GLOSSARY OF SYMBOLS

$\bar{A}$	Complement of set A
ANOVA	Analysis of variance
α(alpha)	Probability of a type I error
β(beta)	Probability of a type II error
$1 - \beta$	Power of a statistical test
β_0	y-intercept of the true linear relationship
β_1	Slope of the true linear relationship
b_0	y-intercept for the line of best fit for the sample data
b_1	Slope for the line of best fit for the sample data
c_j	Column total
c	Column number or class width
d	Difference in value between two paired pieces of data
$d(\ \ )$	Depth of
df or df$(\ \ )$	Number of degrees of freedom
d_i	Difference in the rankings of the ith element
E	Expected value or maximum error of estimate
$e = y - \hat{y}$	Error (observed)
ε(epsilon)	Experimental error
ε_{ij}	Amount of experimental error in the value of the jth piece of data in the ith row
F	F distribution statistic
$F(\mathrm{df}_n, \mathrm{df}_d, \alpha)$	Critical value for the F distribution
f	Frequency
H	Value of the largest-valued piece of data in a sample
H_a	Alternative hypothesis
H_0	Null hypothesis
i	Index number when used with Σ notation
i	Position number for a particular data
k	Identifier for the kth percentile
k	Number of cells or variables
L	Value of the smallest-valued piece of data in a sample
LH	Lower hinge
m	Number of classes
MAD	Mean absolute deviation
MS$(\ \ )$	Mean square
MSE	Mean square error
μ(mu)	Population mean
μ_d	Mean value of the paired differences

Elementary Statistics

THE DUXBURY SERIES IN STATISTICS AND DECISION SCIENCES

Elementary Statistics
Fifth Edition

ROBERT JOHNSON
MONROE COMMUNITY COLLEGE

PWS–KENT Publishing Company
Boston

Publishing Company

20 Park Plaza
Boston, Massachusetts 02116

Library of Congress Cataloging-in-Publication Data

Johnson, Robert Russell, 1939–
  Elementary statistics/Robert Johnson. —5th ed.
    p. cm.
  Includes index.
  ISBN 0-534-91719-4
  1. Statistics. I. Title.
QA276.12.J64 1988   87-19830
519.5–dc19   CIP

Printed in the United States of America.
  89 90 91 92—10 9 8 7 6 5 4 3

SPONSORING EDITOR   Michael Payne
PRODUCTION COORDINATOR   Elise Kaiser
PRODUCTION   Del Mar Associates
INTERIOR/COVER DESIGN   Elise Kaiser
COMPOSITION   Polyglot Pte. Ltd.
COVER PRINTER   New England Book Components
TEXT PRINTER/BINDER   R. R. Donnelley & Sons Company

Cover photo ©Bill Binzen/Photo Researchers, Inc.

To my mother
and to the
memory of my father

Preface

The primary objective of *Elementary Statistics, Fifth Edition* is to present a truly readable introduction to statistics—one that will motivate students by presenting statistics in a context that relates to their personal experiences and that is organized to promote learning and understanding.

Statistics is a practical discipline that responds to the changing needs of our society. Today's student is a product of a particular cultural environment and, as such, is motivated differently from the student of a few years ago. Statistics is presented in this text as a very useful tool in learning about the world around us. While studying descriptive and inferential concepts, students will become aware of the practical application of these concepts to such fields as business, biology, engineering, industry, and the social sciences.

This book was written for use in an introductory course for nonmathematics major students who need a working knowledge of statistics but do not have a strong mathematics background. Statistics requires the use of many formulas, so those students who have not had intermediate algebra should complete at least one semester of college mathematics before beginning this course.

CHANGES IN THE FIFTH EDITION

The teaching objectives of this edition are the same as those of the previous editions. The following significant changes made in this revision should be helpful in attaining these objectives:

1. A discussion of box plots has been added to Chapter 2.

2. Over 30 case studies have been added throughout the text to demonstrate the application of the statistical topics being studied.

3. Chapter 8, Introduction to Statistical Inferences, has been substantially revised to increase the statistical precision of its exposition without sacrificing its clarity of explanation.

4. The exercise sets have been extensively revised. More than 30 percent of the exercises are new. Attention has been paid to ensure that the problems give adequate practice in mastering the techniques of statistical problem solving and provide relevant applications to engage student interest. Another new feature of the exercise sets is that some problems now require analysis of MINITAB output in order to solve them.

5. The entire manuscript has been carefully reviewed to ensure that the writing style is unambiguous, accurate, and very accessible to the student.

6. Additional output from MINITAB has been added where appropriate to place the presentation within a modern computing environment.

TO THE INSTRUCTOR: THE TEXT AS A TEACHING TOOL

As stated earlier, one of the primary objectives in writing this book was to produce a truly readable presentation of elementary statistics. The chapters are designed to interest and involve students and guide them step by step through the material in a logical manner. Each chapter includes the following main features:

A **chapter outline** that shows students what to look for in the chapter.

A **news article** that illustrates how statistics are actually applied in the real world and demonstrates the types of statistics to be studied in that chapter.

Chapter objectives that tell students the specific information to be learned in that chapter.

Worked-out examples with solutions to illustrate concepts as well as demonstrate the applications of statistics in real-world situations.

Case studies that are brief versions of actual newspaper and magazine articles specifically focusing on the use of statistics in everyday life.

End-of-section exercises to facilitate practice in the use of concepts as they are presented.

An **in retrospect** section provides a summary of the material in the chapter and relates the material to the chapter objectives.

Chapter exercises that give students further opportunity to master conceptual and computational skills.

A **vocabulary list** to help students review key terms.

A **chapter quiz** that helps students evaluate their mastery of the material.

At the end of each part is a **working with your own data** section that has been included for student exploration. These data sections provide a more personalized learning experience by directing students to collect their own data and apply the techniques they have been studying.

The first three chapters are introductory in nature. Chapter 3 is a descriptive (first-look) presentation of bivariate data. This material is presented at this point in the book because students often ask about the relationship between two sets of data (such as heights and weights).

In the chapters on probability (4 and 5), the concepts of permutations and combinations are deliberately avoided. They are of no help in understanding statistics. Thus only the binomial coefficient is introduced in connection with the binomial probability distribution.

The instructor has several options in the selection of topics to be studied in a given course. Chapters 1 through 9 are considered to be the basic core of a course (some sections of Chapters 2, 3, and 6 may be omitted without affecting continuity). Following the completion of Chapter 9, any combination of Chapters 10 through 14 may be studied. There are two restrictions, however: Chapter 3 must be studied prior to Chapter 13, and Chapter 10 must precede Chapter 12.

The suggestions of instructors using the previous editions have been invaluable in helping me improve the text for the present revision. Should you, in teaching from this edition, have comments or suggestions, I would be most grateful to receive them. Please address such communications to me at Monroe Community College, Rochester, New York 14623.

TO THE STUDENT: THE TEXT AS A LEARNING TOOL

Plain talk and a stress on common sense are this book's main merits as a learning tool. Such a treatment should allow you, provided that you have the necessary basic mathematics skills, to work your way through the course with relative ease. Examples of this procedure are (1) Illustration 1–4 (p. 10), which is used to reemphasize the meanings of the eight basic definitions presented in Chapter 1, and (2) the use of real-life situations to introduce a new concept (see the introduction of hypothesis testing on page 284; and Illustration 13-4 on regression analysis, p. 489).

It is the aim of this book to motivate and involve you in the statistics that you are learning. The chapter format reflects these aims and can best promote learning if each part of each chapter is used as indicated.

1. Read the annotated **chapter outline** to gain an initial familiarity with several of the basic terms and concepts to be presented.

2. Read the **news article**, which puts to practical use some of the concepts to be learned in the chapter.

3. Use the **chapter objectives** as a guide to map out the direction and scope of the chapter.

4. Learn and practice using the concepts of each section by doing the **exercises** at the end of the respective section. Answers and partial solutions are provided at the back of the book to complement the study illustrations and to enable you to work independently. While working within a chapter, it will be helpful to save the results of the exercises, since some results will be used again in later exercises in the chapter. When this occurs, the later use has been cross-referenced.

5. Use the **in retrospect** section to reflect on the concepts you have just learned and the relationship of the material in this chapter to the material of previous chapters. At this point it would be meaningful to reread the news article.

6. The **chapter exercises** at the end of the chapter offer additional learning experiences, since in these exercises you must now identify the technique to be used and must be able to apply it. The exercises are graded—everyone will be able to complete the first exercises with reasonable ease, but succeeding exercises become more challenging. Occasionally the results of an exercise should be saved for use in later exercises.

7. The **vocabulary list** and **quiz** are provided as self-testing devices. You are encouraged to use them. The correct responses for the quiz may be found in the back of the book.

8. The **working with your own data** sections direct you to collect a set of data, often of your own interest, and to apply the techniques you have studied. This opportunity should (a) reinforce the concepts studied and (b) result in an interesting and informative statistical experience with real data.

SUPPLEMENTS

The complete instructional package that accompanies this book includes the following:

1. The **Study Guide with Self-Correcting Exercises** offers students an alternative approach to mastering difficult concepts. For each chapter in the text, the study guide provides alternative explanations of difficult, numerous, worked-out examples, and self-correcting exercises. There is a review of elementary algebra.

2. The **Solutions Manual** shows at least one complete solution to each exercise in the textbook. Occasionally some parenthetical comments have been added to aid the teacher in such areas as when to assign specific problems and how some problems can be of greatest use.

3. A **Partial Solutions Manual**, new to this edition, is available for students. It contains the complete solutions to all selected exercises whose answers appear in the back of the book.

4. **Computerized Test Items**, providing a series of multiple-choice questions and problems, are available on disk for the Apple and IBM PC computers, along with the printed versions of the test items.

5. **Transparency Masters** highlighting important illustrations and pertinent examples are available for this edition.

6. The **Minitab Student Supplement** by James Scott and Kenneth Bond of Creighton University is provided for those interested in teaching or learning the course interactively with the computer. This supplement is a text-specific introduction to the MINITAB statistical analysis system and is keyed to text discussion and examples.

ACKNOWLEDGMENTS

I owe a debt to many other books. Many of the ideas, principles, examples, and developments that appear in this text stem from thoughts provoked by these sources.

It is a pleasure to acknowledge the aid and encouragement I have received throughout the development of this text from my students and colleagues at Monroe Community College. Special thanks go also to those who read and offered suggestions about the previous editions. I also want to acknowledge and thank the reviewers for this edition: Shirley Dowdy, West Virginia University; James E. Holstein, University of Missouri; Robert Hoyt, Southwestern Montana State; Mark Anthony McComb, Mississippi College; Jeffrey Mock, Diablo Valley College; and Larry J. Ringer, Texas A & M University. In addition, I would especially like to thank Larry Stephens of the University of Nebraska—Omaha for his invaluable assistance in helping me to develop this fifth edition.

I would also like to express my appreciation for the quality work that Elise Kaiser and Nancy Sjoberg have put into the production of the book. To Michael Payne, Editor, I would like to say thanks for all your assistance and encouragement.

Thanks also to the many authors and publishers who so generously extended reproduction permissions for the news articles and tables used in the text. These acknowledgments are specified individually thoughout the text.

Robert R. Johnson

Contents

Part 4   *More Inferential Statistics*   419

Part 1

Descriptive Statistics

When one embarks on a statistical solution to a problem, a sequence of events must develop. The order in which these events occur should be: (1) the situation investigated is carefully and fully defined, (2) a sample of data is collected from the appropriate population following an established and appropriate procedure, (3) the sample data are converted into usable information (this usable information, either numerical or pictorial, is called the *descriptive statistics*), and (4) the theories of statistical inference are applied to the sample information in order to draw conclusions about the sampled population (these conclusions or answers are called *inferences*).

The first part of this textbook, Chapters 1 through 3, will concentrate on the first three of the four events identified above. The second part will deal with probability theory, the theory on which statistical inferences rely. The third and fourth parts will survey the various types of inferences that can be made from sample information.

Chapter 1

Statistics

It's
Your
Life

Statistics about
the American teen

What's your biggest problem? If it's school, you're not alone. Dr. Myron Harris and Jane Norman have discovered that school weighs heaviest on the minds of young people. They surveyed over 160,000 teenagers and published the results in *The Private Life of the American Teenager* (Rawson, Wade). Here's a list of some of their findings:

* More than eight out of ten teens with a working mother feel happy and proud of her.
* Eighty percent of the girls want a career.
* Ninety percent of the teens surveyed believe in marriage.
* Seventy-four percent say they would live with someone before marrying him or her.
* Seven out of ten high school students have tried marijuana.
* Sixty percent study only to pass tests, not to learn. Fifty-five percent admit that they cheat.
* Eighty-three percent say they can usually tell one or both parents how they feel about an issue.
* Almost sixty percent fear their parents' death—even more than they fear their own.
* Seventy-five percent of the teens believe that divorce is justified if parents argue often, if physical violence is involved, or if one or both parents are unfaithful.

Erdice Court

Average Gain on Today's Stockmarket Was 4.14

4 out of 5 doctors recommend . . .

"Oh, I was just an average guy . . . . . . "

USA SNAPSHOTS

A look at statistics that shape our lives

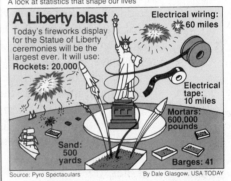

A Liberty blast

Today's fireworks display for the Statue of Liberty ceremonies will be the largest ever. It will use:

Rockets: 20,000

Electrical wiring: 60 miles

Electrical tape: 10 miles

Mortars: 600,000 pounds

Sand: 500 yards

Barges: 41

Source: Pyro Spectaculars   By Dale Glasgow, USA TODAY

USA SNAPSHOTS

A look at statistics that shape the nation

Why we like our jobs

Seven of 10 adults say they work outside the home and 91 percent like their jobs, a new poll says. What they like most:

Don't know 1%
Other 16%
Boss 3%
The work itself 32%
Company benefits 6%
Hours 7%
Money 12%
People at work 23%

Source: Media General-Associated Press poll of 1,464 adults

By Heidi E. Capousis, USA TODAY

Chapter
Objectives

The purpose of this introductory chapter is to present (1) an initial image of the field of statistics, (2) several of the basis vocabulary words used in studying statistics, and (3) the basic ideas and concerns about the processes used to obtain sample data.

1.1   WHAT IS STATISTICS?

Statistics is the universal language of the sciences. Statistics is more than just a "kit of tools." As potential users of statistics, we need to master the "art" of using these tools correctly. Careful use of statistical methods enables us to (1) accurately describe the findings of scientific research, (2) make decisions, and (3) make estimations.

Statistics involves information, numbers to summarize this information, and their interpretation. The word "statistics" has different meanings to people of varied backgrounds and interests. To some people it is a field of "hocus-pocus" whereby a person in the know overwhelms the layperson. To other people it is a way of collecting and displaying large amounts of numerical information. And to still another group it is a way of "making decisions in the face of uncertainty." In the proper perspective, each of these points of view is correct.

descriptive and inferential statistics

The field of statistics can be roughly subdivided into two areas: descriptive statistics and inferential statistics. **Descriptive statistics** is what most people think of when they hear the word "statistics." It includes the collection, presentation, and description of data. The term **inferential statistics** refers to the technique of interpreting the values resulting from the descriptive techniques and then using them to make decisions.

Statistics is more than just numbers—it is what is done to or with numbers. Let's use the following definition:

statistics

Statistics
The science of collecting, classifying, presenting, and interpreting data.

Before we begin our detailed study, let's look at a few examples of how and when statistics can be applied.

Illustration 1-1   State University is planning an expansion program of its physical facilities. To draw up an effective course of action, the board of trustees decides that it needs to answer this question: How many college students will we need to accommodate over the next ten years? This question can immediately be broken down into a number of smaller questions: How many college students will there be in the U.S.? How many will want to attend State U? and so on. To answer these questions the researcher will need to obtain data that will indicate the proportion of future high school graduates that will want to attend State U. Then he or she will somehow need to project, or predict, the number of high school graduates that will be available over the next ten years.

Let's consider the question of the proportion of high school graduates that will wish to attend State U. The best way to answer this question is to find out what proportion have attended in the past. Note that this "best" answer assumes that there is a relationship between past and future. This does not always hold true, however. Wars, economic depressions, and other events will alter the "natural" progression of events.   ■

As you can see, when accurate answers are desired, many problems must be overcome. One rather obvious problem is how to obtain historical data. Do we need to consider every student who has graduated from high school for the last several years? Do we need to account for all schools within a 500-mile radius? No, we do not; it would be impossible to research these questions completely. So we will obtain
sample   information about only some of this population—in other words, we will **sample** the population.

There are other considerations, too. How accurate are our results? What is the probability that there will be a larger proportion of students wanting to attend State U? and so on.

We have not begun to exhaust the questions that may be relevant. At this time I only wish to start you on your way to considering some of the problems involved in answering this type of question.

Illustration 1-2   "Everybody loves Pickadilly Pete!" Well, that is what Pickadilly Pete claims, so he has decided to run for mayor of Skunk Hollow (population 279). But his campaign manager is not sure that he understands exactly what Pickadilly Pete means. Does Pete mean that all 279 inhabitants of Skunk Hollow love him, or that the majority of the 279 (at least 140) love him, or that at least half of the inhabitants of voting age love him? Pete could have meant any or all of these, and what he means could make a big difference in the campaign. (Suppose, for example, that the 140 who love him are all children and can't vote.) If you were hired as an independent research agent, what would you do to test the "accuracy" of Pickadilly Pete's claim?   ■

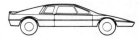

Illustration 1-3   How tall are sports car drivers? This is the question posed by the owners of Custom Sport Coupe, a local manufacturer of the world's finest sports car. They want to design and build a new model that is truly comfortable for the driver. Their present model is designed for people between 5 feet 2 inches and 5 feet 8 inches tall. The manufacturers are concerned because they have heard rumors that their car is uncomfortable for a large proportion of sports car enthusiasts. (If you have ever ridden in a CSC, you will understand the rumor "built for short people with very short necks.") How might you go about obtaining an answer to the original question? What special considerations would you give to the process of obtaining sample information?   ■

Each of these illustrations poses questions that should make you think about the situation and at the same time give you a feeling for statistics.

1.2   USES AND MISUSES OF STATISTICS

The uses of statistics are unlimited. It is much harder to name a field in which statistics is not used than it is to name one in which statistics plays an integral part. The following are a few examples of how and where statistics is used:

1. In education, descriptive statistics are frequently used to describe test results.

2. In science, the data resulting from experiments must be collected and analyzed.

3. In government, many kinds of statistical data are collected all the time. In fact, the U.S. government is probably the world's greatest collector of statistical data.

Misuses of statistics are often colorful and sometimes troublesome. Many people are concerned about the indifference of statistical descriptions; other people believe all statistics are lies. Most statistical lies are innocent, however, and result from using an inappropriate statistic, an open, nonspecific statement (such as Pickadilly Pete's claim), or data derived from a faulty sample. All these lead to a common result: misunderstanding of the information on the part of the consumer. Specific illustrations of the misuse of statistics will be given in Chapter 2.

Case Study 1-1

"What we throw away,"
Copyright 1986, USA TODAY. Reprinted with permission.

WHAT WE THROW AWAY

The July 11, 1986 issue of *USA Today* presented a graphic on the statistics of what is normally thrown away daily by every person in the U.S. What information was collected? What unit of measure appears to have been used to collect that information? How is the information reported on the graphic? (*See* Exercise 1.1.)

Case Study 1-2

"How we pay for things,"
Copyright 1986, USA TODAY. Reprinted with permission.

HOW WE PAY FOR THINGS

The July 16, 1986 issue of *USA Today* presented a graphic on the statistics of "How we pay for things" we purchase (check, credit, or cash). The reader receives the message that most purchases are paid for by check and that cash is used the least. The *Snapshot* reports figures of 57 percent, 36 percent, and 6 percent. The trouble is, percent of what? These percentages could be based on (1) the dollar value of items purchased, (2) the number of items purchased, or (3) the number of purchases. (*See* Exercise 1.2.)

Case Study 1-3

"Many say no to nursing homes,"
Copyright 1987, USA TODAY. Reprinted with permission.

MANY SAY NO TO NURSING HOMES

The January 30, 1987 issue of *USA Today* presented a graphic that shows how adults of different age groups feel about the possibility of having to live in a nursing home. Sixty-five percent of the 18- to 34-year-olds said they were not willing to be cared for in a nursing home. Fifty-nine percent of the 35- to 54- year-olds and 48% of those 55 and older expressed the same feeling. This information is descriptive in nature. An inferential statement is also made. What is it? (*See* Exercise 1.3.)

EXERCISES

1.1 Refer to the *USA Snapshot* "What we throw away" (Case Study 1-1).

a. If you were planning to collect similar information about your neighborhood, what information would you collect?

b. What unit of measure would you use in collecting the information?

c. What information did the National Solid Wastes Management Association collect for the graphic?

d. What unit of measure appears to have been used to collect that information?

e. How is the information reported on the graphic?

1.2 Refer to the *USA Snapshot* "How we pay for things" (Case Study 1-2).

a. How do you think the ratio 57 percent was obtained?

b. If you had to describe, in a similar manner, how customers at a local department store paid for their purchases, what information would you want to collect?

c. Suppose you had collected your information about every customer at that store for a given time period. How would you determine the percentages you would report?

1.3 Refer to the *USA Snapshot* "Many say no to nursing homes" (Case Study 1-3).

a. What descriptive statistics are cited?

b. Which statement is inferential?

1.3 INTRODUCTION TO BASIC TERMS

To study statistics we need to be able to speak its language. Let's first define a few basic terms that will be used throughout this text. (These definitions are descriptive in nature and are not necessarily mathematically complete.)

population

Population

A collection, or set, of individuals or objects whose properties are to be analyzed.

The population is the complete collection of individuals or objects that are of interest to the sample collector. The concept of a population is the most fundamental idea in statistics. The population of concern must be carefully defined and is considered fully defined only when its membership list of elements is specified. The set of "all students who have ever attended a U.S. college" is an example of a well-defined population.

Typically we think of a population as a collection of people. However, in statistics the population could be a collection of animals, of manufactured objects, or whatever. For example, the set of all redwood trees in California could be a population.

sample

Sample

A subset of a population.

A sample consists of the individuals, objects, or measurements selected by the sample collector from the population.

response variable

Response Variable (or, simply, Variable)

A characteristic of interest about each individual element of a population or sample.

A student's age at entrance into college, the color of her hair, her height, weight, and so on, are all response variables.

data (singular)

Data (singular)

The value of the response variable associated with one element of a population or sample.

For example, Bill Jones entered college at the age of "23," his hair is "brown," he is "6 feet 1 inch" tall, and he weighs "183 pounds." These four pieces of data are single values for the four response variables as applied to Bill Jones.

data (plural)

Data (plural)

The set of values collected for the response variable from each of the elements belonging to the sample.

The set of 25 heights collected from the 25 students is an example of a set of data.

experiment

Experiment

A planned activity whose results yield a set of data.

parameter

Parameter

A numerical characteristic of an entire population.

The "average" age at time of admission of all students who have ever attended our college or the proportion of students who were over 21 years of age when they entered college are examples of two different population parameters. A parameter is a value that describes the entire population. It is a common practice in statistics to use a Greek letter to symbolize the names of the various parameters. These symbols will be assigned as we study individual parameters.

statistic

Statistic

A numerical characteristic of a sample.

The "average" height found by using the set of 25 heights is an example of a sample statistic. A statistic is a value that describes a sample. Most sample statistics are found with the aid of formulas and are assigned symbolic names using letters of the English alphabet (for example, $\bar{x}$, s, and r).

Illustration 1-4   A statistics student is interested in finding out something about the dollar value of the typical car owned by the faculty members of a college. Each of the eight terms just described can be identified in this situation.

1. The *population* is the collection of all cars owned by the faculty members.

2. A *sample* is any part of that population. For example, the cars owned by members of the mathematics department would be a sample.

3. The *response variable* is the actual value of each individual car.

4. One *data* would be the value of a particular car. Mr. Jones's car, for example, is valued at $5400.

5. The *data* would be the set of values that correspond to the sample obtained (5400, 8700, 5950, . . .).

6. An *experiment* would be the methods used to select the cars forming the sample and determining the value of each car in the sample. It could be carried out by questioning each member of the mathematics department, or in other ways.

7. The *parameter* about which we are seeking information is the "average" value in the population.

8. The *statistic* that will be found is the "average" value in the sample. ■

There are basically two kinds of data: (1) data obtained from qualitative information and (2) data obtained from quantitative information.

attribute or qualitative data

Attribute or Qualitative Data

Results from a process that categorizes or describes an element of a population.

For example, color is an attribute of a car. A sample of the colors of the cars in a nearby parking lot would result in such data as blue, red, yellow, and so on. In general, the resulting data are a collection of word responses (as opposed to number responses).

numerical or quantitative data

Numerical or Quantitative Data

Result from a process that quantifies—that is, counts (of how many) or measurements (length, weight, and so on).

discrete

continuous

Numerical data can be subdivided into two classifications: (1) **discrete** numerical data and (2) **continuous** numerical data. In most cases the two can be distinguished by deciding whether the data result from a count or from a measurement. A count will always yield discrete numerical data. The number of bugs on the right headlight of an automobile will be 0, 1, 2, ...,* but it cannot be 1.9 or 3.25. (The partial remains of any bug will be counted as a whole bug.) The idea of discontinuous numerical values is somewhat synonymous with that of discrete numerical values.

A measure of a quantity will usually be continuous. If John says, for example, that he weighs 162 pounds, to the nearest pound, all we can be sure of is that his weight is some value between 161.5 and 162.5 pounds. He could actually weigh 162.000, 161.789, or any value in the interval 161.5–162.5. This is the basic concept of a continuous variable. There are many continuous variables that receive responses that appear to be discrete: for example, a person's age. When asked how old he was, Steve responded "19." He was 19 years old on his last birthday, but he is 19 plus some part of a year.

The use of fractions or decimals does not necessarily imply that data are continuous. The scoring of competitive diving is an example of a case in which decimals appear and the variable is actually discrete. A contestant can receive scores by halves only (5.5, 6.0, 6.5, and so on). These scores are actually discrete, because values *between* 6.0 and 6.5 cannot occur. There are other illustrations of this situation, but in this text a *count* or a *measurement* is about the only distinction that we will need to make.

There are situations in which data are collected in numerical form and reported and discussed in attribute form. Two such situations are (1) the measurement of air pollution, as commonly reported in the newspaper (measured in amount and reported as low, medium, or high), and (2) the Richter scale and the measurement of an earthquake.

---

* Note: The notation "..." means that the listing of numbers continues indefinitely.

Let's explore the difference between these terms and the types of data. If I were to go to a nearby parking lot right now and obtain some data, I would most likely think of it as a sample. However, it could be considered to be the population if I defined my population of concern to be the cars in that parking lot right now. It is more likely that it would be considered to be only part of a larger collection of cars, say of all cars that ever park in this lot or a sample of all cars parked in all parking lots at this time. (You must define your population and your sample carefully.) If I were to observe the make of each car, I would be collecting attribute data. If I were to record the number of people riding in each car as it left the parking lot, I would be collecting discrete data. (This clearly is a count of people.) If each car were to be weighed, continuous data would result, because all the various fractional parts of weight units could result from the weighing (*see* Section 1.4).

Don't let the appearance of the data fool you in regard to their type. For example, suppose that after surveying this parking lot, I summarized the sample by reporting five red, eight blue, six green, and two yellow cars. These data appear to be the count of cars; therefore, one would think them to be discrete. This is not the case, however, they are attribute data. **You must look at each individual source to determine the kind of variable being used**. One specific car was red; "red," then is the data for that one car. And red is an attribute. Thus this collection (five red, eight blue, and so on) is a *summary* of the *attribute* data.

Another example of a deceiving variable is the license plate number (number part only) of each car. The numbers appear to be a discrete variable since only whole number values occur; however, these numbers are merely identification numbers. Identification numbers are not variables because they do not measure anything; they serve only to identify. Consider the weights of the cars as recorded on the registration: 3485, 3860, 2091, 4175, and so on. They are all whole numbers, but that does not make the variable discrete. The variable is "weight," and these data are measured to the nearest pound. Thus the variable "weight" is continuous. As you can see, the appearance of the data *after* they are recorded can be misleading in respect to their type. Remember to inspect an individual piece of data and you should have little trouble in distinguishing among attribute data and discrete and continuous data.

EXERCISES

1.4 A drug manufacturer is interested in the proportion of persons who have hypertension (elevated blood pressure) whose condition can be controlled by a new drug the company has developed. A study involving 5000 individuals with hypertension is conducted and it is found that 80 percent of the individuals are able to control their hypertension with the drug. Assuming that the 5000 individuals are representative of the group who have hypertension, answer the following questions.

   a. What is the population?

   b. What is the sample?

    c. Identify the parameter of interest.

    d. Identify the statistic and give its value.

    e. Do we know the value of the parameter?

1.5 Perform the "first-ace" experiment five times (instructions follow) and observe the value of three different variables each time:

    Variable 1: The color of the first ace to appear.

    Variable 2: The longest distance across the dealt pile of cards.

    Variable 3: The count of the card on which the first ace appears.

To perform the first-ace experiment, shuffle an ordinary deck of 52 playing cards containing 4 aces. Deal the cards one at a time onto a pile, stopping when the first ace appears. After the first ace has been placed on the pile, record its color. Now measure the longest distance across the scattered pile, before it is disturbed, and record it. Count and record the number of cards in the pile. Repeat the experiment four more times.

Trial	x = Color of first ace	d = Distance across pile	y = Count of cards, including first ace
1	$x_1 =$	$d_1 =$	$y_1 =$
2	$x_2 =$	$d_2 =$	$y_2 =$
3	$x_3 =$	$d_3 =$	$y_3 =$
4	$x_4 =$	$d_4 =$	$y_4 =$
5	$x_5 =$	$d_5 =$	$y_5 =$
		$\sum\limits_{i=1}^{5} d_i =$	$\sum\limits_{i=1}^{5} y_i =$

Note: See Appendix A for information about the $\sum$ notation.

1.6 Each of the three variables in the first-ace experiment produces a different type of data. What type of data results from each variable?

1.7 A quality-control technician selects assembled parts from an assembly line and records the following information concerning each part:

    a. defective or nondefective

    b. the employee number of the individual who assembled the part

    c. the weight of the part

Classify the responses for each part as either attribute data, discrete variable data, or continuous variable data.

1.8 We want to investigate the cost of education. One of the expenses that a student must contend with is the cost of textbooks. Let x be the cost of all textbooks purchased by each student this semester at your college.

a. Carefully describe the population.

b. Carefully describe the variable.

1.9 The admissions office wishes to estimate the average cost of textbooks per student per semester. (*See* Exercise 1.8.)

a. Describe the population parameter.

b. The admissions office identified 50 students who registered and asked each one to keep track of his or her textbook expenses and to report the total amount. The resulting 50 amounts form a sample. Describe the sample statistic that is of interest to the admissions office.

c. Describe how you would use the 50 data in the sample to calculate the value of the sample statistic discussed in (b).

1.10 The "first-seven" experiment consists of rolling a pair of dice repeatedly until a seven (sum of numbers is seven) occurs. The response variable of interest is the number of rolls necessary to obtain the first seven.

a. What are the possible values for the response variable?

b. Is this variable a discrete or a continuous variable?

1.11 Identify each of the following as examples of (1) attribute, (2) discrete, or (3) continuous variables.

a. the breaking strength of a given type of string

b. the hair color of children auditioning for the musical *Annie*

c. the number of stop signs in towns of less than 500 people

d. whether or not a faucet is defective

e. the number of questions answered correctly on a standardized test

f. the length of time required to answer a telephone call at a certain real estate office

1.12 Identify each of the following as examples of (1) attribute, (2) discrete, or (3) continuous variables.

a. you are polling registered voters as to which candidate they support

b. length of time required for a wound to heal when using a new medicine.

c. the number of telephone calls arriving at a switchboard per 10-minute period

d. the distance first-year college women can kick a football

e. the number of pages per job coming off a computer printer

f. the kind of tree used as a Christmas tree

1.13 Suppose a twelve-year-old asked you to explain the difference between a sample and a population.

a. What information should your answer include?

b. What reasons would you give him or her for why one would take a sample instead of surveying every member of the population?

1.4   MEASURABILITY AND VARIABILITY

Within a set of experimental data, we *always* expect variation. If little or no variation is found, we would guess that the measuring device is not calibrated with a small enough unit. For example, we take a carton of a favorite candy bar (24) and weigh each bar individually. We observe that all 24 candy bars weigh $\frac{7}{8}$ of an ounce, to the nearest $\frac{1}{8}$ ounce. Does this mean that they are all identical in weight? Not really. Suppose that we were to weigh them on an analytical balance that weighs to the nearest milligram. Now their weights will be variable.

variability
It does not matter what the response variable is; there will be **variability** in the numerical response if the tool of measurement is precise enough. A primary objective in statistical analysis is that of measuring variability. For example, in the study of quality control, measuring variability is an absolute essential. Controlling (or reducing) the variability in a manufacturing process is a field all its own (for example, statistical process control).

EXERCISES

1.14 Suppose we measure the weights (in pounds) of individuals contained in each of the following groups:

Group 1: one-year-old male children

Group 2: adult males who are over 20 years of age

For which group would you expect the data to have more variability?

1.15 Suppose you were trying to decide which of two machines to purchase. Furthermore, suppose the length to which the machines cut a particular product part was important. If both machines produced parts that had the same length on the average, what other consideration regarding the lengths would be important? Why?

1.5   DATA COLLECTION

One of the first problems a statistician faces is obtaining data. Data doesn't just happen. It must be collected. We must realize the importance of good sampling techniques, because the inferences we ultimately make will be based on the statistics calculated from our sample data.

The collection of data for statistical analysis is an involved process and includes the following important steps:

1. Defining the objectives of the survey or experiment.

Examples: (a) comparing the effectiveness of a new drug to the effectiveness of the standard drug, (b) estimating the average household income in our county.

2. Defining the variable and the population of interest.
Examples: (a) length of recovery time for patients suffering from a particular disease, (b) total income for households in our county.

3. Defining the data-collection and data-measuring schemes.
This includes sampling procedures, sample size, and the data-measuring device (questionnaire, telephone, and so on).

4. Determining the appropriate descriptive or inferential data-analysis techniques.

Very often an analyst is stuck with data already collected, possibly even data collected for other purposes, which makes it impossible to determine whether or not the data are "good." Collecting the data yourself using approved techniques is much preferred. Although this test will be chiefly concerned with various data-analysis techniques, you should be aware of the problems of data collection.

The following illustrations present populations defined for specific investigations.

Illustration 1-5   The admissions office wishes to estimate the cost of college education at a local college. One of the components of total cost per semester is the cost of textbooks. Specifically, the admissions office would like to estimate the current "average" cost of textbooks per semester per student. The population of interest is the currently enrolled student body.   ■

Illustration 1-6   The cost of a one-minute television advertisement varies drastically according to the station, the day, and the time of day. This rate is determined by the proportion of potential viewers who are tuned to a particular station at a particular time. A national department store chain is opening a new store in a nearby city and is in the process of selecting time slots for advertising on WRQC TV. They want to choose the time slots that will be most cost effective in attracting new customers. The population of interest is the daily television viewing habits of all persons who live within the broadcasting area.   ■

The two methods used to collect data are *experiments* and *surveys*. In an experiment, the investigator controls or modifies the environment and observes the effect on the response variable. We often read about the results obtained by laboratories using white rats to test some new product, such as different doses of a new medication and its effect on blood pressure. The experimental treatments used were designed specifically to obtain the data needed to study the effect on the response variable. In a survey, data are obtained by sampling some population of interest. The investigator, however, does not modify the environment. The remainder of this section deals with some of the simpler methods that might be used in order to obtain sample data from surveys.

census   If every element in the population is listed, or enumerated, and observed, then a **census** is compiled. A census is a 100 percent survey. A census for the population in

Illustration 1-5 could probably be obtained by contacting each student on the registrar's computer printout listing all registered students. However, censuses are seldom used because they are difficult to compile, very time-consuming, and therefore expensive. Imagine the task of compiling a census of every person who lives within the broadcasting area of WRQC TV (Illustration 1-6). Instead of a census, a sample survey is usually conducted.

When selecting a sample for a survey, it is necessary to construct a *sampling frame*.

sampling frame

Sampling Frame

A list of the elements belonging to the population from which the sample will be drawn.

The sampling frame should be identical to the population. Ideally, the frame will have every element of the population listed once and only once. However, this is not always possible because it may be impractical or impossible to select directly from the total population. Since only the elements in the frame have a chance representative sampling to be selected as part of the sample, **it is important that the sampling frame be** frame **representative of the population**.

In Illustration 1-5, the registrar's computer list will serve as a sampling frame for the admissions office. In this case, the census uses the entire sampling frame. In other situations a census may not be so easy to conduct, because a complete frame is not available. In such cases, lists of registered voters or the telephone directory are sometimes used as sampling frames of the general public. One of these might be used in Illustration 1-6. Depending on the nature of the information sought, the list of registered voters or the telephone directory may or may not serve as a good sampling frame.

Once a representative sampling frame has been established, we are prepared to proceed with selecting the sample elements from the sampling frame. This selection process is defined by the *sample design*.

sample design, or
sampling plan

Sample Design, or Sampling Plan

The procedures used to select the elements of the sample.

There are many different types of sample designs. However, all sample designs fit into two categories: *judgment samples* and *probability samples*.

judgment samples

Judgment Samples

Samples that are selected on the basis of being "typical."

When a judgment sample is drawn, the person selecting the sample chooses items that he or she thinks are representative of the population. The validity of the results from a judgment sample reflects the soundness of the collector's judgment.

probability samples

Probability Samples

Samples in which the elements to be selected are drawn on the basis of probability. Each element in a population has a certain probability of being selected as part of the sample.

Note   **Statistical inference** (tests of hypothesis and confidence interval estimates) **requires that the sample design be a probability sample**.

Let's look at a few of the simpler and easier-to-use sample designs.

One of the most common methods used to collect data for a probability sample is the *simple random sample*.

simple random sample

Simple Random Sample

A random sample selected in such a way that every element in the population has an equal probability of being chosen. Equivalently, all samples of size n have an equal chance of being selected.

When a simple random sample is drawn, an effort must be made to ensure that each element has an equal probability of being selected. Mistakes are frequently made because the term *random* (equal chance) is confused with *haphazard* (without pattern). The proper procedure for selecting a simple random sample is to use a random number generator or a table of random numbers.

To select a simple random sample, first assign a number to each element in the sampling frame. This is usually done sequentially using the same number of digits for each element. Then go to a table of random numbers and select as many numbers with that number of digits as are needed for the sample size desired. (*See* Appendix B for specific information regarding use of the random number table.) Each numbered element in the sampling frame that corresponds to a selected random number is chosen for the sample.

Illustration 1-7   Let's return to Illustration 1-5. Mr. Clar, who works in the admissions office, has obtained a computer list of this semester's full-time enrollment. There are 4265 student names on the list. He numbered the students 0001, 0002, 0003, and so on, up to 4265. Then, using four-digit random numbers, he identified a sample. (*See* Appendix B for a discussion of the use of the random number table.)   ■

One of the easiest-to-use methods for approximating a simple random sample is the *systematic sampling method*.

systematic sample

Systematic Sample

A sample in which every kth item in the sampling frame is selected after a random start among the first k elements.

This method of selection uses the random number table only once, to find the starting point (first element to observe). When an x percent sample is desired, it is not necessary to number the elements in the sampling frame or to know the total count of items in the sampling frame. Thus this is a good procedure for sampling a percentage of a large population. However, there are dangers inherent in using the systematic sampling technique. When the population is repetitive or cyclical in nature, systematic sampling should not be used, because the results will not approximate a simple random sample.

To select an x percent systematic sample, we will need to select 1 element from every $\frac{100}{x}$ elements in the sampling frame. The systematic method requires that we select 1 element from the first $\frac{100}{x}$ elements and then every $\frac{100}{x}$th item after that. For example, if we desire a 3 percent systematic sample, we would locate the position of the first element to observe in our sampling frame by randomly selecting an integer between 1 and 33 ($\frac{100}{3} = 33.33$, which when rounded becomes 33). Suppose the integer determined was 27. This means that the data associated with the element in the 27th position in the sampling frame is our first piece of data. Then, starting at the 27th position, every 33d element is selected. The second piece of data comes from the element in the 60th position ($27 + 33 = 60$), the third is in the 93d position ($60 + 33 = 93$), and so on, until the entire sample is collected.

strata

When sampling very large populations, sometimes it is possible to divide the population into subpopulations on the basis of some characteristic. These subpopulations are called **strata**. These smaller, easier-to-work-with strata are sampled separately. One of the sample designs that starts by stratifying the sampling frame is the stratified sampling method.

stratified sample

Stratified Sample

A sample obtained by stratifying the sampling frame and then selecting a fixed number of items from each of the strata by means of simple random sampling.

When a stratified sample is to be drawn, the population is subdivided into the various strata and then a subsample is drawn from each stratum. These subsamples may be drawn from each stratum randomly or systematically. Then the subsamples are summarized separately and this information is combined to draw conclusions about the whole population.

Another sampling method that starts by stratifying results is a cluster sample.

cluster sample

Cluster Sample

A sample obtained by stratifying the sampling frame and then selecting all of the items from some, but not all, of the strata.

The cluster sample is obtained by using either random numbers or a systematic method to first identify the strata (clusters) to be sampled and then to use all the items from within these strata. Then the subsamples are summarized separately and this information is combined.

The sampling plan used in a particular situation depends on several factors: (1) the nature of the population and the variable, (2) the ease (or difficulty) of sampling, and (3) the cost of sampling. These and other factors must be weighed, with trade-offs usually occurring, before the exact sampling technique is determined.

In this text, all of the statistical methods presume that simple ramdom sampling has been used to collect the data.

EXERCISES

1.16 a. What is a sampling frame?

b. What did Mr. Clar use for a sampling frame in Illustration 1-7?

1.17 Consider a simple population consisting of only the numbers 1, 2, and 3 (an unlimited number of each). There are nine different samples of size two that could be drawn from this population: (1, 1), (1, 2), (1, 3), (2, 1), (2, 2), (2, 3), (3, 1), (3, 2), (3, 3).

a. If the population consists of the numbers 1, 2, 3, and 4, list all the samples of size two that could possibly be selected.

b. If the population consists of the numbers 1, 2, and 3, list all the samples of size three that could possibly be selected.

c. If the population consists of the numbers 1, 2, 3, and 4, list all the samples of size three that could possibly be selected.

1.18 The computer system for a large hospital contains the records for 30,000 patients. The records are sequentially numbered from 1 to 30,000. If we sampled the records by choosing patients numbered 100, 200, 300,.., 30,000, thus obtaining a sample of 300 patients, what type of sampling would this be called?

1.19 A wholesale food distributor in a large metropolitan area would like to test the demand for a new food product. He distributes food through five large supermarket chains. The food distributor selects a sample of stores from each chain and tests his new product in these stores. What type of sampling does this represent?

1.20 A simple random sample could be very difficult to obtain. Why?

1.6 COMPARISON OF PROBABILITY AND STATISTICS

Probability and statistics are two separate but related fields of mathematics. It has been said that "probability is the vehicle of statistics." That is, if it were not for the laws of probability, the theory of statistics would not be possible.

Let's illustrate the relationship and the difference between these two branches of mathematics by looking at two boxes. We know the probability box contains five blue, five red, and five white poker chips. The subject of probability tries to answer

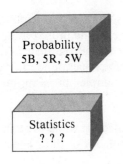

questions such as: If one chip is drawn from this box, what is the chance that it is blue? On the other hand, in the statistics box we don't know what the combination of chips is. We draw a sample and, based on the findings in the sample, we make conjectures about what we believe to be in the box. Note the difference: Probability asks about the chance that something specific (a sample) will happen when you know the possibilities (that is, you know the population). Statistics, on the other hand, asks you to draw a sample, describe the sample (descriptive statistics), and then make inferences about the population based on the information found in the sample (inferential statistics).

EXERCISES

1.21 Classify each of the following as a probability or a statistics problem.

  a. determining whether a new drug shortens the recovery time from a certain illness

  b. determining the chance that a head will result when a coin is flipped

  c. determining the amount of waiting time required to check out at a certain grocery store

  d. determining the chance that you will be dealt a "blackjack"

1.22 Classify each of the following as a probability or a statistics problem.

  a. determining how long it takes to handle a typical telephone inquiry at a real estate office

  b. determining the length of life for the 100-watt light bulbs a company produces

  c. determining the chance that a blue ball will be drawn from a bowl that contains 15 balls of which 5 are blue

  d. determining the shearing strength of the rivets that your company just purchased for building airplanes

1.7   STATISTICS AND THE COMPUTER

In recent years, the computer has had a tremendous effect on almost every aspect of life. The field of statistics is no exception. As you will see, the field of statistics uses many techniques that are repetitive in nature: formulas used to calculate descriptive statistics, procedures for constructing graphic displays of data, and procedures that are followed to formulate statistical inferences. The computer is very good at performing these repetitive operations. It is fairly common for someone with a set of sample data that needs to be analyzed to seek out someone who knows how to use and has access to a computer. If that computer has one of the standard statistical packages on line, it will make the analysis easy to complete. Your local computer center can provide you with a list of what is available. Some of the more readily available packaged programs are: MINITAB, Biomed (Biomedical Programs), SAS (Statistical Analysis System), IBM Scientific Subroutine Packages, and SPSS (Statistical Package for the Social Sciences). You will see examples of computer output throughout this textbook.

Illustration 1-8   Figure 1-1 shows two computer-generated drawings of stem-and-leaf diagrams. The two different diagrams depict the same set of data. Stem-and-leaf diagrams are studied in Section 2.1.

Figure 1-1

Weights of 50 College Students (lbs.)

Stem	Leaves
09	8
10	8 1 8
11	2 8 8 5 5 0 6
12	0 8 0 0 9
13	5 7 2
14	8 3 5 2
15	0 8 7 4 5 4
16	2 7 8 2 1 5
17	0 7 6 0 6
18	6 8 4 3
19	1 5 0 5
20	5
21	5

Weights of 50 College Students (lbs.)

Female		Male
8	09	
8 1 8	10	
6 0 5 5 8 8 2	11	
9 0 0 8 0	12	
2 7 5	13	
2	14	8 3 5
	15	0 8 7 4 5 4
	16	2 7 8 2 1 5
	17	0 7 6 0 6
	18	6 8 4 3
	19	1 5 0 5
	20	5
	21	5

EXERCISES

1.23 How have computers increased the usefulness of statistics for professionals such as researchers, government workers who analyze data, statistical consultants, and so on?

In Retrospect

You should now have a general feeling of what statistics is about—an image that will grow and change as you work your way through this book. You know what a sample and a population are, the distinction between attribute and numerical data, and the difference between continuous and discrete variables. You even know the difference between statistics and probability (although we will not study probability in detail until Chapter 4). You should also have an appreciation for and a partial understanding of how sample data are obtained.

You should realize that the news articles at the beginning of the chapter represent various aspects of statistics. All of these articles, in one way or another, are the matter of statistics. The *USA Snapshots* picture a variety of information.

The "It's Your Life" article reports several statistics about the American teen. The intent in reporting the values is to estimate the level of concern all American teenagers feel, not just the sampled ones. The sign in the cartoon reports another statistic, "1.3 kids."

As members of a computer-age society, we are constantly bombarded by "statistical" information, by way of advertisements and public opinion polls. Information involving numbers can be found in nearly every daily newspaper. However, the fact that the information is numerical does not make it statistical, even though the terms *data* and *statistics* are sometimes used in referring to such information. For example, "20,000 rockets" is not a statistic as it is used in the *USA Snapshot*.

After some exercises using the material in this chapter, we will consider the analysis of data in detail in the next chapters.

Chapter Exercises

1.24 We want to describe the so-called typical student at your college. Describe a variable that measures some characteristic of a student and results in

    a. attribute data   **b.** discrete data   **c.** continuous data

1.25 A candidate for a political office claims that he will win the election. A poll is conducted and 35 of 150 voters indicate that they will vote for the candidate, 100 voters indicate that they will vote for his opponent, and 15 voters are undecided.

    a. What is the population parameter of interest?

    b. What is the value of the sample statistic that might be used to estimate the population parameter?

    c. Would you tend to believe the candidate based on the results of the poll?

1.26 A short survey consists of three questions:

    a. Is your religious preference Christian, Jewish, Moslem, or other?

    b. How many religious services do you attend per year?

    c. How much money did you contribute to religious organizations this last year?
    Classify the responses to a, b, and c as attribute data, discrete variable data, or continuous variable data.

1.27 In "A liberty blast," one of *USA Snapshots* at the beginning of this chapter, the information shown is called statistics. However, it might better be described as "facts and figures." Explain why the number 20,000 (or any of the other values reported) does not satisfy the definition of a "statistic."

1.28 Refer to the information shown in the news article "It's Your Life" at the beginning of this chapter.

    a. What is the population of interest?

    b. Name one variable discussed. [*Hint*: information asked of each person sampled.]

    c. What type of data results from the variable named in (b)?

    d. One of the numerical values reported is "eighty percent." Is this numerical value a piece of data or a statistic? Explain.

1.29 It is impossible for any family to have "1.3 kids," as the sign in the cartoon states. Explain what "1.3 kids" means.

1.30 A researcher studying consumer buying habits asks every twentieth person entering Publix Supermarket, how many times per week he or she goes grocery shopping. He then records the answer as *T*.

    a. Is $T = 3$ an example of (1) a sample, (2) a variable, (3) a statistic, (4) a parameter, or (5) a piece of data?

    Suppose the researcher questions 427 shoppers during the survey.

    b. Give an example of a question that can be answered using the tools of descriptive statistics.

    c. Give an example of a question that can be answered using the tools of inferential statistics.

1.31 A researcher studying the attitudes of parents of preschool children interviews a random sample of 50 mothers, each having one preschool child. She asks each mother, "How many times did you compliment your child yesterday?" She records the answer as *C*.

    a. Is *C* a (1) piece of data, (2) statistic, (3) parameter, (4) variable, or (5) sample?

    b. Give an example of a question that can be answered using the tools of descriptive statistics.

    c. Give an example of a question that can be answered using the tools of inferential statistics.

1.32 A study is designed to estimate the average income per household in a large city. The city is divided into blocks and a simple random sample of blocks is selected.

For the blocks selected, each household is interviewed. What type of sampling does this represent?

1.33 Describe, in your own words, and give an example of the following terms. Your example should not be one given in class or in the textbook.

    a. variable   **b.** data   **c.** sample

    d. population   **e.** statistic   **f.** parameter

1.34 Find an article or a newspaper or magazine ad that exemplifies the use of statistics.

    a. Describe and identify one population of interest (if there is more than one, identify just one for your answer).

    b. Describe and identify one variable used in the article.

    c. Describe and identify one statistic reported in the article.

Vocabulary List

Be able to define each term. In addition, describe, in your own words, and give an example of each term. Your examples should not be ones given in class or in the textbook.

attribute data	probability sample
census	random
cluster sample	representative sampling frame
continuous data	response variable
data	sample
descriptive statistics	sampling frame
discrete data	sampling plan
experiment	simple random sample
inferential statistics	statistic
judgment sample	statistics
numerical data	strata
parameter	stratified sample
population	systematic sample
probability	variability

Quiz

Answer "True" if the statement is always true. If the statement is not always true, replace the words shown in italics with words that make the statement always true.

1.1 *Inferential* statistics is the "study" and description of data that result from an experiment.

1.2 *Descriptive* statistics is the "study" of a sample that enables us to make projections or estimates about the population from which the sample was drawn.

1.3 A *population* is typically a very large collection of individuals or objects about which we desire information.

1.4 A statistic is the calculated measure of some characteristic of a *population*.

1.5 A parameter is the measure of some characteristic of a *sample*.

1.6 As a result of surveying 50 freshmen, it was found that 16 had participated in interscholastic sports, 23 had served as officers of classes and clubs, and 18 had been in school plays during their high school years. This is an example of *discrete numerical data*.

1.7 The number of rotten apples in a shipping crate is an example of *continuous numerical* data.

1.8 The thickness of the sheet metal that a company uses in its manufacturing process is an illustration of *attribute* data.

1.9 A *representative* sample is a sample obtained in such a way that all individuals had an equal chance to be selected.

1.10 The basic objective of *statistics* is that of obtaining a sample, inspecting this sample, and then making inferences about the unknown characteristics of the population from which the sample was drawn.

Chapter 2

Descriptive Analysis and Presentation of Single-Variable Data

HOUSEHOLD EARNINGS RISE FOR 3RD YEAR

USA household incomes are digging out from a slump—and could reach record levels by decade's end, *American Demographics* magazine says.

Its March issue says 1985's median household income of $23,600—up a significant third year after a time of stagnation—could top $24,800 by 1990.

"What this means is the middle class is still strong. The middle class should remain healthy," senior editor Thomas Exter said Sunday.

Inflation and recession contributed to an income decline between 1978–82. Since then, inflation has slowed and the economy has grown.

Fueling the growth:

• Baby Boomers. Our biggest generation is entering peak earning years of middle age. Oldest boomers: 41. They are now half of the work force.

• Working couples. More than half of all married couples hold down two jobs. The number likely will increase.

• Education. One in five adult householders has a college degree. The proportion will grow. Income increases with schooling.

• Baby Bust generation. It's entering the work force now. Fewer in number than boomers, bust workers face less job competition from each other, perhaps forcing up wages as bosses bid for their services.

The forecast doesn't take into account the likelihood of another recession, cautions Robert Greenstein of the Center on Budget and Policy Priorities, a liberal Washington-based think tank.

Exter said 20 percent of our 88 million households had incomes below $10,000. Most were headed by women.

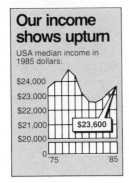

Our income shows upturn

USA median income in 1985 dollars:

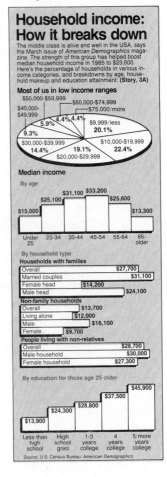

Household income: How it breaks down

The middle class is alive and well in the USA, says the March issue of *American Demographics* magazine. The strength of this group has helped boost median household income in 1985 to $23,600. Here's the percentage of households in various income categories, and breakdowns by age, household makeup and education attainment: (Story, 3A)

Source: U.S. Census Bureau; *American Demographics*

**Chapter
Objectives**

In this chapter we will learn how to present and describe single-variable data. **Single variable** means that we are going to deal with only one numerical response value from a given element of the population. In Chapter 3 we will work with two values from a common element of the population.

The basic idea of descriptive statistics is to describe a set of data. Primarily we will try to describe a set of numerical values in a variety of abbreviated ways. Suppose your instructor just returned an exam paper to you. It carries a grade of 78, and you wish to interpret its value with respect to the grades in the rest of the class. Students usually are interested in three different things. What is the first thing you would ask your instructor?

Most students immediately ask: "What was the average exam grade?" The information being sought here is the location of the *middle* for the set of all the grades. *Measures of central tendency* are used to describe this concept. Your instructor replies that the class average was 68. So you know that your paper was above average by 10 points. Now you would like an indication of the *spread* of the grades. Since 78 is 10 points above the average, you ask: "How close to the top is it?" Your instructor replies that the grades ranged from 42 to 87 points. Pictorially, the information we have so far looks like the accompanying figure.

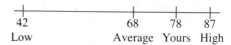

The third question is usually less important, since by now you have a fairly good idea of how your grade compares with those of your classmates. However, the third concept still has some importance. It answers the question "How are the data distributed?" You may have heard about the *normal curve*, but what is it? As you will see later, the idea of distribution describes the data by telling you whether the values are evenly distributed or clustered (bunched) around a certain value. For example, your instructor says that half the class had grades between 65 and 75.

When very large sets of data are involved, measures of *position* are useful. When college board exams are taken, for example, there are thousands of scores that result. Your concern is the position of your score with respect to all the others. Averages, ranges, and so on, are not enough. We would like to say that your specific score is better than a certain percentage of scores. For example, your score is better than 75 percent of all the scores.

The four concepts necessary to describe sets of single-variable data are (1) measures of central tendency, (2) measures of dispersion (spread), (3) measures of position, and (4) types of distribution.

GRAPHIC PRESENTATION OF DATA

2.1 GRAPHS AND STEM-AND-LEAF DISPLAYS

Once we collect the sample of data, we must "get acquainted" with it. One of the most helpful ways to become acquainted is to use an initial exploratory technique that will result in a pictorial representation of the data. The resulting displays visually reveal patterns of behavior of the variable being studied. There are several graphic (pictorial) ways to describe data. The method used is determined by the type of data and the idea to be presented.

Note   Realize from the start that there is no single correct answer when constructing a graphic display. The analyst's judgment and the circumstances surrounding the problem will play a major role in the development of the graphic.

CIRCLE AND BAR GRAPHS

Circle graphs (pie diagrams) and bar graphs are often used to summarize attribute data. You have seen many examples of these before.

Table 2-1 lists the number of cases of each type of operation performed at General Hospital last year. These data are displayed on a circle graph in Figure 2-1,

Table 2-1

Operations performed at General Hospital

Type of Operation	Number of Cases
Thoracic	20
Bones and joints	45
Eye, ear, nose, and throat	58
General	98
Abdominal	115
Urologic	74
Proctologic	65
Neurosurgery	23
	498

Figure 2-1

Circle graph: Operations performed at General Hospital last year

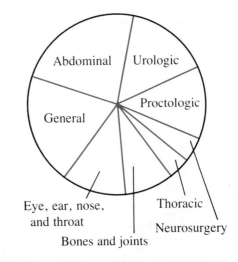

Figure 2-2

Bar graph: Operations performed at General Hospital last year

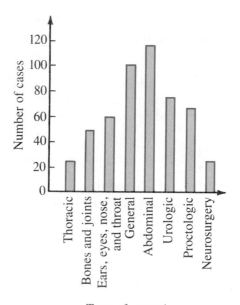

with each type being represented by its proportion and reported in percentages. Figure 2-2 displays the same set of data but in the form of a bar graph.

STEM-AND-LEAF DISPLAYS

stem-and-leaf display

In recent years a technique known as the **stem-and-leaf display** has become very popular for summarizing numerical data. These displays are well-suited for computer application. This technique, very simple to create and use, is a combination of a graphic technique and a sorting technique. (To **sort** data is to make a listing of the data in rank order according to numerical value.) The data values themselves are used to do this sorting. The **stem** is the leading digit(s) of the data, and the **leaf** is the trailing digit(s). For example, the numerical data value 458 might be split 45—8 as shown:

Leading Digits	Trailing Digits
45	8
Used in sorting	Shown in display

Illustration 2-1
20 test scores.

Let's construct a stem-and-leaf display of the following set of

82   74   88   66   58   74   78   84   96   76
62   68   72   92   86   76   52   76   82   78

At a quick glance we see that there are scores in the 50s, 60s, 70s, 80s, and 90s. Let's use the first digit of each score as the stem and the second digit as the leaf. Typically, the display is constructed in a vertical position. Draw a vertical line and place the stems, in order, to the left of it.

```
5 |
6 |
7 |
8 |
9 |
```

Next we place each leaf on its stem. This is done by placing the trailing digit on the right side of the vertical line opposite its corresponding leading digit. Our first data value is 82; 8 is the stem and 2 is the leaf. Thus we place a 2 opposite the stem 8.

```
8 | 2
```

The next data value is 74, so a leaf of 4 is placed on the 7 stem.

```
7 | 4
8 | 2
```

We continue until each of the other 18 leaves is placed on the display. Figure 2-3 shows the resulting stem-and-leaf display.

Figure 2-3

20 Exam Scores							
5	8	2					
6	6	2	8				
7	4	4	8	6	2	6	6	8
8	2	8	4	6	2		
9	6	2					

Once the data have been entered into the computer, reworking it to create a display with different stems is usually easy to accomplish.

In Figure 2-3 all scores with the same tens digit are placed on the same branch. But, this may not always be desired. Suppose we reconstruct the display; this time instead of grouping ten possible values on each stem, let's group the values so that only five possible values could fall on each stem. Do you notice a difference in the appearance of Figure 2-4? The general shape, approximately symmetrical about the 70s, is very much the same. It is fairly typical of many variables to display a

Figure 2-4

20 Exam Scores		
(50–54) 5	2	
(55–59) 5	8	
(60–64) 6	2	
(65–69) 6	6	8
(70–74) 7	4	4	2
(75–79) 7	8	6	6	6	8
(80–84) 8	2	4	2
(85–89) 8	8	6
(90–94) 9	2	
(95–99) 9	6	

distribution that is concentrated (mounded) about a central value and then in some manner dispersed in both directions. ∎

Often a graphic display reveals something that the analyst may or may not have anticipated. Illustration 2-2 demonstrates what generally occurs when two populations are sampled together.

Illustration 2-2　　A random sample of 50 college students was selected. Their weights were obtained from their medical records. The resulting data are listed in Table 2-2.

Table 2-2

Student	1	2	3	4	5	6	7	8	9	10
Male/Female	F	M	F	M	M	F	F	M	M	F
Weight	98	150	108	158	162	112	118	167	170	120

Student	11	12	13	14	15	16	17	18	19	20
Male/Female	M	M	M	F	F	M	F	M	M	F
Weight	177	186	191	128	135	195	137	205	190	120

Student	21	22	23	24	25	26	27	28	29	30
Male/Female	M	M	F	M	F	F	M	M	M	M
Weight	188	176	118	168	115	115	162	157	154	148

Student	31	32	33	34	35	36	37	38	39	40
Male/Female	F	M	M	F	M	F	M	F	M	M
Weight	101	143	145	108	155	110	154	116	161	165

Student	41	42	43	44	45	46	47	48	49	50
Male/Female	F	M	F	M	M	F	F	M	M	M
Weight	142	184	120	170	195	132	129	215	176	183

Figure 2-5

```
Weights of 50 College Students (lbs.)

09 | 8
10 | 8 1 8
11 | 2 8 8 5 5 0 6
12 | 0 8 0 0 9
13 | 5 7 2
14 | 8 3 5 2
15 | 0 8 7 4 5 4
16 | 2 7 8 2 1 5
17 | 0 7 6 0 6
18 | 6 8 4 3
19 | 1 5 0 5
20 | 5
21 | 5
```

Figure 2-6

Weights of 50 College Students (lbs.)		
Female		Male
8	09	
8 1 8	10	
6 0 5 5 8 8 2	11	
9 0 0 8 0	12	
2 7 5	13	
2	14	8 3 5
	15	0 8 7 4 5 4
	16	2 7 8 2 1 5
	17	0 7 6 0 6
	18	6 8 4 3
	19	1 5 0 5
	20	5
	21	5

Notice that the weights range from 98 to 215 pounds. Let's group the weights on stems of ten units using the hundreds and the tens digits as stems and the units digit as the leaf. (*See* Figure 2-5.)

Close inspection of Figure 2-5 suggests that there may be two overlapping distributions involved. That is exactly what we have: a distribution of female weights and a distribution of male weights. Figure 2-6, which shows a "back-to-back" stem-and-leaf display of this set of data, makes it quite obvious that we do in fact have two distinct distributions. ■

EXERCISES

2.1 A computer-anxiety questionnaire was given to 200 students in a course in which computers are used. One of the questions was "I enjoy using computers." The responses to this particular question were:

Response	Number
strongly agree	50
agree	75
slightly agree	25
slightly disagree	15
disagree	15
strongly disagree	20

Construct a bar graph that shows these responses.

2.2 The Madison Gas & Electric Company uses various forms of energy for generating its electricity. The sources and the percentages of energy derived from each source for last year and ten years ago are shown in the following table.

Source of Energy	10 Years Ago	Last Year
Coal	16.8%	62.6%
Nuclear	26.9%	34.1%
Gas	55.9%	2.9%
Other (including oil)	0.4%	0.4%
Total	100.0%	100.0%

Prepare a horizontal bar graph for these data. Put percentages along the horizontal axis and the sources of energy along the vertical axis.

2.3 The distribution of the number of Washington Gas Light Company shares owned by stockholders is as follows:

Number of Shares Owned	Proportion of Total
1–99	0.43
100–249	0.41
250–999	0.14
1,000 or more	0.02
	1.00

Prepare a pie chart that shows the distribution of the number of shares owned.

2.4 A random sample, drawn for the records at a chemical-dependency unit, of the number of days spent by a patient in treatment is shown on the following stem-and-leaf display.

```
1 | 5 1 3 5
2 | 0 2 5 0 3 0 4
3 | 4 4
4 | 1
5 | 0
```

a. How many patients are represented?

b. What was the shortest treatment time?

c. What was the longest treatment time?

d. What length of time occurred most often?

2.5 Construct a stem-and-leaf display of the following data:

13.7   13.7   15.6   11.3   11.7   11.2   13.9
14.0   10.5   12.4   11.2   12.8   11.4   15.1

2.6 An electronic Geiger counter was used to count the number of detectable emissions during a 10-second period. The experiment was repeated 22 times, and the following counts resulted:

$$
\begin{array}{ccccccccccc}
8 & 23 & 18 & 22 & 22 & 15 & 21 & 23 & 25 & 18 & 24 \\
22 & 21 & 37 & 19 & 22 & 22 & 12 & 27 & 16 & 26 & 32
\end{array}
$$

Construct a stem-and-leaf display of these data.

2.7 On the first day of class last semester, 50 students were asked for their one-way travel times from home to college (to the nearest 5 minutes). The resulting data were as follows:

$$
\begin{array}{cccccccccc}
20 & 20 & 30 & 25 & 20 & 25 & 30 & 15 & 10 & 40 \\
35 & 25 & 15 & 25 & 25 & 40 & 25 & 30 & 5 & 25 \\
25 & 30 & 15 & 20 & 45 & 25 & 35 & 25 & 10 & 10 \\
15 & 20 & 20 & 20 & 20 & 25 & 20 & 20 & 15 & 20 \\
5 & 20 & 20 & 10 & 5 & 20 & 30 & 10 & 25 & 15
\end{array}
$$

Construct a stem-and-leaf display of these data.

2.8 A term often used in solar energy research is *heating-degree-days*. This concept is related to the difference between an indoor temperature of 65°F and the average outside temperature for a given day. If the average outside temperature is 5°F, this would give 60 heating-degree-days. The annual heating-degree-days normals for several Nebraska locations follow.

$$
\begin{array}{cccccccccc}
6726 & 6796 & 6946 & 6197 & 6368 & 6437 & 6434 & 6740 & 6886 & 6197 \\
6582 & 6297 & 6811 & 6102 & 6261 & 6919 & 6086 & 6320 & 6139 & 6420 \\
6684 & 6070 & 6169 & 6955 & 6545
\end{array}
$$

Construct a stem-and-leaf display for these data. (Use the leading two digits as stems and the trailing two digits as leaves.)

2.2 FREQUENCY DISTRIBUTIONS, HISTOGRAMS, AND OGIVES

Listing a large set of data does not present much of a picture to the reader. Sometimes we want to condense the data into a more manageable form. This can be accomplished with the aid of a **frequency distribution**.

frequency distribution

FREQUENCY DISTRIBUTIONS

To demonstrate the concept of a frequency distribution, let's use the following set of data:

$$
\begin{array}{ccccc}
3 & 2 & 2 & 3 & 2 \\
4 & 4 & 1 & 2 & 2 \\
4 & 3 & 2 & 0 & 2 \\
2 & 1 & 3 & 3 & 1
\end{array}
$$

If we use x to represent these data values, a frequency distribution can be used to represent this set of data by listing the x values with their frequencies. For example, the value 1 occurs in the sample three times; therefore, the frequency for $x = 1$ is 3. The set of data is represented by the frequency distribution shown in Table 2-3. The **frequency** f is the number of times the value x occurs in the sample. This is an **ungrouped frequency distribution**. We say "ungrouped" because each value of x in the distribution stands alone.

ungrouped frequency distribution

Table 2-3
Frequency distribution

x	f
0	1
1	3
2	8
3	5
4	3

classes

When a large set of data has many different x values instead of a few repeated values, as in the previous example, we can group the values into a set of **classes** and construct a frequency distribution. The stem-and-leaf display in Figure 2-3 shows, in picture form, a grouped frequency distribution. Each stem represents a class. The number of leaves on each stem is the same as the frequency for that same class. The data represented in Figure 2-3 are listed as a frequency distribution in Table 2-4.

Table 2-4

Class	Frequency
50–59	2
60–69	3
70–79	8
80–89	5
90–99	2
	20

The stem-and-leaf process can be used to construct a frequency distribution; however, the stem representation is not compatible with all class widths. For example, class widths of 3, 4, or 7 are awkward to use. Thus, sometimes we will find it advantageous to have a separate procedure for constructing a grouped frequency distribution. To illustrate this grouping (or classifying) procedure, let's use a sample of 50 final exam scores. Table 2-5 lists the 50 scores in ranked order.

The two basic guidelines that should be followed in constructing a grouped frequency distribution are:

1. Each class should be of the same width.

2. Classes should be set up so that they do not overlap and so that each piece of data belongs to exactly one class.

Table 2-5

Ranked data for statistics exam score

27	68	79	91	107
43	71	80	91	108
43	71	81	93	108
44	71	82	94	116
47	73	82	94	120
49	73	84	94	120
50	74	84	96	122
54	75	86	97	123
58	76	88	103	127
65	77	88	106	128

Three additional helpful (but not necessary) guidelines are:

3. Five to 12 classes are most desirable for the exercises given in this textbook.

4. When it is convenient, an odd class width is often advantageous.

5. Use a system that takes advantage of a number pattern, to guarantee accuracy. (This will be demonstrated in the following example.)

Procedure

1. Identify the high and the low scores ($H = 128$, $L = 27$) and find the range. Range $= H - L = 128 - 27 = $ **101**.

2. Select a number of classes ($m = 10$) and a class width ($c = 11$) so that the product ($mc = 110$) is a bit larger than the range (range $= 101$).

3. Pick a starting point. This starting point should be a little smaller than the lowest score L. Suppose that we start at 22; counting from there by elevens (the class width), we get 22, 33, 44, 55, ..., 132. These are called the lower class limits. (They are all multiples of eleven, an easily recognized pattern.)

lower and upper class limits

The **lower class limit** is the smallest piece of data that can go into each class. The **upper class limits** are the largest values fitting into each class: 32, 43, 54, and so on, in our example. Our *classes* for this example are

22–32	77–87
33–43	88–98
44–54	99–109
55–65	110–120
66–76	121–131

Notes

1. At a glance you can check the number pattern to determine whether the arithmetic used to form the classes was correct.

class width

2. The **class width** is the difference between a lower class limit and the next lower class limit. (It is *not* the difference between the lower and upper limit of the same class.)

class boundary

3. Class boundaries are numbers that do not occur in the sample data but are halfway between the upper limit of one class and the lower limit of the next class. In the previous example, the class boundaries are 21.5, 32.5, 43.5, 54.5,..., 120.5, and 131.5. The difference between the upper and lower class boundaries will also give you the class width.

When classifying data it helps to use a **standard chart** (*see* Table 2-6).

Table 2-6

Standard chart for
frequency distribution

Class Number	Class Limits	Tallies	Frequency (f)
1	22–32	|	1
2	33–43	||	2
3	44–54	⅏	5
4	55–65	||	2
5	66–76	⅏ ||||	9
6	77–87	⅏ ||||	9
7	88–98	⅏ ⅏	10
8	99–109	⅏	5
9	110–120	|||	3
10	121–131	||||	4
			50

Once the classes are set up, we need to tally the data (Table 2-6). If the data have been ranked, this tallying is unnecessary; if they are not ranked, be careful as you tally them. The frequency f for each class is the number of pieces of data that belong in that class. The sum of the frequencies should be exactly equal to the number of pieces of data n ($n = \sum f$). This summation serves as a good check.

class mark

In Table 2-7, there is a column for the class mark, x. The **class mark** (sometimes called *class midpoint*) is the numerical value that is exactly in the middle of each class.

Table 2-7

Frequency table with
class marks, a grouped
frequency distribution

Class Number	Class Limits	f	Class Mark (x)
1	22–32	1	27
2	33–43	2	38
3	44–54	5	49
4	55–65	2	60
5	66–76	9	71
6	77–87	9	82
7	88–98	10	93
8	99–109	5	104
9	110–120	3	115
10	121–131	4	126
		50	

In Table 2-7, the class marks are

$$x_1 = \frac{22 + 32}{2} = \mathbf{27}, \qquad x_2 = \frac{33 + 43}{2} = \mathbf{38},$$

and so on. As a check of your arithmetic, successive class marks should be a class width apart, which is 11 in this illustration.

Note   Now you can see why it is helpful to have an odd class width; the class marks are whole numbers.

grouped frequency distribution

Once the class marks are determined, we have a **grouped frequency distribution**. Note that when we classify data in this fashion, we lose some information. Only when we have all the raw data before us do we know the exact values that were actually observed in each class. For example, we put a 58 and a 65 into class number 4, with class limits of 55 and 65. Once they are placed in the class, their values are lost to us and we use the class mark, 60, as their representative value. This loss of identity costs some accuracy, but the computations are made easier.

HISTOGRAMS

histogram

The **histogram** is a type of bar graph representing an entire set of data. The distribution of frequencies from Table 2-7 appears in histogram form in Figure 2-7. A histogram is made up of the following components:

1. A title, which identifies the population of concern.

2. A vertical scale, which identifies the frequencies in the various classes. Figure 2-7 is a frequency histogram.

Figure 2-7

Frequency histogram: 50 final exam scores in elementary statistics

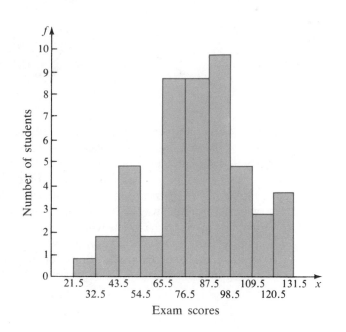

3. A horizontal scale, which identifies the variable x. Values for the class boundaries, class limits, or class marks may be labeled along the x-axis. Use whichever *one* of these sets of class numbers best presents the variable.

[*Note*: Be sure to identify both scales so that the histogram tells the complete story.]

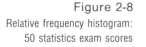

By changing the vertical scale to $\frac{0}{50}, \frac{1}{50}, \frac{2}{50}$, and so on, the histogram would become a **relative frequency histogram**. The **relative frequency** is a proportional measure of the frequency of an occurrence. It is found by dividing the class frequency by the total number of observations. For example, in this illustration the frequency associated with the seventh class (88–98) is 10. The relative frequency is $\frac{10}{50}$ (or 0.20), since these values occurred 10 times in 50 observations. Relative frequency can often be useful in a presentation. They are particularly useful when comparing the frequency distributions of two different sets of data. Figure 2-8 is a relative frequency histogram of the sample of 50 scores. Compare and contrast Figures 2-7 and 2-8.

Figure 2-8

Relative frequency histogram:
50 statistics exam scores

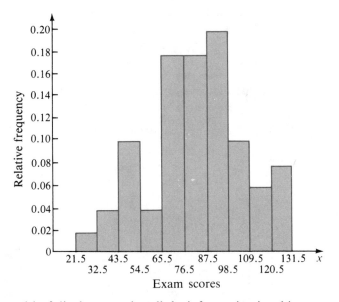

A stem-and-leaf display contains all the information in a histogram. Figure 2-3 shows the stem-and-leaf display constructed in Illustration 2-1. Figure 2-3 has been rotated 90° and labels have been added to form the histogram shown in Figure 2-9.

Histograms are valuable tools. For example, the histogram of the sample should have a distribution shape that is very similar to that of the population from which the sample was drawn. If the reader of a histogram is at all familiar with the variable involved, he or she will usually be able to interpret several important facts. Figure 2-10 presents histograms with descriptive labels resulting from their geometric shape. Can you think of populations whose samples might yield histograms like these?

Figure 2-9
20 exam scores

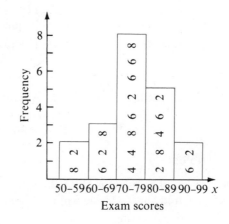

Exam scores

Figure 2-10
Shapes of histograms

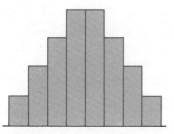

Symmetrical, normal, or triangular

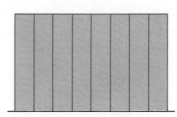

Uniform or rectangular

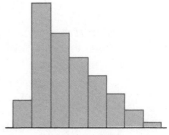

Skewed to right

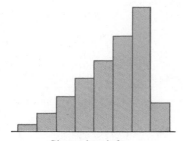

Skewed to left

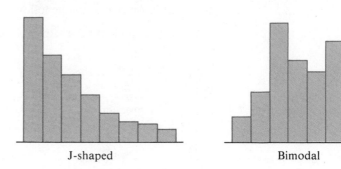

J-shaped

Bimodal

Briefly, the terms used to describe histograms are as follows:

Symmetrical: Both sides of this distribution are identical.

Uniform (rectangular): Every value appears with equal frequency.

Skewed: One tail is stretched out longer than the other. The
                            direction of skewness is on the side of the longer tail.

J-shaped: There is no tail on the side of the class with the highest frequency.

Bimodal: The two most populous classes are separated by one or more classes. This situation often implies that two populations are being sampled.

Normal: A symmetrical distribution that is mounded up about the mean and becomes sparse at the extremes. (Additional properties are discussed later.)

Notes

mode

1. The **mode** is the value of the piece of data that occurs with the greatest frequency. (Mode will be discussed in Section 2.3.)

modal class

2. The **modal class** is the class with the highest frequency.

bimodal frequency distribution

3. A **bimodal frequency distribution** has the two highest frequency classes separated by classes with lower frequencies.

OGIVE

cumulative, and cumulative relative frequency distributions

A frequency distribution can very easily be converted to a **cumulative frequency distribution** by replacing the frequencies with cumulative frequencies. This is done by placing a subtotal of the frequencies next to each class, as in Table 2-8. The **cumulative frequency** for any given class is the sum of the frequency for that class and the frequencies of all classes of smaller values.

The same information can be presented by using a **cumulative relative frequency distribution** (*see* Table 2-9). This combines the cumulative frequency idea and the relative frequency idea.

ogive

An **ogive** (pronounced o'jīv) is a cumulative frequency or cumulative relative frequency graph. An ogive has the following components:

1. A title, which identifies the population.

Table 2-8
A cumulative frequency distribution

Frequency Distribution		Cumulative Frequency Distribution	
Class Limits	Frequencies	Class Boundaries	Cumulative Frequencies
22–32	1	21.5–32.5	1
33–43	2	32.5–43.5	3 (1 + 2)
44–54	5	43.5–54.5	8 (1 + 2 + 5)
55–65	2	54.5–65.5	10 (1 + 2 + 5 + 2)
66–76	9	65.5–76.5	19 (1 + 2 + 5 + 2 + 9)
77–87	9	76.5–87.5	28 ⋮
88–98	10	87.5–98.5	38
99–109	5	98.5–109.5	43
110–120	3	109.5–120.5	46
121–131	4	120.5–131.5	50
	50		

Table 2-9

A cumulative relative
frequency distribution

Class Boundaries	Cumulative Relative Frequency
21.5–32.5	$\frac{1}{50} = 0.02$
32.5–43.5	$\frac{3}{50} = 0.06$
43.5–54.5	$\frac{8}{50} = 0.16$
54.5–65.5	$\frac{10}{50} = 0.20$
65.5–76.5	$\frac{19}{50} = 0.38$
76.5–87.5	$\frac{28}{50} = 0.56$
87.5–98.5	$\frac{38}{50} = 0.76$
98.5–109.5	$\frac{43}{50} = 0.86$
109.5–120.5	$\frac{46}{50} = 0.92$
120.5–131.5	$\frac{50}{50} = 1.00$

2. A vertical scale, which identifies either the cumulative frequencies or the cumulative relative frequencies. (Figure 2-11 shows an ogive with cumulative relative frequencies.)

3. A horizontal scale, which identifies the upper class boundaries. Until the upper boundary of a class has been reached, you cannot be sure you have accumulated all the data in that class. Therefore, **the horizontal scale for an ogive is always based on the upper class boundaries**.

Note   Every ogive starts on the left with a relative frequency of zero at the lower class boundary of the first class and ends on the right with a relative frequency of 100% at the upper class boundary of the last class.

All graphic representations of sets of data should be completely self-explanatory. That includes a descriptive and meaningful title and proper identification of the vertical and horizontal scales.

Figure 2-11

Ogive: 50 final exam
scores in statistics

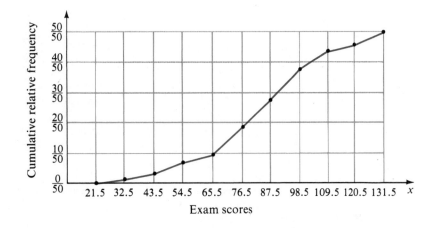

Case Study 2-1

USA SNAPSHOTS
A look at statistics that shape the nation

Cracking the books on campus

The time college freshmen spend studying:

20 - 3.2%
16-20 5.4%
11-15 12.6%
6-10 28.3%
None 2.3%
1-5 48.2%

Hours per week

Source: The American Freshman—National Norms for Fall 1986; December 1986 survey of 204,000 freshmen

"Cracking the books on campus,"
Copyright 1987, USA TODAY. Reprinted with permission.

CRACKING THE BOOKS ON CAMPUS

A February 1987 issue of *USA Today* presented this graphic showing various percentages of 204,000 college freshmen and how many hours the students spent studying. These percentages are *relative frequencies*, as described in this section. This same information could be expressed using either a relative frequency distribution or a relative frequency histogram. (*See* Exercise 2.11.)

EXERCISES

2.9 The monthly percentage changes in the Consumer Price Index for last year were:

0.7 1.0 0.6 0.4 0.7 0.7 1.2 0.8 1.2 0.4 0.5 0.4

a. Prepare a frequency distribution of these changes.

b. Prepare a frequency histogram of these changes.

2.10 The ages of 50 dancers who responded to a call to audition for a musical comedy are:

21 19 22 19 18 20 23 19 19 20
19 20 21 22 21 20 22 20 21 20
21 19 21 21 19 19 20 19 19 19
20 20 19 21 21 22 19 19 21 19
18 21 19 18 22 21 24 20 24 17

a. Prepare an ungrouped frequency distribution of these ages.

b. Prepare an ungrouped relative frequency distribution of the same data.

    c. Prepare a cumulative relative frequency distribution of the same data.

    d. Prepare a relative frequency histogram of these data.

    e. Prepare an ogive of these data.

2.11 Refer to Case Study 2-1, which shows the results of a survey of 204,000 college freshmen.

    a. Express this same information as a relative frequency distribution.

    b. Express this same information as a relative frequency histogram.

2.12 The KSW computer-science aptitude test was given to 50 students. The following frequency distribution resulted from their scores.

KSW Test Score	Frequency
0–3	4
4–7	8
8–11	8
12–15	20
16–19	6
20–23	3
24–27	1

    a. Give all class boundaries associated with this frequency distribution.

    b. Give all class marks associated with this frequency distribution.

    c. What is the class width?

    d. Give the relative frequencies for the classes.

    e. Give the cumulative frequencies for the classes.

    f. Give the cumulative relative frequencies for the classes.

2.13 The speeds of 55 cars were measured by a radar device on a city street:

27	23	22	38	43	24	35	26	28	18	20
25	23	22	52	31	30	41	45	29	27	43
29	28	27	25	29	28	24	37	28	29	18
26	33	25	27	25	34	32	36	22	32	33
21	23	24	18	48	23	16	38	26	21	23

    a. Classify these data into a grouped frequency distribution by using classes of 15–19, 20–24, ..., 50–54. (Retain this solution for use in answering Exercises 2.26 and 2.39.)

    b. Find the class width.

    c. For the class 20–24, find (1) the class mark (2) the lower class limit (3) the upper class boundary.

    d. Construct a frequency histogram of these data.

2.14 The hemoglobin A_{1c} test, a blood test given to diabetics during their periodic checkups, indicates the level of control of blood sugar during the past two to three months. The following data were obtained for 40 different diabetics at a university clinic that treats diabetic patients.

$$
\begin{array}{cccccccccc}
6.5 & 5.0 & 5.6 & 7.6 & 4.8 & 8.0 & 7.5 & 7.9 & 8.0 & 9.2 \\
6.4 & 6.0 & 5.6 & 6.0 & 5.7 & 9.2 & 8.1 & 8.0 & 6.5 & 6.6 \\
5.0 & 8.0 & 6.5 & 6.1 & 6.4 & 6.6 & 7.2 & 5.9 & 4.0 & 5.7 \\
7.9 & 6.0 & 5.6 & 6.0 & 6.2 & 7.7 & 6.7 & 7.7 & 8.2 & 9.0
\end{array}
$$

a. Classify these A_{1c} values into a grouped frequency distribution using the classes 3.7–4.6, 4.7–5.6, and so on.

b. What are the class boundaries for these classes?

c. What are the class marks for these classes?

d. What shape histogram will these data produce?

$ **2.15** The following 40 amounts are the fees that Fast Delivery charged for delivering small freight items last Thursday afternoon:

$$
\begin{array}{cccccccc}
2.03 & 1.56 & 1.10 & 4.04 & 3.62 & 1.16 & 0.93 & 1.82 \\
2.30 & 1.86 & 2.57 & 1.59 & 2.57 & 4.16 & 0.88 & 3.02 \\
3.46 & 1.87 & 4.81 & 2.91 & 1.62 & 1.62 & 1.80 & 1.70 \\
2.15 & 2.07 & 1.77 & 3.77 & 5.86 & 2.63 & 2.81 & 0.86 \\
3.02 & 3.24 & 2.02 & 3.44 & 2.65 & 1.89 & 2.00 & 0.99
\end{array}
$$

a. Classify these data into a grouped frequency distribution by using classes of 0.01–1.00, 1.01–2.00, ..., 5.01–6.00.

b. Find the class width.

c. For the class 2.01–3.00, name the value of (1) the class mark (2) the class limits (3) the class boundaries.

d. Construct a relative frequency histogram of these data.

CALCULATED DESCRIPTIVE STATISTICS

2.3 MEASURES OF CENTRAL TENDENCY

measure of central tendency

Measures of central tendency are numerical values that tend to locate in some sense the *middle* of a set of data. The term *average* is often associated with these measures. Each of the several measures of central tendency can be called the *average* value.

MEAN

mean

To find the **mean**, $\bar{x}$ (read "x bar"), the average with which you are probably most familiar, you will add all the values of the variable x (this sum of x values is symbolized by $\sum x$) and divide by the number of these values, n. We express this in formula form as

$$
\text{sample mean} = \bar{x} = \frac{\sum x}{n} \tag{2-1}
$$

summation   Note   See Appendix A for information about the $\sum$ (read "**sum**") **notation**.

Illustration 2-3   A set of data consists of the five values 6, 3, 8, 5, and 3. Find the mean.

Solution   Using formula (2-1), we find

$$\bar{x} = \frac{6 + 3 + 8 + 5 + 3}{5} = \frac{25}{5} = 5$$

Therefore, the mean of this sample is 5.   ■

A physical representation of the mean can be constructed by thinking of a number line balanced on a fulcrum. A weight is placed on a number on the line corresponding to each number in the sample. In Figure 2-12 there is one weight each on the 6, 8, and 5 and two weights on the 3, since there were two 3s in the sample of Illustration 2-3. The mean is the value that balances the weights on the number line, in this case, 5.

Figure 2-12
Physical representation
of mean

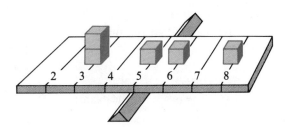

Don't be misled by Illustration 2-3 into believing that the mean value has to be a value in the data set. For example, consider the case in which a salesperson sells $130 worth of merchandise one week and $110 in another. The mean of these values is $120, which is not a value in the data set.

When the sample data has the form of a frequency distribution, we will need to make a slight adaptation in order to find the mean. Consider the frequency distribution of Table 2-10 on page 50. This frequency distribution represents a sample of 28 values, five 1s, nine 2s, eight 3s, and six 4s. To calculate the mean $\bar{x}$ using formula (2-1), we need $\sum x$, the sum of the 28 x-values.

$$\sum x = \underbrace{1 + 1 + \cdots + 1}_{5 \text{ of them}} + \underbrace{2 + 2 + \cdots + 2}_{9 \text{ of them}} + \underbrace{3 + 3 + \cdots + 3}_{8 \text{ of them}} + \underbrace{4 + 4 + \cdots + 4}_{6 \text{ of them}}$$

$$\sum x = (5)(1) + (9)(2) + (8)(3) + (6)(4)$$

$$= 5 + 18 + 24 + 24 = 71$$

We can also obtain this sum by direct use of the frequency distribution. The extensions xf (*see* Table 2-11) are formed for each row by multiplying each x by its corresponding f and then totaled to obtain $\sum xf$. The $\sum xf$ is the sum of the data. Therefore, the mean of a frequency distribution may be found by dividing the sum of the data, $\sum xf$, by the sample size, $\sum f$. We can rewrite formula (2-1) for use with a

Table 2-10

Ungrouped frequency distribution

x	f
1	5
2	9
3	8
4	6
	28

Table 2-11

Extension xf

x	f	xf
1	5	5
2	9	18
3	8	24
4	6	24
Total	28	71

frequency distribution as

$$\bar{x} = \frac{\sum xf}{\sum f} \tag{2-2}$$

The mean for the example is found by using formula (2-2).

Mean:

$$\bar{x} = \frac{\sum xf}{\sum f},$$

$$\bar{x} = \frac{71}{28} = 2.536 = \mathbf{2.5}$$

Notes

1. The extensions in the xf column are the same subtotals as we found previously.

2. The two column totals, $\sum f$ and $\sum xf$, are the same values as were previously known as n and $\sum x$, respectively. That is, $\sum f = n$, the sum of the frequencies is the number of pieces of data. The $\sum xf$ is the sum of all the pieces of data.

3. The f in the summation expression $\sum xf$ indicates that the sum was obtained with the use of a frequency distribution.

Let's return now to the sample of 50 exam scores. In Table 2-7 (see page 40) you will find a grouped frequency distribution for the 50 scores. We will calculate the mean for the sample by using the frequencies and the class marks in the same manner as in the last example. Table 2-12 shows the extensions and totals. [*Note*: The class marks are now being used as representative values for the observed data. Hence the sum of the class marks is meaningless and is therefore not found.]

Table 2-12

Calculations and totals
for exam scores

Class Number	Class Limits	f	Class Marks x	xf
1	22–32	1	27	27
2	33–43	2	38	76
3	44–54	5	49	245
4	55–65	2	60	120
5	66–76	9	71	639
6	77–87	9	82	738
7	88–98	10	93	930
8	99–109	·5	104	520
9	110–120	3	115	345
10	121–131	4	126	504
		50		4,144

The mean: Using formula (2-2) we obtain

$$\bar{x} = \frac{\sum xf}{\sum f} = \frac{4,144}{50} = 82.88 = \mathbf{82.9}$$

To illustrate again what the mean is, consider a balance beam (a real number line) with a cube placed on the value for each response (class mark). That is, nine cubes will be placed on 71, and so on. When the cubes are all placed, one cube for each piece of data, the beam will balance when the fulcrum is at 82.9, the mean. (*See* Figure 2-13.) Formula (2-2) gives an approximation of $\bar{x}$ when working with a grouped frequency distribution. Recall that the value of the class mark is used to represent the value of each piece of data that fell into that class interval.

Figure 2-13

Mean is 82.9

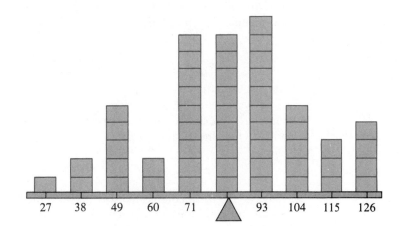

MEDIAN

median   The **median** $\tilde{x}$(read "x tilde" or "median") is the value of the data that occupies the middle position when the data are ranked in order according to size. The data in Illustration 2-3, ranked in order of size, are 3, 3, 5, 6, and 8. The 5 is in the third, or

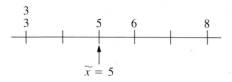

middle, position of the five numbers. Thus, the median is 5. Notice that the median essentially "breaks" the ranked set of data into two subsets.

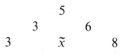

The *depth* (number of positions from either end), or *position*, of the median is determined by the formula

$$\text{depth of median} = d(\tilde{x}) = \frac{n+1}{2}, \tag{2-3}$$

where 1 is the position of the smallest data value, and n is the number of pieces of data, or the position number, of the largest data value. The median's depth (or position) is then found by adding the position numbers of the smallest and largest data values and dividing by 2.

To find the median $\tilde{x}$ you must first have the data in ranked order. Next, determine the depth of the median, and then count over the ranked data, starting from either end, finding the data in the $d(\tilde{x})$th position. The median will be the value of that data and will be the same regardless of which end of the ranked data (high or low) you count from. In fact, counting from both ends will serve as an excellent check.

When n is an odd number, the median will be the exact middle piece of data. In our example, $n = 5$, and therefore the depth of the median is

$$d(\tilde{x}) = \frac{5+1}{2} = 3$$

That is, the median is the third number from either end in the ranked data, or $\tilde{x} = 5$.

However, if n is even, the depth of the median will always be a half-number. For example, let's look at a sample whose ranked data are 6, 7, 8, 9, 9, and 10. Here $n = 6$, and therefore the median's depth is

$$d(\tilde{x}) = \frac{6+1}{2} = 3.5$$

This says that the median is halfway between the third and fourth pieces of data. To find the number halfway between any two values, add the two values together and divide by 2. In this case, add the 8 and the 9, then divide by 2. **The median is 8.5,** a number halfway between the "middle" two numbers. (*See* Figure 2-14.) Notice that the median again "breaks" the ranked set of data into two subsets.

$$
\begin{array}{ccccc}
 & 8 & 8.5 & 9 & \\
 & 7 & \tilde{x} & 9 & \\
6 & & & & 10
\end{array}
$$

Figure 2-14
Median

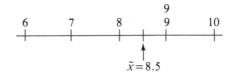

$$\tilde{x} = 8.5$$

OTHER AVERAGES

mode   The **mode** is the value of x that occurs most frequently. In the set of data 3, 3, 5, 6, 8, the mode is 3. The mode in the sample 6, 7, 8, 9, 9, 10 is 9. In both cases these are the only numbers that occur more than once. If it happens that more than one of the values are tied for the highest frequency (number of occurrences), we say there is *no mode*. For example, in the sample 3, 3, 4, 5, 5, and 7, both the 3 and the 5 appear an equal number of times. There is no one value that appears most often. Thus this sample has no mode.

midrange   Another measure of central tendency is the **midrange**. A set of data will always have a lowest value L and a highest value H. The midrange is a number exactly midway between them. It is found by averaging the low and the high values:

$$\text{midrange} = \frac{L + H}{2} \tag{2-4}$$

For the sample 6, 7, 8, 9, 9, and 10, $L = 6$ and $H = 10$. Therefore, the midrange is

$$\frac{6 + 10}{2} = \mathbf{8}$$

The midrange is the numerical value halfway between the two extreme values, the low L and the high H.

The four measures of central tendency represent four different methods of describing the middle. These four values may be the same but more likely they will result in different values. For the sample data shown in Figure 2-14, the mean $\bar{x}$ is 8.2, the median $\tilde{x}$ is 8.5, the mode is 9, and the midrange is 8.

Case Study 2-2 | **"AVERAGE" MEANS DIFFERENT THINGS**

When it comes to convenience, few things can match that wonderful mathematical device called averaging.

How handy it is! With an average you can take a fistful of figures on any subject—temperatures, incomes, velocities, populations, light-years, hairbreadths, anything at all that can be measured—and compute one figure that will represent the whole fistful.

But there is one thing to remember. There are several kinds of measures ordinarily known as averages. And each gives a different picture of the figures it is called on to represent.

Take an example. Here are the annual incomes of ten families:

$54,000	$31,500
$39,000	$31,500
$37,500	$31,500
$36,750	$31,500
$35,250	$25,500

What would this group's "typical" income be? Averaging would provide the answer, so let's compute the typical income by the simpler and most frequently used kinds of averaging.

The arithmetic mean. When anyone cites an average without specifying which kind, you can probably assume that he has the arithmetic mean in mind.

It is the most common form of average, obtained by adding items in the series, then dividing by the number of items. In our example, the sum of the ten incomes divided by 10 is $35,400.

The mean is representative of the series in the sense that the sum of the amounts by which the higher figures exceed the mean is exactly the same as the sum of the amounts by which the lower figures fall short of the mean.

The median. As you may have observed, six families earn less than the mean, four earn more. You might very well wish to represent this varied group by the income of the family that is right smack dab in the middle of the whole bunch.

To do this, you need to find the median. It would be easy if there were 11 families in the group. The family sixth from highest (or sixth from lowest) would be in the middle and have the median income. But with ten families there is no middle family. So you add the two central incomes ($31,500 and $35,250 in this case) and divide by 2. The median works out to $33,375, less than the mean.

The midrange. The median, you will note, is the middle item in the series. Another number that might be used to represent the group is the midrange, computed by calculating the figure that lies halfway between the highest and lowest incomes. To find this figure, add the highest and lowest incomes ($54,000 and $25,500), divide by 2 and you have the amount that lies halfway between the extremes, $39,750.

The mode. So, three kinds of averages, and not one family actually has an income matching any of them.

Say you want to represent the group by stating the income that occurs most frequently. That kind of representativeness is called a mode. In this example $31,500 would be the modal income. More families earn that income than any other. If no two families earned the same income, there would be no mode.

Four different averages, each valid, correct and informative in its way. But how they differ!

arithmetic mean	$35,400
median	$33,375
midrange	$39,750
mode	$31,500

And they would differ still more if just one family in the group were a millionaire—or one were jobless!

So there are three lessons to take away from today's class in averages. First, when you see or hear an average, find out which average it is. Then you'll know what kind of picture you are being given.

Second, think about the figures being averaged so you can judge whether the average used is appropriate.

And third, don't assume that a literal mathematical quantification is intended every time somebody says "average." It isn't. All of us often say "the average person" with no thought of implying a mean, median or mode. All we intend to convey is the idea of other people who are in many ways a great deal like the rest of us.

Reprinted with permission from CHANGING TIMES Magazine, © 1980 Kiplinger Washington Editors, Inc., Mar. 1980. This reprint is not to be altered in any way, except with permission from CHANGING TIMES.

Case Study 2-3

IS YOUR RETURN AUDIT BAIT?

HOW TO AVOID GETTING SWEPT INTO THE IRS' EVER-WIDENING NET

The jaw-rattling visit to the dentist in *The Little Shop of Horrors* aside, few things in life are more harrowing than an Internal Revenue Service audit. Your chances of experiencing this trauma have been mercifully slim up to now: the IRS trains its kliegs on a mere 1.3% of the 96 million or so individual returns filed each year. But here is some chilling news: the IRS is doing its best to tilt those odds more heavily against you.

· · ·

Some factors that determine who gets audited are beyond your control. If you earn more than $50,000, for example, you are eight to 10 times more likely to get an audit notice than someone earning less than $10,000. And if you are self-employed, the odds are even higher that the IRS will call you in for a chat. The IRS has found that self-employed people are more likely than salaried workers to understate their income and overstate their expenses.

Still, by understanding how the IRS audit-selection system works and knowing which deductions are most likely to touch off alarms, you can reduce the likelihood of sending in a tax return loaded with audit bait. Most returns are singled out for an audit via the IRS' discriminant function system, or DIF, a computer program that gives numerical values to such key items on your tax return as deductions, adjustments to income and the number of credits and exemptions you claim. "If a deduction is disproportionate to your income," says Gerald Portney, a tax partner with the accounting firm Peat Marwick in Washington, D.C., "the computer flags it." . . .

This table shows average deductions in six categories for several income groups, based on statistics compiled by the Internal Revenue Service from 1984 tax returns. Tax specialists agree that your chances of being audited increase significantly if your deductions exceed these averages by more than 10%.

Adjusted Gross Income	Employee Business Expense	Medical	Interest	Alimony	Gift	Casualty Theft Loss
$20,000–24,999	$2,054	$1,468	$3,076	$3,223	$767	$2,621
25,000–29,999	2,150	1,421	3,476	4,513	802	5,968
30,000–39,999	2,190	1,749	4,032	5,267	874	3,256
40,000–49,999	2,227	1,876	4,796	4,685	1,104	4,340
50,000–74,999	3,030	3,127	6,492	6,880	1,545	3,964
75,000–99,999	4,062	6,315	9,589	9,958	2,420	2,965
100,000–200,000	6,662	9,494	13,703	14,293	4,234	11,319

Source: From Walter L. Updegrave, "Is Your Return Audit Bait?" *Money*, February 1987, by special permission. Copyright 1987 Time, Inc. All rights reserved.

EXERCISES

2.16 Consider the sample 2, 4, 7, 8, and 9. Find the following:
- **a.** the mean $\bar{x}$
- **b.** the median $\tilde{x}$
- **c.** the mode
- **d.** the midrange

2.17 Consider the sample 6, 8, 7, 5, 3, and 7. Find the following:
- **a.** the mean $\bar{x}$
- **b.** the median $\tilde{x}$
- **c.** the mode
- **d.** the midrange

2.18 Consider the sample 7, 6, 10, 7, 5, 9, 3, 7, 5, and 13. Find the following:
- **a.** the mean $\bar{x}$
- **b.** the median $\tilde{x}$
- **c.** the mode
- **d.** the midrange

2.19 Fifteen randomly selected college students were asked to state the number of hours they slept last night. The resulting data are 5, 6, 6, 8, 7, 7, 9, 5, 4, 8, 11, 6, 7, 8, 7. Find the following:
- **a.** the mean $\bar{x}$
- **b.** the median $\tilde{x}$
- **c.** the mode
- **d.** the midrange

2.20 Recruits for a police academy were required to undergo a test that measures their exercise capacity. The exercise capacity (measured in minutes) was obtained for each of 20 recruits.

25	27	30	33	30	32	30	34	30	27
26	25	29	31	31	32	34	32	33	30

Find the mean, median, mode, and midrange.

2.21 Case Study 2-2 uses a sample of ten annual incomes to discuss the four averages. Calculate the mean, median, mode, and midrange for the ten incomes. Compare your results with those found in the article.

2.22 Use formula (2-2) to find the mean for the following frequency distribution:

x	f
0	1
1	3
2	8
3	5
4	3

2.23 The weight gains (in grams) for chicks fed on a high-protein diet were as follows:

Weight Gain	Frequency
12.5	2
12.7	6
13.0	22
13.1	29
13.2	12
13.8	4

a. Find the mean.   **b.** Find the median.

c. Find the mode.   **d.** Find the midrange.

2.24 Find the mean for the following grouped frequency distribution:

Class Limits	f
3–5	2
6–8	10
9–11	12
12–14	9
15–17	7

2.25 Find the mean for the following grouped frequency distribution:

Class Limits	f
2–5	7
6–9	15
10–13	22
14–17	14
18–21	2

2.26 Find the mean for the set of speeds for the 55 cars given in Exercise 2.13. Use the frequency distribution found in answering Exercise 2.13. (Retain these solutions for use in answering Exercise 2.39.)

2.27 A quality-control technician selected 25 one-pound boxes from a production process and found the following distribution of weights (in ounces).

Weight	Frequency
15.95–15.97	2
15.98–16.00	4
16.01–16.03	15
16.04–16.06	3
16.07–16.09	1

Find the mean weight for this distribution.

2.28 Walter Updegrave cites many averages in the table that appears in Case Study 2-3. Which average (mean, median, midrange, mode) do you believe is used? Explain why.

2.4 MEASURES OF DISPERSION

Once the middle of a set of data has been determined, our search for information immediately turns to the measures of dispersion (spread). The measures of dispersion include the range, variance, and standard deviation. These numerical values describe the amount of spread, or variability, that is found among the data. **Closely grouped data will have relatively small values, and more widely spread-out data will have larger values for these various measures of dispersion.**

RANGE

range   The **range** is the simplest measure of dispersion. It is the difference between the highest- (largest) valued (H) and the lowest- (smallest) valued (L) pieces of data:

$$\text{range} = H - L \tag{2-5}$$

For the sample 3, 3, 5, 6, and 8, the range is $8 - 3 = $ **5**. The range tells us that the five pieces of data all fall within a distance of 5 units on the number line.

The other two measures of dispersion, variance and standard deviation, are actually measures of dispersion about the mean. To develop a measure of dispersion about the mean, let's look first at the concept of **deviation from the mean**. An deviation from the mean   individual value x deviates, or is located, from the mean by an amount equal to $(x - \bar{x})$. This deviation $(x - \bar{x})$ is zero when x is equal to the mean. The deviation $(x - \bar{x})$ is positive if x is larger than $\bar{x}$ and negative if x is less than $\bar{x}$.

Consider the sample 6, 3, 8, 5, and 3. Using formula (2-1), $\bar{x} = (\sum x)/n$, we find that the mean is 5. Each deviation is then found by subtracting 5 from each x value.

x	6	3	8	5	3
$x - \bar{x}$	1	−2	3	0	−2

Figure 2-15

Deviations from the mean

7

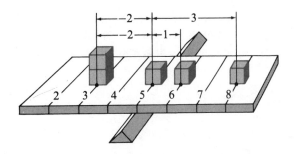

Figure 2-15 shows the deviation from the mean for each value. We might suspect that the sum of all these deviations, $\sum(x - \bar{x})$, would serve as a measure of dispersion about the mean. However, that sum is exactly zero. As a matter of fact, it will always be zero. Why? Think back to the definition of the mean (page 48) and see if you can justify this statement.

If the sum of the deviations, $\sum(x - \bar{x})$, is always zero, it is not going to be of any value in describing a particular set of data. However, we still want to be able to use the idea of deviation about the mean, since the mean is the most common average used. Yet there is a neutralizing effect between the deviations of x values smaller than the mean (negative) and those values larger than the mean (positive). This neutralizing effect can be removed if we make all the deviations positive. By using the absolute value of $x - \bar{x}$, $|x - \bar{x}|$, we can accomplish this. For the previous illustration we now obtain the following absolute deviations:

x	6	3	8	5	3
$|x - \bar{x}|$	1	2	3	0	2

mean absolute deviation

The sum of these deviations is 8. Thus we define a value known as the **mean absolute deviation**:

$$\text{mean absolute deviation} = \frac{\sum|x - \bar{x}|}{n} \qquad (2\text{-}6)$$

For our example the mean absolute deviation then becomes 8/5, or **1.6**. Although this particular measure of spread is not used too frequently, it is a measure of dispersion. It tells us the average distance that a piece of data is from the mean.

There is another way to eliminate the neutralizing effect. Squaring the deviations will cause all these values to be nonnegative (positive or zero). When these values are totaled, the result is positive. The sum of the squares of the deviations from the mean, $\sum(x - \bar{x})^2$, is used to define the variance.

VARIANCE

variance

The **variance**, s^2, of a sample is the numerical value found by applying the following formula to the data:

$$\text{sample variance} = s^2 = \frac{\sum(x - \bar{x})^2}{n - 1} \qquad (2\text{-}7)$$

where n is the sample size, that is, the number of items in the sample. The variance of a sample is a measure of the spread of the data about the mean. The variance of our sample 6, 3, 8, 5, and 3 is found in Table 2-13.

Table 2-13

Computing s^2 for 6, 3, 8, 5, 3 using formula (2-7)

Step 1. x	Step 3. $x - \bar{x}$	Step 4. $(x - \bar{x})^2$
6	1	1
3	−2	4
8	3	9
5	0	0
3	−2	4
25	0	18

Step 2. $\quad \bar{x} = \dfrac{\sum x}{n} = \dfrac{25}{5} = \mathbf{5}$

Step 5. $\quad s^2 = \dfrac{\sum (x - \bar{x})^2}{n - 1} = \dfrac{18}{4} = \mathbf{4.5}$

Let's look at another sample: 1, 3, 5, 6, and 10. The calculations of the totals, the mean, and the variance are shown in Table 2-14.

Table 2-14

Computing s^2 for 1, 3, 5, 6, 10 using formula (2-7)

1. x	3. $x - \bar{x}$	4. $(x - \bar{x})^2$
1	−4	16
3	−2	4
5	0	0
6	1	1
10	5	25
25	0	46

2. $\quad \bar{x} = \dfrac{\sum x}{n} = \dfrac{25}{5} = \mathbf{5}$

5. $\quad s^2 = \dfrac{\sum (x - \bar{x})^2}{n - 1} = \dfrac{46}{4} = \mathbf{11.5}$

Notes

1. The sum of all the x's is used to find $\bar{x}$.

2. The sum of the deviations, $\sum (x - \bar{x})$, is always zero, provided the exact value of $\bar{x}$ is used. Use this as a check in your calculations, as we did in Tables 2-13 and 2-14.

3. If a rounded value of $\bar{x}$ is used, then the $\sum (x - \bar{x})$ will not always be exactly zero. It will, however, be reasonably close to zero.

The last set of data is more dispersed than the previous set, and therefore its variance is larger. A comparison of these two samples is shown in Figure 2-16.

Figure 2-16
Comparison of data

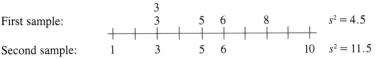

First sample: $s^2 = 4.5$

Second sample: 1   3   5   6   10 $s^2 = 11.5$

STANDARD DEVIATION

standard deviation   The **standard deviation** of a sample, s, is the positive square root of the variance:

$$\text{sample standard deviation} = s = \sqrt{s^2} = \sqrt{\frac{\sum(x - \bar{x})^2}{n - 1}} \qquad (2\text{-}8)$$

For the samples shown in Figure 2-16, the standard deviations are $\sqrt{4.5}$, or **2.1**, and $\sqrt{11.5}$, or **3.4**.

Round-off Rule

When rounding off an answer, let's agree to keep one more decimal place in our answer than was present in our original data. To avoid round-off buildup, round off only the final answer, not the intermediate steps. That is, do not use a rounded variance to obtain a standard deviation. In our previous examples, the data were composed of whole numbers; therefore, those answers that have decimal values should be rounded to the nearest tenth. See Appendix C for specific instructions on how to perform the rounding off.

Note   The numerator for sample variance, $\sum(x - \bar{x})^2$, is often called the *sum of squares for x* and symbolized by $SS(x)$. Thus, formula (2-7) can be expressed

$$s^2 = \frac{SS(x)}{n - 1}, \quad \text{where} \quad SS(x) = \sum(x - \bar{x})^2.$$

The formulas for variance can be modified into other forms for easier use in various situations. For example, suppose that we have the sample 6, 3, 8, 5, and 2. The variance for this sample is computed in Table 2-15.

Table 2-15
Computing s^2 for 6, 3, 8, 5, 2 using formula (2-7)

x	$x - \bar{x}$	$(x - \bar{x})^2$
6	1.2	1.44
3	-1.8	3.24
8	3.2	10.24
5	0.2	0.04
2	-2.8	7.84
24	0	22.80

Using formula (2-1): $\bar{x} = \dfrac{\sum x}{n} = \dfrac{24}{5} = \mathbf{4.8}$

Using formula (2-7): $s^2 = \dfrac{\sum(x - \bar{x})^2}{n - 1} = \dfrac{22.80}{4} = \mathbf{5.7}$

The arithmetic for this example has become more complicated because the mean contains nonzero digits to the right of the decimal point. However, the sum of squares for x can be rewritten:

$$SS(x) = \sum (x - \bar{x})^2 = \sum x^2 - \frac{(\sum x)^2}{n} \tag{2-9}$$

Combining formulas (2-7) and (2-9) yields formula (2-10).

$$s^2 = \frac{SS(x)}{n - 1} = \frac{\sum x^2 - (\sum x)^2/n}{(n - 1)} \tag{2-10}$$

shortcut formula   Formula 2-10 is called the **shortcut formula** because it bypasses the calculation of $\bar{x}$. This shortcut formula can be shown to be equivalent to the defining formula (Exercise 2.44). The computations for s^2 using formula (2-9) are performed as shown in Table 2-16.

Table 2-16
Computing s^2 with the
shortcut formula

x	x^2
6	36
3	9
8	64
5	25
2	4
24	138

Using formula (2-9):

$$SS(x) = 138 - \frac{(24)^2}{5} = 138 - 115.2 = 22.8$$

$$s^2 = \frac{22.8}{4} = \mathbf{5.7}$$

$$s = \sqrt{5.7} = \mathbf{2.4}$$

The units of measure for the standard deviation are the same as the units of measure for the data. For example, if our data are in pounds, then the standard deviation s will also be in pounds. The unit of measure for variance might then be thought of as units squared. In our example of pounds, this would be *pounds squared*. As you can see, this unit has very little meaning.

The reason we have more than one formula to calculate variance is for convenience, not for confusion. As is so common in life, there is an easy way and a hard way to accomplish a goal. In statistics we have several formulas for variance, and if you use the appropriate formula your work will be kept to a minimum.

How do you decide which formula is applicable? With small samples there are only two formulas for variance,

$$\frac{\sum (x - \bar{x})^2}{n - 1} \quad \text{and} \quad \frac{\sum x^2 - (\sum x)^2/n}{n - 1}$$

Notice that the first formula involves the mean. If the mean is unknown or is a decimal (and thereby hard to work with), use the second formula. If the mean $\bar{x}$ is known and is a whole number (and thus convenient to work with), use the first formula. If you need to find both the mean and the variance of a sample, always find the mean first.

In order to use formula (2-10) to calculate the variance when the sample is in the form of a frequency distribution, we must determine the values of n, $\sum x$, and $\sum x^2$ using an extensions table. Recall that in Section 2.3 the values of n and $\sum x$ were found on an extensions table as the values $\sum f$ and $\sum xf$. To obtain $\sum x^2 f$ (it is the value of $\sum x^2$) we need only add a column to our extensions table. Let's look at Table 2-17, which is the same ungrouped frequency distribution used in Section 2-3 (Table 2-10).

To calculate the sum $\sum x^2$ needed, we find the extension $x^2 f$ for each class. These extensions (*see* Tables 2-18 and 2-19) will have the same value as the sum of the squares of the like values of x. For example, $x = 3$ has a frequency of 8, meaning that the value 3 occurs 8 times in the sample. Thus, $x^2 f = (3^2)(8) = 72$, which is the same value obtained by adding the squares of eight 3s. These extensions serve as subtotals. When added, they result in the sum $\sum x^2 f$, the replacement for $\sum x^2$.

Table 2-17

Ungrouped frequency distribution

x	f
1	5
2	9
3	8
4	6
	28

Table 2-18

$\sum x$ and $\sum x^2$

x	x^2	x	x^2
1	1	3	9
1	1	3	9
1	1	3	9
1	1	3	9
1	1	3	9
2	4	3	9
2	4	3	9
2	4	3	9
2	4	4	16
2	4	4	16
2	4	4	16
2	4	4	16
2	4	4	16
2	4	4	16
		71	209

Table 2-19

Extensions xf and x^2f

x	f	xf	x^2f
1	5	5	5
2	9	18	36
3	8	24	72
4	6	24	96
Total	28	71	209

Notes

1. The extension x^2f is found by multiplying x^2 by f or by multiplying xf by x.
2. The three column totals, $\sum f$, $\sum xf$, and $\sum x^2f$, are the same values previously known as n, $\sum x$, and $\sum x^2$, respectively. That is, $\sum f = n$, the sum of the frequencies, is the number of pieces of data. The $\sum xf = \sum x$ and $\sum x^2f = \sum x^2$. The f in the summation expression indicates only that the sum was obtained with the use of a frequency distribution. With these ideas in mind, the formula for the variance (2-10) becomes

$$s^2 = \frac{\sum x^2f - \frac{(\sum xf)^2}{\sum f}}{\sum f - 1} \qquad (2\text{-}11)$$

The standard deviation is the positive square root of variance.

The variance and standard deviation for the example are found by using formula (2-11).

Variance:

$$SS(x) = 209 - \frac{(71)^2}{28}$$

$$= 209 - 180.036 = 28.964$$

$$s^2 = \frac{28.964}{27} = 1.073$$

$$= \mathbf{1.1}$$

Standard deviation:

$$s = \sqrt{s^2} = \sqrt{1.073} = 1.036$$

$$= \mathbf{1.0}$$

Let's return now to the sample of 50 exam scores. In Table 2-7 you will find a grouped frequency distribution for the 50 scores. We will calculate the variance and standard deviation for the sample by using the frequencies and the class marks in the same manner as in the preceding example. Table 2-20 shows the extensions and totals. [*Note:* The class marks are now being used as representative values for the observed data. Hence the sum of the class marks is meaningless and is therefore not found.]

Table 2-20
Calculations and totals
for exam scores

Class Number	Class Limits	f	Class Marks, x	xf	x²f
1	22–32	1	27	27	729
2	33–43	2	38	76	2,888
3	44–54	5	49	245	12,005
4	55–65	2	60	120	7,200
5	66–76	9	71	639	45,369
6	77–87	9	82	738	60,516
7	88–98	10	93	930	86,490
8	99–109	5	104	520	54,080
9	110–120	3	115	345	39,675
10	121–131	4	126	504	63,504
		50		4,144	372,456

The variance: Using formula (2-11) we have

$$s^2 = \frac{\sum x^2 f - \frac{\left(\sum xf\right)^2}{\sum f}}{\sum f - 1}$$

$$= \frac{372{,}456 - \frac{(4{,}144)^2}{50}}{49} = \mathbf{591.86}$$

The standard deviation is

$$s = \sqrt{s^2} = \sqrt{591.86} = 24.33 = \mathbf{24.3}$$

EXERCISES

2.29 Consider the sample 2, 4, 7, 8, and 9. Find the following:
   a. range
   b. variance s^2; use formula (2-7)
   c. standard deviation s

2.30 Consider the sample 6, 8, 7, 5, 3, and 7. Find the following:
   a. range
   b. variance s^2; use formula (2-7)
   c. standard deviation s

2.31 Given the sample 7, 6, 10, 7, 5, 9, 3, 7, 5, 13, find the following:
   a. variance s^2; use formula (2-7)
   b. variance s^2; use formula (2-10)
   c. standard deviation s

2.32 Using the information from Exercise 2.19 concerning the sample of number of hours slept, find the following:

    a. variance s^2; use formula (2-7)

    b. variance s^2; use formula (2-10)

    c. standard deviation s

2.33 A company specializing in the manufacture of shafts is considering the purchase of a computer-controlled cutting machine. The company engineer tests machines from two different manufacturers. The diameters (in centimeters) of the shafts cut by the two machines were found to be:

    Manufacturer 1:   2.001, 2.000, 2.004, 1.998, and 1.997

    Manufacturer 2:   2.002, 2.008, 1.995, 1.990, and 2.005

Compute the mean and standard deviation for each and comment on the results obtained from the two machines.

2.34 During the last few years Portland General Electric Company has requested rate increases several times. As a result of the granted rate increases, the following revenue amounts will be earned:

$34.5 million	$62.3 million
13.3	83.8
22.0	58.3
41.5	10.8

What is the variance for the revenue amounts realized?

2.35 Use formula (2-11) to find the variance and the standard deviation for the following frequency distribution:

x	f
0	1
1	3
2	8
3	5
4	3

2.36 Find the variance and the standard deviation for the following grouped frequency distribution:

Class Limits	f
3–5	2
6–8	10
9–11	12
12–14	9
15–17	7

2.37 Find the mean and variance for the following grouped frequency distribution:

Class Limits	f
2–5	7
6–9	15
10–13	22
14–17	14
18–21	2

2.38 The following distribution of commuting distances was obtained for a sample of Mutual of Nebraska employees.

Distance (Miles)	Frequency
1.0–2.9	2
3.0–4.9	6
5.0–6.9	12
7.0–8.9	50
9.0–10.9	35
11.0–12.9	15
13.0–14.9	5

Find the standard deviation for the commuting distances.

2.39 Find the standard deviation for the set of speeds for the 55 cars given in Exercise 2.13. Use the frequency distribution found in answering Exercise 2.13.

2.40 A sheet metal firm employs several troubleshooters to make emergency repairs of furnaces. Typically, the troubleshooters take many short trips. For the purpose of estimating travel expenses for the coming year a sample of 20 travel-expense vouchers related to troubleshooting was taken. The following information resulted:

Dollar Amount on Voucher	Number of Vouchers
$0.01–10.00	2
10.01–20.00	8
20.01–30.00	7
30.01–40.00	2
40.01–50.00	1
Total in Sample	20

Calculate the mean and the standard deviation for these travel account dollar amounts.

2.41 Adding (or subtracting) the same number from each value in a set of data does not affect the measures of variability for that set of data.

   a. Find the variance of the following set of annual heating-degree-days data.

6017, 6173, 6275, 6350, 6001, and 6300

b. Find the variance of the following set of data [obtained by subtracting 6000 from each value in (a)]:

$$17, 173, 275, 350, 1, \text{ and } 300$$

2.42 Consider the following two sets of data:

| Set 1: | 46 | 55 | 50 | 47 | 52 |
| Set 2: | 30 | 55 | 65 | 47 | 53 |

Both sets have the same mean, which is 50. Compare these measures for both sets: $\sum (x - \bar{x})$, $\sum |x - \bar{x}|$, SS(x), and range. Comment on the meaning of these comparisons.

2.43 Comment on the following statement: "The mean loss for customers at First State Bank (which was not insured) was \$150. The standard deviation of the losses was $-\$125$."

2.44 Show that the shortcut formula for the sum of squares for x, SS(x), and $\sum x^2 - (\sum x)^2/n$ [formula (2-9)] is equivalent to $\sum (x - \bar{x})^2$, which is the sum of squares by definition.

2.5 MEASURES OF POSITION

measure of position   **Measures of position** are used to describe the location of a specific piece of data in relation to the rest of the sample. Quartiles and percentiles are the two most popular measures of position.

QUARTILES

quartile   **Quartiles** are numbers that divide the ranked data into quarters; each set of data has three quartiles. The **first quartile**, Q_1, is a number such that at most one-fourth of the data are smaller in value than Q_1 and at most three-fourths are larger. The **second quartile** is the median. The **third quartile**, Q_3, is a number such that at most three-fourths of the data are smaller in value than Q_3 and at most one-fourth are larger. (*See* Figure 2-17.)

The procedure for determining the value of the quartiles is the same as that for percentiles and is shown in the following description of percentiles.

Figure 2-17
Quartiles

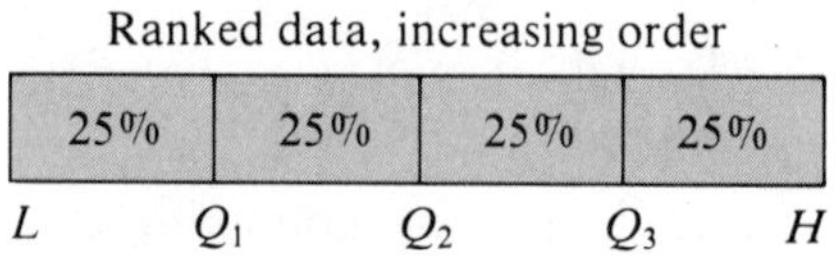

Ranked data, increasing order

| 25% | 25% | 25% | 25% |

L $\quad$ Q_1 $\quad$ Q_2 $\quad$ Q_3 $\quad$ H

PERCENTILES

percentile   **Percentiles** are numbers that divide a set of ranked data into 100 equal parts; each set of data has 99 percentiles. (*See* Figure 2-18.) The kth percentile, P_k, is a number

Figure 2-18

Percentiles

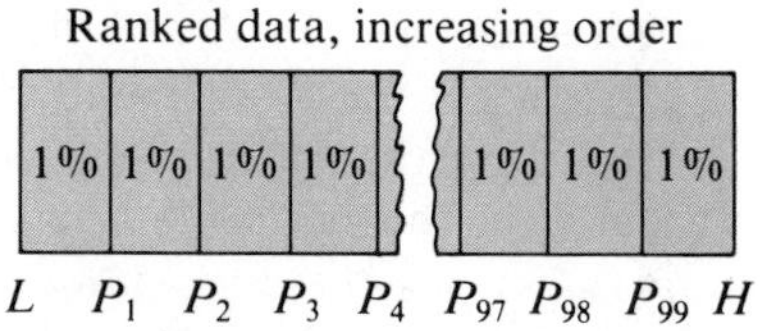

Ranked data, increasing order

$$1\% \mid 1\% \mid 1\% \mid 1\% \quad\quad 1\% \mid 1\% \mid 1\%$$

$$L \quad P_1 \quad P_2 \quad P_3 \quad P_4 \quad P_{97} \ P_{98} \ P_{99} \ H$$

Figure 2-19

kth percentile

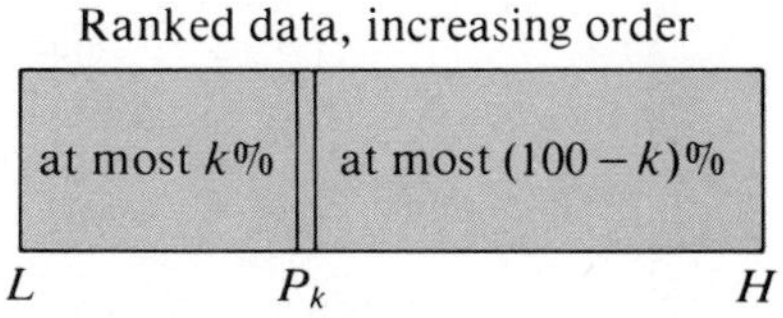

Ranked data, increasing order

$$\text{at most } k\% \quad\quad \text{at most } (100-k)\%$$

$$L \quad\quad\quad P_k \quad\quad\quad\quad\quad H$$

value such that at most k percent of the data are smaller in value than P_k and at most $(100 - k)$ percent of the data are larger. (*See* Figure 2-19.)

Notes

1. The 1st quartile and the 25th percentile are the same; that is, $Q_1 = P_{25}$. Also, $Q_3 = P_{75}$.

2. The median, the 2d quartile, and the 50th percentile are all the same: $\tilde{x} = Q_2 = P_{50}$. Therefore, when asked to find P_{50} or Q_2, use the procedure for finding the median.

The procedure for determining the value of any kth percentile (or quartile) involves three basic steps.

Step 1: The data must be ranked: start at the lowest-valued piece of data and proceed toward the larger-valued data.

Step 2: The depth of the kth percentile must be determined. If k is less than 50: The depth is found by first calculating the value of $\frac{nk}{100}$, where n is the sample size. If the calculated value is not an integer (that is, it contains a fraction), then the depth is equal to the next larger integer. For example, if $\frac{nk}{100} = 17.2$, then $d(P_k) = 18$. If the calculated value is an integer, then the depth is equal to $\frac{nk}{100} + 0.5$. For example, if $\frac{nk}{100} = 23$, then $d(P_k) = 23.5$.

If k is greater than 50:
Subtract k from 100 and use the value of $100 - k$ in place of k in Step 2. For example, if $k = 80$, then $100 - k = 100 - 80 = 20$. The depth of this kth percentile is then determined using $k = 20$.

Step 3: Locate the value of P_k. To do this when $k < 50$, count the data starting from the lowest-valued piece of data, finding the data in the $d(P_k)$th position. (If $k > 50$, start the counting process from the highest-valued piece of data.) If $d(P_k)$ is an integer, the value of P_k will be the value of the data located. If $d(P_k)$ is not an integer, that is, it contains the fraction $\frac{1}{2}$, and the value of P_k is halfway between the $\frac{nk}{100}$th and the $(\frac{nk}{100} + 1)$th piece of data, then add the two values and divide by 2.

Illustration 2-4   A sample of 50 raw scores was taken from a population of raw scores for an elementary statistics final exam. These scores are listed in Table 2-21 in the order of collection. Find the first quartile Q_1, the 56th percentile P_{56}, and the third quartile Q_3.

Table 2-21
Raw scores for elementary
statistics exam

75	96	74	120	71
97	58	73	94	106
71	94	68	79	86
65	43	54	80	108
84	116	50	82	84
27	123	49	71	93
108	91	81	88	77
91	120	128	88	107
122	94	103	47	44
82	43	76	73	127

Solution   The first step is to rank the data. The resulting ranked data are listed in Table 2-22 in uniform columns (that is, each column has the same number of items).

Table 2-22
Ranked data

27	68	79	91	107
43	71	80	91	108
43	71	81	93	108
44	71	82	94	116
47	73	82	94	120
49	73	84	94	120
50	74	84	96	122
54	75	86	97	123
58	76	88	103	127
65	77	88	106	128

To find Q_1 we need to determine the location of Q_1 by using $\frac{nk}{100}$.

$$n = 50 \quad \text{since there are 50 pieces of data}$$
$$k = 25 \quad \text{since } Q_1 = P_{25}$$

$$\frac{nk}{100} = \frac{(50)(25)}{100} = 12.5$$

Thus $d(Q_1) = 13$. So Q_1 is the thirteenth value, counting from L. Thus $Q_1 = 71$.
To find P_{56} we need to determine its location, using $\frac{nk}{100}$.

$$n = 50 \quad \text{since there are 50 pieces of data}$$
$$k = 56 \quad \text{from } P_{56}; \text{ therefore use } 44 \ (100 - 56) \text{ for } k$$

$$\frac{nk}{100} = \frac{(50)(44)}{100} = 22$$

Thus $d(P_{56}) = 22.5$ from H. So P_{56} is the value halfway between the 22d and the

23d pieces of data counting from H down. Therefore,

$$P_{56} = \frac{86 + 88}{2} = 87$$

To find Q_3 we need to determine its location, using $\frac{nk}{100}$.

$$n = 50 \quad \text{and} \quad k = 25 \quad \text{since } Q_3 = P_{75}$$

(Notice that these calculations are the same ones that were completed for Q_1). Thus $d(Q_3) = 13$. So Q_3 is the thirteenth value, counting from H. Thus $Q_3 = \mathbf{97}$.   ■

An additional measure of central tendency, the midquartile, can now be defined.
midquartile   The **midquartile** is the numerical value midway between the first quartile and the third quartile.

$$\text{midquartile} = \frac{Q_1 + Q_3}{2} \tag{2-12}$$

Illustration 2-5   Find the midquartile for the set of 50 exam scores given in Illustration 2-4.

Solution $Q_1 = 71$ and $Q_3 = 97$, as found in Illustration 2-4. Thus,

$$\text{midquartile} = \frac{71 + 97}{2} = \frac{168}{2} = \mathbf{84}$$ ■

Note   The median, the midrange, and the midquartile are not necessarily the same value. They are each *middle* values, but by different definitions of the middle.

5-NUMBER SUMMARIES

5-number summary   A **5-number summary** is very effective in describing a set of data. It is easy information to obtain and is very informative to the reader. The 5-number summary is composed of

1. L, the smallest value in the data set
2. Q_1, the first quartile (also called P_{25}, twenty-fifth percentile)
3. $\tilde{x}$, the median
4. Q_3, the third quartile (also called P_{75}, seventy-fifth percentile)
5. H, the largest value in the data set

The 5-number summary for the set of 50 exam scores in Illustration 2-4 is

27	71	83	97	128
L	Q_1	$\tilde{x}$	Q_3	H

Notice that these five numerical values divide the set of data into four subsets, with one-quarter of the data in each subset. From this information we can observe how much the data is spread out in each of the fourths. An additional measure of
interquartile range   dispersion can now be defined. The **interquartile range** is the difference between the first and third quartiles. It is *the range of the middle 50 percent of the data*.

This 5-number summary is even more informative when it is displayed on a diagram drawn to scale. One of the modern computer-generated graphic displays that accomplishes this is known as the *box-and-whisker* display.

BOX-AND-WHISKER DISPLAYS

box-and-whisker display

The **box-and-whisker display** is a graphic 5-number summary of a set of data. Five numerical values (smallest, *lower hinge*, median, *upper hinge*, and largest) are located on a scale that can be either vertical or horizontal. The **box** is used to depict the middle half of the data that lies between the two hinges. The **whiskers** are line segments used to depict the other half of the data; one line segment represents the quarter of the data that is smaller in value than the lower hinge, and a second line segment represents the quarter of the data that is larger in value than the upper hinge. The hinges and the median break the ranked set of data into four subsets in much the same way that the median breaks the data into two subsets. The values

hinges

used as *hinges* are often the same as the values of the quartiles, but, depending on the number of pieces of data, the value of a quartile and its corresponding hinge may be slightly different. This difference, when it occurs, results because the method used to determine the hinge is slightly different from that used for the quartiles.

The following procedure for finding the hinges is, however, quite similiar to that for finding percentiles.

Step 1: Rank the data.

Step 2: The depth of the hinge, the number of positions from extreme values, is calculated by adding 1 to the *integer value* of the depth of the median and dividing by 2.

$$\text{depth of hinge} = d(H) = \frac{[d(\tilde{x})] + 1}{2}$$

integer value

Note   The *integer value* of a number, symbolized by [ ], is the integer part only. Any fractional part is discarded. For example, 9.0 remains the value 9, and 9.5 becomes 9.

Let's consider the following set of 19 quiz scores (already ranked): 52, 62, 66, 68, 72, 74, 74, 76, 76, 76, 78, 78, 82, 82, 84, 86, 88, 92, 96

$$\text{Depth of the median is } 10 \left(d(\tilde{x}) = \frac{n+1}{2} = \frac{19+1}{2} = 10 \right)$$

$$\text{Depth of the hinge is } 5.5 \left(d(\text{hinge}) = \frac{[d(\tilde{x})] + 1}{2} = \frac{10+1}{2} = 5.5 \right)$$

The *lower hinge*, *LH*, is the value in the 5.5th position when you start counting from the *L*, the smallest value. (Recall that the half position number means that the value is halfway between two data values.) Thus *LH* = 73, the sum of the 5th and 6th values divided by 2. The *upper hinge*, *UH*, is the value in the 5.5th position when you start counting from the *H*, the largest value. Thus *UH* = 83, the sum of the 5th and 6th values in from *H* divided by 2.

The 5-number summary is $L = 52, LH = 73, \tilde{x} = 76, UH = 83, H = 96$. Notice the way the ranked set of data is "broken" into four subsets.

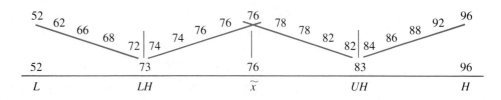

The box-and-whisker display for this set follows.

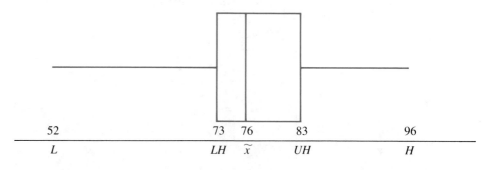

Examine the box-and-whisker display. Notice the amount of spread in each of the fourths: L to LH, LH to $\tilde{x}$, $\tilde{x}$ to UH, UH to H. One-fourth of the data is between 73 and 76 in value. One-half of the data is between 73 and 83.

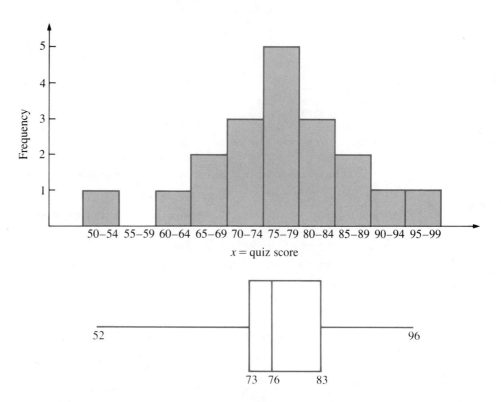

Compare the box-and-whisker display on page 73 to a histogram of the same data. The figure contains both the histogram and the box-and-whisker diagram drawn on the same horizontal scale.

Now let's consider a different set of 19 quiz scores.

52, 56, 58, 60, 64, 66, 68, 70, 72, 74, 76, 76, 78, 82, 84, 86, 86, 92, 96

The following figure shows this second set of quiz scores. Notice how the histogram is flatter than the histogram for the first set. Notice also the difference in appearance of the box-and-whisker display. The data in this second set are much more dispersed.

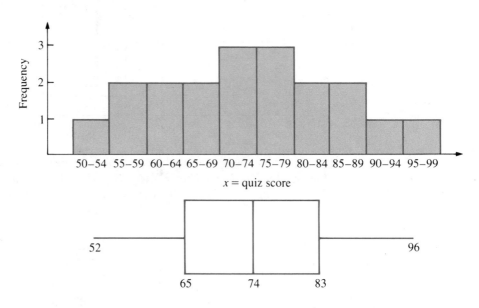

z SCORE

z, or standard, score   The **z score**, also called the **standard score**, is the position a particular value of *x* has in terms of the number of standard deviations it is from the mean of the set of data to which it is being compared. The *z* score is found by the formula

$$z = \frac{\text{piece of data} - \text{mean of the data}}{\text{standard deviation of the data}} = \frac{x - \bar{x}}{s} \qquad (2\text{-}13)$$

Illustration 2-6   Find the standard scores for (a) 92 and (b) 72 with respect to a sample of exam grades that have a mean score of 75.9 and a standard deviation of 11.1.

Solution

a. $x = 92$, $\bar{x} = 75.9$, and $s = 11.1$. Thus

$$z = \frac{92 - 75.9}{11.1} = \frac{16.1}{11.1} = \mathbf{1.45}$$

b. $x = 72$, $\bar{x} = 75.9$, and $s = 11.1$. Thus

$$z = \frac{72 - 75.9}{11.1} = \frac{-3.9}{11.1}$$

$$= -0.35$$

This means that the score 92 is approximately one and one-half standard deviations above the mean, while the score 72 is approximately one-third of a standard deviation below the mean. ■

Notes

1. Typically, the calculated value of z is rounded to the nearest hundredth.
2. z scores typically range in value from approximately -3.00 to $+3.00$.

Since the z score is a measure of relative position with respect to the mean, it can be used to help make a comparison of two raw scores that come from separate populations. For example, suppose that you want to compare a grade that you received on a test with a friend's grade on a comparable exam in her course. You received a raw score of 45 points; she obtained 72 points. Is her grade better? We need more information before we can draw such a conclusion. Suppose that the mean on the exam you took was 38 and the mean on her exam was 65. Your grades are both 7 points above the mean, so we still can't draw a definite conclusion. However, the standard deviation on the exam you took was 8 points, and it was 14 points on your friend's exam. This means that your score is $\frac{7}{8}$ of a standard deviation above the mean ($z = \frac{7}{8}$), whereas your friend's grade is only $\frac{1}{2}$ of a standard deviation above the mean ($z = \frac{1}{2}$). Since your score has the "better" relative position, we would conclude that your score was slightly higher than your friend's score. (Again, this is speaking from a relative point of view.)

EXERCISES

2.45 The following data are the yields, in pounds, of hops:

3.9	3.4	5.1	2.7	4.4
7.0	5.6	2.6	4.8	5.6
7.0	4.8	5.0	6.8	4.8
3.7	5.8	3.6	4.0	5.6

a. Find the first and the third quartiles of the yield.
b. Find the midquartile.
c. Find the following percentiles: (1) P_{15}, (2) P_{33}, (3) P_{90}.

2.46 A research study of manual dexterity involved determining the time required to complete a task. The time required for each of 40 handicapped individuals is as follows (data are ranked):

7.1	7.2	7.2	7.6	7.6	7.9	8.1	8.1	8.1	8.3
8.3	8.4	8.4	8.9	9.0	9.0	9.1	9.1	9.1	9.1
9.4	9.6	9.9	10.1	10.1	10.1	10.2	10.3	10.5	10.7
11.0	11.1	11.2	11.2	11.2	12.0	13.6	14.7	14.9	15.5

Find **a.** Q_1,   **b.** Q_2,   **c.** Q_3,   **d.** P_{95},   **e.** the 5-number summary.
f. Draw the box-and-whisker display.

2.47 Consider the following set of ignition times that were recorded for a synthetic fabric:

30.1	30.1	30.2	30.5	31.0	31.1	31.2	31.3	31.3	31.4
31.5	31.6	31.6	32.0	32.4	32.5	33.0	33.0	33.0	33.5
34.0	34.5	34.5	35.0	35.0	35.6	36.0	36.5	36.9	37.0
37.5	37.5	37.6	38.0	39.5					

Find **a.** the median,   **b.** the midrange,   **c.** the midquartile,
d. the 5-number summary,   **e.** Draw the box-and-whisker display.

2.48 Find the following percentiles for the set of 50 final exam scores shown in Table 2-22.

    a. P_{35} **b.** P_{60} **c.** P_{95}

2.49 A sample has a mean of 50 and a standard deviation of 4.0. Find the z score for each value of x.

    a. $x = 54$ **b.** $x = 50$ **c.** $x = 59$ **d.** $x = 45$

2.50 A nationally administered test has a mean of 500 and a standard deviation of 100. If your standard score on this test was 1.8, what was your test score?

2.51 A sample has a mean of 120 and a standard deviation of 20.0. Find the value of x that corresponds to each of these standard scores:

    a. $z = 0.0$ **b.** $z = 1.2$ **c.** $z = -1.4$ **d.** $z = 2.05$

2.52 a. What does it mean to say that $x = 152$ has a standard score of $+1.5$?

    b. What does it mean to say that a particular value of x has a z score of -2.1?

    c. In general, the standard score is a measure of what?

2.53 Which x value has the higher value relative to the set of data from which it comes?

    $A: x = 85$, where mean $= 72$ and standard deviation $= 8$
    $B: x = 93$, where mean $= 87$ and standard deviation $= 5$

2.54 Which x value has the lower relative position with respect to the set of data from which it comes?

    $A: x = 28.1$, where $\bar{x} = 25.7$ and $s = 1.8$
    $B: x = 39.2$, where $\bar{x} = 34.1$ and $s = 4.3$

2.6 INTERPRETING AND UNDERSTANDING STANDARD DEVIATION

Standard deviation is a measure of fluctuation (dispersion) in the data. It has been defined as a value calculated with the use of formulas. But you may wonder what it really is. It is a kind of yardstick by which we can compare one set of data with another. This particular "measure" can be understood further by examining two statements, Chebyshev's theorem and the empirical rule.

Chebyshev's theorem

Chebyshev's Theorem

The proportion of any distribution that lies within k standard deviations of the mean is at least $1 - (\frac{1}{k^2})$, where k is any positive number larger than 1. This theorem applies to any distribution of data.

This theorem says that within two standard deviations of the mean ($k = 2$) you will always find at least 75 percent (that is, 75 percent *or more*) of the data.

$$1 - \frac{1}{k^2} = 1 - \frac{1}{2^2} = 1 - \frac{1}{4} = \frac{3}{4} = 0.75$$

Figure 2-20 shows a mounded distribution that illustrates this.

If we consider the interval enclosed by three standard deviations on either side of the mean ($k = 3$), the theorem says that we will find at least $\frac{8}{9}$, or 89 percent, of the data $[1 - (\frac{1}{k^2}) = 1 - (\frac{1}{3^2}) = 1 - (\frac{1}{9}) = \frac{8}{9}]$, as shown in Figure 2-21.

Figure 2-20
Chebyshev's theorem
with $k = 2$

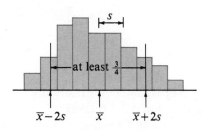

Figure 2-21
Chebyshev's theorem
with $k = 3$

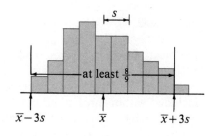

Empirical Rule

If a variable is normally distributed, then within one standard deviation of the mean there will be approximately 68 percent of the data. Within two standard deviations of the mean there will be approximately 95 percent of the data, and within three standard deviations of the mean there will be approximately 99.7 percent of the data. [This rule applies specifically to a normal (bell-shaped) distribution, but it is frequently applied as an interpretive guide to any mounded distribution.]

Figure 2-22 shows the intervals of one, two, and three standard deviations about the mean of an approximately normal distribution. Usually these proportions will not occur exactly in a sample, but your observed values will be close when a large sample is drawn from a normally distributed population.

Figure 2-22
Empirical rule

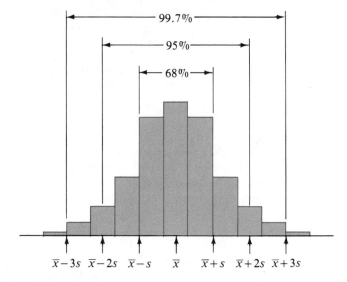

The empirical rule can be used to determine whether or not a set of data is approximately normally distributed. Let's demonstrate this by working with the distribution of final exam scores that we have been using throughout this chapter. The mean $\bar{x}$ was found to be 82.9, and the standard deviation s was 24.3. The interval from one standard deviation below the mean, $\bar{x} - s$, to one standard deviation above the mean, $\bar{x} + s$, is $82.9 - 24.3 = \mathbf{58.6}$ to $82.9 + 24.3 = \mathbf{107.2}$. This interval includes 59, 60, 61, ..., 106, 107. Upon inspection of the ranked data (Table 2-5), we see that 32 of the 50 data pieces, or 64 percent, lie within one standard deviation of the mean. Further, $\bar{x} - 2s = 82.9 - (2)(24.3) = 82.9 - 48.6 = \mathbf{34.3}$, $\bar{x} + 2s = 82.9 + 48.6 = \mathbf{131.5}$, and the interval from 35 to 131, two standard deviations about the mean, includes 49 of the 50 data pieces, or 98 percent. All 50 data, or 100 percent, are included within three standard deviations of the mean (from 10.0 to 155.8). This information can be placed in a table for comparison with the values given by the empirical rule (*see* Table 2-23).

Table 2-23

Observed percentages versus the empirical rule

Interval	Empirical Rule Percentage	Percentage Found
$\bar{x} - s$ to $\bar{x} + s$	≈ 68	64
$\bar{x} - 2s$ to $\bar{x} + 2s$	≈ 95	98
$\bar{x} - 3s$ to $\bar{x} + 3s$	≈ 99.7	100

These percentages are reasonably close to those found by using the empirical rule. By combining this evidence with the shape of the histogram of the data, we can safely say that the sample data are approximately normally distributed.

If a distribution is approximately normal, it will be nearly symmetrical and the mean (the mean and the median are the same in a symmetrical distribution) will divide the distribution in half. This allows us to refine the empirical rule. Figure 2-23 shows this refinement.

Figure 2-23

Refinement of empirical rule

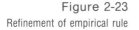

There is also a graphic way to test for normality. This is accomplished by drawing a relative frequency ogive of the grouped data on **probability paper** (which can be purchased at your college bookstore). On this paper the vertical scale is measured in percentages and is placed on the right side of the graph paper. All the directions and guidelines given on pages 44–45 for drawing an ogive must be followed. An ogive of the statistics exam scores is drawn and labeled on a piece of probability paper in Figure 2-24. To **test for normality**, first draw a straight line from the lower-left corner to the upper-right corner of the graph. Then, if the ogive lies close to this straight line, the distribution is said to be approximately normal. The ogive for an exactly normal distribution will trace the straight line.

The dashed line in Figure 2-24 is the straight-line test for normality. The ogive suggests that the distribution of exam scores is approximately normal. (*Warning*: This graphic technique is very sensitive to the scale used along the horizontal axis.)

The ogive is a very handy tool to use for finding percentiles of a frequency distribution. If the frequency distribution is drawn on probability paper, as in Figure 2-24, it is a very simple matter to determine any kth percentile. Locate the value k on the vertical scale along the right-hand side, and follow the horizontal line for k until it intersects the line of the ogive. Then follow the vertical line, that passes through this point of intersection, to the bottom of the graph. Read the value of x from the horizontal scale. This value of x is the value of the kth percentile. For example, let's find the value of P_{40} for the data shown on Figure 2-24. Then locate the

probability paper

test for normality

Figure 2-24

Ogive on probability paper:
Final exam scores in statistics

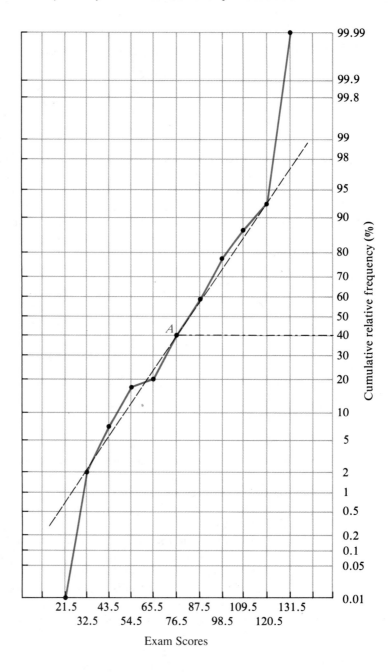

value of k (40) along the right-hand side of the graph. Follow the horizontal line at k (40) until it intersects the graph of the ogive. Then determine the value of x that corresponds to this point of intersection (point A on Figure 2-24) by reading it from the scale on the x-axis: $P_{40} = 76.5$. This method can be used to find percentiles, quartiles, or the median. (*Note*: Because of the grouping into classes, the results may differ slightly from the answers obtained when using the ranked data.)

Case Study 2-4

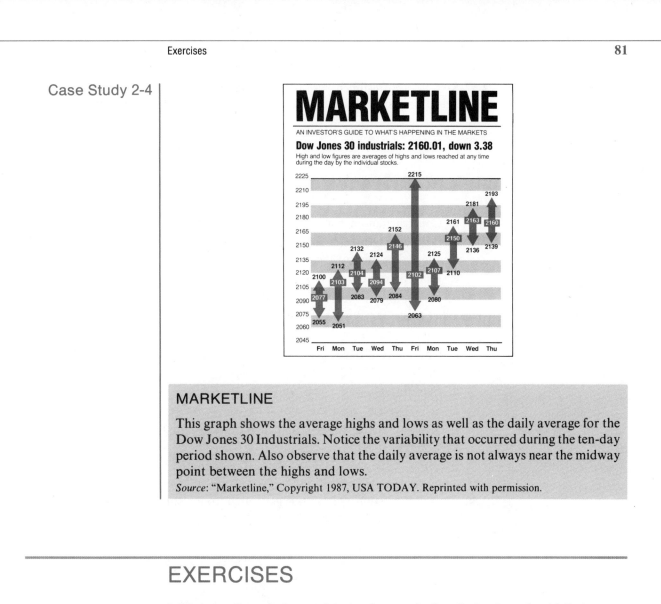

MARKETLINE

This graph shows the average highs and lows as well as the daily average for the Dow Jones 30 Industrials. Notice the variability that occurred during the ten-day period shown. Also observe that the daily average is not always near the midway point between the highs and lows.

Source: "Marketline," Copyright 1987, USA TODAY. Reprinted with permission.

EXERCISES

2.55 According to the empirical rule, practically all the data should lie between $(\bar{x} - 3s)$ and $(\bar{x} + 3s)$. The range also accounts for all the data.

   a. What relationship should hold (approximately) between the standard deviation and the range?

   b. How can you use the results of (a) to estimate the standard deviation?

2.56 Chebyshev's theorem can be stated in a form equivalent to that given on page 77. For example, to say that "At least 75 percent of the data fall within two standard deviations of the mean" is equivalent to stating that "At most 25 percent will be more than two standard deviations away from the mean."

   a. At most what percent of a distribution will be three or more standard deviations from the mean?

   b. At most what percent of a distribution will be four or more standard deviations from the mean?

2.57 Sixty college freshmen were asked to give the number of children in their family (number of their brothers and sisters plus 1). The data collected follow:

```
1  6  3  5  5  3  4  1  2  7  3  2
3  4  5  3  1  3  2  1  4  4  2  2
3  9  4  3  3  5  3  5  7  3  1  1
3  5  2  6  4  3  3  3  3  3  2  3
4  3  5  7  3  2  1  2  3  2  4  3
```

a. Construct an ungrouped frequency distribution of these data.

b. Use the ungrouped frequency distribution found in (a) to find the mean and standard deviation of the data.

c. Find the values of $\bar{x} - s$ and $\bar{x} + s$.

d. How many of the 60 pieces of data have values between $\bar{x} - s$ and $\bar{x} + s$? What percentage of the sample is this?

e. Find the values of $\bar{x} - 2s$ and $\bar{x} + 2s$.

f. How many of the 60 pieces of data have values between $\bar{x} - 2s$ and $\bar{x} + 2s$? What percentage of the sample is this?

g. Find the values of $\bar{x} - 3s$ and $\bar{x} + 3s$.

h. What percentage of the sample has values between $x - 3s$ and $x + 3s$?

i. Compare the answers found in (f) and (h) to the results predicted by Chebyshev's theorem.

j. Compare the answers found in (d), (f), and (h) to the results predicted by the empirical rule. Does the result suggest an approximately normal distribution?

2.58 The following information was generated from a study of stocks listed on the New York Stock Exchange and the American Stock Exchange.

	New York Exchange	American Exchange
Number of Stocks	24	25
Mean 52-week Price Difference	$ 5.57	$11.79
Standard Deviation	3.57	10.55

a. At least 89 percent of the American Stock Exchange stocks had 52-week price differences within what range?

b. At least 75 percent of the New York Stock Exchange stocks had 52-week price differences within what range?

2.59 The average clean-up time for a crew of a medium-size firm is 84.0 hours and the standard deviation is 6.8 hours. Assuming that the empirical rule is appropriate,

a. what proportion of the time will it take the clean-up crew 97.6 or more hours to clean the plant?

b. the total clean-up time will fall within what interval 95 percent of the time?

2.60 On the first day of class last semester, 50 students were asked for the one-way distance from home to college (to the nearest mile). The resulting data were:

6	5	3	24	15	15	6	2	1	3
5	10	9	21	8	10	9	14	16	16
10	21	20	15	9	4	12	27	10	10
3	9	17	6	11	10	12	5	7	11
5	8	22	20	13	1	8	13	4	18

a. Construct a grouped frequency distribution of the data by using $1-3$ as the first class.

b. On probability paper, draw and label carefully the ogive depicting the distribution of one-way distances.

c. Does this distribution appear to have an approximately normal distribution? Explain.

d. Estimate the values of P_{10}, Q_1, $\tilde{x}$, Q_3, and P_{90} from the ogive drawn in (b).

2.7 THE ART OF STATISTICAL DECEPTION

"There are three kinds of lies—lies, damned lies, and statistics." These remarkable words spoken by Disraeli (nineteenth-century British prime minister) represent the cynical view of statistics held by many people. Most people are on the consumer end of statistics and therefore have to "swallow" them.

GOOD ARITHMETIC, BAD STATISTICS

Let's explore an outright statistical lie. Suppose that a small business firm employs eight people who earn between $200 and $240 per week. The owner of the business pays himself $800 per week. He reports to the general public that the average wage paid to the employees of his firm is $280 per week. That may be an example of good arithmetic, but it is also an example of bad statistics. It is a misrepresentation of the situation, since only one employee, the owner, receives more than the mean salary. The public will think that most of the employees earn about $280 per week.

TRICKY GRAPHS

Graphic representation can be tricky and misleading. The frequency scale (which is usually the vertical axis) should start at zero in order to present a total picture. Usually, graphs that do not start at zero are used to save space. Nevertheless, this can be deceptive. Graphs in which the frequency scale starts at zero tend to emphasize the size of the numbers involved, whereas graphs that are *chopped off*

Figure 2-25
A graph that doesn't start at zero

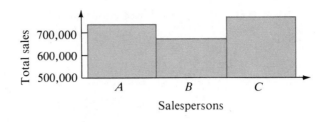

may tend to emphasize the variation in the number without regard to the actual size of the number.

The graph in Figure 2-25 was presented in an annual report of a small business to show the amount of sales made by the company's three salespersons. Without careful study, a viewer of the graph might falsely conclude that B sold only a little more than half of what C sold. In reality, the graph should look like the one in Figure 2-26. The two graphs do not show any different information, they merely create different impressions. The variability among the three amounts of sales is the same in both figures. However, in Figure 2-25 the reader tends to *see* the variation, whereas in Figure 2-26 the reader tends to see that all the numbers are relatively large, and this de-emphasizes the variability that exists.

Figure 2-26
The same graph, but starting at zero

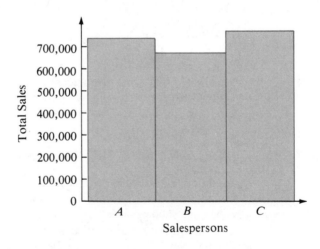

Figure 2-27 represents a daily report of air pollution as reported in a local newspaper. Notice that the attribute scales (Low, Medium, High) are approximately equal, but the numerical scale is not uniform. (The nonuniform scale is due to the pollutants' effect on a person's breathing.) Likewise, the arrow indicator points to the middle of the attribute level and is not in a position relative to the numerical scale. This seems to be a two-way deception.

The histogram in Figure 2-28 shows the number of drivers fatally injured in automobile accidents in New York State in 1968. Notice the "normal-appearing" distribution of ages of fatally injured drivers. The reason for grouping the ages in this manner is completely unclear. Certainly the age of fatally injured automobile drivers is not expected to be normally distributed. Notice too that the class width

Figure 2-27

Air pollution chart: Two-way deception?

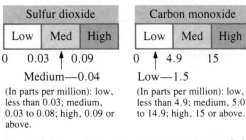

Air pollution

Reading represents average air-pollution levels for a 24-hour period ending at midnight this morning as measured at the Environmental Conservation Department's automatic monitoring station on Farmington Road near the Rochester-Irondequoit line.

Sulfur dioxide		
Low	Med	High

0   0.03 ↑ 0.09
Medium—0.04

(In parts per million): low, less than 0.03; medium, 0.03 to 0.08; high, 0.09 or above.

Carbon monoxide		
Low	Med	High

0   ↑ 4.9   15
Low—1.5

(In parts per million): low, less than 4.9; medium, 5.0 to 14.9; high, 15 or above.

Soiling or dust		
Low	Med	High

0   ↑ 0.4   0.8
Low—0.3

(In reflective units of dirt shade): low, less than 0.4; medium, 0.4 to 0.7; high, 0.8 or above.

From *The Times-Union*, Rochester, N.Y.; by permission

Figure 2-28

Histogram: Fatal automobile accidents

varies considerably. The histogram seems to indicate that 30- to 39-year-olds are the "poorest drivers." This may or may not be the case. Perhaps this age group is larger than any of the other groups. The group with the most drivers might be expected to have the highest number of driver fatalities. A vertical scale that showed the number of driver fatalities per thousand drivers might be more representative. An age grouping that created classifications with the same number of drivers in each class might also be more representative.

There are situations in which a change in the class width might be appropriate for the graphic representation. However, one must be extremely careful when calculating the values of the numerical measures. The annual salaries paid to the employees of a large firm is a situation for which a change in class width might be appropriate. A large firm has employees with salaries ranging from $8,000 to $80,000. Ten or 12 classes of equal width in this case would probably put 95 percent of the employees into the first two classes. To present a more accurate picture, one might consider a classification system similar to 8,000—9,000—10,000—12,000—15,000—18,000—24,000—30,000—50,000 and up. This system should allow for more accurate description for the lower salaries and will also allow the higher salaries to be lumped together. There are, generally speaking, usually only a few very high salaries.

INSUFFICIENT INFORMATION

A realtor describes a mountainside piece of land as a perfect place for a nice summer home and tells you that the average high temperature during the months of July and August is 77 degrees. If the standard deviation is 5 degrees, this is not too bad. Assuming that these daily highs are approximately normally distributed, the temperature will range from three standard deviations below to three standard deviations above the mean temperature. The range would be from 62 to 92 degrees, with only a few days above 87 and a few days below 67. However, what she didn't tell you is that the standard deviation of temperature highs is 10 degrees, which means the daily high temperatures range from 47 degrees to 107 degrees. These extremes are not too pleasant!

The familiar advertisement that shows the "higher level of pain reliever" is a tricky graph with no statistics. It's statistical appearance is a result of the clever use of a graphic presentation. No variable, no units—just implications.

What it all comes down to is that statistics, like all languages, can be and is abused. In the hands of the careless, the unknowledgeable, or the unscrupulous, statistical information can be as false as "damned lies."

In Retrospect   You have been introduced to some of the more common techniques of descriptive statistics. There are far too many specific types of statistics used in nearly every specialized field of study for us to review here. We have outlined the uses of only the most universal statistics. Specifically, you have seen several basic graphic techniques (circle and bar graphs, stem-and-leaf displays, histograms, and ogives) that are used to present sample data in picture form. You have also been introduced to some of the more common measures of central tendency (mean, median, mode, midrange, and midquartile), measures of dispersion (range, variance, and standard deviation), and measures of position (quartiles, percentiles, and z score).

You should now be aware that an average can be any one of five different statistics, and you should understand the distinctions among the different types of averages. The article "Average Means Different Things" discusses four of the averages studied in this chapter. You might reread it now and find that it has more meaning and is of more interest. It will be time well spent!

You should also have a feeling for, and an understanding of, the concept of a standard deviation. You were introduced to Chebyshev's theorem and the empirical rule for this purpose.

The exercises in this chapter are extremely important; they will help you nail down the concepts studied before you go on to learn how to use these ideas in later chapters.

Chapter Exercises

2.61 Samples A and B are shown in the following table. Notice that the two samples are the same except that the 8 in A has been replaced by a 9 in B.

$$A:\quad 2,\ 4,\ 5,\ 5,\ 7,\ 8$$

$$B:\quad 2,\ 4,\ 5,\ 5,\ 7,\ 9$$

What effect does changing the 8 to a 9 have on each of the following statistics? Explain why.

a. mean   **b.** median   **c.** mode   **d.** midrange

e. range   **f.** variance   **g.** standard deviation

2.62 The following figures are the "average family income needed to live moderately" in the selected cities:

Atlanta	$20,797	Dallas-Fort Worth	$20,219
Baltimore	22,439	Houston	21,028
Boston	27,029	Los Angeles	21,954
Buffalo	23,393	San Diego	22,185
Chicago	22,717	Seattle-Everett	23,392

a. Determine the median income needed to live moderately for this sample of ten cities.

b. List the cities that fall below the first quartile.

2.63 The average daily occupancy rate for the Metropolitan Hospital for each month last year is given below:

Month	Jan.	Feb.	Mar.	Apr.	May	June	July	Aug.	Sept.	Oct.	Nov.	Dec.
Patients	265	259	258	242	245	249	234	222	226	254	215	190

a. Determine the median average daily occupancy rate for this information.

b. Determine the third quartile for these data.

c. What is the mean monthly occupancy rate?

2.64 The acreage of each offshore Louisiana lease acquired by Kerr-McGee last year is:

Location	Gross Acres
Main Pass	4,995
Ship Shoal	3,327
South Timbalier	3,772
Vermilion	5,000
West Cameron	5,000
West Cameron	2,500
West Cameron	5,000
West Delta	1,250

a. What is the range of the acreage of these acquisitions?

b. What is the mean?

c. What is the median?

d. What is the mode?

e. Which measure do you think is most representative of the plots acquired? Explain.

2.65 The percent earned on average shareholders' equity (the rate of return on common stock) for the Pacific Lighting Corporation for the last several years has been:

$$8.6\% \quad 9.8\% \quad 10.9\% \quad 11.8\% \quad 13.7\% \quad 13.9\% \quad 14.6\%$$

Determine:

a. the median rate of return for these years

b. the mean rate of return for these years

c. the mean absolute deviation for these data

2.66 Ask one of your instructors for a list of exam grades (15 to 25 grades) from a class.

a. Find five measures of central tendency.

b. Find the three measures of dispersion.

c. Construct a stem-and-leaf display. Does this diagram suggest that the grades are normally distributed?

d. Find the following measures of location: (1) Q_1 and Q_3 (2) P_{15} and P_{60} (3) the standard score z for the highest grade.

2.67 The following figures show the amount of sales per person recorded for 25 salespeople who work for a large company.

150	312	988	750	500	713	650	92	117	835
750	919	400	550	670	803	414	275	850	700
525	435	820	435	535					

Construct a stem-and-leaf display for these data.

2.68 A survey of 32 workers at building 815 of Eastman Kodak Company was taken last May. Each worker was asked: "How many hours of television did you watch yesterday?" The results were as follows:

$$
\begin{array}{cccccccc}
0 & 0 & \frac{1}{2} & 1 & 2 & 0 & 3 & 2\frac{1}{2} \\
0 & 0 & 1 & 1\frac{1}{2} & 5 & 2\frac{1}{2} & 0 & 2 \\
2\frac{1}{2} & 1 & 0 & 2 & 0 & 2\frac{1}{2} & 4 & 0 \\
6 & 2\frac{1}{2} & 0 & \frac{1}{2} & 1 & 1\frac{1}{2} & 0 & 2
\end{array}
$$

a. Construct a stem-and-leaf display.   **b.** Find the mean.

c. Find the median.   **d.** Find the mode.

e. Find the midrange.

f. Which one of the measures of central tendency would "best" represent the average viewer if you were trying to portray the typical television viewer? Explain.

g. Which measure of central tendency would best describe the amount of television watched? Explain.

h. Find the range.   **i.** Find the variance.

j. Find the standard deviation.

2.69 A time-study analyst observed a packaging operation and collected the following times (in seconds) required for the operation to fill packages of a fixed volume:

10.8	14.4	19.6	18.0	8.4	15.2	11.0	13.3	23.1	17.2
17.1	16.1	12.6	14.6	9.1	12.0	11.6	14.7	12.4	17.3

a. Find the mean.   **b.** Find the variance.

c. Find the standard deviation.

2.70 The stopping distance on a wet surface was determined for 25 cars each traveling at 30 miles per hour. The data (in feet) are shown on the following stem-and-leaf display.

$$
\begin{array}{r|llllllll}
6 & 3 & 7 & 6 & 3 & 9 \\
7 & 4 & 2 & 0 & 1 & 1 & 2 & 0 & 5 \\
8 & 5 & 4 & 5 & 5 & 6 \\
9 & 4 & 1 & 0 & 0 & 5 \\
10 & 5 & 4
\end{array}
$$

Find the mean and standard deviation of these stopping distances.

2.71 The following set of data gives the ages of 118 known offenders who committed an auto theft last year in Garden City, Michigan.

11	14	15	15	16	16	17	18	19	21	25	36
12	14	15	15	16	16	17	18	19	21	25	39
13	14	15	15	16	17	17	18	20	22	26	43
13	14	15	15	16	17	17	18	20	22	26	46
13	14	15	16	16	17	17	18	20	22	27	50
13	14	15	16	16	17	17	19	20	23	27	54
13	14	15	16	16	17	18	19	20	23	29	59
13	15	15	16	16	17	18	19	20	23	30	67
14	15	15	16	16	17	18	19	21	24	31	
14	15	15	16	16	17	18	19	21	24	34	

a. Find the mean.   b. Find the median.   c. Find the mode.

d. Find Q_1 and Q_3.   e. Find P_{10} and P_{95}.

2.72 The numbers of pigs born per litter last year at Circle J Pig Farm are given in the following list.

11	8	13	14	11	8	14	7	11	13	10	8
7	4	9	11	10	12	6	12	5	10	9	10
12	10	3	6	9	10	10	13	12	9	12	7
9	5	12	7	11	9	4	8	13	12	11	
13	11	12	8	6	13	11	6	12	7	5	

a. Construct an ungrouped frequency distribution of these data.

b. How many litters of pigs were there last year?

c. What is the meaning and value of $\sum f$?

2.73 The distribution of credit hours, per student, taken this semester at a certain college was found to be:

Credit Hours	Frequency
3	75
6	150
8	30
9	50
12	70
14	300
15	400
16	1050
17	750
18	515
19	120
20	60

Find: **a.** mean   **b.** median   **c.** mode   **d.** midrange

2.74 Given the following frequency distribution of ages, x, in years, of cars found in a parking lot,

x	1	2	3	4	5	6	7	8	9	10	11
f	16	20	18	12	9	7	6	3	5	3	1

a. draw a histogram of the data.

b. find the five measures of central tendency.

c. find Q_1 and Q_3.   d. find P_{15} and P_{12}.

e. find the three measures of dispersion (range, s^2, s).

2.75 Compute the mean and standard deviation for the following set of data. Then find the percentage of the data that is within two standard deviations of the mean.

```
1 | .4  .7  .1
2 | .4  .5
3 | .5  .0  .4  .1
4 | .4
5 | .5  .8  .7
6 | .8  .8  .2  .8  .6
7 | .5
8 |
9 | .4
```

2.76 A survey was conducted to determine how many hours elementary school children spend viewing television per week. The results are as follows.

Hours Per Week	Frequency
5–9	2
10–14	16
15–19	54
20–24	112
25–29	64
30–34	10

Find the mean and standard deviation for this distribution.

2.77 A set of 75 measurements has a mean equal to 40 and a variance equal to 100. Find the $\sum x$ and $\sum x^2$.

2.78 A survey was conducted to determine the number of radios per household. The results are summarized as follows (x represents the number of radios per household):

x	1	2	3	4	5	6	7
f	20	35	100	90	65	40	5

Find: **a.** Q_3 **b.** P_{90}

2.79 The grouped frequency distribution in the following table represents a sample of the ages of 120 randomly selected patients who were admitted to Memorial Hospital last June.

Class Limits	Frequency
0–8	17
9–17	14
18–26	10
27–35	14
36–44	10
45–53	16
54–62	9
63–71	11
72–80	8
81–89	11

a. Construct a relative frequency histogram that shows the ages of the 120 patients.

b. Construct an ogive that shows the cumulative distribution of the ages.

c. Calculate the mean age of the patients.

d. Calculate the standard deviation of the ages.

2.80 Classify the following data into a frequency distribution with classes of 3.0–5.1, 5.2–7.3, and so on.

3	.1	.6	.3				
4	.0	.1	.0	.6			
5	.7	.6	.5	.5	.8		
6	.7	.0	.0	.5	.0	.5	.9
9	.1	.3	.1	.4			
11	.5	.7	.5				
13	.5						

Find: **a.** all class marks   **b.** all relative frequencies

c. all cumulative frequencies

2.81 The lengths (in millimeters) of 100 brown trout in pond 2-B at Happy Acres Fish Hatchery on June 15 of last year were as follows:

```
15.0   15.3   14.4   10.4   10.2   11.5   15.4   11.7   15.0   10.9
13.6   10.5   13.8   15.0   13.8   14.5   13.7   13.9   12.5   15.2
10.7   13.1   10.6   12.1   14.9   14.1   12.7   14.0   10.1   14.1
10.3   15.2   15.0   12.9   10.7   10.3   10.8   15.3   14.9   14.8
14.9   11.8   10.4   11.0   11.4   14.3   15.1   11.5   10.2   10.1
14.7   15.1   12.8   14.8   15.0   10.4   13.5   14.5   14.9   13.9
10.1   14.8   13.7   10.9   10.6   12.4   14.5   10.5   15.1   15.8
12.0   15.5   10.8   14.4   15.4   14.8   11.4   15.1   10.3   15.4
15.0   14.0   15.0   15.1   13.7   14.7   10.7   14.5   13.9   11.7
15.1   10.9   11.3   10.5   15.3   14.0   14.6   12.6   15.3   10.4
```

a. Find the mean   **b.** Find the median.

c. Find the mode.   **d.** Find the midrange.

e. Find the range.   **f.** Find Q_1 and Q_3.

g. Find the midquartile.   **h.** Find P_{35} and P_{64}.

i. Construct a grouped frequency distribution that uses 10.0–10.5 as the first class.

j. Construct a histogram of the frequency distribution.

k. Construct a cumulative relative frequency distribution.

l. Construct an ogive of the cumulative relative frequency distribution.

m. Find the mean of the frequency distribution.

n. Find the standard deviation of the frequency distribution.

2.82 The length of life of 220 incandescent 60-watt lamps was obtained and yielded the frequency distribution shown in the following table.

Class Limit	Frequency
500–599	3
600–699	7
700–799	14
800–899	28
900–999	64
1000–1099	57
1100–1199	23
1200–1299	13
1300–1399	7
1400–1499	4

a. Construct a histogram of these data, using a vertical scale for relative frequencies.

b. Find the mean length of life.   **c.** Find the standard deviation.

2.83 Earnings per share for 40 firms in the radio and transmitting equipment industry follow:

4.62	0.25	1.07	5.56	0.10	1.34	2.50	1.62
1.29	2.11	2.14	1.36	7.25	5.39	3.46	1.93
6.04	0.84	1.91	2.05	3.20	−0.19	7.05	2.75
9.56	3.72	5.10	3.58	4.90	2.27	1.80	0.44
4.22	2.08	0.91	3.15	3.71	1.12	0.50	1.93

a. Prepare a frequency distribution and a frequency histogram for these data.

b. Which class of your frequency distribution contains the median?

2.84 The following table shows the age distribution of heads of families.

Age of Head of Family (years)	Number
20–24	23
25–29	38
30–34	51
35–39	55
40–44	53
45–49	50
50–54	48
55–59	39
60–64	31
65–69	26
70–74	20
75–79	16
	450

a. Find the mean age of the heads of families.

b. Find the standard deviation.

2.85 For a normal (or mound-shaped) distribution, a z score of 2 corresponds to approximately what percentile rank?

2.86 Bill and Bob are good friends, although they attend different high schools in their city. The city school system uses a battery of fitness tests to test all high school students. After completing the fitness tests, Bill and Bob are comparing their scores to see who did better in each event. They need help.

	Sit-ups	Pull-ups	Shuttle Run	50 Yard Dash	Softball Throw
Bill	$z = -1$	$z = -1.3$	$z = 0.0$	$z = 1.0$	$z = 0.5$
Bob	61	17	9.6	6.0	179 ft.
Mean	70	8	9.8	6.6	173 ft.
Standard Deviation	12	6	0.6	0.3	16 ft.

Bill received his test results in z scores, whereas Bob was given raw scores. Since both boys understand raw scores, convert Bill's z scores to raw scores in order to make an accurate comparison.

2.87 Given the following information, compare an individual's achievement in relation to the rest of his class on a midterm exam and a final exam.

	His Score	Class Mean	Class Standard Deviation
Midterm	70	75	10
Final	50	55	15

On which exam did he perform better?

2.88 Mrs. Adam's twin daughters, Jean and Joan, are in fifth grade (different sections) and the class has been given a series of ability tests.

	Results	
Skill	Jean: z Score	Joan: Percentile
Fitness	2.0	99
Posture	1.0	69
Agility	1.0	88
Flexibility	-1.0	35
Strength	0.0	50

If the scores for these ability tests are approximately normally distributed, which girl has the higher relative score on each of the skills listed? Explain your answers.

2.89 Chebyshev's theorem guarantees what proportion of a distribution to be included between the following?

  a. $\bar{x} - 2s$ and $\bar{x} + 2s$ **b.** $\bar{x} - 3s$ and $\bar{x} + 3s$

2.90 The empirical rule indicates that we can expect to find what proportion of the sample to be included between the following?

  a. $\bar{x} - s$ and $\bar{x} + s$ **b.** $\bar{x} - 2s$ and $\bar{x} + 2s$ **c.** $\bar{x} - 3s$ and $\bar{x} + 3s$

2.91 Why is it that the z score for a value belonging to a normal distribution usually lies between -3 and 3?

2.92 The mean mileage per tire is 30,000 miles and the standard deviation is 2,500 miles for a certain tire.

a. If we assume that the mileage is normally distributed, approximately what percent of all such tires will give between 22,500 and 37,500 miles?

b. If we assume nothing about the shape of the distribution, approximately what percent of all such tires will give between 22,500 and 37,500 miles?

Vocabulary List

Be able to define each term. In addition, describe, in your own words, and give an example of each term. Your examples should not be ones given in class or in the textbook.

bell-shaped distribution
bimodal distribution
box-and-whisker plot
Chebyshev's theorem
class
class boundary
class limits (lower and upper)
class mark
class width
cumulative frequency distribution
cumulative relative frequency
   distribution
depth
deviation from mean
distribution
empirical rule
five-number summary
frequency
frequency distribution
frequency histogram
grouped frequency distribution
histogram
interquartile range
lower hinge
mean
mean absolute deviation
measure of central tendency
measure of dispersion
measure of position

median
midquartile
midrange
modal class
mode
normal distribution
ogive
percentile
probability paper
quartile
range
rectangular distribution
relative frequency
relative frequency distribution
relative frequency histogram
sample
single-variable data
skewed distribution
standard deviation
standard score
stem-and-leaf display
summation
tally
variability
ungrouped frequency distribution
upper hinge
variance
x bar $(\bar{x})$
z score

Quiz   *Answer "True" if the statement is always true. If the statement is not always true, replace the italicized words with words that make the statement always true.*

2.1 The *mean* of a sample always divides the data into two equal halves—half larger and half smaller in value than itself.

2.2 A measure of *central tendency* is a quantitative value that describes how widely the data are dispersed about a central value.

2.3 The sum of the squares of the deviations from the mean, $\sum(x - \bar{x})^2$, will *sometimes* be negative.

2.4 For any distribution, the sum of the deviations from the mean equals *zero*.

2.5 The standard deviation for the set of values 2, 2, 2, 2, and 2 is *2*.

2.6 On a test John scored at the 50th percentile and Jim scored at the 25th percentile; therefore, John's test score was *twice* Jim's test score.

2.7 The frequency of a class is the number of pieces of data whose values fall within the *boundaries* of that class.

2.8 *Frequency distributions* are used in statistics to present large quantities of repeating values in a concise form.

2.9 The unit of measure for the standard score is always in *standard deviations*.

2.10 For a bell-shaped distribution, the range will be approximately equal to *six standard deviations*.

Chapter 3

Descriptive Analysis and Presentation of Bivariate Data

AGES OF MATED PAIRS OF CALIFORNIA GULLS

Several studies of long-lived seabirds have shown that mated pairs are usually the same or nearly the same age. For example, Coulson and Horobin (1976) report ages of 29 pairs of Arctic terns ranging in age from 3 to 18 years old. Eighty-three percent of mated pairs differed in age by no more than 2 years. Mating of similar-aged individuals has been reported for a number of other species.... Here we describe ages of mated pairs of California Gulls (*Larus californicus*) and discuss possible explanations that may account for high correlation in ages of mated pairs.

• **Methods** Data were gathered in 1979 and 1980 on California gulls breeding on an island on Bamforth Lake, Wyoming. Detailed descriptions of the breeding colony and nesting habitat are in Pugesek and Diem (1983).

Gulls from the Bamforth colony have been banded and wing-marked as nearly fledged chicks almost every year since 1958. Gulls were watched on the breeding island in 1979 and 1980 as they defended nesting territories. Their leg band or wing marker numbers were read with the aid of a spotting scope and the nesting territory marked with a coded stake (8 × 2 × 40 cm). We returned to staked nest sites at various intervals to determine whether mates were also banded. Gulls were aged by banding records. Sex was determined by bill characteristics (bill length, head-bill length, and bill depth) (Fox et al. 1981, Diem, unpubl. data) recorded at the time of banding. Gender was further verified by visual comparisons of mates at the nest site.

Males are usually larger than females and can be distinguished from females in side-by-side comparison of bill characteristics. In all cases, gender, as determined by such comparisons, agreed with that determined by previous bill measurements.

• **Results and Discussion** Breeding adults in the Bamforth colony range in age from 3 to 20 years old (Pugesek and Diem, unpubl. data) Despite the wide variation in ages of potential mates, gulls nearly always mated with individuals of the same age (Fig. 1). In 13 of 18 pairs (72%), mates were born in the same year. Mates were within 3 years of age in 16 of 18 pairs (89%). Results, therefore, are consistent with those reported for other seabird species.

The similarities in ages of mates could occur for a number of reasons. In gulls, timing of reproductive readiness varies with age. For example Mills (1973) found that older Redbilled Gulls come into reproductive condition before younger gulls. A similar condition is likely in the California Gulls at Bamforth Lake as oldest members of the population are first to arrive in the breeding area, and establish nesting territories (Pugesek 1983). Thus if variation in the date at which gulls are physiologically ready to accept mates differs between age-classes, gulls would only find mates of the same age. The process of pairing similar-aged individuals may be further augmented by retention of mates from year to year (Coulson 1966).

It is unlikely, however, that timing of reproduction and retention of mates are sufficient to explain the high degree of correlation in ages of mated pairs observed during our study. Considerable variation occurs within age classes in behavior and timing of reproductive events (Pugesek 1983). Many young gulls establish territories, incubate, and hatch eggs before older gulls. For example, mean hatching date of the youngest gulls (3–6 years old) was only 4 days later than that of gulls 11 years and older. Standard deviations in hatching dates were 5.0 and 5.5 days, respectively, indicating a high degree of overlap between age groups.

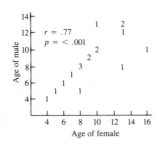

FIG. 1. Age of male and female mated California Gulls with Pearson's Correlation Coefficient and significance level. Numbers designate pairs found with a particular age combination (N = 18 pairs)

$r = .77$
$p = < .001$

Age of male

Age of female

Source: The Wilson Bulletin, Vol. 98, No. 4, December 1986.

Chapter
Objectives

In the field of statistics there are many problems that require the analysis of more than one variable. In business, in education, and in many other fields, we often try to obtain answers to the following questions: Are these two variables related? If so, how are they related? Are these variables correlated? The relationships discussed here are not cause-and-effect relationships. They are only mathematical relationships that predict the behavior of one variable from knowledge about a second variable.

Let's look at a few specific illustrations.

Illustration 3-1   As a person grows taller, he or she usually gains weight. Someone might ask, "Is there a relationship between height and weight?"   ■

Illustration 3-2   Students are constantly spending their time studying and taking examinations while going to school. We could ask, "Is it true that the more you study the higher your grade will be?"   ■

Illustration 3-3   Research doctors test new drugs (old ones, too) by prescribing different amounts and observing the responses of their patients. One question we could ask here is: "Does the amount of drug prescribed determine the amount of recovery time needed by the patient?"   ■

Illustration 3-4   A high school guidance counselor would like to predict the academic success that students graduating from her school will have in college. In cases like this, the predicted value (grade point average at college) depends on many traits of the students: (1) how well they did in high school, (2) their intelligence, (3) their desire to succeed at college, and so on.

■

These illustrations all require correlation or regression analysis to obtain answers. In this chapter we will take a first look at the very simplest form of correlation and regression analysis. The basic objectives of this chapter are (1) to gain an understanding of the distinction between the basic purposes of correlation analysis and regression analysis, and (2) to become familiar with the ideas of descriptive presentation. With these objectives in mind, we will restrict our discussion to the simplest and most basic form of correlation and regression analysis—the bivariate linear case.

3.1   BIVARIATE DATA

bivariate data

Bivariate data consist of the values of two different response variables that are obtained from the same population element.

ordered pair

Expressed mathematically, bivariate data are composed of **ordered pairs**—let's call them x and y, where x is the value of the first variable and y is the value of the second variable. We write an ordered pair as (x, y). The data are said to be *ordered* because one value, x in this case, is always written first. They are said to be *paired* because for each x value there is a corresponding y value. For example, if x is height and y is weight, a height and corresponding weight are recorded for each person.

input and output
variables

It is customary to call the **input variable** (sometimes called the *independent* variable) x and the **output variable** (sometimes called the *dependent* variable) y. The input variable x is measured or controlled in order to predict the output variable y. For example, in Illustration 3-3 the researcher can control the amount of drug prescribed. Therefore, the amount of drug would be referred to as x. In the case of height and weight, either variable could be treated as input, the other as output, depending on the question being asked. However, different results will be obtained from the regression analysis, depending on the choice made.

scatter diagram

In problems that deal with both correlation and regression analysis, we will present our sample data pictorially on a scatter diagram. A **scatter diagram** is a plot of all the ordered pairs of bivariate data on a coordinate axis system. The input variable x is plotted on the horizontal axis and the output variable y is plotted on the vertical axis.

Note   When constructing a scatter diagram, it is convenient to construct scales so that the range of data along the vertical axis is equal to or slightly shorter than the range along the horizontal axis.

Illustration 3-5   In Mr. Chamberlain's physical fitness course, several fitness scores were taken. The following sample is the number of push-ups and sit-ups done by ten randomly selected students:

$$(27, 30), (22, 26), (15, 25), (35, 42), (30, 38),$$
$$(52, 40), (35, 32), (55, 54), (40, 50), (40, 43)$$

Table 3-1 shows these sample data, and Figure 3-1 shows a scatter diagram of these data.

Table 3-1
Data for push-ups and sit-ups

	Student									
	1	2	3	4	5	6	7	8	9	10
Push-ups (x)	27	22	15	35	30	52	35	55	40	40
Sit-ups (y)	30	26	25	42	38	40	32	54	50	43

Figure 3-1
Scatter diagram: Push-ups versus sit-ups for ten students

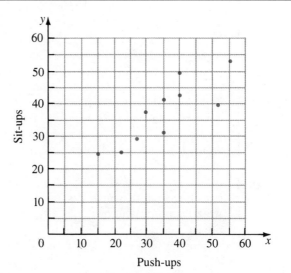

Case Study 3-1

BREAST CANCER: THE DIET-CANCER CONNECTION

Doctor Kimberly Kline discusses how a person's diet may contain factors that correlate with cancer. Included in her report is an interesting scatter diagram showing values from many different countries.

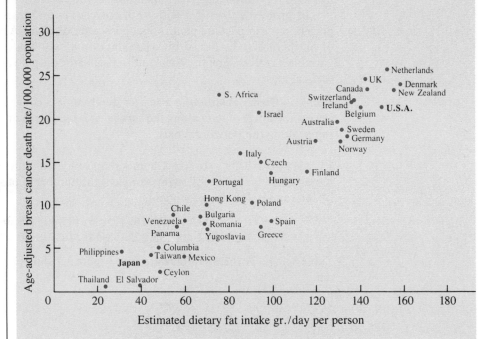

FIG. 1. Daily dietary fat intake in relation to death rates from breast cancer

Epidemiologic Research That Supports the High-Fat Diet-Breast Cancer Connection. One way that epidemiologists identify cancer risk factors is to identify geographic correlations. As illustrated in Figure 1, when the age-adjusted breast cancer death rate of a country is plotted against the estimated dietary fat intake in grams of fat per day per person in that country, a strikingly strong positive correlation between number of breast cancer deaths and high dietary fat intake is seen. In support of correlation data linking high dietary fat intake with high mortality rates from breast cancer are studies of women who migrate from a country with a low incidence of breast cancer to a country with a high incidence of breast cancer.... These immigrants rapidly acquire the breast cancer risks of women from their adopted country. One example is a study that analyzed the incidence of breast cancer in Japanese women who immigrated to the United States. In Japan, where the estimated daily fat intake is 20% of total calories consumed, death from breast cancer is low (Figure 1). By comparison the average American woman consumes 40% of total calories from fat and the American breast cancer mortality rate is high. It is of interest and significant that third generation Japanese/American women exhibit the same high breast cancer mortality rate as American women. Other migrating population studies have shown that women migrating from countries with low rates of breast cancer deaths to countries with high rates of breast cancer deaths acquire the rates of breast cancer deaths of their host environment more rapidly. It is possible that the generation lag observed in the Japanese to American migration may be accounted for by a slower adoption of an American life-style by the original as well as the first generation Japanese/American immigrants.

Although all of the epidemiologic data are not entirely consistent with a high-fat diet-breast cancer connection, it is widely recognized that it is extremely difficult to determine with any precision what people eat, especially over long periods of time. And although various confounding variables might exist that would weaken or invalidate the conclusions drawn from epidemiologic data; nevertheless, epidemiologic data have focused attention on the high-fat diet-breast cancer connection and has resulted in the formulation of a hypothesis that has been tested in the laboratory: *High levels of dietary fat increase the risk of breast cancer* . (As an alternative explanation critics of the "fat" theory suggest that these same women also have higher calorie diets and that "calories" rather than specifically fat may be the relevant variable.)

Source: Kimberly Kline, *Nutrition Today*, May/June 1986.

EXERCISES

3.1 Refer to Doctor Kline's discussion of the diet-cancer connection in "Breast Cancer: The Diet-Cancer Connection" (Case Study 3.1). Fig. 1 shows a scatter diagram in which the points plotted represent countries and their related values.
    a. What variable is plotted along the horizontal axis?
    b. What variable is plotted along the vertical axis?
    c. Explain how this is bivariate data.

3.2 The accompanying data show the number of hours x studied for an exam and the grade y received on the exam. (y is measured in 10s; that is, $y = 8$ means that the grade, rounded to the nearest 10 points, is 80.) Draw the scatter diagram. (Retain this solution for use in answering Exercises 3.9 and 3.21.)

x	2	3	3	4	4	5	5	6	6	6	7	7	7	8	8
y	5	5	7	5	7	7	8	6	9	8	7	9	10	8	9

3.3 An experimental psychologist asserts that the older a child is the fewer irrelevant answers he or she will give during a controlled experiment. To investigate this claim, the following data were collected. Draw a scatter diagram. (Retain this solution for use in answering Exercises 3.10 and 3.22.)

Age (x)	2	4	5	6	6	7	9	9	10	12
Number of Irrelevant Answers (y)	12	13	9	7	12	8	6	9	7	5

3.4 An educational study was conducted to investigate the relationship between mathematical competency and computer-science aptitude. The Backmann-Beal mathematical competency test and the KSW computer-science aptitude test were used. The results for 20 students were as follows:

Math Competency (x)	28	35	42	41	44	42	36
Computer-science Aptitude (y)	4	16	20	13	22	21	15

x	44	39	36	40	40	33	27	44	45	41	31	41	43
y	20	19	16	18	17	8	6	5	20	18	11	19	22

Prepare a scatter diagram for these data.

3.5 A sample of 15 upperclassmen who commute to classes was selected at registration. They were asked to estimate the distance (x) and the time (y) required to commute each day to class. (*See* the following table.) Construct a scatter diagram depicting these data.

Distance, x (nearest mile)	Time, y (nearest 5 minutes)	Distance, x (nearest mile)	Time, y (nearest 5 minutes)
18	20	2	5
8	15	15	25
20	25	16	30
5	20	9	20
5	15	21	30
11	25	5	10
9	20	15	20
10	25		

3.6 A clinic that specializes in the treatment of hypertension (high blood pressure) designed an experiment to investigate the relationship between blood pressure readings obtained by medical personnel using the standard technique and readings obtained by patients using an electronic device. For a given patient, let x represent the diastolic blood pressure as determined by the standard technique and let y represent the diastolic blood pressure obtained using the electronic device.

x	72	84	96	90	85	105	98	102	87	80
y	73	80	95	90	88	101	96	100	90	82

Draw the scatter diagram for these data. Does there appear to be a consistent bias for either technique?

3.2 LINEAR CORRELATION

correlation analysis

independent and
dependent variables

linear correlation

The primary purpose of linear correlation analysis is to measure the strength of a linear relationship between two variables. Let's examine some scatter diagrams demonstrating different relationships between input, or **independent variables** x and output, or **dependent variables** y. If as x increases there is no definite shift in the values of y, we say there is **no correlation**, or no relationship, between x and y. If as x increases there is a shift in the values of y, there is a **correlation**. The correlation is positive when y tends to increase and negative when y tends to decrease. If the values of x and y tend to follow a straight-line path, there is a **linear** correlation. The preciseness of the shift in y as x increases determines the strength of the linear correlation. The scatter diagrams in Figure 3-2 demonstrate these ideas.

Perfect linear correlation occurs when all the points fall exactly along a straight line, as shown in Figure 3-3. This can be either positive or negative, depending on whether y increases or decreases as x increases. If the data form a straight horizontal or vertical line, there is no correlation, since one variable has no effect on the other.

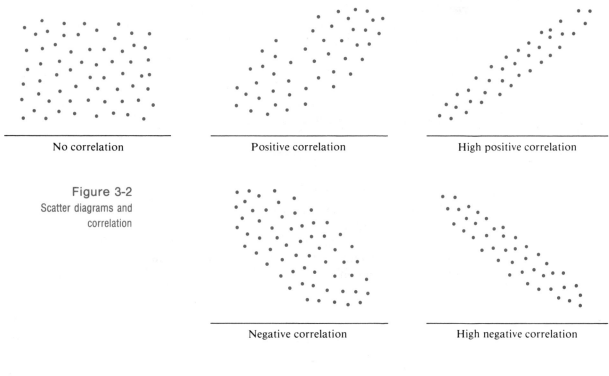

No correlation   Positive correlation   High positive correlation

Figure 3-2
Scatter diagrams and
correlation

Negative correlation   High negative correlation

Figure 3-3
Perfect positive correlation

Figure 3-4
No linear correlation

Scatter diagrams do not always appear in one of the five forms shown in Figure 3-2. Sometimes they will suggest relationships other than linear. Such is the case in Figure 3-4. There appears to be a definite pattern; however, the two variables are not related linearly, and therefore they are not linearly correlated.

coefficient of
linear correlation

The **coefficient of linear correlation r** is the measure of the strength of the linear relationship between two variables. The coefficient reflects the consistency of the effect that a change in one variable has on the other.

The value of linear correlation coefficient helps us to answer the question "Is there a linear correlation between the two variables under consideration?" The linear correlation coefficient r always has a value between -1 and $+1$. A value of $+1$ signifies a perfect **positive correlation**, and a value of -1 shows a perfect **negative correlation**.

positive negative
correlation

If as x increases there is a general increase in the value of y, then r will indicate a positive linear correlation. For example, a positive value of r would be expected for age and height of children, because as children grow older, they grow taller. Also, consider the age x and relative value y of an automobile. As the car ages, its value decreases. Since as x increases, y decreases, their inverse relationship is denoted by a negative value.

The value of r for a sample is obtained from the formula

$$r = \frac{\sum (x - \bar{x})(y - \bar{y})}{(n - 1)s_x s_y} \tag{3-1}$$

Note s_x and s_y are the standard deviations of the x and y variables.

Pearson's product
moment

This formula is called **Pearson's product moment r**. The development of this formula is discussed in Chapter 13.

To calculate r, we will use an alternative formula that is equivalent to formula (3-1). We will calculate three separate sums of squares and then substitute them into formula (3-2) to obtain r.

$$r = \frac{SS(xy)}{\sqrt{SS(x)SS(y)}} \tag{3-2}$$

where $SS(x) = \sum x^2 - ((\sum x)^2/n)$, from formula (2-9) on page 62,

$$SS(y) = \sum y^2 - \frac{(\sum y)^2}{n}, \tag{3-3}$$

and

$$SS(xy) = \sum xy - \frac{(\sum x)(\sum y)}{n} \tag{3-4}$$

Table 3-2
Computations needed to
calculate r

Student	Push-ups (x)	x^2	Sit-ups (y)	y^2	xy
1	27	729	30	900	810
2	22	484	26	676	572
3	15	225	25	625	375
4	35	1,225	42	1,764	1,470
5	30	900	38	1,444	1,140
6	52	2,704	40	1,600	2,080
7	35	1,225	32	1,024	1,120
8	55	3,025	54	2,916	2,970
9	40	1,600	50	2,500	2,000
10	40	1,600	43	1,849	1,720
Total	351	13,717	380	15,298	14,257

Let's return to Illustration 3-5 and find the linear correlation coefficient. First we need to construct a table (Table 3-2) listing all the pairs of values (x, y) and the extensions for x^2, xy, and y^2, and the column totals. Then we use the totals from the table and formulas (2-9), (3-3), (3-4), and (3-2) to calculate r.

Using formula (2-9),

$$SS(x) = 13,717 - \frac{(351)^2}{10} = 1396.9$$

Using formula (3-3),

$$SS(y) = 15,298 - \frac{(380)^2}{10} = 858.0$$

Using formula (3-4),

$$SS(xy) = 14,257 - \frac{(351)(380)}{10} = 919.0$$

Using formula (3-2), we find r to be

$$r = \frac{919.0}{\sqrt{(1396.9)(858.0)}} = 0.8394$$

$$= \mathbf{0.84}$$

Note   Typically, r is rounded to the nearest hundredth.

The value of the calculated linear correlation coefficient is supposed to help us answer the question "Is there a linear correlation between the two variables under consideration?" When the calculated value of r is close to zero, we conclude that there is very little or no linear correlation. When the calculated value of r is close to $+1$ or -1, we suspect that there is a linear correlation between the two variables. Further discussion involving the interpretation of the calculated value of r is found in Chapter 13.

ESTIMATING THE LINEAR CORRELATION COEFFICIENT

With a formula as complex as formula (3-2), it would be very convenient to be able to inspect the scatter diagram of the data and estimate the calculated value of r. This would serve as a check on the calculations. The following **method for estimating r** is quick and generally yields a reasonable estimate.

1. Lay two pencils on your scatter diagram. Keeping them parallel, move them so that they are as close together as possible but yet have all the points on the scatter diagram between them. (*See* Figure 3-5.)

2. Visualize a rectangular region that is bounded by the two pencils and that ends just beyond the points on the scatter diagram. (*See* the shaded proportion of Figure 3-6.)

3. Estimate how many times longer the rectangle is than it is wide. An easy way to do this is to mentally mark off squares in the rectangle. (*See* Figure 3-7.) Call this number of multiples k.

Figure 3-5

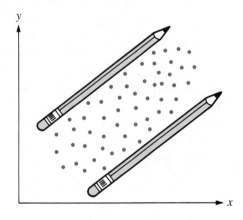

Figure 3-6

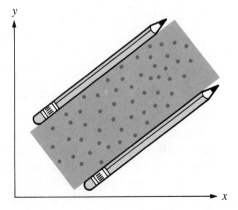

Figure 3-7

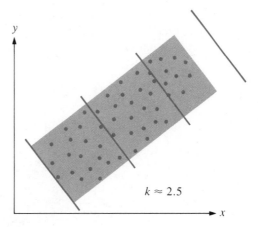

$k \approx 2.5$

4. The value of r may be estimated as

$$\pm\left(1 - \frac{1}{k}\right)$$

The sign assigned to r is determined by the general position of the length of the rectangular region. If it lies in an increasing position (*see* Figure 3-8), r will be positive; if it lies in a decreasing position, r will be negative. If the rectangle is in either a horizontal or a vertical position, then r will be zero, regardless of the length-width ratio.

Figure 3-8

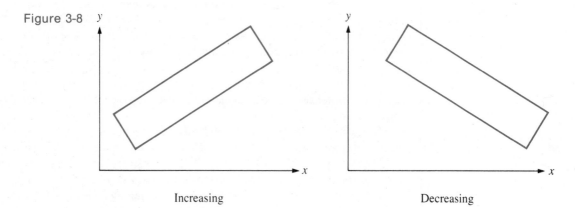

Increasing   Decreasing

Note   *This estimating technique is very crude. It is very sensitive to the "spread" of the diagram.* However, if the range of the x values and the range of the y values are approximately equal, this approximation will be helpful. *This technique should be used only as a mental check.*

Let's use this method to estimate the value of the linear correlation coefficient for the relationship between the number of push-ups and sit-ups shown in

Figure 3-9. From the figure, we find that the rectangle is approximately 3.5 times longer than it is wide; that is, $k = 3.5$, and the rectangle lies in an increasing position. Therefore, our estimate for r is $+0.7$ $[1 - (1/3.5) = 0.7]$.

Figure 3-9
Push-ups versus sit-ups
for ten students

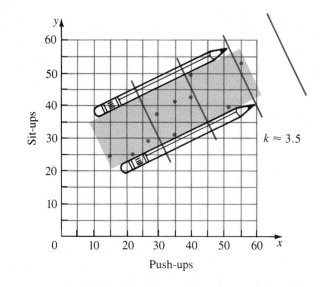

Case Study 3-2

MEDIA AND RAPE STUDY

Professor Larry Baron, a Yale sociologist, claims that a "moderate correlation" was found between media habits and rape rates in a 1980 study. States that had high readership rates for certain magazines also had relatively high rates of reported rape. Professor Baron also points out that evidence of correlation does not show cause and effect.

States with high numbers of viewers of violent television programs and readers of violence-related magazines are also high on the list of reported cases of rape, a Yale sociologist claims.

Studying figures from 1980, Professor Larry Baron reports what he calls a "moderate correlation" between high viewership rates for programs such as "Charlie's Angels," "Incredible Hulk," "Hart to Hart" and "Dukes of Hazzard" and high rates of reported rape per 100,000 population.

States that had high readership rates of such magazines as Easyriders, Guns & Ammo, Heavy Metal Times and Shooting Times also had relatively high rates of reported rape, the study found.

Baron cautioned that such reading and viewing do not cause people to commit rape but rather are part of a social climate making rape more likely.

He presented the results at a New York Academy of Sciences meeting. Coresearchers were Murray Straus of the University of New Hampshire and David Jaffree of the State University of New York College at New Paltz.

Source: Insight, February 9, 1987.

EXERCISES

3.7 How would you interpret the findings of a correlation study that reported a linear correlation coefficient of -1.34?

3.8 Consider the following data, which give the weight (in thousands of pounds) x and gasoline mileage (miles per gallon) y for ten different automobiles.

x	2.5	3.0	4.0	3.5	2.7	4.5	3.8	2.9	5.0	2.2
y	40	43	30	35	42	19	32	39	15	44

Find **a.** $SS(x)$ **b.** $SS(y)$ **c.** $SS(xy)$ **d.** Pearson's product moment r

3.9 a. Use the scatter diagram you draw in answering Exercise 3.2 to estimate r for the sample data relating the number of hours studied and exam grade.

b. Calculate r.

(Retain these solutions for use in answering Exercise 3.21.)

3.10 a. Use the scatter diagram you draw in answering Exercise 3.3 to estimate r for the sample data involving age and irrelevant answers.

b. Calculate r.

(Retain these solutions for use in answering Exercise 3.22).

3.11 A marketing firm wished to determine whether or not the number of television commercials broadcast were linearly correlated to the sales of its product. The data, obtained from each of several cities, are shown in the following table.

City	A	B	C	D	E	F	G	H	I	J
No. TV Commercials (x)	12	6	9	15	11	15	8	16	12	6
Sales Units (y)	7	5	10	14	12	9	6	11	11	8

a. Draw a scatter diagram. **b.** Estimate r. **c.** Calculate r.

3.12 A study was designed to investigate the relationship between the weight x and diastolic blood pressure y of adult males between the ages of 19 to 30.

x	173	178	145	146	157	175	173	137
y	76	76	74	70	80	68	90	70

x	199	131	152	171	163	170	135	159
y	96	80	90	72	76	80	68	72

Calculate r, the linear correlation between weight and diastolic blood pressure.

3.13 A high positive correlation coefficient is reported as a result of a study about the ages of mated pairs of California gulls. The news article at the beginning of this chapter, page 99, contains part of the report.

a. What population was being studied?

b. How large was the sample?

c. What variables were found to have a high correlation?

3.14 Refer to the Media and Rape Study from the February 9, 1987 *Insight* (Case Study 3-2).

a. Professor Baron reports what he calls a "moderate correlation" between high viewership rates for programs such as "Charlie's Angels," "Incredible Hulk," "Hart to Hart," and "Dukes of Hazzard" and high rates of reported rape per 100,000 population. What does this mean?

b. Later in the article, Baron cautioned that "such reading and viewing do not cause people to commit rape but rather are part of a social climate making rape more likely." Explain what he means.

3.15 An experiment was conducted to study the relationship between the corn yield *y* and the amount of fertilizer applied *x* per plot.

x	1.3	2.5	1.8	3.1	2.7
y	60	72	61	70	68

The Pearson product moment *r* was computed using the **MINITAB** statistical package:

```
          READ DATA INTO C1, C2
DATA> 1.3, 60
DATA> 2.5, 72
DATA> 1.8, 61
DATA> 3.1, 70
DATA> 2.7, 68
DATA> END DATA
       5 ROWS READ
MTB > CORR C1, C2
Correlation of C1 and C2 = 0.879
MTB > STOP
```

The statement "Read data into C1, C2" is used to enter the data. The statement "Corr C1, C2" instructs the computer to calculate *r*. The value calculated was 0.879. Verify this value.

3.16 Verify, algebraically, that formula (3-2) for calculating *r* is equivalent to the definition formula (3-1).

3.3 LINEAR REGRESSION

Although correlation tells us the strength of a linear relationship, it does not tell us the exact numerical relationship. For example, the correlation coefficient calculated for the data in Table 3-2 implies that there is a linear relationship between the number of push-ups and the number of sit-ups done by a student. That means that we should be able to use the number of push-ups to predict the number of sit-ups. But correlation analysis does not show us how to determine a *y* value given an *x* value. This is done by regression analysis. **Regression analysis calculates an equation that provides values of *y* for given values of *x*.** One of the primary objectives of regression analysis is to **make predictions**—for example, predicting the success a student will have in college based on results of high school, or predicting the distance required to stop a car based on its speed.

regression analysis

Generally, the exact value of *y* is not predicted. We are usually satisfied if the predictions are reasonably close. The statistician seeks an equation to express the relationship between the two variables. The equation he or she chooses is the one that *best fits* the scatter diagram.

Here are some examples of various relationships, called **prediction equations**:

prediction equation

$$y = b_0 + b_1 x \text{ (linear)}$$
$$y = a + bx + cx^2 \text{ (quadratic)}$$
$$y = a(b^x) \text{ (exponential)}$$
$$y = a \log_b x \text{ (logarithmic)}$$

Figures 3-10 through 3-13 illustrate some of these relationships. These graphs might result if we were to sample the following bivariate populations:

The hourly wages earned by store clerks *x* and their weekly pay *y*.

Figure 3-10
Linear regression with positive slope

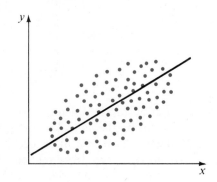

The age of a used American economy car *x* and its resale value as a used car *y*.

Figure 3-11
Linear regression with negative slope

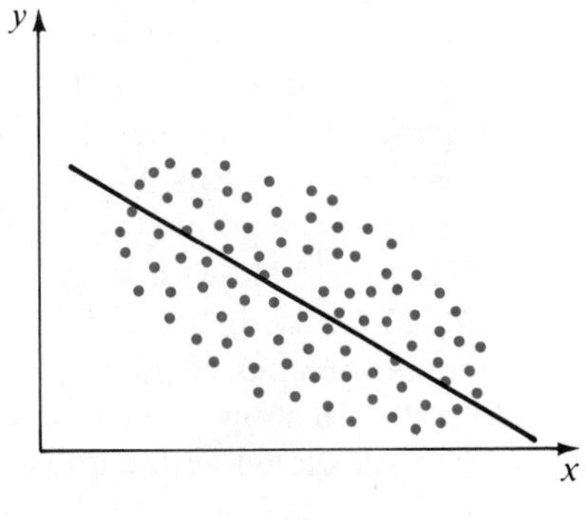

The week of the semester x and the number of hours spent studying statistics y.

Figure 3-12
Curvilinear regression
(looks quadratic)

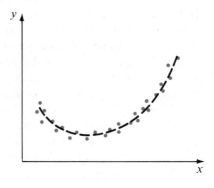

The number of the row that a student sat in for a large lecture course x and the final grade received for the course y.

Figure 3-13
No relationship

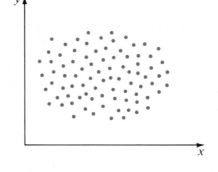

The relationship between these two variables will be an algebraic expression describing the mathematical relationship between x and y. In Figures 3-10, 3-11, and 3-12 there does appear to be a relationship. In Figure 3-13, however, the values of x and y do not seem to be related.

Given a set of bivariate (x, y) data, how do we find the line of best fit? If a straight-line relationship seems appropriate, the best-fitting straight line is found by using the **method of least squares**. Suppose that $\hat{y} = b_0 + b_1 x$ is the equation of a straight line, where $\hat{y}$ (read "y hat") represents the **predicted value** of y that corresponds to a particular value of x. The **least squares criterion** requires that we find the constants b_0 and b_1 such that the **sum $\sum(y - \hat{y})^2$ is as small as possible**.

method of least squares
predicted value
least squares criterion

Figure 3-14 shows the distance of an observed value of y from a predicted value of y. The length of this distance represents the value $(y - \hat{y})$. Note that $(y - \hat{y})$ is positive when the point (x, y) is above the line and negative when (x, y) is below the line. Figure 3-15 shows a scatter diagram with what might appear to be the line of best fit, along with ten individual $(y - \hat{y})$'s. The sum of the squares of these differences is **minimized** (made as small as possible) if the line is indeed the line of best fit. Figure 3-16 shows the same data points as Figure 3-15 with the ten individual $(y - \hat{y})$'s associated with a line that is definitely *not* the line of best fit.

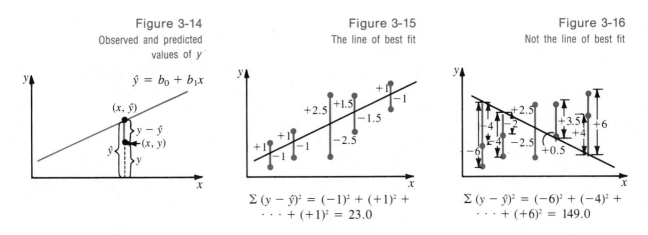

Figure 3-14
Observed and predicted values of y

$\hat{y} = b_0 + b_1 x$

$(x, \hat{y})$

$y - \hat{y}$

(x, y)

$\hat{y}$

y

Figure 3-15
The line of best fit

$\Sigma (y - \hat{y})^2 = (-1)^2 + (+1)^2 +$
$\cdots + (+1)^2 = 23.0$

Figure 3-16
Not the line of best fit

$\Sigma (y - \hat{y})^2 = (-6)^2 + (-4)^2 +$
$\cdots + (+6)^2 = 149.0$

line of best fit

The **equation of the line of best fit** is determined by its slope (b_1) and its y-intercept (b_0). (See the Study Guide for a review of the concepts of slope and intercept of a straight line.) The values of the constants that satisfy the least squares criterion are found by using these formulas:

slope

$$\textbf{slope:} \quad b_1 = \frac{\Sigma (x - \bar{x})(y - \bar{y})}{\Sigma (x - \bar{x})^2} \tag{3-5}$$

intercept

$$\textbf{intercept:} \quad b_0 = \frac{1}{n}\left(\Sigma y - b_1 \cdot \Sigma x\right) \tag{3-6}$$

(The derivation of these formulas is beyond the scope of this text.)

We will use a mathematical equivalent of formula (3-5) that is easier to apply. We will calculate the slope b_1 by using the formula

$$b_1 = \frac{SS(xy)}{SS(x)} \tag{3-7}$$

and formulas (2-9), page 62, and (3-4), page 106. Notice that the numerator of formula (3-7) is the same as the numerator of formula (3-2) for the linear correlation coefficient. Notice also that the denominator is the first radicand (expression under a radical sign) in the denominator of formula (3-2). Thus if you calculate the linear correlation coefficient by using formula (3-2), you can easily find the slope of the line of best fit. If you did not calculate r, set up a table similar to Table 3-2 and find the necessary values.

Now let's consider the data in Illustration 3-5 and the question of predicting a student's sit-ups based on the number of push-ups. We want to find the line of best fit, $\hat{y} = b_0 + b_1 x$. The necessary calculations and totals for determining b_1 have already been obtained in Table 3-2. Recall that $SS(xy) = 919.0$ and $SS(x) = 1396.9$ (see page 107). Then, using formula (3-7), the slope is calculated to be

$$b_1 = \frac{919.0}{1396.9} = 0.6579 = \textbf{0.66}$$

Using formula (3-6), the y-intercept is calculated to be

$$b_0 = \frac{1}{10}[380 - (0.6579)(351)] = \frac{1}{10}[380 - 230.9229] = 14.9077 = \mathbf{14.9}$$

Remember to keep the extra decimal places in the calculations to ensure an accurate answer.

Thus the equation of the line of best fit is

$$\hat{y} = \mathbf{14.9 + 0.66}x$$

Note   When rounding off the calculated values of b_0 and b_1, always keep at least two significant digits in the final answer, but *more* as calculations are being done.

Figure 3-17 shows a scatter diagram of the data and the line of best fit.

Figure 3-17
Line of best fit for push-ups
versus sit-ups

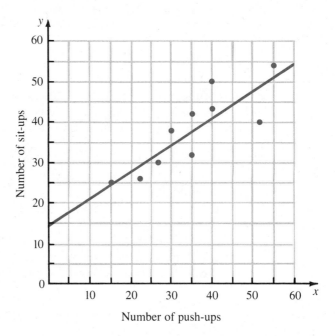

Number of push-ups

There are some additional facts about the least squares method that we need to discuss:

1. The slope b_1 represents the predicted change in y per unit increase in x. In our example, $b_1 = 0.66$; thus if a student had done an additional ten push-ups (x), we would predict that he or she would have done an additional seven (0.66×10) sit-ups (y).

2. The y-intercept is the value of y where the line of best fit intersects the y-axis, that is, where $x = 0$. However, in interpreting b_0 you first must consider whether $x = 0$ is a reasonable value before you conclude that you would predict $y = b_0$ if $x = 0$. To predict that if a student did no push-ups he or she would still do approximately 15 sit-ups $(b_0 = 14.9)$ is probably incorrect. Second, all x value of zero is outside the domain of the data on which the regression line is based. **In predicting y based on an x value, you should check to be sure that the x value is within the domain of the x values observed.**

3. The line of best fit will always pass through the point $(\bar{x}, \bar{y})$. When drawing the line of best fit on your scatter diagram, you should use this point as a check. It must lie on the line of best fit.

Illustration 3-6   In a random sample of eight college women, each was asked for her height (to the nearest inch) and her weight (to the nearest 5 pounds). The data obtained are shown in Table 3-3. Find an equation to predict the weight of a college woman based on her height (the equation of the line of best fit) and draw it on the scatter diagram.

Table 3-3
Data for college women's heights and weights

	Woman							
	1	2	3	4	5	6	7	8
Height (x)	65	65	62	67	69	65	61	67
Weight (y)	105	125	110	120	140	135	95	130

Solution   Before we find the equation for the line of best fit, we need to decide whether or not the two variables appear to be linearly related. One way to check this is to draw the scatter diagram and observe whether the plot of points suggests a linear relationship. If the scatter diagram suggests a linear relationship, we will calculate the equation for the line of best fit. The scatter diagram for the data on the height and weight of college women is shown in Figure 3-18. The scatter diagram suggests that a linear line of best fit is appropriate.

Figure 3-18
Scatter diagram: College women's heights versus weights

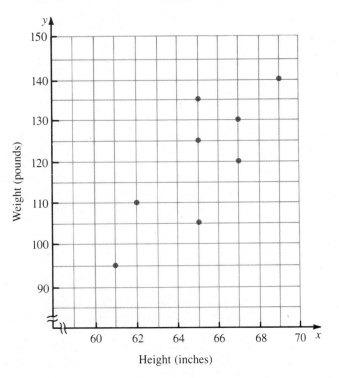

Height (inches)

Table 3-4
Calculations needed
to find b_1 and b_0

Coed	Height (x)	x_2	Weight (y)	xy
1	65	4,225	105	6,825
2	65	4,225	125	8,125
3	62	3,844	110	6,820
4	67	4,489	120	8,040
5	69	4,761	140	9,660
6	65	4,225	135	8,775
7	61	3,721	95	5,795
8	67	4,489	130	8,710
Total	521	33,979	960	62,750

Let's calculate the equation for the line of best fit using the summations in Table 3-4. By formulas (3-7), (3-4), and (2-9), we find

$$SS(xy) = 62,750 - \frac{(521)(960)}{8} = 230.0$$

$$SS(x) = 33,979 - \frac{(521)(521)}{8} = 48.875$$

$$b_1 = \frac{230.0}{48.875} = 4.706 = \mathbf{4.71}$$

By formula (3-6), we find

$$b_0 = \frac{1}{n}\left[\sum y - b_1 \cdot \sum x\right]$$

$$= \frac{1}{8}[960 - (4.706)(521)]$$

$$= -186.478 = \mathbf{-186.5}$$

Therefore,

$$\hat{y} = \mathbf{-186.5 + 4.71}x$$

To draw the line of best fit on the scatter diagram, we need to locate two points. Substitute two values for x, for example, 60 and 70, into the regression equation and obtain two corresponding values for y:

$$\hat{y} = -186.5 + (4.71)(60) = -186.5 + 282.6 = 96.1 = 96$$

and

$$\hat{y} = -186.5 + (4.71)(70) = -186.5 + 329.7 = 143.2 = 143$$

The values (60, 96) and (70, 143) represent two points (designated by a + in Figure 3-19) that enable us to draw the line of best fit.

Note   In Figure 3-19, $(\bar{x}, \bar{y}) = (65.1, 120)$ is also on the line of best fit. It is the colored circle. Use $(\bar{x}, \bar{y})$ as a check on your work.

Figure 3-19

Line of best fit for college women's heights versus weights

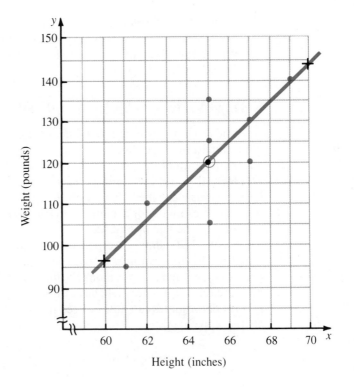

Height (inches)

One of the main purposes for obtaining a regression equation is for making predictions. Once a linear relationship has been established and the value of the input variable x is known, we can predict a value of y ($\hat{y}$). For example, in the physical fitness illustration, the equation was found to be $\hat{y} = 14.9 + 0.66x$. If student A can do 25 push-ups, how many sit-ups do you therefore predict that A will be able to do? (This predicted value will be the average number of sit-ups that you would expect from a sample taken of students who could do exactly 25 push-ups.) The predicted value is

$$\hat{y} = 14.9 + (0.66)(25) = 14.9 + 16.5 = 31.4 = \mathbf{31}$$

When making predictions based on the line of best fit, you must observe the following restrictions:

1. The equation should be used to make predictions only about the population from which the sample was drawn. For example, using the relationship between the height and weight of college women to predict the weight of professional athletes given their height would be questionable.

2. The equation should be used only over the sample domain of the input variable. For example, for Illustration 3-6 the prediction that a college woman of height zero weighs -186.5 pounds is nonsense. You should not use a height outside the sample domain of 60 to 70 inches to predict weight.

3. If the sample was taken in 1987, do not expect the results to have been valid in 1929 or to hold in 1999. The women of today may be different from the women of 1929 and the women of 1999. On occasion you might wish to use the line of best fit to estimate values outside the domain interval of the sample. This can be done, but you should do it with caution and only for values close to the domain interval.

ESTIMATING THE LINE OF BEST FIT

It is possible to estimate the line of best fit and its equation. As with the approximation of r, it should be used mostly as a mental check of your work. On the scatter diagram of the data, draw the straight line that appears to be the line of best fit. [*Hint*: If you draw a line parallel to and halfway between your pencils (whose location was described in Section 3.2, p. 108, for estimating r) you will have a reasonable estimate for the line of best fit.] Then use this line to approximate the equation of the line of best fit.

Figure 3-20, shows the pencils and the resulting estimated line for Illustration 3-6. This line can now be used to approximate the equation. First, locate two points, (x_1, y_1) and (x_2, y_2), along the line and determine their coordinates. Two such points, circled in Figure 3-20, have the coordinates (59, 85) and (66, 125). These two pairs of coordinates can now be used in the following formula to obtain an estimate for the slope b_1.

$$\text{estimate for } b_1 = \frac{y_2 - y_1}{x_2 - x_1} = \frac{125 - 85}{66 - 59} = \frac{40}{7} = 5.7$$

Figure 3-20

Estimating the line of best fit
for the college coed data

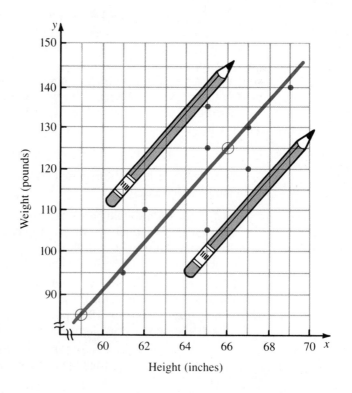

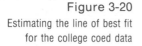

Using this result, the coordinates of one of the points, and the following formula, we can determine an estimation for the y-intercept b_0.

$$\text{estimate for } b_0 = y - b_1 \cdot x = 85 - (5.7)(59)$$
$$= 85 - 336.3 = -251.3 \text{ or } -250$$

We now estimate the equation for the line of best fit to be $\hat{y} = -250 + 5.7x$. This should serve as a crude estimate. [*Note*: The graph's y-intercept is -250, not approximately 80, as might be read. Why?]

The techniques presented in this section presume that the scatter diagram suggests a linear relationship.

Case Study 3-3

POINTS AND FOULS IN BASKETBALL

Albert Shulte discusses the relationship between the number of personal fouls committed and total points scored during a season by the members of a junior-varsity basketball team. It appears, from Fig. 1, that the players score about 3 points for every personal foul committed. The data for the second year, Table 2, seem to indicate approximately the same relationship. Draw a scatter diagram for the data in Table 2. Does the relationship seem to be about the same as for the first-year data? (See Exercise 3.20)

Sometimes you may think that two different sets of numbers are related in some way, although not perfectly. How can you decide if they are related? If they are related, is there any reason why they should be, or is it purely accidental? Does a change in one of the variables cause a change in the other?

The data come from the records of a junior-varsity basketball team...for two separate years.... A quick look at Table 2 makes one feel that a strong relationship exists between the number points... and the number of personal fouls.... In Fig. 1, the data for the first year have been plotted....

TABLE 2. Second year

Player	Total Points	Personal Fouls
Brummett	2	1
Cooper	75	24
Felice	0	1
Hook	59	18
Hurd	9	9
Kampsen	7	3
McPartlin	35	5
Pointer	46	20
Schuback	0	1
Wilson	2	3
Zuelch	57	22

Now let's think a bit more about the relationship that seems to exist.... The table and the figure both indicate that the more fouls he commits the more points

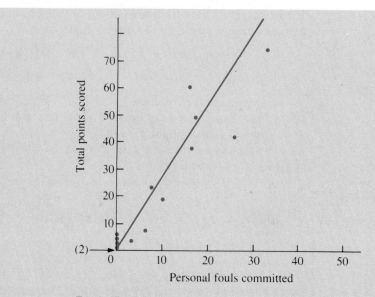

FIG. 1. First year

he scores.... Surely no coach would coach a player to make lots of fouls in the hope that he would therefore score more points. The crucial word in the previous sentence is "therefore." Is the fact that a player commits more fouls the reason that he scores more points? Of course not. But maybe both of these... are the results of some other act. Perhaps... more game time.

A graph such as that in Fig. 1 can show that a high degree of correlation exists, but it does not tell us why.

Source: Albert P. Shulte, "Points and Fouls in Basketball," in *Exploring Data*, from *Statistics by Example*. Edited and prepared by the Joint Committee on the Curriculum in Statistics and Probability of the American Statistical Association and the National Council of Teachers of Mathematics. Copyright 1973 by Addison-Wesley Publishing Co., Inc. Reprinted by permission.

EXERCISES

3.17 Draw a scatter diagram for these data:

x	2	12	4	6	9	4	11	3	10	11	3	1	13	12	14	7	2	8
y	4	8	10	9	10	8	8	5	10	9	8	3	9	8	8	11	6	9

Would you be justified in using the techniques of linear regression on these data to find the line of best fit? Explain.

3.18 Explain why the y-intercept is not 80 for the line of best fit in Figure 3-20.

3.19 A study was conducted to investigate the relationship between the cost y (in tens of thousands of dollars) per unit of equipment manufactured and the number of units produced per run x. The equation of the line of best fit was found to be $\hat{y} = 7.31 - 0.01x$, and x was observed for values between 10 and 200. If a production run is scheduled to produce 50 units, what would you predict the cost per unit to be?

3.20 Use the second-year data from Case Study 3-3

    a. Construct the scatter diagram.

    b. Estimate the correlation coefficient.

    c. Estimate the equation for the line of best fit.

    d. Does there seem to be a relationship between the two variables?

    e. Does this mean that an increase in the number of fouls (x) will cause an increase in number of points (y)? Explain.

3.21 **a.** Draw the line of best fit by eye and find an estimate of its equation by using the scatter diagram found in answering Exercise 3.2.

    b. Calculate the equation of the line of best fit. (Use the work you completed in answering Exercise 3.9.)

    c. Compare your answers for (a) and (b).

    d. Using the answer to (b), what grade would you expect to receive if you had studied for 6 hours?

    e. What exactly does your answer to (d) mean?

3.22 **a.** Calculate the equation of the line of best fit for the psychologist's problem in Exercise 3.3. (Use the work you completed in answering Exercise 3.10.)

    b. Using the evidence found [answers to Exercises 3.3, 3.10, and part (a) of this exercise], does the psychologist's claim seem to be valid?

3.23 A biological study of a minnow called the blacknose dace was conducted. The length x in millimeters and the age y to the nearest year were recorded.

x	25	80	45	40	36	75	50	95	30	15
y	0	3	2	2	1	3	2	4	1	1

    a. Calculate the correlation coefficient.

    b. Find the equation of the line of best fit.

3.24 A record of maintenance costs is kept for each of several cash registers throughout a department store chain. A sample of 14 registers gave the following data.

Age, x (years)	Maintenance Cost, y (dollars)	Age, x (years)	Maintenance Cost, y (dollars)
6	92	2	49
7	181	1	64
1	23	9	141
3	40	3	110
6	126	8	105
4	82	9	181
5	117	8	152

   a. Draw a scatter diagram that shows these data.

   b. Calculate the equation of the line of best fit.

   c. A particular cash register is eight years old. How much maintenance (cost) do you predict it will require this year?

   d. Interpret your answer to (c).

   e. Calculate r.

3.25 Verify, algebraically, that formula (3-7) is equivalent to formula (3-5).

In Retrospect

To sum up what we have just learned, there is a distinct difference between the purpose of regression analysis and the purpose of correlation. In regression analysis, we seek a relationship between the variables. The equation that represents this relationship may be the answer desired or it may be the means to the prediction that is desired. In correlation analysis, on the other hand, we simply ask: Is there a significant linear relationship between the two variables?

The article at the beginning of this chapter and the articles that form the case studies show a wide variety of applications for the techniques of correlation and regression. These articles are worth reading again. When bivariate data appear to fall along a straight line on the scatter diagram, they suggest a linear relationship. But, as noted in two of the articles, this is not proof of cause and effect. Clearly, if a basketball player commits too many fouls, he will not be scoring more points. He will be in foul trouble and riding the bench. It also seems reasonable that the more game time he has, the more points he will score and the more fouls he will commit. Thus, a positive correlation and a positive regression relationship will exist between these two variables.

The linear methods we have studied thus far, the basic concepts, have been presented for the purpose of a first, descriptive look. More details must, by necessity, wait until additional developmental work has been completed. After reading this chapter you should have a basic understanding of bivariate data, how they are different from just two sets of data, how to present them, what correlation and regression analysis are, and how each is used. The evaluation and the interpretation of the meaning and the meaningfulness of these results are studied in Chapter 13. After a few more exercises to practice the methods of this chapter, we will study some basic probability.

Chapter Exercises

3.26 Determine whether each of the following questions requires correlation analysis or regression analysis to obtain an answer.

   a. Is there a correlation between the grades a student attained in high school and the grades he or she attained in college?

   b. What is the relationship between the weight of a package and the cost of mailing it first class?

   c. Is there a linear relationship between a person's height and shoe size?

   d. What is the relationship between the number of worker hours and the number of units of production completed?

e. Is the score obtained on a certain aptitude test linearly related to a person's ability to perform a certain job?

3.27 An automobile owner records the number of gallons of gasoline x required to fill the gasoline tank and the number of miles traveled y between fill-ups.

a. If he does a correlation analysis on the data, what would be his purpose and what would be the nature of his result?

b. If he does a regression analysis on the data, what would be his purpose and what would be the nature of his results?

3.28 The following data were generated using the equation $y = 2x + 1$.

x	0	1	2	3	4
y	1	3	5	7	9

A scatter plot of these data results in five points that fall perfectly on a straight line. Find the correlation coefficient and the equation of the line of best fit.

3.29 The following equation gives a relationship that exists between b_1 and r.

$$r = b_1 \sqrt{\frac{SS(x)}{SS(y)}}$$

a. Verify this equation for the following data.

x	4	3	2	3	0
y	11	8	6	7	4

b. Verify this equation using formulas (3-2) and (3-7).

💲 **3.30** A study of beef prices showed that as the demand for beef increased by 1000 units, the price tended to increase by 10¢ a pound.

a. What type of analysis would produce this result?

b. Is the correlation coefficient for the demand for beef and the price of beef positive or negative?

💲 **3.31** A regression line for estimating the yearly food cost $\hat{y}$ for a family of five based on yearly income x is given by $\hat{y} = 350 + 0.3x$.

a. What is the average yearly expenditure for families with incomes of $6,000? $12,000?

b. Why would this formula not be appropriate for use with incomes of $0 or $100,000?

3.32 Using the data presented in Fig. 1 of the article about California gulls, page 99,

a. calculate the linear correlation coefficient.

b. calculate the line of best fit using *age of female* as the independent variable and *age of male* as the dependent variable.

3.33 A survey was conducted at a large metropolitan university campus. Twenty-four students were interviewed. Two of the questions asked were: "How many hours per week are you employed?" and "How many credit hours are you currently registered for?"

Hours Employed (x)	20	40	35	15	40	20	20	0	20	40
Credit Hours (y)	6	3	6	9	6	6	3	15	6	9

x	10	20	30	40	15	0	0	0	10	40	0	0	30	25
y	9	3	6	6	3	12	15	18	6	6	21	12	6	9

Is there a correlation between x and y?

3.34 The following data are the ages and the asking prices for 19 used foreign compact cars.

Age, x (years)	Price, y (× $100)	Age, x (years)	Price, y (× $100)
3	34	6	21
5	26	8	11
3	31.5	5	25
6	12	6	18
4	30	5	23
4	30	7	18
6	14	4	24
7	18	7	10
2	34	5	18
2	32		

a. Draw a scatter diagram.

b. Calculate the linear correlation coefficient r.

c. Is there evidence of correlation? Explain.

d. Calculate the equation for the line of best fit.

e. Graph the line of best fit on the scatter diagram.

f. Predict the average asking price for all such foreign cars that are five years old.

3.35 The following data resulted from a psychological experiment.

x	y	x	y
310	3.2	250	2.5
280	3.0	360	3.4
120	2.6	215	2.6
230	3.5	365	3.8
175	2.8	230	3.0
295	3.8	325	3.8
135	2.2	280	3.3

a. Draw a scatter diagram.

b. Calculate the linear correlation coefficient r.

c. Is there evidence of correlation? Explain.

d. Calculate the equation for the line of best fit.

e. Graph the line of best fit on the scatter diagram.

f. If x equals 300, what value can be predicted for y?

3.36 The number of one-family building permits issued and the number of one-family housing starts for the last ten months in a certain town are recorded in the following table:

Month	1	2	3	4	5	6	7	8	9	10
Permits Issued	528	494	453	398	413	454	450	436	460	444
Starts	696	614	623	507	554	550	592	568	627	564

Determine whether there is a relationship between x, the number of permits issued, and y, the number of starts. The best relationship will probably be obtained by relating permits to starts one month later, since it takes time to get construction started after a permit is issued. [*Hint*: There will be only nine data points; (528, 614), (494, 623), ...]

3.37 Consider the following set of bivariate data.

x	8	4	2	5	6
y	6	4	3	5	4

The MINITAB statistical package was used to find the line of best fit. The MINITAB statements follow.

```
          READ DATA INTO C1, C2
     DATA> 6 8
     DATA> 4 4
     DATA> 3 2
     DATA> 5 5
     DATA> 4 6
     DATA> END DATA
         5 ROWS READ
     MTB > REGRESS Y IN C1 ON 1 PREDICTOR IN C2
     The regression equation is
     C1 = 2.15 + 0.450 C2
```

Verify that $\hat{y} = 2.15 + 0.45x$ is the line of best fit.

Vocabulary List

Be able to define each term. In addition, describe, in your own words, and give an example of each term. Your examples should not be ones given in class or in the textbook.

bivariate data

coefficient of linear correlation

correlation analysis

dependent variable

independent variable	Pearson's product moment r
input variable	positive correlation
least squares criterion	predicted value
linear	prediction equation
line of best fit	regression analysis
method of least squares	scatter diagram
negative correlation	slope, b_1
ordered pair	y-intercept, b_0
output variable	zero correlation
paired data	

Quiz   *Answer "True" if the statement is always true. If the statement is not always true, replace the words shown in italics with words that make the statement always true.*

3.1 *Correlation* analysis is a method of obtaining the equation that represents the relationship between two variables.

3.2 The linear correlation coefficient is used to determine the *equation* that represents the relationship between two variables.

3.3 A correlation coefficient of *zero* means that the two variables are perfectly correlated.

3.4 Whenever the slope of the regression line is zero, the *correlation coefficient* will also be zero.

3.5 When r is positive, b_1 will always be *negative.*

3.6 The *slope* of the regression line represents the amount of change that is expected to take place in y when x increases by one unit.

3.7 When the calculated value of r is positive, the calculated value of b_1 will be *negative.*

3.8 Correlation coefficients range between *0 and +1.*

3.9 The value that is being predicted is called the *input variable.*

3.10 The line of best fit is used to predict the *average value of y* that can be expected to occur at a given value of x.

Working with Your Own Data

Each semester, new students enter your college environment. You may have wondered, "What will the student body be like this semester?"

As a beginning statistics student, you have just finished studying three chapters of basic descriptive statistical techniques. Let's use some of these techniques to describe some characteristic of your college's student body.

A | SINGLE VARIABLE DATA

1. Define the population to be studied.

2. Choose a variable to define. (You may define your own variable or you may use one of the variables in the accompanying table if you are not able to collect your own data. Ask your instructor for guidance.)

3. Collect 35 pieces of data for your variable.

4. Construct a stem-and-leaf display of your data. Be sure to label it.

5. Calculate the value of the measure of central tendency that you believe best answers the question "What is the average value of your variable?" Explain why you chose this measure.

6. Calculate the sample mean for your data (unless you used the mean in Question 5).

7. Calculate the sample standard deviation for your data.

8. Find the value of the 85th percentile, P_{85}.

9. Construct a graphic display (other than a stem-and-leaf) that you believe "best" displays your data. Explain why the graph best presents your data.

B | BIVARIATE DATA

1. Define the population to be studied.

2. Choose and define two variables that will produce bivariate data. (You may define your own variables or you may use two of the variables in the accompanying table if you are not able to collect your own data. Ask your instructor for guidance.)

3. Collect 15 ordered pairs of data.

4. Construct a scatter diagram of your data. (Be sure to label it completely.)

5. Using a table to assist with the organization, calculate the extensions x^2, xy, and y^2, and the summations of x, y, x^2, xy, and y^2.

6. Calculate the linear correlation coefficient r.
7. Calculate the equation of the line of best fit.
8. Draw the line of best fit on your scatter diagram.

The following table of data was collected on the first day of class last semester. You may use it as a source for your data if you are not able to collect your own.

Variable A = student's sex (male/female)

Variable B = student's age at last birthday

Variable C = number of completed credit hours toward degree

Variable D = "Do you have a job (full/part-time)?" (yes/no)

Variable E = number of hours worked last week, if D = yes

Variable F = wages (before taxes) earned last week, if D = yes

Student	A	B	C	D	E	F	Student	A	B	C	D	E	F
1	M	21	16	No			31	F	24	45	No		
2	M	18	0	Yes	10	34	32	M	34	4	No		
3	F	23	18	Yes	46	206	33	M	29	48	No		
4	M	17	0	No			34	M	22	80	Yes	40	336
5	M	17	0	Yes	40	157	35	M	21	12	Yes	26	143
6	M	40	17	No			36	F	18	0	No		
7	M	20	16	Yes	40	300	37	M	18	0	Yes	13	65
8	M	18	0	No			38	F	42	34	Yes	40	244
9	F	18	0	Yes	20	70	39	M	25	60	Yes	60	503
10	M	29	9	Yes	8	32	40	M	39	32	Yes	40	500
11	M	20	22	Yes	38	146	41	M	29	13	Yes	39	375
12	M	34	0	Yes	40	340	42	M	19	18	Yes	51	201
13	M	19	31	Yes	29	105	43	M	25	0	Yes	48	500
14	M	18	0	No			44	F	18	0	No		
15	M	20	0	Yes	48	350	45	M	32	68	Yes	44	473
16	F	27	3	Yes	40	130	46	F	21	0	No		
17	M	19	10	Yes	40	202	47	F	26	0	Yes	40	320
18	F	18	16	Yes	40	140	48	M	24	11	Yes	45	330
19	M	19	4	Yes	6	22	49	F	19	0	Yes	40	220
20	F	29	9	No			50	M	19	0	Yes	10	33
21	F	21	0	Yes	20	80	51	F	35	59	Yes	25	88
22	F	39	6	No			52	F	24	6	Yes	40	300
23	M	23	34	Yes	42	415	53	F	20	33	Yes	40	170
24	F	31	0	Yes	48	325	54	F	26	0	Yes	52	300
25	F	22	7	Yes	40	195	55	F	17	0	Yes	27	100
26	F	27	75	Yes	20	130	56	M	25	18	Yes	41	355
27	F	19	0	No			57	M	24	0	No		
28	M	22	20	Yes	40	470	58	M	21	0	Yes	30	150
29	F	60	0	Yes	40	390	59	M	30	12	Yes	48	555
30	M	25	14	No			60	F	19	0	Yes	38	169

Student	A	B	C	D	E	F	Student	A	B	C	D	E	F
61	M	32	45	Yes	40	385	81	F	27	0	Yes	40	550
62	M	26	90	Yes	40	340	82	F	42	47	Yes	37	300
63	M	20	64	Yes	10	45	83	F	41	21	Yes	40	250
64	M	24	0	Yes	30	150	84	M	36	0	Yes	40	400
65	M	20	14	No			85	M	25	16	Yes	40	480
66	M	21	70	Yes	40	340	86	F	18	0	Yes	45	189
67	F	20	13	Yes	40	206	87	M	22	0	Yes	40	385
68	F	33	3	Yes	32	246	88	F	27	9	Yes	40	260
69	F	25	68	Yes	40	330	89	F	26	3	Yes	40	240
70	F	29	48	Yes	40	525	90	F	23	9	Yes	40	330
71	F	40	0	Yes	40	400	91	M	41	3	Yes	23	253
72	F	36	3	Yes	40	300	92	M	39	0	Yes	40	110
73	F	35	0	Yes	40	280	93	M	21	0	Yes	40	246
74	F	28	0	Yes	40	350	94	F	32	0	Yes	40	350
75	M	40	64	Yes	40	390	95	F	48	58	Yes	40	714
76	F	31	0	Yes	40	200	96	F	26	0	Yes	32	200
77	F	32	0	Yes	40	270	97	F	27	0	Yes	40	350
78	F	37	0	Yes	24	150	98	F	52	56	Yes	40	390
79	F	35	0	Yes	40	350	99	F	34	27	Yes	8	77
80	M	21	72	Yes	45	470	100	F	49	3	Yes	24	260

The required calculations are accomplished most easily with the assistance of an electronic calculator or a canned program on a computer. Your local computer center will have a list of the available canned programs and will also provide you with the necessary information and assistance to run your data on the computer.

Part 2

Probability

Before continuing our study of statistics, we must make a slight detour and study some basic probability. Probability is often called the "vehicle" of statistics; that is, the probability associated with chance occurrences is the underlying theory for statistics. Recall that in Chapter 1 we described probability as the science of making statements about what will occur when samples are drawn from known populations. Statistics was described as the science of selecting a sample and making inferences about the unknown population from which it is drawn. To make these inferences, we need to study sample results in situations where the population is known, so that we will be able to understand the behavior of chance occurrences.

In Part 2 we will study the basic theory of probability (Chapter 4), probability distributions of discrete variables (Chapter 5), and probability distributions for continuous random variables (Chapter 6). Following this brief study of probability, we will study the techniques of inferential statistics in Part 3.

Chapter 4
Probability

ANOTHER WAY ALCOHOL CAN TRULY HURT YOU

"I learned on my mother's knee that God takes care of babies, fools, and drunks," says Patricia Waller, associate director for driver studies at the University of North Carolina Highway Safety Research Center. "But I think you're going to have to remove drunks from the list."

Since 1959, several controlled studies of inebriated dogs, cats, and rats have appeared, showing that, in response to a blow to the spinal cord or brain, there's more swelling and hemorrhaging if alcohol is present. To find out whether this holds for humans, Waller and her colleagues analyzed data on 1,136,507 drivers in automobile crashes on North Carolina's roads. They published their findings in the *Journal of the American Medical Association*.

Of course, drunks have worse judgment and coordination than sober people, and they're more likely to be driving fast, at night, and with seat belts unfastened. That means they're likely to get in more serious accidents. So the researchers grouped accidents into type, speed, and degree of vehicle deformation. Even when these variables are controlled, she says, alcohol "renders organisms more vulnerable to injury from any given impact"—the higher the level of alcohol, the greater the chance of being injured or killed.

However, the study showed that the worse the car was damaged the less difference there was between a drunk and a sober driver's chances of survival. In minor crashes, drunk drivers were more than four times as likely to be killed as sober ones; in average crashes, more than three times as likely; in the worst ones, almost twice as likely. Overall, drunks were more than twice as likely to die.

Chapter
Objectives

You may already be familiar with some ideas of probability, because probability is part of our everyday culture. We constantly hear people making probability-oriented statements such as

"Our team will probably win the game tonight."

"There is a 40 percent chance of rain this afternoon."

"I will most likely have a date for the winter weekend."

"If I park in the faculty parking area, I will probably get a ticket."

"I have a 50-50 chance of passing today's chemistry exam."

Everyone has made or heard these kinds of statements. What exactly do they mean? Do they, in fact, mean what they say? Some statements may be based on scientific information and others on subjective prejudice. Whatever the case may be, they are probabilistic inferences—not fact, but conjectures.

In this chapter we will study the basic concept of probability and the rules that apply to the probability of both simple and compound events.

CONCEPTS OF PROBABILITY

4.1 THE NATURE OF PROBABILITY

Let's consider an experiment in which we toss two coins simultaneously and record the number of heads that occur. 0H (zero heads), 1H (one head), and 2H (two heads) are the only possible outcomes. Let's toss the two coins 10 times and record our findings.

$$2H, \quad 1H, \quad 1H, \quad 2H, \quad 1H, \quad 0H, \quad 1H, \quad 1H, \quad 1H, \quad 2H$$

Summary:

Outcome	Frequency
2H	3
1H	6
0H	1

Suppose that we repeat this experiment 19 times. Table 4-1 shows the totals for 20 sets of 10 tosses. (Trial 1 shows the totals from our first experiment.) The total of 200 tosses of the pair of coins resulted in 2H on 53 occasions, 1H on 104 occasions, and

Table 4-1
Experimental results of
tossing two coins

Outcome	Trial																				Total
	1	2	3	4	5	6	7	8	9	10	11	12	13	14	15	16	17	18	19	20	
2H	3	3	5	1	4	2	4	3	1	1	2	5	6	3	1	4	1	0	3	1	53
1H	6	5	5	5	5	7	5	5	5	5	8	4	3	7	5	1	5	4	5	9	104
0H	1	2	0	4	1	1	1	2	4	4	0	1	1	0	4	5	4	6	2	0	43

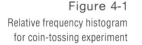

Figure 4-1

Relative frequency histogram
for coin-tossing experiment

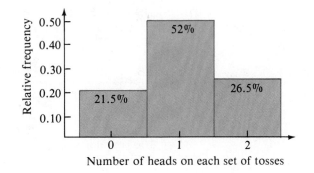

OH on 43 occasions. We can express these results in terms of relative frequencies and show the results using a histogram, as shown in Figure 4-1.

What would happen if this experiment were repeated or continued? Would the relative frequencies change? If so, by how much? If we look at the individual sets of 10 tosses, we notice a large variation in the number of times each of the events (2H, 1H, and 0H) occurred. In both the 0H and the 2H categories, there were as many as 6 occurrences and as few as 0 occurrences in a set of 10 tosses. In the 1H category there were as few as 1 occurrence and as many as 9 occurrences.

If we were to continue this experiment for several hundred more tosses, what would you expect to happen in terms of the relative frequencies of these three events? It looks as if we have approximately a 1:2:1 ratio in the totals of Table 4-1. We might therefore expect to find the relative frequency for 0H to be approximately $\frac{1}{4}$, or 25 percent; the relative frequency for 1H to be approximately $\frac{1}{2}$, or 50 percent; and the relative frequency for 2H to be approximately $\frac{1}{4}$, or 25 percent. These relative frequencies accurately reflect the concept of probability.

EXERCISES

4.1 Toss a single coin 10 times and record H (head) or T (tail) after each toss. Using your results, find the relative frequency of

    a. heads   **b.** tails

4.2 Roll a single die 20 times and record a 1, 2, 3, 4, 5, or 6 after each roll. Using your results, find the relative frequency of

    a. 1   **b.** 2   **c.** 3   **d.** 4   **e.** 5   **f.** 6

4.3 Place three coins in a cup, shake and dump them out, and observe the number of heads showing. Record 0H, 1H, 2H, or 3H after each trial. Repeat the process 25 times. Using your results, find the relative frequency of

    a. 0H   **b.** 1H   **c.** 2H   **d.** 3H

4.4 MINITAB can be used to simulate certain experiments such as flipping a coin or rolling a die. The following MINITAB output gives a simulated outcome for rolling a die 50 times and tossing a coin 100 times.

```
MTB > IRANDOM 50 INTEGERS BETWEEN 1 AND 6, PUT IN C1
        50 random integers between 1 and 6
MTB > PRINT C1
C1
2 5 5 6 1 1 2 4 4 6 4 5 2 3 5
3 3 5 2 1 2 1 4 2 5 5 4 6 1 1
3 4 3 6 2 3 2 1 6 5 5 2 5 4 4
4 2 6 2 5
MTB > IRANDOM 100 INTEGERS BETWEEN 1 AND 2, PUT IN C2
        100 random integers between 1 and 2
MTB > PRINT C2
C2
2 2 2 2 1 2 2 1 2 2 1 1 1 1 1
1 2 2 2 1 1 2 1 1 2 2 2 2 1 1
1 2 2 2 1 1 2 2 2 1 1 1 1 2 1
1 1 2 1 2 2 1 2 2 2 2 1 1 1 2
2 2 2 1 1 2 2 1 2 1 2 1 2 2 1
1 2 1 2 1 2 2 1 1 1 1 2 1 1 2
2 2 1 1 1 1 2 2 1 1
MTB > STOP
```

a. Give the relative frequency for the outcomes 1, 2, 3, 4, 5, and 6 based on the MINITAB output.

b. Give the relative frequency for a head (1) and a tail (2) based on the MINITAB output.

4.2 PROBABILITY OF EVENTS

We are now ready to define what is meant by probability. Specifically, we talk about "the probability that a certain event will occur."

probability of an event

> ### Probability That an Event Will Occur
> The relative frequency with which that event can be expected to occur.

experimental or empirical probability

The probability of an event may be obtained in three different ways: (1) empirically, (2) theoretically, and (3) subjectively. The first method was illustrated in the experiment in Section 4-1 and might be called **experimental**, or **empirical**, **probability**. This is nothing more than the **observed relative frequency with which an event occurs**. In our coin-tossing illustration we observed exactly one head (1H) on 104 of the 200 tosses of the pair of coins. The observed empirical probability for the occurrence of 1H was 104/200, or 0.52.

When the value assigned to the probability of an event results from experimental data, we will identify the probability of the event with the symbol $P'(\ \ )$. The name of the event is placed in parentheses:

$$P'(1H) = 0.52, \ P'(0H) = 0.215, \ \text{and} \ P'(2H) = 0.265$$

Thus the *prime notation* is used to denote empirical probabilities.

The value assigned to the probability of event A as a result of experimentation can be found by means of the formula

$$P'(A) = \frac{n(A)}{n} \tag{4-1}$$

where $n(A)$ is the number of times that event A actually is observed and n is the number of times the experiment is attempted.

$P'(A)$ represents the relative frequency with which event A occurred. This may or may not be the relative frequency with which that event "can be expected" to occur. $P'(A)$, however, represents our best *estimate* of what "can be expected" to occur. But what is meant by the relative frequency that "can be expected" to occur?

Consider the rolling of a single die. Define event A as the occurrence of a 1. In a single roll of a die, there are six possible outcomes. Assuming that the die is symmetrical, each number should have an equal likelihood of occurring. Intuitively, we see that the probability of A, or the expected relative frequency of a 1, is $\frac{1}{6}$. (Later we will formalize this calculation.)

What does this mean? Does it mean that once in every six rolls a 1 will occur? No, it does not. Saying that the probability of a 1, $P(1)$, is $\frac{1}{6}$ means that in the long run the proportion of times that a 1 occurs is approximately $\frac{1}{6}$. How close to $\frac{1}{6}$ can we expect the observed relative frequency to be?

Table 4-2 shows the number of 1's observed in each set of six rolls of a die (Column 1), an observed relative frequency for each set of six rolls (Column 2), and a cumulative relative frequency (Column 3). Each trial is a set of six rolls.

Figure 4-2a (page 140) shows the fluctuation of the observed probability for event A on each of the 20 trials (Column 2, Table 4-2). Figure 4-2b shows the

Table 4-2

Experimental results of rolling a die six times in each trial

Trial	(1) Number of 1s Observed	(2) Relative Frequency	(3) Cumulative Relative Frequency
1	1	1/6	1/6 = 0.17
2	2	2/6	3/12 = 0.25
3	0	0/6	3/18 = 0.17
4	2	2/6	5/24 = 0.21
5	1	1/6	6/30 = 0.20
6	1	1/6	7/36 = 0.19
7	2	2/6	9/42 = 0.21
8	2	2/6	11/48 = 0.23
9	0	0/6	11/54 = 0.20
10	0	0/6	11/60 = 0.18
11	2	2/6	13/66 = 0.20
12	0	0/6	13/72 = 0.18
13	2	2/6	15/78 = 0.19
14	1	1/6	16/84 = 0.19
15	1	1/6	17/90 = 0.19
16	3	3/6	20/96 = 0.21
17	0	0/6	20/102 = 0.20
18	1	1/6	21/108 = 0.19
19	0	0/6	21/114 = 0.18
20	1	1/6	22/120 = 0.18

Figure 4-2
Fluctuations found in the
die-tossing experiment

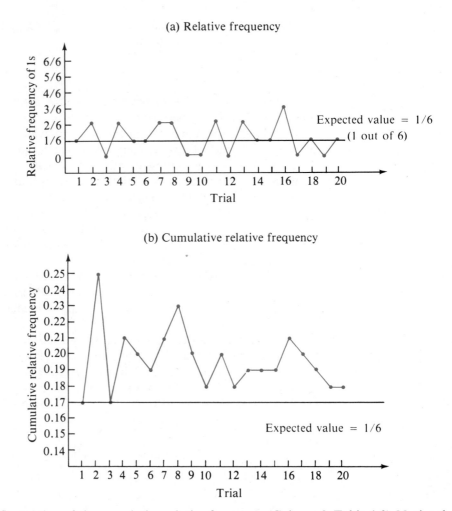

Figure 4-2
Fluctuations found in the
die-tossing experiment

(a) Relative frequency

(b) Cumulative relative frequency

fluctuation of the cumulative relative frequency (Column 3, Table 4-2). Notice that the observed relative frequency on each trial of six rolls of a die tends to fluctuate about $\frac{1}{6}$. Notice also that the observed values on the cumulative graph seem to become more stable; in fact, they become relatively close to the expected $\frac{1}{6}$.

A cumulative graph such as Figure 4-2b demonstrates the idea of long-term average. When only a few rolls were observed (as on each trial), the probability $P'(A)$ fluctuated between 0 and $\frac{1}{2}$ (*see* the relative frequency of each trial, Column 2, Table 4-2). As the experiment was repeated, however, the cumulative graph suggests a stabilizing effect on the observed cumulative probability. This stabilizing effect, or **long-term average** value, is often referred to as the *law of large numbers*.

long-term average

law of large numbers

Law of Large Numbers

If the number of times that an experiment is repeated is increased, the ratio of the number of successful occurrences to the number of trials will tend to approach the theoretical probability of the outcome for an individual trial.

The law of large numbers can be applied in a different manner. Consider the following experiment. A thumbtack is tossed and the result is observed and recorded as a point up (⌃) or a point down (⌀). What is the probability that the tack will land point up? That is, what is $P(⌃)$? When we worked with the die in the preceding example, we could easily see the true probability. With the tack, we have no idea. Is $P(⌃)$ equal to $\frac{1}{2}$? Since the tack is not symmetrical, we would suspect not. More than $\frac{1}{2}$? Less than $\frac{1}{2}$? About the only way we can tell is to conduct an experiment. And the law of large numbers can help us here. After many trials of the experiment, $P'(⌃)$ should indicate a value that is very close to the true value of $P(⌃)$, even though the true value cannot be proven mathematically.

Table 4-3 lists the data recorded after repeated tosses of a thumbtack. After each 10 tosses the number of times the tack landed point up was recorded. The cumulative value of the number of times it landed point up was computed and reported as a fraction of the total tosses. Data on 500 tosses are recorded. Figure 4-3 illustrates the value of $P(⌃)$ that we are attempting to estimate. The figure illustrates the tendency for the observed probability $P'(⌃)$ to be zeroing in on the value 0.605. We can conclude that $P(⌃)$ is very close to 0.605.

An important point is illustrated in Figure 4-3 and Table 4-3.

Table 4-3 Experimental results of tossing a thumbtack					

Multiples of 10 Tosses	$n(⌃)$ in Each Set of 10 Tosses	Cumulative $P'(⌃)$	Multiples of 10 Tosses	$n(⌃)$ in Each Set of 10 Tosses	Cumulative $P'(⌃)$
1	5	$5/10 = 0.500$	26	3	$156/260 = 0.600$
2	7	$12/20 = 0.600$	27	5	$161/270 = 0.596$
3	10	$22/30 = 0.733$	28	7	$168/280 = 0.600$
4	6	$28/40 = 0.700$	29	7	$175/290 = 0.603$
5	9	$37/50 = 0.740$	30	9	$184/300 = 0.613$
6	5	$42/60 = 0.700$	31	7	$191/310 = 0.616$
7	6	$48/70 = 0.686$	32	7	$198/320 = 0.619$
8	6	$54/80 = 0.675$	33	7	$205/330 = 0.621$
9	2	$56/90 = 0.622$	34	6	$211/340 = 0.621$
10	4	$60/100 = 0.600$	35	7	$218/350 = 0.623$
11	4	$64/110 = 0.582$	36	6	$224/360 = 0.622$
12	7	$71/120 = 0.592$	37	4	$228/370 = 0.616$
13	6	$77/130 = 0.592$	38	6	$234/380 = 0.616$
14	8	$85/140 = 0.607$	39	4	$238/390 = 0.610$
15	6	$91/150 = 0.607$	40	5	$243/400 = 0.608$
16	3	$94/160 = 0.588$	41	5	$248/410 = 0.605$
17	7	$101/170 = 0.594$	42	7	$255/420 = 0.607$
18	9	$110/180 = 0.611$	43	7	$262/430 = 0.609$
19	7	$117/190 = 0.616$	44	6	$268/440 = 0.609$
20	3	$120/200 = 0.600$	45	4	$272/450 = 0.604$
21	5	$125/210 = 0.595$	46	7	$279/460 = 0.607$
22	8	$133/220 = 0.605$	47	5	$284/470 = 0.604$
23	8	$141/230 = 0.613$	48	8	$292/480 = 0.608$
24	6	$147/240 = 0.613$	49	5	$297/490 = 0.606$
25	6	$153/250 = 0.612$	50	6	$303/500 = 0.606$

Figure 4-3
Experimental probability of
thumbtack pointing up

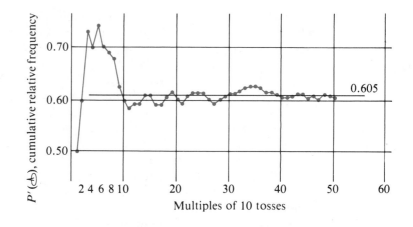

Figure 4-3
Experimental probability of
thumbtack pointing up

> The larger the number of experimental trials n, the closer the experimental probability $P'(A)$ is expected to be to the true probability $P(A)$.

Many situations are analogous to the thumbtack experiment. We'll look at one example in Illustration 4-1.

Illustration 4-1   The key to establishing proper life insurance rates is using the probability that the insureds will live one, two, or three years, and so forth, from the time they purchase policies. These probabilities are derived from actual life and death statistics and hence are experimental probabilities. They are published by the government and are extremely important to the life insurance industry.   ∎

Case Study 4-1

RATINGS-RACE ROUNDUP

The ratings race is not a horse race. It is, however, an application of empirical probabilities that affects all of us. That is, it has a great influence on what shows we see on television. The Nielsen Ratings are relied on very heavily by the television industry. What does a rating of 31.4 mean? (See Exercise 4.10.)

To the networks, the annual ratings race lasts 30 weeks. Then it's time to declare a winner and begin summer reruns while preparing for next season's race. That means that on Jan. 4 the ABC, CBS and NBC steeds thundered past the halfway pole—the 15-week mark—of the 1986-87 ratings derby. Here's a look at how they're doing, based on Nielsen's numbers. The ratings figures refer to the percentage of TV households tuned to a show; one ratings point equals 874,000 TV households.

Program Types

What types of prime-time programs drew the strongest ratings? Sports events scored highest overall, but there were only 38 of them among the 1046 prime-time programs shown by the networks during the first half of the season.

Category	Number Shown	Ratings Average
Sports Events	38	19.7
Comedy Series	340	18.1
Nighttime Soap Operas	67	16.8
Movies	108	15.2
Crime Series	238	14.9
News and Documentary Series	51	14.6
Drama Series	133	13.7
Specials and Miniseries	71	13.6

The Bottom 11

Eleven series ranked at the bottom of the chart for the first half of the season, and six, marked [C], have already been canceled or put on the shelf. Nine of the 11 are new shows (marked with asterisks). It's worth noting that the last-place show. *Our World*, has a great many viewers. With each ratings point equaling 874,000 households. *Our World* reaches an average of 5,418,800 households each week.

*	Downtown [C] (CBS)	10.4
	1986 [C] (NBC)	10.4
*	Starman (ABC)	10.1
*	Sidekicks (ABC)	10.1
*	Dads [C] (4 shows) (ABC)	10.1
*	Life with Lucy [C] (ABC)	9.7
	The Twilight Zone [C] (CBS)	9.6
*	Gung Ho [C] (4 shows) (ABC)	8.5
*	Ellen Burstyn [C] (ABC)	8.3
*	Heart of the City [C] (ABC)	7.3
*	Our World (ABC)	6.2

Top 25 Series

Bill Cosby continued his winning ways in the first half of the season with ratings numbers that played a large role in NBC's continuing success. Six new series (marked with asterisks) made the Top 25.

1.	Cosby (NBC)	34.0
2.	Family Ties (NBC)	31.4
3.	Cheers (NBC)	27.4
4.	Murder She Wrote (CBS)	25.5
5.	The Golden Girls (NBC)	25.1
6.	Night Court (NBC)	24.7
7.	60 Minutes (CBS)	23.4
8.	Moonlighting (ABC)	22.5
9.	Dallas (CBS)	21.7
	Growing Pains (ABC)	21.7
11.	Who's the Boss? (ABC)	21.5
12.	*Outlaws (2 shows) (CBS)	20.6
13.	*Amen (NBC)	19.7
	Newhart (CBS)	19.7
15.	227 (NBC)	18.8
16.	Kate & Allie (CBS)	18.2
17.	*My Sister Sam (CBS)	17.8
18.	Falcon Crest (CBS)	17.4
	*Matlock (NBC)	17.4
20.	Miami Vice (NBC)	17.3
21.	Hunter (NBC)	17.2
22.	Highway to Heaven (NBC)	16.9
	Dynasty (ABC)	16.9
24.	*L.A. Law (NBC)	16.7
25.	*The Cavanaughs (5 shows) (CBS)	16.6

ABC:4   CBS:9   NBC:12

The Race Tightens

The 15-week averages show NBC in the lead by two points over CBS and almost four points over ABC (3.80)

NBC 17.9
CBS 15.9
ABC 14.1

In the first five weeks NBC beat CBS by 3.16 ratings points and ABC by 4.44. In the second five-week period, NBC's lead was 2.00 points over CBS and 4.02 over ABC. In the last five weeks. CBS came within a ratings point of NBC and ABC cut is deficit to 2.88.

EXERCISES

4.5 Explain what is meant by the statement "The probability of a 1 when a single die is rolled is $\frac{1}{6}$."

4.6 a. In Exercise 4.1, you were asked to toss a single coin 10 times. What was the empirical probability of heads after 10 tosses?

b. Toss the coin 10 more times so that you have a total of 20 tosses. What is $P'(H)$ after 20 tosses?

c. Toss the coin 30 more times so that you have a total of 50 tosses. What is $P'(H)$ after 50 tosses?

d. Toss the coin 50 more times. What is $P'(H)$ after 100 tosses?

4.7 Draw a card from a shuffled standard deck and observe whether or not the card is a face card (that is, a jack, a queen, or a king). Shuffle the deck and repeat the draw. Then continue until you have made 50 draws.

a. Find the empirical probability of obtaining a face card.

b. Do the same experiment 100 times and use your results to compute the probability of a face card.

c. Comment on your answers to (a) and (b).

4.8 Take two dice (one white and one black) and roll them 50 times, recording the results as ordered pairs [for example, (3.5) represents 3 on the white die and 5 on the black die]. Then calculate the following observed probabilities.

a. P'(white die is an odd number)

b. P'(sum is 6)

c. P'(both dice show odd numbers)

d. P'(number on black die is larger than number on white die)

4.9 Using a coin, perform the die and tack experiment discussed on pages 139–141. Toss a single penny 10 times, observing the number of heads (or use a cup containing 10 pennies and use one toss for each block of 10), and record the results. Repeat until you have made 200 tosses. Chart and graph the data. Do your data tend to support the claim that $P(\text{head}) = \frac{1}{2}$? If not, what do you think caused the difference?

4.10 Explain how a Nielsen rating of 31.4, Case Study 4-1, is a probability.

4.3 SIMPLE SAMPLE SPACES

Let's return to an earlier question: What values might we expect to be assigned to the three events (0H, 1H, 2H) associated with our coin-tossing experiment? As we inspect these three events, we see that they do not tend to happen with the same relative frequency. Why? Suppose that the experiment of tossing two pennies and observing the number of heads had actually been carried out using a penny and a nickel—two distinct coins. Would this have changed our results? No, it would have had no effect on the experiment. However, it does show that there are more than three possible outcomes.

When a penny is tossed, it may land as heads or tails. When a nickel is tossed, it may also land as heads or tails. If we toss them simultaneously, we see that there are actually four different possible outcomes of each toss. Each observation would be one of the following possibilities: (1) heads on the penny and heads on the nickel, (2) heads on the penny and tails on the nickel, (3) tails on the penny and heads on the nickel, or (4) tails on the penny and tails on the nickel. Notice that the previous events (0H, 1H, and 2H) match up with these four in the following manner: event (1) is the same as 2H, event (4) is the same as 0H, and events (2) and (3) together make up what was previously called 1H.

ordered pair   In this experiment with the penny and the nickel, let's use an **ordered pair** notation. The first listing will correspond to the penny and the second will correspond to the nickel. Thus (H, T) represents the event that a head occurs on the penny and a tail occurs on the nickel. Our listing of events for the tossing of a penny and a nickel looks like this:

$$(H, H), (H, T), (T, H), (T, T)$$

What we have accomplished here is a listing of what is known as the "sample space" for this experiment.

experiment

Experiment

Any process that yields a result or an observation.

outcome

Outcome

A particular result of an experiment.

sample space

Sample Space

The set of all possible outcomes of an experiment. The sample space is typically called S and may take any number of forms: a list, a tree diagram, a lattice grid *sample points* system, and so on. The individual outcomes in a sample space are called **sample points**. $n(S)$ is the number of sample points in sample space S.

event

Event

Any subset of the sample space. If A is an event, then $n(A)$ is the number of sample points that belong to event A.

Regardless of the form in which they are presented, the outcomes in a sample space can never overlap. Also, all possible outcomes must be represented. These *mutually exclusive* characteristics are called **mutually exclusive** and **all inclusive**, respectively. A more *all inclusive* detailed explanation of these characteristics will be presented later; for the moment, however, an intuitive grasp of their meaning is sufficient.

Now let's look at some illustrations of probability experiments and their associated sample spaces.

Experiment 4-1   A single coin is tossed once and the outcome—a head (H) or a tail (T)—is recorded.

$$\text{Sample space: } S = \{H, T\}$$

Experiment 4-2   Two coins, one penny and one nickel, are tossed simultaneously and the outcome for each coin is recorded using ordered pair notation: (penny, nickel).

The sample space is shown here in two different ways.

tree diagram   Tree diagram representation:

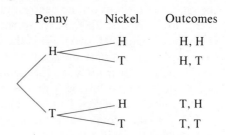

(four branches, each branch
shows a possible outcome)

Listing:

$$S = \{(H, H), (H, T), (T, H), (T, T)\}$$

Notice that both representations show the same four possible outcomes. For example, the top branch on the tree diagram shows heads on both coins, as does the first ordered pair in the listing.   ■

Experiment 4-3   A die is rolled one time and the number of spots on the top face observed.

The sample space is

$$S = \{1, 2, 3, 4, 5, 6\}$$ ■

Experiment 4-4   A box contains three poker chips (one red, one blue, one white), and two are drawn *with replacement*. (This means that one chip is selected, its color is observed, and then the chip is replaced in the box.) The chips are scrambled before a second chip is selected and its color observed.

The sample space is shown in two different ways.

Tree diagram representation:

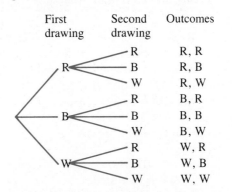

* See the *Study Guide* for information about tree diagrams.

Listing:

$$S = \{(R, R), (R, B), (R, W), (B, R), (B, B),$$
$$(B, W), (W, R), (W, B), (W, W)\}$$ ∎

Experiment 4-5   This experiment is the same as Experiment 4-4 except that the first chip is not replaced before the second selection is made.

The sample space is shown in two ways.

Tree diagram representation:

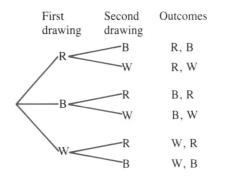

Listing:

$$S = \{(R, B), (R, W), (B, R), (B, W), (W, R), (W, B)\}$$ ∎

Experiment 4-6   A white die and a black die are each rolled one time and the number of dots showing on each die is observed.

The sample space is shown by a chart representation.

Chart representation:

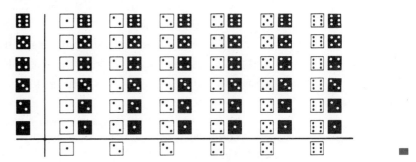

∎

Experiment 4-7   Two dice are rolled and the sum of their dots is observed.

The sample space is

$$S = \{2, 3, 4, 5, 6, 7, 8, 9, 10, 11, 12\}$$

(or the 36-point sample space listed in Experiment 4-6). ∎

You will notice that two different sample spaces are suggested for Experiment 4-7. Both of these sets satisfy the definition of a sample space and thus either could be used. We will learn later why the 36-point sample space is more useful than the other.

Experiment 4-8   A single marble is drawn from a box that contains 10 red and 90 blue marbles and its color is recorded.

The sample space is

$$S = \{R, B\}$$ ∎

Experiment 4-9   A thumbtack is tossed and the outcome is recorded [whether it lands point up (⌣) or point down (⌢)].

The sample space is

$$S = \{⌣, ⌢\}$$ ∎

Experiment 4-10   A weather forecaster predicts that there will be a measurable amount of precipitation or no precipitation on a given day.

The sample space is

$$S = \{\text{precipitation, no precipitation}\}$$ ∎

Special attention should always be given to the sample space. Like the statistical population, the sample space must be well defined. Once the sample space is defined, you will find the remaining work much easier.

EXERCISES

4.11 The face cards are removed from a regular deck and then 1 card is selected from this set of 12 face cards. List the sample space for this experiment.

4.12 An experiment consists of drawing one marble from a box that contains a mixture of red, yellow, and green marbles.

   a. List the sample space.

   b. Can we be sure that each outcome in the sample space is equally likely?

4.13 Two marbles are drawn from the box in Exercise 4.12. List the sample space.

4.14 a. A balanced coin is tossed twice. List a sample space showing the possible outcomes.

   b. A biased coin (it favors heads in a ratio of 3 to 1) is tossed twice. List a sample space showing the possible outcomes.

4.15 An experiment consists of two trials. The first is tossing a penny and observing heads or tails; the second is rolling a die and observing a 1, 2, 3, 4, 5, or 6. Construct the sample space.

4.16 A computer generates (in random fashion) pairs of integers. The first integer is between 1 and 5, inclusive, and the second is between 1 and 4, inclusive. Represent the sample space on a coordinate axis system where x is the first number and y is the second number.

4.17 A coin is tossed and a head or a tail observed. If a head results, the coin is tossed a second time. If a tail results on the first toss, a die is rolled.

a. Construct the sample space for this experiment.

b. What is the probability that a die is rolled in the second stage of this experiment?

4.18 A box stored in a warehouse contains 100 identical parts of which 10 are defective and 90 are nondefective. Three parts are selected without replacement. Construct a tree diagram representing the sample space.

4.4   RULES OF PROBABILITY

Let's return now to the concept of probability and relate the sample space to it. Recall that the probability of an event was defined as the relative frequency with which the event could be expected to occur.

In the sample space associated with Experiment 4-1, the tossing of one coin, we find two possible outcomes: heads (H) and tails (T). We have an "intuitive feeling" that these two events will occur with approximately the same frequency. The coin is a symmetrical object and therefore would not be expected to favor either of the two outcomes. We would expect heads to occur $\frac{1}{2}$ of the time. Thus the probability that a head will occur on a single toss of a coin is thought to be $\frac{1}{2}$.

This description is the basis for the second technique for assigning the probability of an event. In a sample space containing **sample points that are equally likely to occur**, the probability $P(A)$ of an event A is the ratio of the number $n(A)$ of points that satisfy the definition of event A to the number $n(S)$ of sample points in the entire sample space. That is,

equally likely events

$$P(A) = \frac{n(A)}{n(S)} \tag{4-2}$$

theoretical probability

This formula gives a **theoretical probability** value of event A's occurrence. The *prime symbol* of formula (4-1) *is not used* with theoretical probabilities.

The probability of event A is still a relative frequency, but it is now based on all the possible outcomes regardless of whether they actually occur. We can now see that the probability of obtaining two heads, $P(2H)$, when the penny and nickel are tossed is $\frac{1}{4}$. The probability of exactly one head, $P(1H)$ [(H, T) or (T, H)], is $\frac{2}{4}$, or $\frac{1}{2}$. And $P(0H) = \frac{1}{4}$. These are the values that were indicated by the experiment.

The use of formula (4-2) requires the existence of a sample space in which each outcome is equally likely. Thus when dealing with experiments that have more than one possible sample space, it is helpful to construct a sample space in which the sample points are equally likely.

Consider Experiment 4-7, where two dice were rolled. If you list the sample space as the 11 sums, the sample points are not equally likely. If you use the 36-point sample space, all the sample points are equally likely. The 11 sums in the first sample space represent combinations of the 36 equally likely sample points. For example, the sum of 2 represents $\{(1, 1)\}$; the sum of 3 represents $\{(2, 1), (1, 2)\}$; and the sum of 4 represents $\{(1, 3), (3, 1), (2, 2)\}$. Thus we can use formula (4-2) and the 36-point sample space to obtain the probabilities for the 11 sums.

$$P(2) = \tfrac{1}{36}, \qquad P(3) = \tfrac{2}{36}, \qquad P(4) = \tfrac{3}{36},$$

and so forth.

In many cases the assumption of equally likely events does not make sense. Experiments 4-1 through 4-6 are examples in which the sample space elements are all equally likely. The sample points in Experiment 4-7 are not equally likely, and there is no reason to believe that the sample points in Experiments 4-8, 4-9, and 4-10 are equally likely.

What do we do when the sample space elements are not equally likely or not a combination of equally likely events? We could use empirical probabilities. But what do we do when no experiment has been done or can be performed?

Let's look again at Experiment 4-10. The weather forecaster often assigns a probability to the event "precipitation." For example, "there is a 20 percent chance of rain today," or "there is a 70 percent chance of snow tomorrow." In such cases the only method available for assigning probabilities is personal judgment. These probability assignments are called **subjective probabilities**. The accuracy of subjective probabilities depends on the individual's ability to correctly assess the situation.

Often, personal judgment of the probability of the possible outcomes of an experiment is expressed by comparing the likelihood between the various outcomes. For example, the weather forecaster's personal assessment might be that "it is five times more likely to rain (R) tomorrow than not rain (NR)"; $P(R) = 5 \cdot P(NR)$. If this is the case, what values should be assigned to $P(R)$ and $P(NR)$? To answer this question, we need to review some of the ideas about probability that we've already discussed.

subjective probability

1. Probability represents a relative frequency.

2. $P(A)$ is the ratio of the number of times an event can be expected to occur divided by the number of trials.

3. The numerator of the probability ratio must be a positive number or zero.

4. The denominator of the probability ratio must be a positive number (greater than zero).

5. The number of times an event can be expected to occur in n trials is always less than or equal to the total number of trials.

Thus it is reasonable to conclude that a probability is always a numerical value between zero and one.

Property 1 $0 \le P(A) \le 1$

Notes

1. The probability is zero if the event cannot occur.

2. The probability is one if the event occurs every time.

Property 2 $\displaystyle\sum_{\text{all outcomes}} P(A) = 1$

Property 2 states that if we add up the probabilities of each of the sample points in the sample space, the total probability must equal one. This makes sense because when we sum up all the probabilities, we are really asking, "What is the probability the experiment will yield an outcome?" and this will happen every time.

Now we are ready to assign probabilities to $P(R)$ and $P(NR)$. The events R and NR cover the sample space, and the weather forecaster's personal judgment was

$$P(R) = 5 \cdot P(NR)$$

From Property 2, we know that

$$P(R) + P(NR) = 1$$

By substituting $5 \cdot P(NR)$ for $P(R)$, we get

$$5 \cdot P(NR) + P(NR) = 1$$
$$6 \cdot P(NR) = 1$$
$$P(NR) = \tfrac{1}{6}$$
$$P(R) = 1 - P(NR) = 1 - \tfrac{1}{6} = \tfrac{5}{6}$$

EXERCISES

4.19 Suppose that a box of marbles contains an equal number of red marbles and yellow marbles but twice as many green marbles as red marbles. Draw one marble from the box and observe its color. Assign probabilities to the elements in the sample space.

4.20 A transportation engineer in charge of a new traffic control system expresses the subjective probability that the system functions correctly 99 times as often as it malfunctions.

> **a.** Based on this belief, what is the probability that the system functions properly?
>
> **b.** Based on this belief, what is the probability that the system malfunctions?

4.21 Events A, B, and C are defined on sample space S. Their corresponding sets of sample points do not intersect and their union is S. Further, event B is twice as likely to occur as event A, and event C is twice as likely to occur as event B. Determine the probability of each of these three events.

4.22 Three coins are tossed and the number of heads observed is recorded. Find the probability for each of the possible results, 0H, 1H, 2H, and 3H.

4.23 Let x be the success rating of a new television show. The accompanying table lists the subjective probabilities assigned to each x for a particular new

	Judge		
Success Rating (x)	A	B	C
Highly successful	0.5	0.6	0.3
Successful	0.4	0.5	0.3
Not successful	0.3	−0.1	0.3

show as assigned by three different media people. Which of these sets of probabilities are inappropriate because they violate a basic rule of probability? Explain.

4.24 Two dice are rolled (Experiment 4-6). Find the probabilities in parts (b) through (e). Use the sample space given on page 147.

    a. Why is the set {2, 3, 4,..., 12} not a useful sample space?

    b. P(white die is an odd number)

    c. P(sum is 6)

    d. P(both dice show odd numbers)

    e. P(number on black die is larger than number on white die)

4.25 A group of files in a medical clinic classifies the patients by gender and by type of diabetes (I or II). The groupings may be shown as follows. The table gives the number in each classification.

		Type of Diabetes	
		I	II
Gender	Male	25	20
	Female	35	20

If one file is selected at random, find the probability that

    a. the selected individual is female.

    b. the selected individual is a Type II.

4.26 A lottery is conducted and 500 tickets are sold. The stubs to the tickets are well shuffled and the winner is chosen by randomly selecting one stub. If you bought 25 tickets, what is the probability that you will win?

CALCULATING PROBABILITIES OF COMPOUND EVENTS

compound events   **Compound events** are formed by combining several simple events. We will study the following four compound events in the remainder of this chapter:

    1. the probability of the complementary event, $P(\bar{A})$

    2. the probability that either event A or event B will occur, $P(A \text{ or } B)$

    3. the probability that both events A and B will occur, $P(A \text{ and } B)$

    4. the probability that event A will occur given that event B has occurred, $P(A|B)$

Note   In determining which compound probability we are seeking, it is not enough to look for the words "not," "either/or," "and," or "given" in the question. We must carefully examine the question asked to determine what combination of events is called for.

4.5   COMPLEMENTARY EVENTS, MUTUALLY EXCLUSIVE EVENTS, AND THE ADDITION RULE

COMPLEMENTARY EVENTS

complementary event

> ### Complement of an Event
>
> The set of all sample points in the sample space that do not belong to event A. The complement of event A is denoted by $\bar{A}$(read "A complement").

For example, the complement of the event "success" is "failure"; the complement of "heads" is "tails" for the tossing of one coin; the complement of "at least one head" on 10 tosses of a coin is "no heads."

By combining the information in the definition of complement with Property 2 (page 150), we can say that

$$P(A) + P(\bar{A}) = 1.0 \qquad \text{for any event A}$$

It then follows that

$$P(\bar{A}) = 1 - P(A) \qquad (4\text{-}3)$$

Note   Every event A has a complementary event $\bar{A}$. Complementary probabilities are very useful when the question asks for the probability of "at least one." Generally this represents a combination of several events, but the complementary event "none" is a single outcome. It is easier to solve for the complementary event and get the answer by using formula (4-3).

Illustration 4-2   Two coins are tossed. What is the probability that at least one head appears?

Solution   Let event A be the occurrence of no heads; then $\bar{A}$ represents the occurrence of one or more heads, that is, at least one head. The sample space is {(HH), (HT), (TH), (TT)}.

$$P(A) = \tfrac{1}{4} = \mathbf{0.25} \qquad [\text{using formula (4-2)}]$$
$$P(\bar{A}) = 1 - P(A) = 1 - \tfrac{1}{4} = \tfrac{3}{4} = \mathbf{0.75} \qquad [\text{using formula (4-3)}] \qquad \blacksquare$$

MUTUALLY EXCLUSIVE EVENTS

mutually exclusive events

> ### Mutually Exclusive Events
>
> Events defined in such a way that the occurrence of one event precludes the occurrence of any of the other events. (In short, if one of them happens, the others cannot happen.)

The following illustrations give examples of events to help understand the concept of mutually exclusive.

Illustration 4-3   A group of 200 college students is known to consist of 140 full-time (80 female and 60 male) students and 60 part-time (40 female and 20 male) students.

	200 College Students		
	Full-time	Part-time	Total
Female	80	40	120
Male	60	20	80
Total	140	60	200

Consider the experiment in which one of these students is to be selected at random. Two events are defined. Event A is "the student selected is full-time" and event B is "the student selected is a part-time male." Since no student is both "full-time" and "part-time male," the two events A and B are mutually exclusive events. A third event, event C, is defined to be "the student selected is female." Now let's consider the two events A and C. Since there are 80 students that are "full-time" and "female," the two events A and C are not mutually exclusive events.

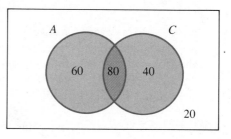

Illustration 4-4   Consider an experiment in which two dice are rolled. Three events are defined:

    A: The sum of the numbers on the two dice is 7.

    B: The sum of the numbers on the two dice is 10.

    C: Each of the two dice shows the same number.

Let's determine whether these three events are mutually exclusive.

    Are events A and B mutually exclusive? Yes, they are, since the sum on the two dice cannot be both 7 and 10 at the same time. If a sum of 7 occurs, it is impossible for the sum to be 10.

    Figure 4-4 presents the sample space for this experiment. This is the same sample space shown in Experiment 4-6, except that ordered pairs are used in place of the pictures. The ovals, diamonds, and rectangles show the ordered pairs that are in

intersection   events A, B, and C, respectively. We can see that events A and B do not **intersect** at a common sample point. Therefore, they are mutually exclusive. Point (5, 5) satisfies

Figure 4-4

Sample space for the roll of two dice

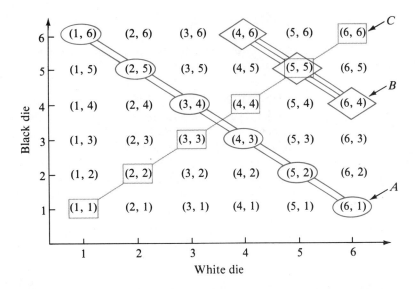

both events B and C. Therefore, B and C are not mutually exclusive. Two dice can each show a 5, which satisfies C; and the total of the two 5s satisfies B.

Venn diagram

Figure 4-5 is a **Venn diagram** that shows the relationships between the events defined in Illustration 4-4. Venn diagrams are simple "area," or "region," diagrams. The basic concept of a Venn diagram is that those points (and only those points) belonging to a given event are shown inside a circle. (For further background information, see the Study Guide.)

In Figure 4-5, events A and B are mutually exclusive. There are six sample points that belong to event A and there are three sample points in event B. The two events do not have a point in common, that is, they are mutually exclusive.

Similarly, events A and C do not intersect; therefore, they too are mutually exclusive. Events C and B, however, do share a common sample point, (5, 5). Notice that the point (5, 5) is in the intersection of the two Venn diagram circles. Consequently, events C and B are not mutually exclusive.

Figure 4-5

Venn diagram events A, B, and C

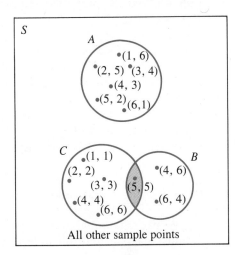

ADDITION RULE

Let us now consider the compound probability $P(A$ or $B)$, where A and B are mutually exclusive events.

Illustration 4-5   In Illustration 4-3 we considered an experiment in which one student was to be selected at random from a group of 200. Event A was "student selected is full-time," and event B was "student selected is a part-time male." The probability of event A, $P(A)$, is $\frac{140}{200}$, or 0.7, and the probability of event B, $P(B)$, is $\frac{20}{200}$, or 0.1. Let's find the probability of A or B, $P(A$ or $B)$. It seems reasonable to add the two probabilities 0.7 and 0.1 to obtain an answer of $P(A$ or $B) = 0.8$.

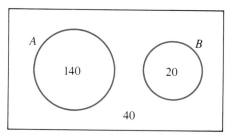

This is further justified by looking at the sample space. We see that there is a total of 160 that are either full-time or are male part-time students ($\frac{160}{200} = 0.8$). Recall that these two events, A and B, are mutually exclusive.   ∎

When events are **not mutually exclusive**, we cannot find the probability that one or the other occurs by simply adding the individual probabilities, as shown in Illustration 4-5. Why not? Let's look at an illustration and see what happens when events are not mutually exclusive.

Illustration 4-6   Using the sample space and the events defined in Illustration 4-3, find the probability that the student selected is "full-time" or "female," $P(A$ or $C)$.

Solution   If we look at the sample space, we see that $P(A) = \frac{140}{200}$, or 0.7, and that $P(C) = \frac{120}{200}$, or 0.6. If we add the two numbers 0.7 and 0.6 together, we will get 1.3, a number larger than one. We also know, from the basic properties of probability, that probability numbers are never larger than one. So what happened? If we take another look at the sample space we will see that 80 of the 200 students have been counted twice if we add $\frac{140}{200}$ and $\frac{120}{200}$. There are only 180 students that are "full-time" or "female." Thus the probability of A or C is

$$P(A \text{ or } C) = \tfrac{180}{200} = 0.9$$

(*See* the figure in Illustration 4-3.)   ∎

We can add probabilities to find the probability of an "or" compound event, but we must make an adjustment in situations such as the previous example.

addition rule

General Addition Rule

Let A and B be two events defined in a sample space S.

$$P(A \text{ or } B) = P(A) + P(B) - P(A \text{ and } B) \qquad (4\text{-}4a)$$

Special Addition Rule

Let A and B be two events defined in a sample space. If A and B are **mutually exclusive** events, then

$$P(A \text{ or } B) = P(A) + P(B) \qquad (4\text{-}4b)$$

This can be expanded to consider **more than two** mutually exclusive events:

$$P(A \text{ or } B \text{ or } C \text{ or} \dots \text{or } E) = P(A) + P(B) + P(C) + \dots + P(E) \qquad (4\text{-}4c)$$

The key to this formula is the property "mutually exclusive." If two events are mutually exclusive, there is no double counting of sample points. If events are not mutually exclusive, then when probabilities are added, the double counting will occur. Let's look at some examples.

In Figure 4-6, events A and B are mutually exclusive. Simple addition is justified, since the total probability of the shaded regions is sought. [In a Venn diagram, probabilities are often represented by enclosed areas. In Figure 4-6, for example, we've shaded the areas representing the probabilities $P(A)$ and $P(B)$.]

In Figure 4-7, events A and B are not mutually exclusive. The probability of the event "A and B," $P(A \text{ and } B)$, is represented by the area contained in region II. The probability $P(A)$ is represented by the area of circle A. That is, $P(A) = P(\text{region I}) + P(\text{region II})$. Also, $P(B) = P(\text{region II}) + P(\text{region III})$. And $P(A \text{ or } B)$ is

Figure 4-6
Mutually exclusive events

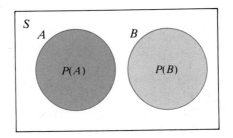

Figure 4-7
Nonmutually exclusive events

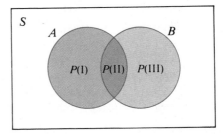

the sum of the probabilities associated with the three regions:

$$P(A \text{ or } B) = P(I) + P(II) + P(III)$$

However, if $P(A)$ is added to $P(B)$, we have

$$P(A) + P(B) = [P(I) + P(II)] + [P(II) + P(III)]$$
$$= P(I) + 2P(II) + P(III)$$

This is the double count previously mentioned. However, if we subtract one measure of region II from this total, we will be left with the correct value.

The addition formula, (4-4b), is a special case of the more general rule stated in formula (4-4a). If A and B are mutually exclusive events, $P(A \text{ and } B) = 0$. (They cannot both happen at the same time.) Thus the last term in formula (4-4a) is zero when events are mutually exclusive.

Illustration 4-7   One white die and one black die are rolled. Find the probability that the white die shows a number smaller than 3 or the sum of the dice is greater than 9.

Solution 1   A = white die shows a 1 or a 2; B = sum of both dice is 10, 11, or 12.

$$P(A) = \frac{12}{36} = \frac{1}{3} \quad \text{and} \quad P(B) = \frac{6}{36} = \frac{1}{6}$$

$$P(A \text{ or } B) = P(A) + P(B) - P(A \text{ and } B)$$

$$= \frac{1}{3} + \frac{1}{6} - 0 = \frac{1}{2}$$

($P(A \text{ and } B) = 0$, since the events do not intersect.)

Solution 2

$$P(A \text{ or } B) = \frac{n(A \text{ or } B)}{n(S)} = \frac{18}{36} = \frac{1}{2}$$

(Look at the sample space, Figure 4-4, and count.)   ■

Illustration 4-8   A pair of dice is rolled. Event T is defined as the occurrence of a "total of 10 or 11," and event D is the occurrence of "doubles." Find the probability $P(T \text{ or } D)$.

Solution   Look at the sample space of 36 ordered pairs for the rolling of two dice in Figure 4-4. Event T occurs if any one of 5 ordered pairs occurs; (4, 6), (5, 5), (6, 4), (5, 6), (6, 5). Therefore, $P(T) = \frac{5}{36}$. Event D occurs if any one of 6 ordered pairs occurs; (1, 1), (2, 2), (3, 3), (4, 4), (5, 5), (6, 6). Therefore, $P(D) = \frac{6}{36}$. Notice, however, that these two events are not mutually exclusive. The two events "share" the point (5, 5). Thus the probability $P(T \text{ and } D) = \frac{1}{36}$. As a result, the probability $P(T \text{ or } D)$ will be found using formula (4-4a).

$$P(\text{T or D}) = P(\text{T}) + P(\text{D}) - P(\text{T and D})$$
$$= \tfrac{5}{36} + \tfrac{6}{36} - \tfrac{1}{36} = \tfrac{10}{36} = \tfrac{5}{18}.$$

(Look at the sample space in Figure 4-4.)   ■

EXERCISES

4.27 Determine whether or not each of the following pairs of events is mutually exclusive.

a. Five coins are tossed: "one head is observed," "at least one head is observed."

b. A salesperson calls on a client and makes a sale: "the sale exceeds $100," "the sale exceeds $1000."

c. One student is selected at random from a student body: the person selected is "male," the person selected is "over 21 years of age."

d. Two dice are rolled: the total showing is "less than 7," the total showing is "more than 9."

4.28 Determine whether or not each of the following sets of events is mutually exclusive.

a. Five coins are tossed: "no more than one head is observed," "two heads are observed," "three or more heads are observed."

b. A salesperson calls on a client and makes a sale: the amount of the sale is "less than $100," is "between $100 and $1000," is "more than $500."

c. One student is selected at random from the student body: the person selected is "female," is "male" or is "over 21."

d. Two dice are rolled: the number of dots showing on each die are "both odd," are "both even," are "total seven" or "total eleven."

4.29 If $P(A) = 0.3$ and $P(B) = 0.4$, and if A and B are mutually exclusive events, find the following:

    a. $P(\bar{A})$ **b.** $P(\bar{B})$ **c.** $P(A \text{ or } B)$ **d.** $P(A \text{ and } B)$

4.30 Explain why $P(A \text{ and } B) = 0$ when events A and B are mutually exclusive.

4.31 One student is selected at random from a student body. Suppose the probability that this student is female is 0.5 and the probability that this student is working a part-time job is 0.6. Are the two events "female" and "working" mutually exclusive events? Explain.

4.32 If $P(A) = 0.4$, $P(B) = 0.5$, and $P(A \text{ and } B) = 0.1$, find $P(A \text{ or } B)$.

4.33 The following table summarizes the teaching experience and educational background of the teachers in a public school.

Educational Background	Teaching Experience	
	Less Than 5 Years	5 Years or More
Less than a master's degree	75	40
Master's degree or more	55	30

Let A be the event that a teacher, selected at random, "has less than a master's degree" and let B represent "the teacher has less than 5 years experience." Find:

    a. $P(A \text{ and } B)$ **b.** $P(A \text{ or } B)$ **c.** $P(\bar{B})$

4.34 A parts store sells both new and used parts. Sixty percent of the parts in stock are used. Sixty-one percent are used or defective. If 5 percent of the store's parts are defective, what percent are both used and defective?

4.35 Union officials report that 60 percent of the workers at a large factory belong to the union, 90 percent make over $5 per hour, and 40 percent belong to the union and make over $5 per hour. Do you believe these percentages? Explain.

4.6 INDEPENDENCE, MULTIPLICATION RULE, AND CONDITIONAL PROBABILITY

Next we will study the compound event "A and B." For example, what is the probability that two heads occur when two coins (one penny and one nickel) are tossed? Let A represent the occurrence of a head on the penny and B represent the occurrence of a head on the nickel. What is $P(A \text{ and } B)$? $P(A \text{ and } B)$ is found by using the definition $n(A \text{ and } B)/n(S)$. The sample space for Experiment 4-2 suggests that this value should be $\frac{1}{4}$. How can we obtain this value by using $P(A)$ and $P(B)$? Both $P(A)$ and $P(B)$ have values of $\frac{1}{2}$. If $P(A)$ is multiplied by $P(B)$, we obtain $\frac{1}{4}$. Thus we might suspect that $P(A \text{ and } B)$ equals $P(A) \cdot P(B)$.

Consider this example. The event that a 2 shows on a white die is A, and the event that a 2 shows on a black die is B. If both dice are rolled once, what is the probability that two 2s occur?

$$P(A) = \frac{1}{6} \quad \text{and} \quad P(B) = \frac{1}{6}$$

$$P(A \text{ and } B) = \frac{n(A \text{ and } B)}{n(S)} = \frac{1}{36}$$

However, note that $\frac{1}{6}$ multiplied by $\frac{1}{6}$ is also $\frac{1}{36}$. In this case, multiplication yields the correct answer. Multiplication does not always work, however. For example, $P(\text{sum of 7 and double})$ when two dice are rolled is zero (as seen in Figure 4-4). However, if $P(7)$ is multiplied by $P(\text{double})$, we obtain $(\frac{1}{6})(\frac{1}{6}) = \frac{1}{36}$.

Multiplication does not work for $P(\text{sum of 10 and double})$, either. By definition and by inspection of the sample space, we know that $P(10 \text{ and double}) = \frac{1}{36}$ (the point (5, 5) is the only element). However, if we multiply $P(10)$ by $P(\text{double})$, we obtain $(\frac{3}{36})(\frac{6}{36}) = \frac{1}{72}$. The probability of this event cannot be both values.

independence

The property that is required for multiplying probabilities is **independence**. Multiplication worked in the two foregoing examples because the events were independent. In the other two cases, the events were not independent and multiplication gave us incorrect answers.

Note   There are several situations that result in the compound event "and." Some of the more common ones are: (1) A followed by B, (2) A and B occurred simultaneously, (3) the intersection of A and B, (4) both A and B and (5) A but not B (equivalent to A and not B).

INDEPENDENCE AND CONDITIONAL PROBABILITIES

independent events

Independent Events

Two events A and B are independent events if the occurrence (or nonoccurrence) of one does not affect the probability assigned to the occurrence of the other.

Sometimes independence is quite easy to determine, for example, if the two events being considered have to do with unrelated trials, such as the tossing of a penny and a nickel. The results on the penny in no way effect the probability of heads or tails on the nickel. Similarly, the results on the nickel have no effect on the probability of heads or tails on the penny. Therefore, the results on the penny and the results on the nickel are *independent*. Also, if a coin and a die are tossed, either simultaneously or one following the other, the results on either one are independent of the results on the other. The coin and the die can each be thought of as a separate trial. However, if *events* are defined as combinations of outcomes from the separate trials, the independence of the events may or may not be so easy to determine. The separate results of each trial (dice in the next illustration) may be independent, but the compound events defined using both trials (both dice) may or may not be independent.

dependent events

Lack of independence, called dependence, is demonstrated by the following illustration. Reconsider the experiment of rolling two dice and observing the two events "sum of 10" and "double." As stated previously, $P(10) = \frac{3}{36} = \frac{1}{12}$ and $P(\text{double}) = \frac{6}{36} = \frac{1}{6}$. Does the occurrence of 10 affect the probability of a double? Think of it this way. A sum of 10 has occurred; it must be one of the following: $\{(4, 6), (5, 5), (6, 4)\}$. One of these three possibilities is a double. Therefore, we must conclude that the $P(\text{double}, \textit{knowing } 10 \text{ has occurred})$, written $P(\text{double}|10)$, is $\frac{1}{3}$. Since $\frac{1}{3}$ does not equal the original probability of a double, $\frac{1}{6}$, we can conclude that the event "10" has an effect on the probability of a double. Therefore, "double" and "10" are dependent events.

Whether or not events are independent often becomes clear by examining the events in question. Rolling one die does not affect the outcome of a second roll. However, in many cases, independence is not self-evident, and the question of independence itself may be of special interest. Consider the events "having a checking account at a bank" and "having a loan account at the same bank." Having a checking account at a bank may increase the probability that the same person has a loan account. This has practical implications. For example, it would make sense to advertise loan programs to checking-account clients if they are more likely to apply for loans than are people who are not customers of the bank.

One approach to the problem is to *assume* independence or dependence. The correctness of the probability analysis depends on the truth of the assumption. In practice, we often assume independence and compare *calculated* probabilities with *actual frequencies* of outcomes in order to infer whether the assumption of independence is warranted.

conditional probability

> The symbol $P(A|B)$ represents the probability that A will occur given that B has occurred. This is called a **conditional probability**.

The previous definition of independent events can now be written in a more formal manner.

> ## Independent Events
>
> Two events A and B are independent events if
> $P(A|B) = P(A)$ or if $P(B|A) = P(B)$.   (4-5)

Let's consider conditional probability. Take, for example, the experiment in which a single die is rolled; $S = \{1, 2, 3, 4, 5, 6\}$. Two events that can be defined for this experiment are B = "an even number occurs" and A = "a 4 occurs." Then $P(A) = \frac{1}{6}$. Event A is satisfied by exactly one of the six equally likely sample points in S. The conditional probability of A given B, $P(A|B)$, is found in a similar manner, but S is no longer the sample space. Think of it this way. A die is rolled out of your sight, and you are told that the number showing is even, that is, event B has occurred. That is the given condition. Knowing this condition, you are asked to assign a probability to the event that the even number is a 4. There are only three possibilities in the new (or reduced) sample space, $\{2, 4, 6\}$. Each of the three outcomes is equally likely; thus $P(A|B) = \frac{1}{3}$.

We can write this as

$$P(A|B) = \frac{P(A \text{ and } B)}{P(B)}$$ (4-6)

Thus, for our example,

$$P(A|B) = \frac{\frac{1}{6}}{\frac{1}{2}} = \frac{1}{3}$$

Illustration 4-9   In a sample of 150 residents, each person was asked if he or she favored the concept of having a single countywide police agency. The county is composed of one large city and many suburban townships. The residence (city or outside the city) and the responses of the residents are summarized in Table 4-4. If

Table 4-4

Sample results for

Illustration 4-9

Residence	Opinion		Total
	Favor (F)	Oppose ($\bar{F}$)	
In city (C)	80	40	120
Outside of city ($\bar{C}$)	20	10	30
Total	100	50	150

one of these residents was to be selected at random, what is the probability that the person will (a) favor the concept? (b) favor the concept if the person selected is a city resident? (c) favor the concept if the person selected is a resident from outside the city? (d) Are the events F (favor the concept) and C (reside in city) independent?

Solution

(a) $P(F)$ is the proportion of the total sample that favor the concept. Therefore,

$$P(F) = \frac{n(F)}{n(S)} = \frac{100}{150} = \frac{2}{3}$$

(b) $P(F|C)$ is the probability that the person selected favors the concept given that he or she lives in the city. The sample space is reduced to the 120 city residents in the sample. Of these, 80 favored the concept; therefore,

$$P(F|C) = \frac{n(F \text{ and } C)}{n(C)} = \frac{80}{120} = \frac{2}{3}$$

(c) $P(F|\bar{C})$ is the probability that the person selected favors the concept, knowing that the person lives outside the city. The sample space is reduced to the 30 noncity residents; therefore,

$$P(F|\bar{C}) = \frac{n(F \text{ and } \bar{C})}{n(\bar{C})} = \frac{20}{30} = \frac{2}{3}$$

(d) All three probabilities have the same value, $\frac{2}{3}$. Therefore, we can say that the events F (favor) and C (reside in city) are independent. The location of residence did not affect $P(F)$. ∎

MULTIPLICATION RULE

general multiplication rule

General Multiplication Rule

Let A and B be two events defined in sample space S. Then

$$P(A \text{ and } B) = P(A) \cdot P(B|A) \tag{4-7a}$$

or

$$P(A \text{ and } B) = P(B) \cdot P(A|B) \tag{4-7b}$$

If events A and B are independent, then the general multiplication rule [formula (4-7)], reduces to the *special multiplication rule*, formula (4-8).

special multiplication rule

Special Multiplication Rule

Let A and B be two events defined in sample space S. If A and B are **independent events**, then

$$P(A \text{ and } B) = P(A) \cdot P(B) \qquad (4\text{-}8a)$$

This formula can be expanded. If A, B, C,..., G are independent events, then

$$P(A \text{ and } B \text{ and } C \text{ and} \ldots \text{and } G) = P(A) \cdot P(B) \cdot P(C) \cdots P(G) \qquad (4\text{-}8b)$$

Illustration 4-10   One student is selected at random from a group of 200 known to consist of 140 full-time (80 female and 60 male) students and 60 part-time (40 female and 20 male) students. (*See* Illustration 4-3). Event A is "the student selected is full-time" and event C is "the student selected is female."

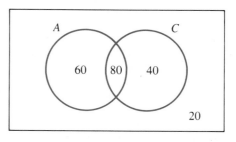

(a) Are events A and C independent? (b) Find the probability $P(A \text{ and } C)$ using the multiplication rule.

Solution 1

(a) First find the probabilities $P(A)$, $P(C)$, and $P(A|C)$.

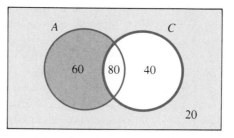

$P(A) = \frac{140}{200} = 0.7$ $P(A|C) = \frac{80}{120} = 0.66$
$P(C) = \frac{120}{200} = 0.6$
A and C are dependent events, $P(A) \neq P(A|C)$.

(b) $P(\text{A and C}) = P(\text{C}) \times P(\text{A}|\text{C})$
$$= \tfrac{120}{200} \times \tfrac{80}{120} = \tfrac{80}{200} = \textbf{0.4}$$

Solution 2

(a) First find the probabilities $P(\text{A})$, $P(\text{C})$, and $P(\text{C}|\text{A})$.

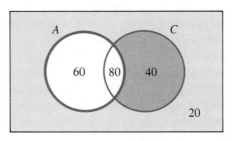

$P(\text{A}) = \tfrac{140}{200} = 0.7$ $P(\text{C}|\text{A}) = \tfrac{80}{140} = 0.57$

$P(\text{C}) = \tfrac{120}{200} = 0.6$

A and C are dependent events, $P(\text{C}) \neq P(\text{C}|\text{A})$.

(b) $P(\text{A and C}) = P(\text{A}) \times P(\text{C}|\text{A})$
$$= \tfrac{140}{200} \times \tfrac{80}{140} = \tfrac{80}{200} = \textbf{0.4} \qquad \blacksquare$$

Illustration 4-11   One white and one black die are rolled. Find the probability that the sum of their numbers is 7 and that the number on the black die is larger than the number on the white die.

Solution   A = "sum is 7"; B = "black number larger than white number." The "and" requires the use of the multiplication rule. However, we do not yet know whether events A and B are independent. (Refer to Figure 4-4 for the sample space of this experiment.) We see that $P(\text{A}) = \tfrac{6}{36} = \tfrac{1}{6}$. Also, $P(\text{A}|\text{B})$ is obtained from the reduced sample space, which includes 15 points above the gray diagonal line. Of the 15 equally likely points, 3 of them—(1, 6), (2, 5), and (3, 4)—satisfy event A. Therefore, $P(\text{A}|\text{B}) = \tfrac{3}{15} = \tfrac{1}{5}$. Since this is a different value than $P(\text{A})$, the events are dependent. So we must use formula (4-7b) to obtain $P(\text{A and B})$.

$$P(\text{A and B}) = P(\text{B}) \cdot P(\text{A}|\text{B}) = \tfrac{15}{36} \cdot \tfrac{3}{15} = \tfrac{3}{36} = \tfrac{1}{12} \qquad \blacksquare$$

Notes

1. Independence and mutual exclusiveness are two different concepts.

2. The term *mutually exclusive* describes whether the events can occur together, whereas *independence* describes the effect that one event has on the probability of the other event's occurrence.

3. If two events are *not* mutually exclusive, then their corresponding sets of sample points intersect.

4. The relationship between independence (or dependence) and mutual exclusiveness (or nonmutual exclusiveness) is summarized by the following four statements:

a. If two events are known to be mutually exclusive, then they are also dependent events.

For example: If $P(A) = 0.3$ and $P(B) = 0.4$ and if A and B are mutually exclusive, then $P(A|B) = 0$. Since $P(A) \neq P(A|B)$, A and B are dependent events.

b. If two events are known to be independent, then they are *not* mutually exclusive events.

For example: If $P(C) = 0.2$ and $P(D) = 0.5$ and if C and D are independent, then $P(C \text{ and } D) = 0.2 \cdot 0.5 = 0.10$. Since $P(C \text{ and } D)$ is greater than zero, C and D intersect. Therefore C and D are not mutually exclusive events.

c. If two events are known to be nonmutually exclusive, then they may be either independent *or* dependent events.

For example: (1) If $P(C) = 0.2$, $P(D) = 0.5$, and $P(C \text{ and } D) = 0.1$ (C and D intersect, that is, they are nonmutually exclusive), then $P(C|D) = 0.1/0.5 = 0.2$. Since $P(C) = P(C|D)$, C and D are independent events.
(2) If $P(E) = 0.2$, $P(F) = 0.5$, and $P(E \text{ and } F) = 0.05$ (E and F intersect, that is, they are nonmutually exclusive), then $P(E|F) = 0.05/0.5 = 0.1$. Since $P(E) \neq P(E|F)$, E and F are dependent events.

d. If two events are known to be dependent, then they may be *either* mutually exclusive *or* nonmutually exclusive.

For example: (1) If $P(E) = 0.2$, $P(F) = 0.5$, and $P(E|F) = 0.1$ [E and F are dependent since $P(E) \neq P(E|F)$], then $P(E \text{ and } F) = 0.5 \cdot 0.1 = 0.05$. Since $P(E \text{ and } F)$ is greater than zero, E and F are intersecting, or nonmutually exclusive, events.
(2) If $P(A) = 0.3$, $P(B) = 0.4$, and $P(B|A) = 0$ [A and B are dependent since $P(B) \neq P(B|A)$], then $P(A \text{ and } B) = 0.3 \cdot 0 = 0$. Since $P(A \text{ and } B) = 0$, A and B do not intersect, or are mutually exclusive events.

Case Study 4-2 | ## PROBABILITY: AN IMPORTANT TOOL IN POLICE SCIENCE

Larry J. Stephens discusses why what might seem to be unconvincing evidence is actually very convincing evidence. Probabilities are the key. Can you verify the probability stated by Professor Stephens? (See Exercise 4.47.)

Probability can be an extremely valuable tool to the police scientist in trying to access the likelihood of certain events or in trying to validate evidence....

The following is a burglary case description as supplied by Chief Richard R. Anderson, Omaha Police Department:

A man was shown a display of thirty-one (31) pieces of Meissen in an attempt to identify some as belonging to him. Within the display of thirty-one (31) pieces, there were eleven (11) that were accounted for; however, the man was unaware of this. Without the knowledge that eleven (11) were accounted for, the man selected nineteen (19) and stated these were his. Out of the nineteen selected, none were any of the eleven (11) that we had verified were not his. The one piece that was left could have been his; however he said he was unsure and would not select it just to say it was his.

All items on display are porcelain-type figurines, all painted differently with different musical instruments. The collection is referred to as a Meissen Monkey Band.

The fact that the man selected 19 pieces and none of them were from the accounted-for group provided strong evidence that the pieces, in fact, did belong to him.

Probability theory provided a way of measuring how strong the evidence, in fact, was....

If the man could not tell the accounted-for from the non-accounted-for pieces, then the odds of selecting 19 and all 19 being from the non-accounted-for set are 20 to 141,120,505 or 1 to 7,056,025. This provides strong evidence that the man could differentiate among the pieces in the collection....

"P[19 of type A]" is the probability that 19 were selected and that all 19 were from those not previously identified, that is, of type A....

$$P[19 \text{ of type A}] = \frac{20}{141,120,525}$$

Reproduced from the *Journal of Police Science and Administration*, Vol. 8, No. 3, pp. 353–354, with permission of the International Association of Chiefs of Police, P. O. Box 6010, 13 Firstfield Road, Gaithersburg, Maryland 20878.

EXERCISES

4.36 Determine whether or not each of the following pairs of events is independent:

a. rolling a pair of dice and observing a "1" on the first die and a "1" on the second die

b. drawing a "spade" from a regular deck of playing cards and then drawing another "spade" from the same deck without replacing the first card

c. same as (b) except the first card is returned to the deck before the second drawing

d. owning a red automobile and having blonde hair

e. owning a red automobile and having a flat tire today

f. studying for an exam and passing the exam

4.37 If $P(A) = 0.3$ and $P(B) = 0.4$, and A and B are independent events, what is the probability of each of the following?

a. $P(A \text{ and } B)$ **b.** $P(B|A)$ **c.** $P(A|B)$

4.38 Suppose that $P(A) = 0.3$, $P(B) = 0.4$, and $P(A \text{ and } B) = 0.12$.

a. What is $P(A|B)$? **b.** What is $P(B|A)$? **c.** Are A and B independent?

4.39 Suppose that $P(A) = 0.3$, $P(B) = 0.4$, and $P(A \text{ and } B) = 0.20$.

a. What is $P(A|B)$? **b.** What is $P(B|A)$? **c.** Are A and B independent?

4.40 Suppose that A and B are events and that the following probabilities are known.

$$P(A) = 0.3, \qquad P(B) = 0.4, \quad \text{and} \quad P(A|B) = 0.2.$$

Find $P(A \text{ or } B)$.

4.41 A single card is drawn from a standard deck. Let A be the event that "the card is a face card" (a jack, a queen, or a king), B be the occurrence of a "red card," and C represent "the card is a heart." Check to determine whether the following pairs of events are independent or dependent.

    a. A and B   **b.** A and C   **c.** B and C

4.42 In a survey that questioned high school students about their attitudes toward business, 0.85 said "Honesty is the best policy" and 0.28 stated "Success in business requires some dishonesty." Assuming that the survey results both accurately reflect the attitudes of all high school students and that the answers to the two attitudes are independent of one another, what is the probability that

    a. a student did not say "Honesty is the best policy" and agreed that "Success in business requires some dishonesty?"

    b. a student said "Honesty is the best policy" or agreed "Success in business requires some dishonesty?"

4.43 A box contains four red and three blue poker chips. What is the probability when three are selected randomly that all three will be red if we select each chip

    a. with replacement?   **b.** without replacement?

4.44 In the United States, the proportion of all births that involve Caesarean section deliveries is 15 out of 100. When Caesarean deliveries are made, 96 out of 100 babies survive. What is the probability that a randomly selected expectant woman will have a Caesarean section and that the baby will survive?

4.45 Consider the set of integers 1, 2, 3, 4, and 5.

    a. One integer is selected at random. What is the probability that it is odd?

    b. Two integers are selected at random (one at a time with replacement, so that each of the five is available for a second selection).

Find the probability that (1) neither is odd; (2) exactly one of them is odd; (3) both are odd.

4.46 A box contains 25 parts, of which 3 are defective and 22 are nondefective. If two parts are selected without replacement, find the following probabilities.

    a. P(both are defective)   **b.** P(exactly one is defective)

    c. P(neither is defective)

4.47 In Case Study 4-2, a man was asked to identify the figurines he believed to be his from a set of 31 such figurines. The police knew that 11 of the figurines were not his. If the man had been guessing, what is the probability that he would be able to select 19 of the 20 unknown figurines? Does your answer agree with that of Professor Stephens? Why should this information convince the police that the man is telling the truth?

4.7 COMBINING THE RULES OF PROBABILITY

Many probability problems can be represented by tree diagrams. In these instances, the addition and multiplication rules can be applied quite readily.

To illustrate the use of tree diagrams in solving probability problems, let's use Experiment 4-5. Two poker chips are drawn from a box containing one each of red, blue, and white chips. The tree diagram representing this experiment (Figure 4-8) shows a first drawing and then a second drawing. One chip was drawn on each drawing and not replaced.

After the tree has been drawn and labeled, we need to assign probabilities to each branch of the tree. If we assume that it is equally likely that any chip would be drawn at each stage, we can assign a probability to each branch segment of the tree, as shown in Figure 4-9. Notice that a set of branches that initiate from a single point have a total probability of 1. In this diagram there are four such sets of branch segments. The tree diagram shows six distinct outcomes. Reading down: branch (1) shows (R, B), branch (2) shows (R, W), and so on. [*Note*: Each outcome for the experiment is represented by a branch that begins at the common starting point and ends at the terminal points at the right.]

The probability associated with outcome (R, B), P(R on 1st drawing and B on 2d drawing), is found by multiplying P(R on 1st drawing) by P(B on 2d drawing|R on 1st drawing). These are the two probabilities $\frac{1}{3}$ and $\frac{1}{2}$ shown on the two branch segments of branch (1) in Figure 4-9. The $\frac{1}{2}$ is the conditional probability asked for by the multiplication rule. Thus we will multiply along the branches.

Some events will be made up of more than one outcome from our experiment. For example, suppose that we had asked for the probability that one red chip and

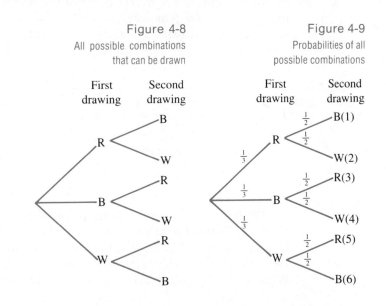

Figure 4-8

All possible combinations
that can be drawn

First drawing / Second drawing

Figure 4-9

Probabilities of all
possible combinations

First drawing / Second drawing

one blue chip are drawn. You will find two outcomes that satisfy this event, branch
(1) or branch (3). With "or" we will use the addition rule (4-4b). Since the branches of
a tree diagram represent mutually exclusive events, we have

$$P(\text{one R and one B}) = \left(\frac{1}{3}\right)\left(\frac{1}{2}\right) + \left(\frac{1}{3}\right)\left(\frac{1}{2}\right) = \frac{1}{6} + \frac{1}{6} = \frac{1}{3}$$

Notes

1. Multiply along the branches (Figure 4-10).
2. Add across the branches.

Figure 4-10
Multiply along the branches

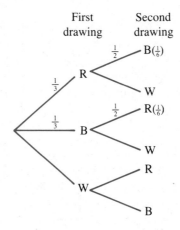

Now let's consider an example that places all the rules in perspective.

Illustration 4-12   A firm plans to test a new product in one randomly selected market area. The market areas can be categorized on the basis of location
and population density. The number of markets in each category is presented in
Table 4-5.

Table 4-5
Number of markets, by
location and by population
density

| Location | Population Density | | Total |
	Urban (U)	Rural (R)	
East (E)	25	50	75
West (W)	20	30	50
Total	45	80	125

What is the probability that the test market selected is in the East, $P(E)$? In the
West, $P(W)$? What is the probability that the test market is in an urban area, $P(U)$?
In a rural area, $P(R)$? What is the probability that the market is a western rural
area, $P(W \text{ and } R)$? What is the probability it is an eastern or urban area, $P(E \text{ or } U)$?
What is the probability that if it is in the East, it is an urban area, $P(U|E)$? Are

"location" and "population density" independent? (What do we mean by independence or dependence in this situation?)

Solution   The first four probabilities, $P(E)$, $P(W)$, $P(U)$, and $P(R)$, represent "or" questions. For example, $P(E)$ means that the area is an eastern urban area or an eastern rural area. Since in this and the other three cases, the two components are mutually exclusive (an area can't be both urban and rural), the desired probabilities can be found by simply adding. In each case the probabilities are added across all the rows or columns of the table. Thus the totals are found in the total column or row.

$$P(E) = \frac{75}{125} \quad \text{(total for East divided by total number of markets)}$$

$$P(W) = \frac{50}{125} \quad \text{(total for West divided by total number of markets)}$$

$$P(U) = \frac{45}{125} \quad \text{(total for urban divided by total number of markets)}$$

$$P(R) = \frac{80}{125} \quad \text{(total for rural divided by total number of markets)}$$

Now we solve for $P(W \text{ and } R)$. There are 30 western rural markets and 125 markets total. Thus

$$P(W \text{ and } R) = \frac{30}{125}$$

Note that $P(W) \cdot P(R)$ does *not* give the right answer $[(\frac{50}{125})(\frac{80}{125}) = \frac{32}{125}.]$ Therefore, "location" and "population density" are dependent events.

$P(E \text{ or } U)$ can be solved in several different ways. The most direct way is to simply examine the table and count the number of markets that satisfy the condition that they are in the East or they are urban. We find $95(25 + 50 + 20)$. Thus

$$P(E \text{ or } U) = \frac{95}{125}$$

Note that the first 25 markets were both in the East and urban; thus E and U are not mutually exclusive events.

Another way to solve for $P(E \text{ or } U)$ is to use the addition formula:

$$P(E \text{ or } U) = P(E) + P(U) - P(E \text{ and } U)$$

which yields

$$\frac{75}{125} + \frac{45}{125} - \frac{25}{125} = \frac{95}{125}$$

A third way to solve the problem is to recognize that the complement of (E or U) is (W and R). Thus $P(E \text{ or } U) = 1 - P(W \text{ and } R)$. Using the previous calculation, we get $1 - \frac{30}{125} = \frac{95}{125}$.

Finally, we solve for $P(U|E)$. Looking at Table 4-5, we see that there are 75 markets in the East. Of the 75 eastern markets, 25 are urban. Thus

$$P(U|E) = \frac{25}{75}$$

The conditional probability formula could also be used:

$$P(U|E) = \frac{P(U \text{ and } E)}{P(E)}$$

$$= \frac{\frac{25}{125}}{\frac{75}{125}} = \frac{25}{75}$$

"Location" and "population density" are not independent events. They are dependent. This means that the probability of these events is affected by the occurrence of the other.   ■

Although each rule for computing compound probabilities has been discussed separately, you should not think they are only used separately. In many cases they are combined to solve problems. Consider the following two illustrations.

Illustration 4-13   A production process produces light bulbs. On the average, 20 percent of all bulbs produced are defective. Each item is inspected before being shipped. The inspector misclasses an item 10 percent of the time; that is,

$$P(\text{classified good}|\text{defective item}) = P(\text{classified defective}|\text{good item})$$
$$= 0.10$$

What proportion of the items will be "classified good"?

Solution   What do we mean by the event "classified good"?

$$G = \text{item good}$$
$$D = \text{item defective}$$
$$CG = \text{item called good by inspector}$$
$$CD = \text{item called defective by inspector}$$

CG consists of two possibilities: "the item is good and is correctly classified good" or "the item is defective and is misclassified good." Thus

$$P(CG) = P[(CG \text{ and } G) \text{ or } (CG \text{ and } D)]$$

Since the two possibilities are mutually exclusive, we can start by using the addition rule, formula (4-4b).

$$P(CG) = P(CG \text{ and } G) + P(CG \text{ and } D)$$

The condition of a bulb and its classification by the inspector are not independent. The multiplication rule for dependent events must be used. Therefore,

$$P(CG) = [P(G) \cdot P(CG|G)] + [P(D) \cdot P(CG|D)]$$

Substituting the known probabilities, we get

$$P(CG) = [(0.8)(0.9)] + [(0.2)(0.1)]$$
$$= 0.72 + 0.02$$
$$= \mathbf{0.74}$$

That is, 74 percent of the items are classified good.   ■

Illustration 4-14   Reconsider Illustration 4-13. Suppose that only items that pass inspection are shipped. Items not classified good are scrapped. What is the quality of the shipped items? That is, what percentage of the items shipped are good, $P(G|CG)$?

Solution   Using the conditional probability formula (4-6),

$$P(G|CG) = \frac{P(G \text{ and } CG)}{P(CG)}$$

In Illustration 4-13 we found $P(CG)$; $P(G \text{ and } CG)$ was also found in the solution of Illustration 4-13. Thus,

$$P(G|CG) = \frac{P(G) \cdot P(CG|G)}{P(CG)}$$

$$= \frac{(0.8)(0.9)}{0.74} = 0.9729 = \mathbf{0.973}$$

In other words, 97.3 percent of all items shipped will be good. Inspection increases the quality of items sold from 80 percent good to 97.3 percent good.   ■

Case Study 4-3

WIN-10

New York State, like many other states and cities, runs several lottery games for the purpose of raising money to finance education. Recently the WIN-10 game was introduced in New York State. The rules for playing and the prizes are described in a pamphlet distributed by New York's Lottery. The pamphlet shows the chances (or probability) of winning at WIN-10, if you bet on 3 numbers, to be approximately 1 in 72. How is the value $\frac{1}{72}$ obtained? To win when you pick 3 numbers, all 3 must be among the 20 winning numbers picked by the lottery.

$$P(\text{picking 3 winning numbers}) = P(A \text{ and } B \text{ and } C)$$
$$= P(A) \cdot P(B, A) \cdot P(C, A, \text{ and } B)$$
$$= \left(\tfrac{20}{80}\right)\left(\tfrac{19}{79}\right)\left(\tfrac{18}{78}\right)$$
$$= \tfrac{6840}{492960}$$
$$= \tfrac{1}{72.07}, \text{ or approximately 1 in 72}$$

Calculate some of the other probabilities of winning. (*See* Exercise 4.55.)

Win up to $200,000 every day!

8 ways to play. 17 ways to win.

Win-10—New York Lottery's newest addition to its exciting game menu. Win-10 allows you more flexibility and excitement than ever.

You may choose 3, 4, 5, 6, 7, 8, 9, or 10 numbers between 1 and 80. Each day the Lottery will randomly select 20 winning numbers from a field of 80.

You decide how large a prize you want to play for by the number of selections you make.

For example, you may want to choose 5 numbers. If your 5 numbers match *any* 5 of the 20 winning numbers drawn—you win $500 for your $1 bet.

Or you might prefer a choice of 10 numbers. This bet allows you to win 5 different ways:
- match *any* 10 winning numbers to win or share the Jackpot
- match *any* 9 winning numbers to win $6,000
- match *any* 8 winning numbers to win $300
- match *any* 7 winning numbers to win $25
- and, if you *don't match any* of the 20 winning numbers you win $4. (Imagine— a game where even when you lose, you can win!)

So play Win-10 to win the prize of your choice. It's easy to play and it's fun!

NTIN 6/7364 / 11 86

WIN-10 PRIZES AND CHANCES OF WINNING

The chart below shows the Win-10 prize levels, prize amounts, and chances of winning. For example, if you select 3 numbers and they match any 3 of the 20 winning numbers drawn, you win $25.

If you select 6 or more numbers, you then have more than one way to win. For example, if you select 9 numbers you win:
- $100,000 if your 9 numbers match any 9 of the 20 winning numbers.
- $3,000 if only 8 of your 9 numbers match any 8 of the 20 winning numbers.
- $250 if only 7 of your 9 numbers match any 7 of the 20 winning numbers.

TYPE OF BET	WINNING NUMBERS MATCHED	PRIZES	APPROXIMATE CHANCES OF WINNING
PICK 3	3	$25	1 in 72
PICK 4	4	$100	1 in 326
PICK 5	5	$500	1 in 1,551
PICK 6	6	$1,500	1 in 7,753
	5	$50	1 in 323
Overall chances of winning a pick 6 bet: 1 in 310			
PICK 7	7	$7,500	1 in 40,979
	6	$250	1 in 1,366
Overall chances of winning a pick 7 bet: 1 in 1,322			
PICK 8	8	$25,000	1 in 230,115
	7	$1,500	1 in 6,232
Overall chances of winning a pick 8 bet: 1 in 6,068			
PICK 9	9	$100,000	1 in 1,380,688
	8	$3,000	1 in 30,682
	7	$250	1 in 1,690
Overall chances of winning a pick 9 bet: 1 in 1,600			
PICK 10	10	JACKPOT*	1 in 8,911,711
	9	$6,000	1 in 163,381
	8	$300	1 in 7,384
	7	$25	1 in 621
	0	$4	1 in 22
Overall chances of winning a pick 10 bet: 1 in 21			

*NOTE: For the Pick-10 Jackpot Prize only, $200,000 will be won or shared by all players matching 10 of the 20 winning numbers drawn in any drawing.

From "Win-10," New York Lottery, November 1986. Reprinted with permission.

EXERCISES

4.48 If $P(A) = 0.4$ and $P(B) = 0.5$, and if A and B are independent events, find $P(A$ or $B)$.

4.49 $P(G) = 0.5$, $P(H) = 0.4$, and $P(G$ and $H) = 0.1$ (see the following diagram).
    a. Find $P(G|H)$.   **b.** Find $P(H|G)$.   **c.** Find $P(\bar{H})$.
    d. Find $P(G$ or $H)$.   **e.** Find $P(G$ or $\bar{H})$.
    f. Are events G and H mutually exclusive? Explain.
    g. Are events G and H independent? Explain.

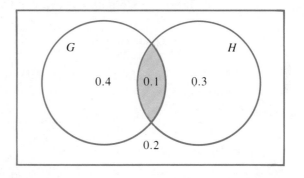

4.50 $P(R) = 0.5$, $P(S) = 0.3$, and events R and S are independent.
    a. Find $P(R$ and $S)$.   **b.** Find $P(R$ or $S)$.   **c.** Find $P(\bar{S})$.
    d. Find $P(R|S)$.   **e.** Find $P(\bar{S}|R)$.
    f. Are events R and S mutually exclusive? Explain.

4.51 $P(M) = 0.3$, $P(N) = 0.4$, and events M and N are mutually exclusive.
    a. Find $P(M$ and $N)$.   **b.** Find $P(M$ or $N)$.   **c.** Find $P(M$ or $\bar{N})$.
    d. Find $P(M|N)$.   **e.** Find $P(M|\bar{N})$.
    f. Are events M and N independent? Explain.

4.52 Two flower seeds are randomly selected from a package that contains five seeds for red flowers and three seeds for white flowers.

    a. What is the probability that both seeds will result in red flowers?
    b. What is the probability that one of each color is selected?
    c. What is the probability that both seeds are for white flowers?

4.53 A, B, and C each in turn toss a balanced coin. The first one to throw a head wins.

    a. What are their respective chances of winning if each tosses only one time?
    b. What are their respective chances of winning if they continue given a maximum of two tosses each?

4.54 The probability that a certain door is locked is 0.6. The key to the door is one of five unidentified keys hanging on a key rack. Two keys are randomly selected before approaching the door. What is the probability that the door may be opened without returning for another key? [*Hint:* Draw a tree diagram.]

4.55 a. Refer to Case Study 4-3. If you were to buy one ticket for New York's Lottery WIN-10 game and you pick 4 numbers, what is the probability that you will win? Calculate the exact probability and compare it to the chances of winning listed in the pamphlet.

b. If you pick 5 numbers, what is the probability of winning?

c. If you pick 6 numbers, what is the probability of winning with all 6 numbers?

d. If you pick 6 numbers, what is the probability of winning with exactly 5 numbers?

e. If you pick 6 numbers, what is the probability of winning (that is, with either 5 or 6 numbers)?

4.56 Probabilities for events A, B, and C are distributed as shown on the following figure. Find

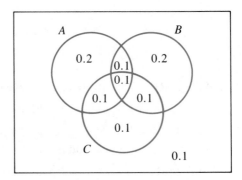

     a. $P(A \text{ and } B)$ **b.** $P(A \text{ or } C)$ **c.** $P(A \mid C)$

4.57 A coin is flipped three times.

     a. Draw a tree diagram that represents all possible outcomes.

     b. Identify all branches that represent the event "exactly one head occurred."

4.58 Box 1 contains two red balls and three green balls, and Box 2 contains four red balls and one green ball. One ball is randomly selected from Box 1 and placed in Box 2. Then one ball is randomly selected from Box 2. What is the probability that the ball selected from Box 2 is green?

4.59 A company that manufactures shoes has three factories. Factory 1 produces 25 percent of the company's shoes, Factory 2 produces 60 percent, and Factory 3 produces 15 percent. One percent of the shoes produced by Factory 1 are mislabeled, 0.5 percent of those produced by Factory 2 are mislabeled, and 2 percent of those produced by Factory 3 are mislabeled. If you purchase one pair of shoes manufactured by this company, what is the probability that the shoes are mislabeled?

In Retrospect You have now studied the basic concepts of probability. These fundamentals need to be understood to allow us to continue our study of statistics. Probability is the vehicle of statistics, and we have begun to see how probabilistic events occur. We have explored theoretical and experimental probabilities for the same event. Does the experimental probability turn out to have the same value as the theoretical? Not exactly, but over the long run we have seen that it does have approximately the same value.

You must, of course, know and understand the basic definition of probability as well as understand the properties of mutual exclusiveness and independence as they apply to the concepts presented in this chapter.

From reading this chapter you should be able to relate the "and" and the "or" of compound events to the multiplication and addition rules. You should also be able to calculate conditional probabilities.

In the next three chapters we will look at distributions associated with probabilistic events. This will prepare us for the statistics that will follow. We must be able to predict the variability that the sample will show with respect to the population before we will be successful at "inferential statistics," in which we describe the population based on the sample statistics available.

Chapter Exercises **4.60** Explain briefly how you would decide which of the following two events is the more unusual:

A: a 90° day in Vermont

B: a 100° day in Florida

4.61 A fair coin is tossed 100 million times. Let B be the total number of heads observed. Identify each of the following statements as true or false. Explain your answer (no computations required).

a. It is very probable that B is very close (within a few thousand) to 50 million.

b. It is very probable that $\frac{B}{100}$ million is very close to $\frac{1}{2}$; maybe between 0.49 and 0.51.

c. Since Hs and Ts fall with complete irregularity, one cannot say anything about what happens in 100 million tosses. It is just as likely that B is equal to one value as to any other value; for example, $P(B = 2) = P(B = 50$ million$)$.

d. Hs and Ts fall about equally often. Therefore, if on the first 10 tosses we get all Hs, it is more probable that the eleventh toss will yield a T than an H.

4.62 Suppose a certain ophthalmic trait is associated with eye color. Three hundred randomly selected individuals are studied with results as follows:

	Eye Color			
Trait	Blue	Brown	Other	Totals
Yes	70	30	20	120
No	20	110	50	180
Totals	90	140	70	300

a. What is the probability that a person selected at random has blue eyes?

b. What is the probability that a person selected at random has the trait?

c. Are events A (has blue eyes) and B (has the trait) independent? Justify your answer.

d. How are the two events A (has blue eyes) and C (has brown eyes) related (independent, mutually exclusive, complementary, or all inclusive)? Explain why or why not each term applies.

4.63 The employees at a large university were classified according to age as well as to whether they belonged to administration, faculty, or staff.

	\multicolumn{4}{c}{Age Group}			
	20–30	31–40	41–50	51 or Over
Administration	2	24	16	17
Faculty	1	40	36	28
Staff	16	20	14	2

For a randomly selected employee, find the probability that he or she

a. was in the administration or was 51 or over.

b. was not a faculty member.

c. was a faculty member given that the individual was 41 or over.

4.64 The following probabilities are known for events R and L: $P(R \text{ and } L) = 0.75$, $P(R) = 0.50$, and $P(L) = 0.50$. Determine whether R and L are independent or dependent events.

4.65 Events R and S are defined on the same sample space. If $P(R) = 0.2$ and $P(S) = 0.5$, explain why each of the following statements is either true or false.

a. If R and S are mutually exclusive, then $P(R \text{ or } S) = 0.10$.

b. If R and S are independent, then $P(R \text{ or } S) = 0.6$.

c. If R and S are mutually exclusive, then $P(R \text{ and } S) = 0.7$.

d. If R and S are mutually exclusive, then $P(R \text{ or } S) = 0.6$.

4.66 Show that if event A is a subset of event B, then $P(A \text{ or } B) = P(B)$.

4.67 Let's assume that there are three traffic lights between your house and a friend's house. As you arrive at each light, it may be red (R) or green (G).

a. List the sample space showing all possible sequences of red and green lights that could occur on a trip from your house to your friend's. (RGG represents red at the first light and green at the other two.) Assuming that each element of the sample space is equally likely to occur,

b. what is the probability that on your next trip to your friend's house you will have to stop for exactly one red light?

c. what is the probability that you will have to stop for at least one red light?

4.68 A fair die is rolled five times. What is the probability that a 6 occurred for the first time on the fifth roll?

4.69 Assuming that a woman is equally likely to bear a boy as a girl, use a tree diagram to compute the probability that a four-child family consists of one boy and three girls.

4.70 Suppose that when a job candidate comes to interview for a job at RJB Enterprises, the probability that he or she will want the job (A) after the interview is 0.68. Also, the probability that RJB wants the candidate (B) is 0.36. The probability $P(A \mid B)$ is 0.88.

    a. Find $P(A \text{ and } B)$.   **b.** Find $P(B \mid A)$.

    c. Are events A and B independent? Explain.

    d. Are events A and B mutually exclusive? Explain.

    e. What would it mean to say A and B are mutually exclusive events in this exercise?

4.71 A traffic analysis at a busy traffic circle in Washington, D.C., showed that 0.8 of the autos using the circle entered from Connecticut Avenue. Of those entering the traffic circle from Connecticut Avenue, 0.7 continued on Connecticut Avenue at the opposite side of the circle. What is the probability that a randomly selected auto observed in the traffic circle entered from Connecticut and will continue on Connecticut?

4.72 In the game of "Craps," you win on the first roll of a pair of dice if the sum of 7 or 11 occurs. You lose on the first roll if the sum 2, 3, or 12 occurs.

    a. What is the probability that you win on the first roll?

    b. What is the probability that you lose on the first roll?

4.73 The probabilities of women being employed in certain occupational groups are as follows.

Occupational Group	Probability
Professional/technical	0.16
Managerial/administrative	0.06
Sales	0.07
Clerical	0.34
Craft	0.02
Operatives, including transport	0.12
Non-farm laborers	0.01
Service, except private households	0.18
Private households	0.03
Farm	0.01
Total	1.00

    a. What is the probability that a randomly selected woman is either a professional or a managerial employee?

    b. What is the probability that two randomly selected women will both be clerical employees?

    c. What is the probability that a randomly selected woman will be both a sales and a clerical employee?

4.74 A store contains 12 aisles. Ten of the aisles are in the grocery section, labeled G1 to G10, and two are in the pharmacy section, labeled P1 and P2. If the store manager randomly selects an aisle to be used for a special display, find the probability that

   a. the aisle is an even-numbered aisle.

   b. the aisle is in the grocery section and has an odd number.

   c. the aisle is in the pharmacy section, given that it is an even-numbered aisle.

4.75 One thousand persons screened for a certain disease are given a clinical exam. As a result of the exam, the sample of 1000 persons is distributed according to height and disease status.

Height	Disease Status				Totals
	None	Mild	Moderate	Severe	
Tall	122	78	139	61	400
Medium	74	51	90	35	250
Short	104	71	121	54	350
Totals	300	200	350	150	1000

Use this information to estimate the probability of being medium or short in height and of having moderate or severe disease status.

4.76 Tires salvaged from a train wreck are on sale at the Getrich Tire Company. Of the 15 tires offered in the sale, 5 tires have suffered internal damage and the remaining 10 are damage free. If you were to randomly select and purchase 2 of these tires,

   a. what is the probability that the tires you purchase are both damage free?

   b. what is the probability that exactly 1 of the tires you purchase is damage free?

   c. what is the probability that at least 1 of the tires you purchase is damage free?

4.77 A shipment of grapefruit arrived containing the following proportions of types: 10 percent pink seedless, 20 percent white seedless, 30 percent pink with seeds, 40 percent white with seeds. A grapefruit is selected at random from the shipment. Find the probability that

   a. it is seedless.   **b.** it is white.

   c. it is pink and seedless.   **d.** it is pink or seedless.

   e. it is pink, given that it is seedless.   **f.** it is seedless, given that it is pink.

4.78 The following table shows the sentiments of 2500 wage-earning employees at the Spruce Company on a proposal to emphasize fringe benefits rather than wage increases during their impending contract discussions.

Employee	Opinion			Total
	Favor	Neutral	Opposed	
Male	800	200	500	1500
Female	400	100	500	1000
Total	1200	300	1000	2500

a. Calculate the probability that an employee selected at random from this group will be opposed.

b. Calculate the probability that an employee selected at random from this group will be female.

c. Calculate the probability that an employee selected at random from this group will be opposed, given that the person is male.

d. Are the events "opposed" and "female" independent? Explain.

4.79 The following information was generated from a recent survey of women employed by a predominantly female company when asked the question "Is college more important for a man than a woman?"

Is College More Important for a Man Than a Woman?

Age	Yes	No	Total
18–24	0.17	0.43	0.60
Over 24	0.26	0.14	0.40
Total	0.43	0.57	1.00

Using this information,

a. what is the probability that a randomly selected woman employee felt college was more important for a man than a woman?

b. what is the probability that a randomly selected woman answered "Yes" or was "over 24"?

c. what is the probability that a randomly selected woman was "over 24" given that she answered "No" to the question?

4.80 The probability that thunderstorms are in the vicinity of a particular midwestern airport on an August day is 0.70. When thunderstorms are in the vicinity, the probability that an airplane lands on time is 0.80. Find the probability that thunderstorms are in the vicinity and the plane lands on time.

4.81 A box contains ten ball-point pens, eight of which write properly and two of which do not write. You are going to randomly select two of these pens.

a. Describe the sample space with a tree diagram.

b. Assign probabilities to each branch of your tree diagram. Consider these three events: A, a good pen is drawn on the first drawing; B, a good pen is drawn on the second drawing; C, at least one defective pen is selected. Find these probabilities:

c. $P(A)$	**d.** $P(B)$	**e.** $P(C)$
f. $P(\bar{A})$	**g.** $P(\bar{B})$	**h.** $P(\bar{C})$
i. $P(A \text{ and } B)$	**j.** $P(A \text{ and } C)$	**k.** $P(B \text{ and } C)$
l. $P(A \text{ or } B)$	**m.** $P(A \text{ or } C)$	**n.** $P(B \text{ or } C)$

o. Are events A and B mutually exclusive? Explain.

p. Are events A and C mutually exclusive? Explain.

q. Are events A and B independent? Explain.

r. Are events A and C independent? Explain.

4.82 According to automobile accident statistics, one out of every six accidents results in an insurance claim of $100 or less in property damage. Three cars insured by an insurance company are involved in different accidents. Consider the following two events:

A: The majority of claims exceed $100.

B: Exactly two claims are $100 or less.

   a. List the sample points for this experiment.

   b. Are the sample points equally likely?

   c. Fine $P(A)$ and $P(B)$.

   d. Are A and B independent? Justify your answer.

4.83 Salesman Adams and saleswoman Jones call on three and four customers, respectively, on a given day. Adams could make 0, 1, 2, or 3 sales, whereas Jones could make 0, 1, 2, 3, or 4 sales. The sample space listing the number of possible sales for each person on a given day is given in the following table.

Adams	Jones 0	1	2	3	4
0	0, 0	1, 0	2, 0	3, 0	4, 0
1	0, 1	1, 1	2, 1	3, 1	4, 1
2	0, 2	1, 2	2, 2	3, 2	4, 2
3	0, 3	1, 3	2, 3	3, 3	4, 3

(3, 1 stands for 3 sales by Jones and 1 sale by Adams.) Assume that each sample point is equally likely.

   Let's define these events:

A = at least one of the salespersons made no sales

B = together they made exactly three sales

C = each made the same number of sales

D = Adams made exactly one sale

Find the following probabilities by *counting* sample points.

   a. $P(A)$ **b.** $P(B)$ **c.** $P(C)$

   d. $P(D)$ **e.** $P(A \text{ and } B)$ **f.** $P(B \text{ and } C)$

   g. $P(A \text{ or } B)$ **h.** $P(B \text{ or } C)$ **i.** $P(A \mid B)$

   j. $P(B \mid D)$ **k.** $P(C \mid B)$ **l.** $P(B \mid \bar{A})$

   m. $P(C \mid \bar{A})$ **n.** $P(A \text{ or } B \text{ or } C)$

Are the following pairs of events mutually exclusive? Explain.

   o. A and B   **p.** B and C   **q.** B and D

Are the following pairs of events independent? Explain.

   r. A and B   **s.** B and C   **t.** B and D

4.84 A testing organization wishes to rate a particular brand of television. Six TVs are selected at random from stock. If nothing is found wrong with any of the six, the brand is judged satisfactory.

   a. What is the probability that the brand will be rated satisfactory if 10 percent of the TVs actually are defective?

   b. What is the probability that the brand will be rated satisfactory if 20 percent of the TVs actually are defective?

   c. What is the probability that the brand will be rated satisfactory if 40 percent of the TVs actually are defective?

4.85 Coin A is loaded in such a way that P(heads) is 0.6. Coin B is a balanced coin. Both coins are tossed. Find the following:

   a. the sample space that represents this experiment and assign a probability measure to each outcome

   b. P(both show heads)

   c. P(exactly one head shows)

   d. P(neither coin shows a head)

   e. P(both show heads | coin A shows a head)

   f. P(both show heads | coin B shows a head)

   g. P(heads on coin A | exactly one head shows)

4.86 On a slot machine there are three reels with digits 0, 1, 2, 3, 4, and 5 and a flower on each reel. When a coin is inserted and the lever pulled, each of the three reels spins independently and comes to rest on one of the seven positions mentioned. Find these probabilities:

   a. that a flower shows on all three reels

   b. that a flower shows on exactly one reel

   c. that a flower shows on exactly two reels

   d. that no flowers show

   e. that a three-digit sequence (in order) appears

   f. that exactly one flower shows and the other two are odd integers

4.87 Professor French forgets to set his alarm with a probability of 0.3. If he sets the alarm it rings with a probability of 0.8. If the alarm rings, it will wake him on time to make his first class with a probability of 0.9. If the alarm does not ring, he wakes in time for his first class with a probability of 0.2. What is the probability that Professor French wakes in time to make his first class tomorrow?

4.88 A two-page typed report contains an error on one of the pages. Two proofreaders review the copy. Each has an 80 percent chance of catching the error. What is the probability that the error will be identified if

   a. each reads a different page?

   b. they each read both pages?

   c. the first proofreader randomly selects a page to read, then the second proofreader randomly selects a page unaware of which page the first selected?

4.89 A computer program generates pairs of random integers. Each integer ranges between 0 and 5, inclusive, so that any one of 36 pairs is equally likely. Let A be the event "the first integer of the pair is a zero" and let B represent the event "the sum of the two integers is an even number." Are A and B independent or dependent events?

4.90 For a particular population, 30 percent are in the age group 0–20 years, 40 percent are in the age group 21–40 years, and the remainder are over 40 years old. Given that an individual from this population is in the 0–20 age group, the probability that the individual has an abnormal result on a glucose tolerance test is 0.05. The conditional probability for an abnormal glucose tolerance test is 0.04 for the 21–40 age group and is 0.10 for the over-40 age group. For a randomly selected individual from this population, what is the probability that he or she has an abnormal glucose tolerance test?

Vocabulary List

Be able to define each term. In addition, describe, in your own words, and give an example of each term. Your examples should not be ones given in class or in the textbook.

The bracketed numbers indicate the chapter in which the term first appeared, but you should define the terms again to show increased understanding of their meaning.

addition rule	listing
all-inclusive events	long-term average
complementary event	multiplication rule
compound event	mutually exclusive events
conditional probability	odds
dependent events	outcome
empirical probability	probability of an event
equally likely events	relative frequency [2]
event	sample point
experiment [1]	sample space
experimental probability	subjective probability
independent events	theoretical probability
intersection	tree diagram
law of large numbers	Venn diagram

Quiz

Answer "True" if the statement is always true. If the statement is not always true, replace the words shown in bold with words that make the statement always true.

4.1 The probability of an event is a **whole number**.

4.2 The concepts of probability and relative frequency as related to an event are very **similar**.

4.3 The **sample space** is the theoretical population for probability problems.

4.4 The sample points of a sample space are **equally likely** events.

4.5 The value found for experimental probability will **always** be exactly equal to the theoretical probability assigned to the same event.

4.6 The probabilities of complementary events always **are equal**.

4.7 If two events are mutually exclusive, they are also **independent**.

4.8 If events A and B are **mutually exclusive**, the sum of their probabilites must be exactly one.

4.9 If the sets of sample points belonging to two different events do not intersect, the events are **independent**.

4.10 A compound event formed by use of the word "and" requires the use of the **addition rule**.

Chapter 5

Probability Distributions (Discrete Variables)

AN ANALYSIS OF ACCIDENTS AT A DAY CARE CENTER

An analysis of 1324 accidents over a 42-month period at a university day care center revealed that toddlers had the highest average number of injuries, most of them self-induced, that accidents peaked in mid-morning, and that September was the month with the highest accident rate. Although accidents were frequent, injuries were minor. Results are contrasted to those of earlier studies.

The dramatic increase in the number of working women with preschool age children has escalated the need for day care. Parents seek warm, caring, stimulating, affordable child care, but recent news reports of injuries from physical and sexual abuse in day care centers have raised questions of safety. Concern is so great that centers report insurance coverage is more difficult to obtain and, where available, rates have risen sharply. The true incidence of injuries from serious abuse and neglect in day care is not known, but injuries due to accidents are likely to be more common. According to Gratz,

> Accidents ... are the leading cause of childhood mortality and rank second only to acute infections as the cause of morbidity and visits to the physician throughout childhood.

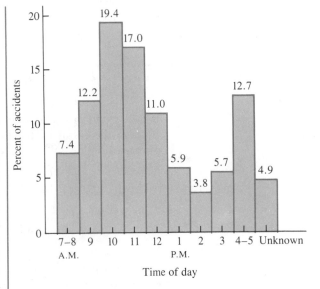

Fig. 1. Percent of accidents by time of day

From Richard Elardo, Ph.D., Hope C. Solomons, Ed.D., Bill C. Snider, Ph.D., *American Journal Orthopsychiatry, 57(1), January 1987.* Reprinted, with permission, from the *American Journal of Orthopsychiatry.* Copyright 1987 by the American Orthopsychiatric Association, Inc.

Chapter
Objectives

Chapter 2 dealt with frequency distributions of data sets, and Chapter 4 dealt with the fundamentals of probability. Now we are ready to combine these ideas to form *probability distributions*, which are much like relative frequency distributions. The basic difference between probability and relative frequency distributions is the use of the random variable. The random variable of a probability distribution corresponds to the response variable of a frequency distribution.

In this chapter we will investigate discrete probability distributions and study measures of central tendency and dispersion for such distributions. Special emphasis will be given to the binomial random variable and its probability distribution, since it is the most important discrete random variable encountered in most fields of application.

5.1   RANDOM VARIABLES

If each event in a probability experiment is assigned a numerical value, then as we observe the result of the experiment we are observing a random variable. This numerical value will be the value of the *random variable* under study.

random variable

Random Variable

A variable that assumes a unique numerical value for each of the outcomes in the sample space of a probability experiment.

In other words, a random variable is used to denote the outcome of a probability experiment. It can take on any numerical value that belongs to the set of all possible outcomes of the experiment. (It is called "random" because the value it assumes is the result of a chance, or random, event.)

Each event in a probability experiment must also be defined in such a way that only one value of the random variable is assigned to it, and each event must have a value assigned to it. Typically, the **discrete random variable** is a count of something.

discrete random variable

The following illustrations demonstrate what we mean by random variable.

Illustration 5-1   We toss five coins and observe the number of heads visible. The random variable x is the number of heads observed and may take on integer values from 0 to 5.   ■

Illustration 5-2   In Experiment 4-7 we rolled two dice and observed the sum of the dots on both dice. If a random variable had been assigned, it would have been the total number of dots showing. In this case, the random variable x could take on integer values from 2 to 12.   ■

Illustration 5-3   Let the number of phone calls received per day by a company be a random variable. It could take on integer values ranging from zero to some very large number.   ■

Illustration 5-4   The "model year" of automobiles owned by the faculty of a college could be used as a random variable. An observed probability distribution would result if the data were presented as a relative frequency distribution.

For example, suppose we find that 0.20 of the faculty-owned cars are 1988 models, 0.25 are 1987 models, 0.28 are 1986, 0.15 are 1985, and so on. (*See* Figure 5-1.) The random variable "model year of automobile" is a discrete random variable since it identifies the year in which the car was manufactured.

Figure 5-1
Model year of automobiles
owned by a college's faculty

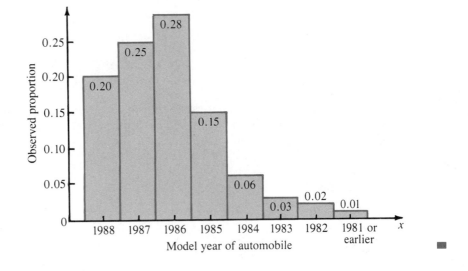

EXERCISES

5.1 A social worker is involved in a study about family structure. She obtains information regarding the number of children per family for a certain community from the census data. Identify the random variable of interest and list its possible values.

5.2 An experiment involves the testing of a new on/off switch. The switch is flipped on and off until it breaks and the flip on which the break occurs is noted. Identify the random variable of interest and list its possible values.

5.3 An archer shoots eight arrows at a bull's-eye of one target. Identify the random variable most likely to be of interest and list its possible values.

5.4 Radios are packed in cartons of 12. A carton of radios is inspected and the number of defective radios found is recorded. Identify the random variable used and list its possible values.

5.2 PROBABILITY DISTRIBUTIONS OF A DISCRETE RANDOM VARIABLE

Recall the coin-tossing experiment we used at the beginning of Section 4-1. Two coins were tossed and no heads, one head, or two heads were observed. If we define the random variable x to be the number of heads observed when two coins are tossed, x can take on the values 0, 1, or 2. The probability of each of these three events is the same as we calculated in Chapter 4.

$$P(x = 0) = P(0H) = \tfrac{1}{4}$$

$$P(x = 1) = P(1H) = \tfrac{1}{2}$$

$$P(x = 2) = P(2H) = \tfrac{1}{4}$$

These probabilities can be listed in any number of ways, but they are best displayed in the form illustrated by Table 5-1. Can you see why the name "probability distribution" is used?

Table 5-1
Probability distribution:
Tossing two coins

x	$P(x)$
0	$\tfrac{1}{4}$
1	$\tfrac{1}{2}$
2	$\tfrac{1}{4}$

probability distribution

Probability Distribution

A distribution of the probabilities associated with each of the values of a random variable. The probability distribution is a theoretical population; it is used to represent empirical populations.

In the experiment in which a single die is rolled and the number of dots on the top surface is observed, the random variable is the number observed. The probability distribution for this random variable is shown in Table 5-2.

Table 5-2
Probability distribution:
Rolling a die

x	$P(x)$
1	$\tfrac{1}{6}$
2	$\tfrac{1}{6}$
3	$\tfrac{1}{6}$
4	$\tfrac{1}{6}$
5	$\tfrac{1}{6}$
6	$\tfrac{1}{6}$

Sometimes it is convenient to write a rule that expresses the probability of an event in terms of the value of the random variable. This expression is typically written in formula form and is called a *probability function*.

probability function

Probability Function

A rule that assigns probabilities to the values of the random variables.

A probability function can be as simple as a list pairing the values of a random variable with their probabilities. Tables 5-1 and 5-2 show two such listings. However, a probability function is most often expressed in formula form.

Consider a die that has been modified so it has one face with one dot, two faces with two dots, and three faces with three dots. Let x be the number of dots observed when this die is rolled. The probability distribution for this experiment is presented in Table 5-3. Each of the probabilities can be represented by the value of x divided by 6. That is, each $P(x)$ is equal to the value of x divided by 6, where $x = 1, 2,$ or 3. Thus $P(x) = \frac{x}{6}$ for $x = 1, 2,$ and 3 is the formula expression of the probability function of this experiment.

Table 5-3

Probability distribution:
Rolling the modified die

x	$P(x)$
1	$\frac{1}{6}$
2	$\frac{2}{6}$
3	$\frac{3}{6}$

The probability function of the experiment of rolling one ordinary die is $P(x) = \frac{1}{6}$ for $x = 1, 2, 3, 4, 5,$ and 6. This particular function is called a **constant function** because the value of $P(x)$ does not change as x changes.

Every probability function must display the two basic properties of probability. These two properties are (1) the probability assigned to each value of the random variable must be between 0 and 1, inclusive, that is,

$$0 \le \text{each } P(x) \le 1$$

and (2) the sum of the probabilities assigned to all the values of the random variable must equal 1, that is,

$$\sum_{\text{all } x} P(x) = 1$$

Is $P(x) = \frac{x}{10}$ for $x = 1, 2, 3,$ and 4 a probability function? To answer this question we need only test the function in terms of the two basic properties. The probability distribution is shown in Table 5-4. Property 1 is satisfied, since $\frac{1}{10}, \frac{2}{10}, \frac{3}{10},$ and $\frac{4}{10}$ are all numerical values between 0 and 1. Property 2 is also satisfied, since the

Table 5-4

Probability distribution for
$P(x) = \frac{x}{10}$

x	$P(x)$
1	$\frac{1}{10}$
2	$\frac{2}{10}$
3	$\frac{3}{10}$
4	$\frac{4}{10}$
Total	$\frac{10}{10} = 1$

sum of all four probabilities is exactly 1. Since both properties are satisfied, we can conclude that $P(x) = \frac{x}{10}$ for $x = 1, 2, 3, 4$ *is* a probability function.

What about $x = 5$ (or any value other than 1, 2, 3, or 4) in the function $P(x) = \frac{x}{10}$ for $x = 1, 2, 3,$ and 4? $P(x = 5)$ is considered to be zero. That is, the probability function provides a probability of zero for all values of x other than the values specified.

Figure 5-2 presents a probability distribution graphically. Regardless of the specific representation, the values of the random variable are plotted on the horizontal scale, and the probability associated with each value of the random variable is plotted on the vertical scale. A discrete random variable should really be

line histogram   presented by a **line histogram**, since the variable can assume only discrete values. Figure 5-2 shows the probability distribution of $P(x) = \frac{x}{10}$ for $x = 1, 2, 3,$ and 4. This line representation makes sense because the variable x can assume *only* the values 1, 2, 3, and 4. The length of the vertical line represents the value of the probability.

Figure 5-2
Line histogram of the probability distribution for $P(x) = \frac{x}{10}$ for $x = 1, 2, 3, 4$

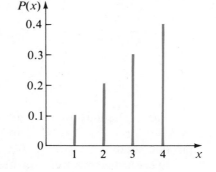

probability histogram   However, a regular **histogram** is more frequently used to present probability distributions. Figure 5-3 shows the same probability distribution of Figure 5-2 but in histogram form. This representation suggests that x can assume all numerical values (fractions, decimals, and so on), but this is not the case. The histogram probability distribution uses the physical area of each bar to represent its assigned probability. The bar for $x = 2$ is 1 unit wide (from 1.5 to 2.5) and is 0.2 unit high. Therefore, its area is $(1)(0.2) = 0.2$, the probability assigned to $x = 2$. The areas of the other bars can be determined in similar fashion. This area representation will be an important concept in Chapter 6 when we begin to work with continuous random variables.

Figure 5-3
Histogram of the probability distribution of $P(x) = \frac{x}{10}$ for $x = 1, 2, 3, 4$

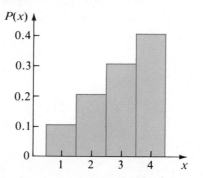

Case Study 5-1

WHO NEEDS THE AMBULANCE?

Robert Giordana used a relative frequency histogram to help him explain how his ambulance services are used. The histogram is of a discrete variable (number of trips per day) and shows, in percents, the relative frequency of days with various numbers of service trips. Explain how this information satisfies the requirements of a discrete probability distribution. (See Exercise 5.7.)

The Austin City Ambulance Company last week appealed to the City Council for additional municipal funding. Mr. Robert Giordana, the company's business manager, stated that while people see ambulances on the streets occasionally, they rarely have any real concept of the frequency with which an ambulance is called upon for assistance.

In surveying the company's records, the Council found that one ambulance responds to between one and six calls for help on a typical day. The records for a recent six-month period showed that an ambulance made three trips on 25% of the days and four trips on 21% of the days. It was further revealed that there was only one day in every three weeks when no trips were made. But on one day in the same three weeks, they made seven or more trips.

Mr. Giordana reminded the Council that the company was working hard for the community all year long.

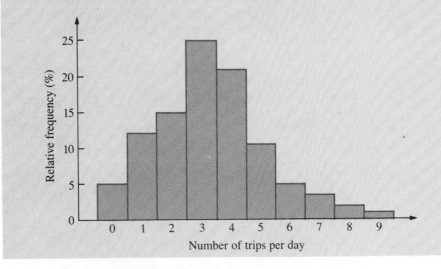

Case Study 5-2

WOES OF THE WEEKEND JOCK

The information about injuries to the recreational athlete is summarized on a chart showing what appear to be many different relative frequency distributions. These numbers are relative frequencies, but they do not always form relative frequency distributions. Explain why. (See Exercise 5.12.)

THE PART-TIME ATHLETE IS NOT IMMUNE TO A HOST OF INJURIES

The pursuit of fitness, a phenomenon of the '70s and '80s, is responsible for generating another phenomenon: Both general practitioners and sportsmedicine specialists are noticing a surge in sports-related injuries among weekend and after-work jocks. Basketball and soccer leagues, ski vacations, evening runs, dance classes, tennis games—all foster injuries once found chiefly among professional and college athletes.

For a comprehensive look at the problems of this new breed of recreational athlete, the Center for Sports Medicine at Saint Francis Memorial Hospital in San Francisco recently compiled statistics on over 10,000 injuries treated at the center. There were some interesting findings:

• Nine activities—basketball, dance, football, gymnastics, running, skiing, tennis, soccer and figure skating—account for nearly three-fourths of the injuries.

• More than two-thirds of the injuries are caused by overuse—problems such as shinsplints and tendinitis that develop from a repetitive trauma to muscle and bone. Tennis, aerobic dance and running frequently cause such problems.

• The remaining injuries are acute ones, incidents that happen instantly, such as a sprained ankle. Not surprisingly, these tend to occur in skiing, football, basketball and soccer.

• Injuries to the knee occasion the most visits to the center, and skiers have the most knee problems.

• Aerobic dance causes more fractures than any other recreational activity.

• Many problems stem from acute injuries that occurred in the past.

• Unlike football injuries, most of the basketball injuries (62 percent) occur in participants over 25 years of age.

Injuries to the Recreational Athlete (age and sex, anatomic site of injury, and injury type (% of total) in 7,458 athletes)

Patients by Sport (No.)	Basketball 377	Dance 2,242	Football 257	Running 3,004	Skiing 425	Soccer 271	Tennis 882
Age and Sex							
0–18	17.5	21.3	33.1	4.8	2.8	40.2	5.2
19–25	20.1	24.2	31.1	11.3	16.5	27.7	6.6
26–39	55.4	45.5	29.6	61.3	58.6	27.7	35.6
40 and over	6.9	8.8	6.2	22.6	22.1	4.4	52.6
Male	87.8	19.8	99.2	55.2	50.8	69.4	63.5
Female	12.3	80.2	0.8	44.8	49.2	30.6	36.5
Anatomic Site of Injury							
Shoulder	4.0	2.3	13.6	0.4	10.8	2.9	15.1
Upper extremity	3.4	1.4	8.2	0.3	14.3	2.2	22.2
Spine	4.8	9.6	9.3	4.9	3.3	4.8	7.9
Hip	4.0	8.7	1.5	7.7	1.2	3.7	3.9
Thigh	2.4	4.0	5.8	4.1	—	4.4	1.9
Knee	42.2	28.6	42.8	38.2	67.3	46.1	24.4
Leg	6.6	12.7	1.9	11.7	3.5	6.6	6.9
Ankle	21.2	12.1	8.6	8.5	4.7	15.5	6.6
Foot	8.5	14.5	0.8	20.7	2.3	9.6	7.9

Patients by Sport (No.)	Basketball 377	Dance 2,242	Football 257	Running 3,004	Skiing 425	Soccer 271	Tennis 882
Injury Types							
Fracture	5.6	9.4	8.9	7.8	3.8	4.8	1.6
Dislocation	4.5	2.4	8.5	1.4	8.7	2.6	2.0
Sprain	36.3	9.5	29.2	5.9	42.6	26.9	10.1
Strain	12.2	23.9	14.0	18.0	11.8	16.6	27.5
Inflammation	15.9	18.6	5.8	36.3	7.1	15.5	31.2
Overuse	8.5	16.4	5.4	19.1	12.2	14.0	12.2
Other	13.2	11.7	18.3	5.8	8.8	14.4	10.0

All percentages don't total 100 because some categories weren't listed and there were multiple injuries in some sports

Amy Wilbur, *Science Digest*, March 1986.

EXERCISES

5.5 Census data are often used to obtain probability distributions for various random variables. Census data for families with a combined income of $50,000 or more in a particular state show that 20 percent have no children, 30 percent have one child, 40 percent have two children, and 10 percent have three children. From this information, construct the probability distribution for x, where x represents the number of children per family for this income group.

5.6 A company manufactures insulin needles and packages them in boxes of 100. Based on historical data obtained from sampling such boxes over a period of years, it is known that 90 percent of all such boxes contain no defective needles, 7 percent contain exactly one defective needle, and 3 percent contain exactly two defective needles. Based on this information, what is the probability distribution for x, where x represents the number of defective needles per box?

5.7 a. List the information shown on the histogram in Case Study 5-1 as a probability distribution.

b. Are these theoretical probabilities? Explain.

c. Explain how this information forms a discrete probability distribution.

5.8 Test the following function to determine whether it is a probability function. If it is not, try to make it into a probability function. List the distribution of probabilities and sketch a histogram.

$$P(x) = \frac{5 - x}{10} \quad \text{for} \quad x = 1, 2, 3, 4$$

5.9 Test the following function to determine whether it is a probability function. If it is not, try to make it into a probability function. List the distribution of probabilities and sketch a histogram.

$$Q(x) = \frac{x^2 - 1}{50} \quad \text{for} \quad x = 2, 3, 4, 5$$

5.10 Test the following function to determine whether it is a probability function. If it is not, try to make it into a probability function. $R(x) = 0.2$ for $x = 0, 1, 2, 3, 4$.

    a. List the distribution of probabilities.

    b. Sketch a histogram.

5.11 Test the following function to determine whether it is a probability function. If it is not, try to make it into a probability function.

$$S(x) = \frac{6 - |x - 7|}{36} \quad \text{for} \quad x = 2, 3, 4, 5, 6, 7, \dots, 11, 12$$

    a. List the distribution of probabilities and sketch a histogram.

    b. Do you recognize $S(x)$? If so, identify it.

5.12 **a.** Why are the numbers reported on the chart in Case Study 5-2 relative frequencies?

    b. Why does the set of numbers reported for skiing and age form a probability distribution?

    c. Why does the set of numbers reported for skiing and anatomic site of injury not form a probability distribution?

5.3 MEAN AND VARIANCE OF A DISCRETE PROBABILITY DISTRIBUTION

Recall that the mean of a frequency distribution is

$$\bar{x} = \frac{\sum xf}{n}$$

which can also be written as

$$\bar{x} = \sum \left(x \cdot \frac{f}{n} \right)$$

We learned in our study of probability that the probability of an event is the expected relative frequency of its occurrence. Thus, if we replace f/n with $P(x)$, we can find the mean of a theoretical probability distribution:

$$\text{mean value of } x = \sum [x \cdot P(x)] \tag{5-1}$$

Notes

1. $\bar{x}$ is the mean of a sample.

2. s is the standard deviation of the individual elements of the sample.

3. $\bar{x}$ and s are called **sample statistics**.

4. μ (Greek letter mu) is the mean of the population under consideration.

5. σ (Greek letter sigma) is the standard deviation of the individual elements of the population under consideration.

6. μ and σ are called **population parameters**. (A parameter is a constant. μ and σ are typically unknown values.)

mean of probability distribution

Recall that the "expected relative frequency" represents what will occur in the long run. The mean of x is therefore the mean of the entire population of experimental outcomes. The symbol for the **mean** value of x (in a probability distribution) is μ, and formula (5-1) can be written as

$$\mu = \sum [x \cdot P(x)] \tag{5-2}$$

[Note: The summation is "over all x-values."]

Illustration 5-5   Let's return to the previous probability function: $P(x) = \frac{x}{10}$ for $x = 1, 2, 3,$ and 4. We can find its mean by using formula (5-2) after we compile a probability distribution table (*see* table 5-5). The mean μ, or the sum of the products of x times $P(x)$, is $\frac{30}{10}$, or **3.0**. [The sum of the probabilities, $\sum P(x)$, must be 1.0. Use this as a check.]

Table 5-5
Probability distribution:
$P(x) = \frac{x}{10}, x = 1, 2, 3, 4$

	$P(x)$	$x \cdot P(x)$
1	$\frac{1}{10}$	$\frac{1}{10}$
2	$\frac{2}{10}$	$\frac{4}{10}$
3	$\frac{3}{10}$	$\frac{9}{10}$
4	$\frac{4}{10}$	$\frac{16}{10}$
Total	$\frac{10}{10} = 1.0$	$\mu = \frac{30}{10} = \mathbf{3.0}$

variance of probability distribution

The **variance** of discrete probability distributions is defined in much the same way as the variance of sample data.

$$\sigma^2 = \sum [(x - \mu)^2 \cdot P(x)] \tag{5-3}$$

The variance of x for the probability distribution discussed in this illustration is found as shown in Table 5-6. The variance σ^2 is $\frac{10}{10}$, or **1.0**. ($\mu = 3$ was found in Table 5-5.)

Formula (5-3) is often not convenient to use; fortunately it can be reworked to appear in the form

$$\sigma^2 = \sum [x^2 \cdot P(x)] - \{\sum [x \cdot P(x)]\}^2 \tag{5-4a}$$

or

$$\sigma^2 = \sum [x^2 P(x)] - \mu^2 \tag{5-4b}$$

Table 5-6
Finding the variance for
$P(x) = \frac{x}{10}, x = 1, 2, 3, 4$

x	$x - \mu$	$(x - \mu)^2$	$P(x)$	$(x - \mu)^2 P(x)$
1	-2	4	$\frac{1}{10}$	$\frac{4}{10}$
2	-1	1	$\frac{2}{10}$	$\frac{2}{10}$
3	0	0	$\frac{3}{10}$	$\frac{0}{10}$
4	1	1	$\frac{4}{10}$	$\frac{4}{10}$
Total				$\sigma^2 = \frac{10}{10}$

It is left for you (Exercise 5.14) to verify that formula (5-4a) and (5-4b) are equivalent to formula (5-3). ■

Illustration 5-6   To find the variance of the probability distribution in Illustration 5-5 by using formula (5-4a), we will need to add two additional columns to Table 5-5. These columns are shown in Table 5-7.

Table 5-7
Calculations needed to find the
variance for $P(x) = \frac{x}{10}$

x	$P(x)$	$x \cdot P(x)$	x^2	$x^2 \cdot P(x)$
1	$\frac{1}{10}$	$\frac{1}{10}$	1	$\frac{1}{10}$
2	$\frac{2}{10}$	$\frac{4}{10}$	4	$\frac{8}{10}$
3	$\frac{3}{10}$	$\frac{9}{10}$	9	$\frac{27}{10}$
4	$\frac{4}{10}$	$\frac{16}{10}$	16	$\frac{64}{10}$
Total	$\frac{10}{10}$	$\frac{30}{10}$		$\frac{100}{10}$

The variance is

$$\sigma^2 = \sum [x^2 \cdot P(x)] - \{\sum [x \cdot P(x)]\}^2$$
$$= \frac{100}{10} - \left(\frac{30}{10}\right)^2 = 10 - (3)^2$$
$$= 10 - 9 = \mathbf{1.0}$$

■

standard deviation of
probability distribution

The **standard deviation** is the square root of variance; therefore, $\sigma = \mathbf{1.0}$.

Illustration 5-7   A coin is tossed three times. Let the number of heads occurring in those three tosses be the random variable x, which can take on the values 0, 1, 2, or 3. There are eight possible outcomes to this experiment: one results in $x = 0$, three in $x = 1$, three in $x = 2$, and one in $x = 3$. Therefore, the probabilities for this random variable are $\frac{1}{8}, \frac{3}{8}, \frac{3}{8}$, and $\frac{1}{8}$. The probability distribution associated with this experiment is shown in Figure 5-4 and in Table 5-8. The necessary extensions and summations for the calculation of its mean and standard deviation are also shown in Table 5-8.

The mean is found with the aid of formula (5-2):

$$\mu = \sum [x \cdot P(x)] = \mathbf{1.5}$$

This result, 1.5, is the mean number of heads expected per experiment.

Figure 5-4
Probability distribution
for Illustration 5-7

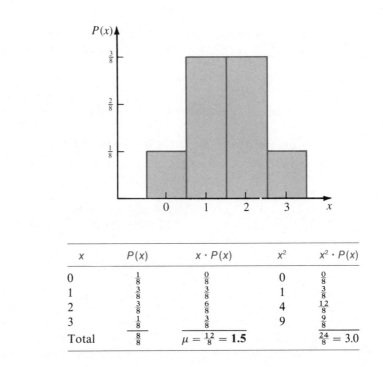

Table 5-8
Probability distribution and
extensions, Illustration 5-7

x	$P(x)$	$x \cdot P(x)$	x^2	$x^2 \cdot P(x)$
0	$\frac{1}{8}$	$\frac{0}{8}$	0	$\frac{0}{8}$
1	$\frac{3}{8}$	$\frac{3}{8}$	1	$\frac{3}{8}$
2	$\frac{3}{8}$	$\frac{6}{8}$	4	$\frac{12}{8}$
3	$\frac{1}{8}$	$\frac{3}{8}$	9	$\frac{9}{8}$
Total	$\frac{8}{8}$	$\mu = \frac{12}{8} = \mathbf{1.5}$		$\frac{24}{8} = 3.0$

The variance is found with the aid of formula (5-4a):

$$\sigma^2 = \sum [x^2 \cdot P(x)] - \{\sum [x \cdot P(x)]\}^2$$
$$= 3.0 - (1.5)^2 = 3.0 - 2.25 = \mathbf{0.75}$$

The standard deviation is the positive square root of the variance:

$$\sigma = \sqrt{0.75} = 0.866 = \mathbf{0.87}$$

That is, 0.87 is the standard deviation expected among the number of heads observed per experiment. ■

EXERCISES

5.13 Given the probability function

$$P(x) = \frac{5 - x}{10} \quad \text{for} \quad x = 1, 2, 3, 4,$$

find the mean and standard deviation.

5.14 Given the probability function

$$R(x) = 0.2 \quad \text{for} \quad x = 0, 1, 2, 3, 4,$$

find the mean and standard deviation.

5.15 The number of calls x to arrive at a switchboard during any 1-minute period is a random variable and has the following probability distribution.

x	0	1	2	3	4
$P(x)$	0.1	0.2	0.4	0.2	0.1

Find the mean and standard deviation of x.

5.16 Based on past history, the distribution of sales of 1-pound containers of cottage cheese at a convenience store are:

Number of Cartons Sold per Day (x)	Probability
10	0.2
11	0.4
12	0.2
13	0.1
14	0.1
Total	1.0

Determine the mean and the standard deviation of x, the number of cartons sold per day.

5.17 A random variable x has the following probability distribution.

x	1	2	3	4	5
$P(x)$	0.6	0.1	0.1	0.1	0.1

How much of the probability distribution is within 2 standard deviations of the mean? That is, find the probability that x is between $\mu - 2\sigma$ and $\mu + 2\sigma$.

5.18 a. Draw a histogram of the probability distribution for the single-digit random numbers (0, 1, 2, ..., 9).

b. Calculate the mean and standard deviation associated with the population of single-digit random numbers.

5.19 Verify that formulas (5-4a) and (5-4b) are equivalent to formula (5-3).

5.4 THE BINOMIAL PROBABILITY DISTRIBUTION

Consider the following probability experiment. You are given a five-question multiple-choice quiz. You have not studied the material to be quizzed and therefore decide to answer the five questions by randomly guessing the answers without reading the question or the answers.

Answer Page to Quiz

Directions: Circle the best answer to each question.

1. A   B   C
2. A   B   C
3. A   B   C
4. A   B   C
5. A   B   C

Circle your answers before continuing.

Before we look at the correct answers to the quiz, let's think about some of the things we might consider when a quiz is answered in this way.

1. How many of the five questions do you think you answered correctly?

2. If an entire class answered the quiz by guessing, what do you think the "average" number of correct answers would be?

3. What is the probability that you selected the correct answers to all five questions?

4. What is the probability that you selected wrong answers for all five questions?

Let's construct the probability distribution associated with this experiment. Let x be the number of correct answers on your paper; x may then take on the values 0, 1, 2,..., 5. Since each individual question has three possible answers and since there is only one correct answer, the probability of selecting the correct answer to an individual question is $\frac{1}{3}$. The probability that a wrong answer is selected is $\frac{2}{3}$. $P(x = 0)$ is the probability that all questions were answered incorrectly.

$$P(x = 0) = \left(\frac{2}{3}\right)\left(\frac{2}{3}\right)\left(\frac{2}{3}\right)\left(\frac{2}{3}\right)\left(\frac{2}{3}\right) = \frac{32}{243} \approx \mathbf{0.132}$$

These events are independent because each question is a separate event. We can therefore multiply the probabilities according to formula (4-8b). Also,

$$P(5) = \left(\frac{1}{3}\right)^5 = \frac{1}{243} \approx \mathbf{0.004}$$

Before we find the other four probabilities, let's look at the events on a tree diagram, Figure 5-5. Notice that the event $x = 0$, "zero correct answers," is shown by the bottom branch, and that event $x = 5$, "five correct answers," is shown by the top branch. The other events, "one correct answer," "two correct answers," and so on, are represented by several branches of the tree.

Figure 5-5 shows 32 branches representing 5 different values of x. We find that the event $x = 1$ occurs on 5 different branches, event $x = 2$ occurs on 10 branches, $x = 3$ occurs on 10 branches, and $x = 4$ on 5 branches. Thus the remaining probabilities can be calculated by finding the probability of a single branch

Figure 5-5

Tree diagram for the
multiple-choice quiz

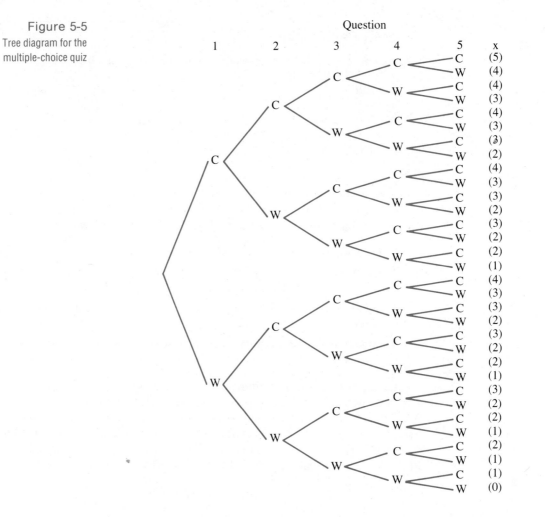

representing the event and multiplying it by the number of branches that represent that event.

$$P(x = 1) = \left(\frac{1}{3}\right)\left(\frac{2}{3}\right)\left(\frac{2}{3}\right)\left(\frac{2}{3}\right)\left(\frac{2}{3}\right)(5) = \frac{80}{243} \approx \mathbf{0.329}$$

$$P(x = 2) = \left(\frac{1}{3}\right)^2\left(\frac{2}{3}\right)^3(10) = \frac{80}{243} \approx \mathbf{0.329}$$

$$P(x = 3) = \left(\frac{1}{3}\right)^3\left(\frac{2}{3}\right)^2(10) = \frac{40}{243} \approx \mathbf{0.165}$$

$$P(x = 4) = \left(\frac{1}{3}\right)^4\left(\frac{2}{3}\right)^1(5) = \frac{10}{243} \approx \mathbf{0.041}$$

The answer to question 1 on page 201 is suggested by the probability distribution in Table 5-9. The most likely occurrence would be to get one or two correct answers. One, two, or three correct answers are expected to result

x	$P(x)$
0	0.132
1	0.329
2	0.329
3	0.165
4	0.041
5	0.004
Total	1.001

approximately 82 percent of the time ($0.329 + 0.329 + 0.165 = 0.823$). We might also argue that $\frac{1}{3}$ of the questions would be expected to be answered correctly. This information provides a reason for answering both questions 1 and 2 page 201 with the value of $\frac{5}{3}$. Questions 3 and 4 were asked with the expectation that you would feel the chance of "all correct" was very slight—and it is: 0.004. "All wrong" is not too likely, but it is not as rare an event as "all correct." It has a probability of 0.132.

The correct answers to the quiz are B, C, B, A, and C. How many correct answers did you have? You might ask several people to answer the quiz for you by guessing at the answers. Then construct an observed relative frequency distribution and compare it to the distribution shown in Table 5-9.

success or failure

Many experiments result in outcomes that can be classified in one of two categories, **success or failure**. Examples of such experiments include the previous experiments of tossing coins or thumbtacks and other more practical experiments, such as determining whether a flashbulb flashes or does not flash, whether a product did its prescribed job or did not do it, whether quiz question answers are right or wrong. There are other experiments that have many outcomes that, under the right conditions, may fit this general description of being classified in one of two categories. For example, when we roll a single die, we usually consider six possible outcomes. However, if we are only interested in knowing whether an ace (a 1) shows or not, there are really only two outcomes: the ace shows or something else shows.

The experiments described above are usually called *binomial probability experiments*.

binomial experiment

trial

binomial random variable

Binomial Probability Experiment

An experiment that is made up of repeated trials of a basic experimental event. The binomial experiment must possess the following properties:

1. Each **trial** has two possible outcomes (success, failure).
2. There are η repeated independent trials.
3. $P(\text{success}) = p$, $P(\text{failure}) = q$, and $p + q = 1$.
4. The **binomial random variable** x is the count of the number of successful trials that occur; x may take on any integer value from zero to n.

[Don't forget: $p = P(\text{success})$ on *each* individual trial and $q = P(\text{failure})$ on *each* trial.]

The binomial experiment *must* demonstrate the two basic properties 1 and 2. Properties 3 and 4 concern notation and the identification of the variables involved.

Note   It is of utmost importance that a probability p be assigned to the particular outcome that is considered to be the "success," since the binomial random variable indicates the number of "successes" that occur.

Illustration 5-8   Consider the roll of a die, where "ace" or "something else" occurs. In this experiment we would count the number of "aces." The random variable x would be assigned a value that is the number of times that an ace is observed in the n trials. Since "ace" is considered "success," $P(\text{ace}) = p$ and $P(\text{not ace}) = q$.   ■

Illustration 5-9   If you were an inspector on a production line in a plant where television sets were manufactured, you would be concerned with identifying the number of defective television sets. You would probably define "success" as the occurrence of a defective television. This is not what we normally think of as success, but if we count defective sets in a binomial experiment, we must define "success" as a "defective." The random variable x indicates the number of defective sets found per lot of n sets. p represents $P(\text{television is defective})$ and q is $P(\text{television is good})$.   ■

Note   *Independent* trials mean that the result of one trial does not affect the probability of success of any other trial in the experiment. In other words, the probability of "success" remains constant throughout the entire experiment.

Each binomial probability experiment has its own specific probability function. However, all such functions have a common format and all such distributions display some similarities. Let's look at a relatively simple binomial experiment and discuss its probability function.

Let's reconsider Illustration 5-7; a coin is tossed three times and we observe the number of heads that occur in the three tosses. This is a binomial experiment because it displays all the properties of a binomial experiment:

1. Each trial (one toss of the coin) has two outcomes: success (heads, what we are counting) and failure (tails).

2. There are n (three) repeated trials that are independent (each is a separate toss and the outcome of any one trial has no effect on the outcome of another.)

3. The probability of success (heads) is p ($\frac{1}{2}$) and the probability of failure (tails) is q ($\frac{1}{2}$), and $p + q = 1$.

4. The random variable x is the number of successes (heads) that occur in n (three) trials. x can assume the value 0, 1, 2, or 3.

Let's consider first the probability of $x = 1$.

$$P(x = 1) = P(\text{exactly one head is observed in three tosses})$$

We want to find this probability. What does it mean to say "exactly one head in three tosses"? The probability that one head appears on one toss is $\frac{1}{2}$. The probability that heads do not occur on the other two tosses is $\frac{1}{2} \times \frac{1}{2} = \frac{1}{4}$. The probability that exactly one head occurs in three tosses, then, must be related to $\frac{1}{2} \times \frac{1}{4} = \frac{1}{8}$ [from the multiplication rule, formula (4-8a)]. But there is one other consideration: In how many different ways can exactly one of the three tosses result in a head?

1. The first toss could be a head and the others tails.

2. The second toss could be a head and the first and last tails.

3. The first two tosses could be tails and the last toss a head.

These are the three ways in which three trials can result in exactly one success. This means that the $\frac{1}{8}$ found above can occur three different ways, resulting in $P(x = 1) = 3(\frac{1}{8}) = \frac{3}{8}$.

If we look at the case $x = 2$, we find a similar situation: there are three ways to obtain exactly two heads. Thus $P(x = 2) = \frac{3}{8}$, also. We can find only one way for $x = 0$ and $x = 3$ to occur—all tails or all heads. In both cases the probability is equal to $(\frac{1}{2})^3 = \frac{1}{8}$. The rest of the probability distribution is exactly as shown in Table 5-8.

If we take a careful look at the probability of each case shown in the tree diagram in Figure 5-5, we can argue that the probability that a random variable x takes on a particular value in a binomial experiment is always the product of three basic factors. These three factors are as follows:

1. the probability of exactly x successes, p^x

2. the probability that failure will occur on the remaining $(n - x)$ trials, q^{n-x}

3. the number of ways that exactly x successes can occur in n trials

The number of ways that exactly x successes can occur in a set of n trials is represented by the symbol

$$\binom{n}{x}$$

binomial coefficient which must always be a positive integer. This is termed the **binomial coefficient**. The binomial coefficient is found by using the formula

$$\binom{n}{x} = \frac{n!}{x!(n - x)!} \tag{5-5}$$

Note $n!$ is an abbreviation for the product of the sequence of integers starting with n and ending with 1. For example, $3! = 3 \cdot 2 \cdot 1 = 6$, and $5! = 5 \cdot 4 \cdot 3 \cdot 2 \cdot 1 = 120$. There is one special case, $0!$, which is defined to be 1. For further information about factorial notation, see the Study Guide.

Also, see the Study Guide for general information on the binomial coefficient. The values for $n!$ when n is equal to or smaller than 20 are found in Table 2 of Appendix E. The values for $\binom{n}{x}$ when n is equal to or smaller than 20 are found in Table 3 of Appendix E.

This information allows us to form a general *binomial probability function*.

binomial probability
function

Binomial Probability Function

If, for a binomial experiment, p is the probability of a success and q is the probability of a failure on a single trial, then the probability $P(x)$ that there will be exactly x successes in n trials is

$$P(x) = \binom{n}{x} \cdot p^x \cdot q^{n-x} \quad \text{for} \quad x = 0, 1, 2, \ldots, n \qquad (5\text{-}6)$$

When this general binomial probability function is applied to the illustration with three coins (Illustration 5-7), we find

$$P(x) = \binom{3}{x} \cdot \left(\frac{1}{2}\right)^x \cdot \left(\frac{1}{2}\right)^{3-x} \quad \text{for} \quad x = 0, 1, 2, 3$$

Let's calculate each of the probabilities and see whether they form a probability distribution:

$$P(x = 0) = \binom{3}{0} \cdot \left(\frac{1}{2}\right)^0 \cdot \left(\frac{1}{2}\right)^3 = 1 \cdot 1 \cdot \frac{1}{8} = \frac{1}{8}$$

$P(x = 0) = \frac{1}{8}$ is the probability of no heads occurring in three tosses of a coin. (*Note*: $(\frac{1}{2})^0 = 1$.)

$$P(x = 1) = \binom{3}{1} \cdot \left(\frac{1}{2}\right)^1 \cdot \left(\frac{1}{2}\right)^2 = 3 \cdot \frac{1}{2} \cdot \frac{1}{4} = \frac{3}{8}$$

$P(x = 1) = \frac{3}{8}$ is the probability of one head occurring in three tosses of a coin.

$$P(x = 2) = \binom{3}{2} \cdot \left(\frac{1}{2}\right)^2 \cdot \left(\frac{1}{2}\right)^1 = 3 \cdot \frac{1}{4} \cdot \frac{1}{2} = \frac{3}{8}$$

$P(x = 2) = \frac{3}{8}$ is the probability of two heads occurring in three tosses of a coin.

$$P(x = 3) = \binom{3}{3} \cdot \left(\frac{1}{2}\right)^3 \cdot \left(\frac{1}{2}\right)^0 = 1 \cdot \frac{1}{8} \cdot 1 = \frac{1}{8}$$

$P(x = 3) = \frac{1}{8}$ is the probability of three heads occurring in three tosses of a coin. Since each of these probabilities is between 0 and 1, and the sum of all the probabilities is exactly 1, this is a probability distribution.

Let's look at another example. Consider an experiment that calls for drawing five cards, one at a time with replacement, from a well-shuffled deck of playing cards. The drawn card is identified as a spade or not a spade, returned to the deck, the deck reshuffled, and so on. The random variable x is the number of spades observed in five drawings. Is this a binomial experiment? Let's identify the various properties.

1. Each drawing is a trial and each drawing has two outcomes, "spade" or "not spade."

2. There are five repeated drawings, so $n = 5$. These individual trials are independent, since the drawn card is returned to the deck and the deck reshuffled before the next drawing.

3. $p = \frac{1}{4}$ and $q = \frac{3}{4}$.

4. x is the number of spades recorded in the five trials. The binomial probability function is

$$P(x) = \binom{5}{x} \cdot \left(\frac{1}{4}\right)^x \cdot \left(\frac{3}{4}\right)^{5-x}$$

for $x = 0, 1, 2, 3, 4,$ and 5. The probabilities are

$$P(0) = \binom{5}{0} \cdot \left(\frac{1}{4}\right)^0 \cdot \left(\frac{3}{4}\right)^5 = 1 \cdot 1 \cdot \left(\frac{3}{4}\right)^5 = \mathbf{0.2373}$$

$$P(1) = \binom{5}{1} \cdot \left(\frac{1}{4}\right)^1 \cdot \left(\frac{3}{4}\right)^4 = 5 \cdot \left(\frac{1}{4}\right) \cdot \left(\frac{3}{4}\right)^4 = \mathbf{0.3955}$$

$$P(2) = \binom{5}{2} \cdot \left(\frac{1}{4}\right)^2 \cdot \left(\frac{3}{4}\right)^3 = 10 \cdot \left(\frac{1}{4}\right)^2 \cdot \left(\frac{3}{4}\right)^3 = \mathbf{0.2637}$$

$$P(3) = \binom{5}{3} \cdot \left(\frac{1}{4}\right)^3 \cdot \left(\frac{3}{4}\right)^2 = 10 \cdot \left(\frac{1}{4}\right)^3 \cdot \left(\frac{3}{4}\right)^2 = \mathbf{0.0879}$$

$$P(4) = \binom{5}{4} \cdot \left(\frac{1}{4}\right)^4 \cdot \left(\frac{3}{4}\right)^1 = 5 \cdot \left(\frac{1}{4}\right)^4 \cdot \left(\frac{3}{4}\right)^1 = \mathbf{0.0146}$$

$$P(5) = \binom{5}{5} \cdot \left(\frac{1}{4}\right)^5 \cdot \left(\frac{3}{4}\right)^0 = 1 \cdot \left(\frac{1}{4}\right)^5 \cdot \left(\frac{3}{4}\right)^0 = \mathbf{0.0010}$$

$$\sum P(x) = \overline{1.0000}$$

The preceding distribution of probabilities indicates that the single most likely value of x is 1, the event of observing exactly one spade in a hand of five cards. What is the least likely number of spades that would be observed?

Let's consider one more binomial probability problem.

Illustration 5-10   The manager of Steve's Food Market guarantees that none of his cartons of one dozen eggs will contain more than one bad egg. If a carton contains more than one bad egg, he will replace the whole dozen and allow the customer to keep the original eggs. The probability that an individual egg is bad is 0.05. What is the probability that Steve will have to replace a given carton of eggs?

Solution   Assuming that this is a binomial experiment, let x be the number of bad eggs found in a carton of a dozen eggs. $p = 0.05$, and let the inspection of each egg be a trial resulting in finding a "bad" or "not bad" egg. To find the probability that the manager will have to make good on his guarantee, we need the probability function associated with this experiment:

$$P(x) = \binom{12}{x} \cdot (0.05)^x \cdot (0.95)^{12-x} \quad \text{for} \quad x = 0, 1, 2, \dots, 12$$

The probability that Steve will replace a dozen eggs is the probability that $x = 2$, 3, 4, 5,..., or 12. Recall that $\sum P(x) = 1$, that is,

$$P(x = 0) + P(x = 1) + P(x = 2) + \cdots + P(x = 12) = 1$$

Therefore,

$$P(x = 2) + P(x = 3) + \ldots + P(x = 12) = 1 - [P(x = 0) + P(x = 1)]$$

Finding $P(x = 0)$ and $P(x = 1)$ and subtracting them from 1 is easier than finding each of the other probabilities.

$$P(x) = \binom{12}{x} \cdot (0.05)^x \cdot (0.95)^{12-x}$$

$$P(0) = \binom{12}{0} \cdot (0.05)^0 (0.95)^{12} = \mathbf{0.540}$$

$$P(1) = \binom{12}{1} \cdot (0.05)^1 (0.95)^{11} = \mathbf{0.341}$$

Note   The value of many binomial probabilities, for small values of n and common values of p, are found in Table 4 of Appendix E. In this example, we have $n = 12$ and $p = 0.05$, and we want the probabilities for $x = 0$ and 1. We need to locate the section of table 4 where $n = 12$, find the column marked $p = 0.05$, and read the numbers opposite $x = 0$ and 1. We find 540 and 341, as shown in Table 5-10. (Look these values up in Table 4.) The decimal point was left out of the table to save space, and we must replace it. It belongs at the front of each table entry. Therefore, our values are 0.540 and 0.341.

Table 5-10

Abbreviated portion of Table 4 in Appendix E, binomial probabilities

n	x	0.01	0.05	0.10	0.20	0.30	0.40	P 0.50	0.60	0.70	0.80	0.90	0.95	0.99	x
(12)	0	886	540	282	069	014	002	0+	0+	0+	0+	0+	0+	0+	0
	1	107	341	377	206	071	017	003	0+	0+	0+	0+	0+	0+	1
	2	006	099	230	283	168	064	016	002	0+	0+	0+	0+	0+	2
	3	0+	017	085	236	240	142	054	012	001	0+	0+	0+	0+	3
	4	0+	002	021	133	231	213	121	042	008	001	0+	0+	0+	4

Now let's return to our illustration:

$$P(\text{replacement}) = 1 - (0.540 + 0.341) = \mathbf{0.119}$$

If $p = 0.05$ is correct, Steve will be busy replacing cartons of eggs. If he replaces 11.9 percent of all the cartons of eggs he sells, he certainly will be giving away a substantial proportion of his eggs. This suggests that he should adjust his guarantee. For example, if he were to replace a carton of eggs only when four or more were found bad, he would expect to replace only 0.003 cartons $[1.0 - (0.540 + 0.341 + 0.099 + 0.017)]$, or 0.3 percent of the cartons sold. Notice that he will be able to control his "risk" (probability of replacement) if he adjusts the value of the random variable stated in his guarantee. ∎

EXERCISES

5.20 Evaluate each of the following:

    a. 4!   **b.** 7!   **c.** 0!   **d.** $\dfrac{6!}{2!}$

    e. $\dfrac{5!}{3!2!}$ **f.** $\dfrac{6!}{4!(6-4)!}$ **g.** $(0.3)^4$ **h.** $\dbinom{7}{3}$

    i. $\dbinom{5}{2}$ **j.** $\dbinom{3}{0}$ **k.** $\dbinom{4}{1}(0.2)^1(0.8)^3$ **l.** $\dbinom{5}{0}(0.3)^0(0.7)^5$

5.21 Four cards are selected, one at a time, from a standard deck of 52 cards. Let x represent the number of aces drawn in the set of 4 cards.

    a. If this experiment is completed without replacement, explain why x is not a binomial random variable.

    b. If this experiment is completed with replacement, explain why x is a binomial random variable.

5.22 The number of branches in a tree diagram representation of a binomial experiment having n trials is equal to 2^n.

    a. For a binomial experiment having two trials, draw the tree diagram and show that it has 2^2, or 4, branches.

    b. For a binomial experiment having three trials, draw the tree diagram and show that it has 2^3, or 8, branches.

    c. How many branches would a tree diagram representation for a binomial experiment have when five trials are made? Ten trials? [Do not draw the tree diagram.]

5.23 Consider a binomial experiment made up of three trials with outcomes of success S and failure F, where $P(S) = p$ and $P(F) = q$.

    a. Complete the accompanying tree diagram. Label all branches completely.

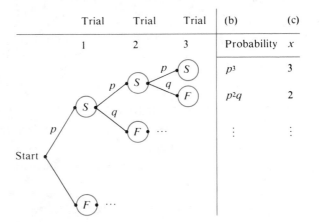

    b. In Column (b) of the tree diagram, express the probability of each outcome represented by the branches as a product of powers of p and q.

c. Let x be the random variable, the number of successes observed. In Column (c), identify the value of x for each branch of tree diagram.

d. Notice that all the products in Column (b) are made up of three factors and that the value of the random variable is the same as the exponent for the number p.

e. Write the equation for the binomial probability function for this situation.

5.24 In reference to the five-question multiple-choice "quiz" on page 201, show that the guessing of the answers is actually a binomial experiment.

a. Specify exactly how this experiment satisfies the four properties of a binomial experiment.

b. Complete the tree diagram (Figure 5-5) showing all possible outcomes for the quiz. There should be a total of 32 outcomes shown. How many outcomes are there for each of the events "0 correct answers" "1 correct answer," "2 correct answers," ..., "5 correct answers"?

c. Write the equation of the binomial probability function for this experiment.

5.25 If x is a binomial random variable, calculate the probability of x for each case.

    a. $n = 4, x = 1, p = 0.3$　　　**b.** $n = 3, x = 2, p = 0.8$

    c. $n = 2, x = 0, p = \frac{1}{4}$　　　**d.** $n = 5, x = 2, p = \frac{1}{3}$

    e. $n = 4, x = 2, p = 0.5$　　　**f.** $n = 3, x = 3, p = \frac{1}{6}$

5.26 If x is a binomial random variable, use Table 4, Appendix E, to determine the probability of x for each case.

    a. $n = 10, x = 8, p = 0.3$　　　**b.** $n = 8, x = 7, p = 0.95$

    c. $n = 15, x = 3, p = 0.05$　　　**d.** $n = 12, x = 12, p = 0.99$

    e. $n = 9, x = 0, p = 0.5$　　　**f.** $n = 6, x = 1, p = 0.01$

    g. Explain the meaning of the symbol $0+$ that appears in Table 4.

5.27 Test the following function to determine whether or not it is a binomial probability function. List the distribution of probabilities and sketch a histogram.

$$T(x) = \binom{5}{x} \cdot \left(\frac{1}{2}\right)^x \cdot \left(\frac{1}{2}\right)^{5-x} \quad \text{for} \quad x = 0, 1, 2, 3, 4, 5$$

5.28 Consider the experiment of tossing a coin five times and counting the number of heads that appear as the random variable (x).

a. Is this a binomial experiment? Describe all the properties and assign the variable.

b. Write the equation of the binomial probability function.

c. Find the various binomial probabilities to form the probability distribution.

5.29 Let x be a random variable with the following probability distribution:

x	0	1	2	3
$P(x)$	0.4	0.3	0.2	0.1

Does x have a binomial distribution? Justify your answer.

5.30 State a very practical reason why the defective item in an industrial situation would be defined to be the "success" in a binomial experiment.

5.31 A manufacturer of matches knows that 0.1 percent of the matches produced by his company are defective. Find the probability that a package of 50 matches contains no defective matches.

5.32 In the biathlon event of the Olympic Games, a participant skis cross-country and on four intermittent occasions stops at a rifle range and shoots a set of five shots. If the center of the target is hit, no penalty points are assessed. If a particular man has a history of hitting the center of the target with 90 percent of his shots, what is the probability that he will hit the center of the target with

a. all five of his next set of five shots?

b. at least four of his next set of five shots? [Assume independence.]

5.33 A machine produces parts of which 0.5 percent are defective. If a random sample of ten parts produced by this machine contains two or more defectives, the machine is shut down for repairs. Find the probability that the machine will be shut down for repairs based on this sampling plan.

5.34 A basketball player has a history of making 80 percent of the foul shots taken during games. What is the probability that he will miss three of the next five foul shots he takes?

5.35 The survival rate during a risky operation for patients with no other hope of survival is 80 percent. What is the probability that exactly four of the next five patients survive this operation?

5.36 According to studies in 1982, 1 out of every 15 individuals entering a department store will attempt to shoplift something before leaving the store. Assuming that the binomial distribution is appropriate, what is the probability that 1 out of 3 randomly selected individuals who have entered the store will attempt to shoplift something before leaving?

5.37 If the binomial $(q + p)$ is squared, the result is $(q + p)^2 = q^2 + 2qp + p^2$. For the binomial experiment with $n = 2$, the probability of no successes in two trials is q^2 (the first term in the expansion), the probability of one success in two trials is $2qp$ (the second term in the expansion), and the probability of two successes in two trials is p^2 (the third term). Find $(q + p)^3$ and compare its terms to the binomial probabilities for $n = 3$ trials.

5.5 MEAN AND STANDARD DEVIATION OF THE BINOMIAL DISTRIBUTION

mean and standard deviation of binomial distribution

The **mean** and **standard deviation** of a theoretical binomial probability distribution can be found by using these two formulas:

$$\mu = np \tag{5-7}$$

$$\sigma = \sqrt{npq} \tag{5-8}$$

The formula for the mean seems appropriate as the number of trials multiplied by the probability that a "success" will occur. (Recall that the mean number of correct answers on the binomial quiz was expected to be $\frac{1}{3}$ of 5, $5 \cdot \frac{1}{3}$, or np.) The formula for the standard deviation is not as easily understood. Thus at this point it is appropriate to look at an example that demonstrates that formulas (5-7) and (5-8) yield the same results as formulas (5-2) and (5-4).

Referring to the case of tossing three coins in Illustration 5-7, $n = 3$ and $p = \frac{1}{2}$. Using formulas (5-7) and (5-8), we find

$$\mu = np = (3)(\tfrac{1}{2}) = \mathbf{1.5}$$

$\mu = 1.5$ is the mean value of the random variable x.

$$\sigma = \sqrt{npq} = \sqrt{(3)(\tfrac{1}{2})(\tfrac{1}{2})} = \sqrt{\tfrac{3}{4}}$$
$$= \sqrt{0.75} = 0.866 = \mathbf{0.87}$$

$\sigma = 0.87$ is the standard deviation of the random variable x. Look back at the solution for Illustration 5-7 and compare the use of formulas (5-2) and (5-4a) with formulas (5-7) and (5-8). Note that the results are the same, regardless of the formula you use. However, formulas (5-7) and (5-8) are much easier to use when x is a binomial random variable.

Illustration 5-11   Find the mean and standard deviation of the binomial distribution where $n = 20$ and $p = \frac{1}{5}$. Recall that the "binomial distribution where $n = 20$ and $p = \frac{1}{5}$" has a probability function

$$P(x) = \binom{20}{x} \cdot (0.2)^x \cdot (0.8)^{20-x}; \quad \text{for} \quad x = 0, 1, 2, \dots, 20$$

and a corresponding distribution with 21 x-values and 21 probabilities as shown in the following distribution chart.

x	$P(x)$	x	$P(x)$
0	0.012	8	0.022
1	0.057	9	0.007
2	0.137	10	0.002
3	0.205	11	0.001
4	0.219	12	0+
5	0.174	13	0+
6	0.109	⋮	⋮
7	0.055	20	0+

Now let's find the mean and the standard deviation of this distribution of x using formulas (5-7) and (5-8).

$$\mu = np = (20)(\tfrac{1}{5}) = \mathbf{4.0}$$
$$\sigma = \sqrt{npq} = \sqrt{(20)(\tfrac{1}{5})(\tfrac{4}{5})} = \sqrt{\tfrac{80}{25}}$$
$$= \frac{(4\sqrt{5})}{5} = \mathbf{1.79}$$

■

EXERCISES

5.38 Given the binomial probability function

$$T(x) = \binom{5}{x} \cdot \left(\frac{1}{2}\right)^x \cdot \left(\frac{1}{2}\right)^{5-x} \quad \text{for} \quad x = 0, 1, 2, 3, 4, 5$$

   a. calculate the mean and standard deviation of the random variable by using formulas (5-2) and (5-4a).

   b. calculate the mean and standard deviation using formulas (5-7) and (5-8).

   c. Compare the results of (a) and (b).

5.39 Find the mean and standard deviation of x for each of the following binomial random variables.

   a. the number of aces seen in 100 draws from a well-shuffled bridge deck (with replacement).

   b. the number of cars found to have unsafe tires among the 400 cars stopped at a road block for inspection. Assume that 6 percent of all cars have one or more unsafe tires.

   c. the number of melon seeds that germinate when a package of 50 seeds is planted. The package states that the probability of germination is 0.88.

   d. the number of defective televisions in a shipment of 125. The manufacturer claimed that 98 percent of the sets were operative.

5.40 A binomial random variable has a mean equal to 200 and a standard deviation of 10. Find the values of n and p.

5.41 Seventy-five percent of the foreign-made autos sold in the United States in 1978 are now falling apart.

   a. Determine the probability distribution of x, the number of these autos that are falling apart in a random sample of five cars.

   b. Draw a histogram of the distribution.

   c. Calculate the mean and the standard deviation of this distribution.

5.42 A binomial random variable x is based upon 15 trials with the probability of success equal to 0.3. Find the probability that this variable will take on a value more than two standard deviations from the mean.

In Retrospect   In this chapter we combined concepts of probability with some of the ideas presented in Chapter 2. We now are able to deal with distributions of probability values and find means, standard deviations, and so on.

   In Chapter 4 we explored the concepts of mutually exclusive events and independent events. The addition and multiplication rules were used on several occasions in this chapter, but very little was said about mutual exclusiveness or independence. Recall that every time we add probabilities together, as we did in each of the probability distributions, we need to know that the associated events are mutually exclusive. If you look back over the chapter, you will notice that the random variable actually requires events to be mutually exclusive; therefore, no real emphasis was placed on this concept. The same basic comment can be made in reference to the multiplication of probabilities and the concept of independent events. Throughout this chapter, probabilities were multiplied together and occasionally independence was mentioned. Independence, of course, is necessary in order to be able to multiply probabilities together.

If we were to take a close look at some of the sets of data in Chapter 2, we would see that several of these problems could be reorganized to form probability distributions. For example: (1) Let x be the number of credit hours that a student is registered for in this semester, with the percentage of the student body being reported for each value of x. (2) Let x be the number of correct passageways through which an experimental laboratory animal passes before taking a wrong one. (3) Let x be the number of trips per day for the ambulance service (Case Study 5-1). (4) Let x be the age of the recreational athlete with an injury as discussed in Case Study 5-2. (This is an example of a continuous random variable.) The list of examples is endless.

We are now ready to extend these concepts to continuous random variables, which we will do in Chapter 6.

Chapter Exercises

5.43 What are the two basic properties of every probability distribution?

5.44 Verify whether or not the following is a probability function. State your conclusion and explain.

$$f(x) = \frac{\frac{3}{4}}{x!(3 - x)!} \quad \text{for} \quad x = 0, 1, 2, 3$$

5.45 Determine which of the following are probability functions.

   a. $f(x) = 0.25$ for $x = 9, 10, 11, 12$

   b. $f(x) = (3 - x)/2$ for $x = 1, 2, 3, 4$

   c. $f(x) = (x^2 + x + 1)/25$ for $x = 0, 1, 2, 3$

5.46 The number of ships to arrive at a harbor on any given day is a random variable represented by x. The probability distribution for x is

x	10	11	12	13	14
$P(x)$	0.4	0.2	0.2	0.1	0.1

Find the probability that on a given day

   a. exactly 14 ships arrive.   **b.** at least 12 ships arrive.

   c. at most 11 ships arrive.

5.47 The number of data-entry mistakes per hour made by an individual entering data at a CRT terminal is a random variable represented by x and has the following probability distribution.

x	0	1	2	3
$P(x)$	0.40	0.30	0.25	0.05

a. Find the mean number of mistakes for 1-hour sessions.

b. Find the probability of at least one mistake during a particular 1-hour session.

5.48 A "wheel of fortune" in a professor's office produces uniformly distributed random integer values between zero and 100. Suppose that each student in this professor's class is assigned his grade by giving the wheel a whirl and that grades are determined by the following distribution.

$$0 \text{ to } 20 \text{ is F}$$
$$21 \text{ to } 40 \text{ is D}$$
$$41 \text{ to } 62 \text{ is C}$$
$$63 \text{ to } 87 \text{ is B}$$
$$88 \text{ to } 100 \text{ is A}$$

a. Find the probability associated with each letter grade.

b. Construct a histogram depicting the random variable x, the random integer generated by the wheel.

5.49 A local elementary school holds four open houses annually. Records show that the probability that a child's parents (one or both) attend from 0 to 4 of the open houses is as shown in the accompanying table.

Number of Open Houses Attended (x)	0	1	2	3	4
Probability	0.12	0.38	0.30	0.12	0.08

a. Is this a probability distribution? Explain.

b. What is the probability that an individual child's parents attend at least one of these functions?

c. Find the mean and the standard deviation for this distribution.

5.50 A coin has a 1 painted on the head and a 2 painted on the tail side. An experiment consists of flipping the coin and rolling a die. The random variable x is defined to be 0 if the coin shows a 1 and the number on the die if the coin shows a 2. Construct the probability distribution of x and calculate its mean.

5.51 A discrete random variable has a standard deviation equal to 10 and a mean equal to 50. Find $\sum x^2 P(x)$.

5.52 A binomial random variable is based on $n = 20$ and $p = 0.4$. Find $\sum x^2 P(x)$.

5.53 For a binomial experiment based on n trials, a tree diagram will have $\binom{n}{r}$ branches, which represent exactly r successes and $(n - r)$ failures. For example, a binomial experiment based on $n = 3$ trials will have $\binom{3}{1} = 3$ branches with exactly one success and two failures, the branches SFF, FSF, and FFS. For a binomial experiment based on four trials, use the foregoing information to determine the number of branches having exactly none, exactly one, exactly two, exactly three, and exactly four successes. Confirm your answers by drawing a tree diagram and listing the branches.

5.54 The ratio of men to women at a college is 2:1. Give the binomial probability function and distribution for the random variable "number of men" in a randomly selected group of four taken from the student body.

5.55 If boys and girls are equally likely to be born, what is the probability that in a randomly selected family of six children, three will be boys? (Find the answer by using a formula.)

5.56 One-fourth of a certain breed of rabbits are born with long hair. What is the probability that in a litter of six rabbits, exactly three will have long hair? (Find the answer by using a formula.)

5.57 Suppose that you take a five-question multiple-choice quiz by guessing. Each question has exactly one correct answer of the four alternatives given. What is the probability that you guess more than one-half of the answers correctly? (Find the answer by using a formula.)

5.58 As a quality-control inspector for toy trucks, you have observed that wooden wheels that are bored off-center occur about 3 percent of the time. If six wooden wheels are used on each toy truck produced, what is the probability that a randomly selected set of wheels has no off-center wheels?

5.59 The town council has nine members. A proposal must have at least $\frac{2}{3}$ of the votes to be accepted. A proposal to establish a new industry in this town has been tabled. If we know that two members of the town council are opposed and that the others randomly vote "in favor" and "against," what is the probability that the proposal will be accepted?

5.60 Ninety percent of the trees planted by a landscaping firm survive. What is the probability that eight or more of the ten trees they just planted will survive? (Find the answer by using a table.)

5.61 In California, 30 percent of the people have a certain blood type. What is the probability that exactly 5 out of a randomly selected group of 14 Californians will have that blood type? (Find the answer by using a table.)

5.62 On the average, 1 out of every 10 boards purchased by a cabinet manufacturer is unusable for building cabinets. What is the probability that 8, 9, or 10 of a set of 11 such boards are usable? (Find the answer by using a table.)

5.63 A local polling organization maintains that 90 percent of the eligible voters have never heard of John Anderson, who was a presidential candidate in 1980. If this is so, what is the probability that in a randomly selected sample of 12 eligible voters 2 or fewer have heard of John Anderson?

5.64 A doctor knows from experience that 10 percent of the patients to whom he gives a certain drug will have undesirable side effects. Find the probabilities that among the ten patients to whom he gives the drug,
    a. at most two will have undesirable side effects.
    b. at least two will have undesirable side effects.

5.65 In a recent survey of women, 90 percent admitted that they had never looked at a copy of *Vogue* magazine. Assuming that this is accurate information, what is the probability that a random sample of three women will show that fewer than two have looked at the magazine?

5.66 Of the Heisman Trophy winners in collegiate football, it is considered that only 0.35 have had successful professional football careers.

   a. What is the probability that none of the next five winners will have successful pro careers?

   b. What is the probability that more than three of the next five winners will have successful pro careers?

5.67 Seventy percent of those seeking a driver's license admitted that they would not report someone if he or she copied some answers during the written exam. You have just entered the room and see ten people waiting to take the written exam. What is the probability that, if the incident happened, five of the ten would not report what they saw?

5.68 The engines on an airliner operate independently. The probability that an individual engine operates for a given trip is 0.95. A plane will be able to complete a trip successfully if at least one-half of its engines operate for the entire trip. Determine whether a four-engine or a two-engine plane has the higher probability of a successful trip.

5.69 CRT operators enter data for a large savings and loan institution. The probability that a character is incorrectly entered is 0.001. For a document containing 10,000 characters, compute the probability that at most one of the characters is incorrectly entered.

5.70 A box contains ten items of which three are defective and seven are nondefective. Two items are selected without replacement and x is the number of defectives in the sample of two. Explain why x is not a binomial random variable.

5.71 A large shipment of radios is accepted upon delivery if an inspection of ten randomly selected radios yields no more than one defective radio.

   a. Find the probability that this shipment is accepted if 5 percent of the total shipment is defective.

   b. Find the probability that this shipment is not accepted if 20 percent of this shipment is defective.

5.72 A business firm is considering two investments. It will choose the one that promises the greater payoff. Which of the investments should it accept? (Let the mean profit measure the payoff.)

Invest in Tool Shop		Invest in Book Store	
Profit	Probability	Profit	Probability
$100,000	0.10	$400,000	0.20
50,000	0.30	90,000	0.10
20,000	0.30	−20,000	0.40
−80,000	0.30	−250,000	0.30
Total	1.00	Total	1.00

5.73 Experiment has shown that a pair of beetles produces black-eyed offspring 30 percent of the time. The eye color of one offspring is not related to that of another, and we are concerned with the "number of black-eyed beetles" in a litter of 12.

    a. Describe how this experiment exhibits the properties of a binomial experiment.

    b. Find the probability that at least one-half of this litter has black eyes.

5.74 A box contains eight identical balls numbered from 1 to 8. An experiment consists of selecting two of these balls, without replacement, and identifying their numbers.

    a. How many different sets of two balls could be selected?

    b. List the sample space for the experiment.

    c. If the balls are selected randomly, what is the probability of each of the possible sets in the sample space?

    d. Define the random variable x to be the total of the numbers on the two balls. Construct the probability distribution for x.

    e. Find the mean and standard deviation of x.

5.75 Bill has completed a ten-question multiple-choice test on which he answered seven questions correctly. Each question had one correct answer to be chosen from five alternatives. Bill says that he answered the test by randomly guessing the answers without reading the questions or answers.

    a. Define the random variable x to be the number of correct answers on this test and construct the probability distribution if the answers were obtained by random guessing.

    b. What is the probability that Bill guessed seven of the ten answers correctly?

    c. What is the probability that anybody can guess six or more answers correctly?

    d. Do you believe that Bill actually randomly guessed as he claims? Explain.

5.76 A random variable that can assume any one of the integer values $1, 2, \ldots, n$ with equal probabilities of $\frac{1}{n}$ is said to have a uniform distribution. The probability function is written $P(x) = \frac{1}{n}$, for $x = 1, 2, 3, \ldots, n$. Show that $\mu = (n + 1)/2$. [*Hint:* $1 + 2 + 3 + \cdots + n = [n(n + 1)]/2$.]

Vocabulary List

Be able to define each term. In addition, describe, in your own words, and give an example of each term. Your examples should not be ones given in class or in the textbook.

    The bracketed numbers indicate the chapter in which the term first appeared, but you should define the terms again to show increased understanding of their meaning.

addition of probabilities [4]
binomial coefficient
binomial experiment

binomial probability function
binomial random variable
constant function

discrete random variable
experiment [1, 4]
failure
independent events [4]
mean of probability distribution
multiplication of probabilities [4]
mutually exclusive events [4]
population parameter [1]
probability distribution
probability function

probability histogram
probability line histogram
random variable
relative frequency [2, 4]
sample statistic [1]
standard deviation of probability
   distribution
success
trial
variance of probability distribution

Quiz *Answer "True" if the statement is always true. If the statement is not always true, replace the words shown in bold with words that make the statement always true.*

5.1 The number of hours that you waited in line to register this semester is an example of a **discrete** random variable.

5.2 The number of automobile accidents that you were involved in as a driver last year is an example of a **discrete** random variable.

5.3 The sum of all the probabilities in any probability distribution is always exactly **two**.

5.4 The various values of a random variable form a list of **mutually exclusive** events.

5.5 A binomial experiment always has **three or more** possible outcomes to each trial.

5.6 The formula $\mu = np$ may be used to compute the mean of a **discrete** population.

5.7 The binomial parameter f is the probability of **one success occurring in n trials** when a binomial experiment is performed.

5.8 A parameter is a statistical measure of some aspect of a **sample**.

5.9 **Sample statistics** are represented by letters from the Greek alphabet.

5.10 The probability of events A or B is equal to the sum of the probability of event A and the probability of event B when A and B are **mutually exclusive events**.

Chapter 6

Normal Probability Distributions

TEACHING BREAST AND TESTICULAR SELF-EXAMS: EVALUATION OF A HIGH SCHOOL CURRICULUM PILOT PROJECT

Introduction

This pilot project, a joint effort of the Cleveland Clinic Foundation and the American Cancer Society (Cuyahoga County Unit), promoted the concept of early detection of cancer to high school students by teaching the topics of breast and testicular self-examination (BSE and TSE). Although the concept for this project evolved from similar work done in Wisconsin, it represents the first comprehensive program in which classroom teachers have been used exclusively to teach students the topics.

Yearly in the United States, there are 115,000 new cases of breast cancer which lead to an estimated 37,300 deaths (American Cancer Society, 1983)....

Evaluation Methodology

Two written questionnaires were used to evaluate the project: a teacher satisfaction questionnaire and a student questionnaire. The student questionnaire included questions about the students: use of self-exams (self-reported); knowledge about BSE and TSE (multiple choice items); and attitudes toward early cancer detection (Likert Scale items)....

Students were placed into one of two study groups, the "Pre-Test Post-Test Study Group" (given the student questionnaire before and after BSE/TSE teaching) and the "Post-Test Only" study group (given the questionnaire once, after BSE/TSE teaching).

Table 1. Mean test scores for knowledge questions

	N	BSE Score (possible = 13) Mean ± SD	TSE Score (possible = 13) Mean ± SD	Total Score (possible = 13) Mean ± SD
Pre-test	532	5.4 ± 2.5	5.6 ± 2.7	11.0 ± 4.7
Post-Test	532	7.7 ± 2.5	8.0 ± 2.6	15.8 ± 4.2

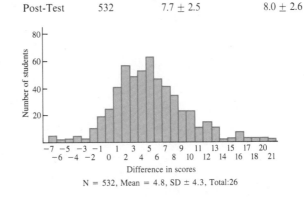

Fig. 1. Distribution of difference scores obtained by subtracting each student's pre-test knowledge score from his or her post-test knowledge score.

Difference in scores

N = 532, Mean = 4.8, SD ± 4.3, Total:26

This article by Stephen L. Luther, Stephen Sroka, Marlene Goormastic, and James E. Montie is reprinted with permission from *Health Education*, February/March 1985, pp. 40–43. *Health Education* is a publication of the American Alliance for Health, Physical Education, Recreation and Dance, 1900 Association Drive, Reston, VA 22091.

Chapter
Objectives

Until now we have considered distributions of discrete variables only. In this chapter we will examine one particular family of probability distributions of major importance whose domain is the set of all real numbers. These distributions are called the normal, the bell-shaped, or the Gaussian distributions. "Normal" is simply the traditional title of this particular type of distribution and is not a descriptive name meaning "typical." Although there are many other types of continuous distributions (rectangular, triangular, skewed, and so on), many variables have an approximately normal distribution. For example, several of the histograms drawn in Chapter 2 suggested a normal distribution. A mounded histogram that is approximately symmetric is an indication of such a distribution.

In addition to learning what a normal distribution is, we will consider (1) how probabilities are found, (2) how they are represented, and (3) how normal distributions are used. Although there are other distributions of continuous variables, the normal distributions are the most important.

6.1 NORMAL PROBABILITY DISTRIBUTIONS

normal distribution

As in all other probability distributions, there are formulas that give us information about the **normal distribution**. Previously, the probability function was the only function of interest. However, for the normal distribution there are two functions that define and describe it. Each formula uses the random variable as an independent variable. Formula (6-1) gives an ordinate (y value) for each point on the graph of the normal distribution for each given abscissa (x value). Formula (6-2) yields the probability associated with x when x is in the interval between the values a and b. Note that the random variable x is a continuous variable. For continuous variables we will discuss the probability that x has a value between the two extreme values of the interval, and we will depict the probability as an area under the graph of the probability distribution.

$$y = f(x) = \frac{1}{\sigma\sqrt{2\pi}} \cdot \exp\left[-\frac{1}{2}\left(\frac{x-\mu}{\sigma}\right)^2 \right] \qquad \text{for all real } x \qquad (6\text{-}1)$$

$$P(a \le x \le b) = \int_a^b f(x)\,dx \qquad (6\text{-}2)$$

If you don't understand these formulas, don't worry about it. We will not be using them in this book. However, a few comments about their meaning will be helpful if you ever use a probability table in any of the standard reference books. The directions for using these tables are often described with the aid of mathematical formulas such as (6-1) and (6-2).

Formula (6-1) and its associated table can be used to graph a normal distribution with a given mean and standard deviation. (The table associated with formula (6-1) is not included in this text, but it is available in many other textbooks and in all standard books of statistical tables.) In a more practical sense, we can draw

bell-shaped curve

a **normal**, or **bell-shaped, curve** and then label it to approximate scale, as indicated in Figure 6-1.

To talk intelligently about formula (6-2), we need to know what the symbols used in the formula represent. The $\int_a^b f(x)\,dx$ is called a "definite integral" and it

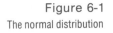

Figure 6-1

The normal distribution

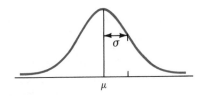

comes from calculus. The definite integral is a number that is the measure of the area under the curve of $f(x)$. This area is bounded by a on the left, b on the right, $f(x)$ at the top, and the x-axis at the bottom. The equation of the normal curve in formula (6-2) gives the measure of the shaded area in Figure 6-2.

Figure 6-2

Shaded area under the curve is $P(a \leq x \leq b)$

Since $P(a \leq x \leq b)$ [read "the probability that x is between a and b"] is given by $\int_a^b f(x)\,dx$, the measure of this probability is also the measure of the probability that a continuous random variable will assume a value within a specified interval. As a student of elementary statistics, you are not expected to have studied calculus, so we will say nothing more about the definite integral in this book.

The values calculated by using formulas (6-1) and (6-2) are also found in a table, with the probabilities expressed to four decimal places. But, before we show you how to read the table, we must point out that the table is expressed in a "standardized" form. It is standardized so that it is not necessary to have a different table for every different pair of values of the mean and standard deviation. For example, the area under the normal curve of a distribution with a mean of 15 and a standard deviation of 3 must be somehow related to the area under the curve of a normal distribution with a mean of 113 and a standard deviation of 38.5 Recall that the empirical rule concerns the percentage of a distribution within a certain number of standard deviations of the mean (see page 78).

percentage—proportion **Note** Percentage and probability are related concepts. **Percentage** is usually used when talking about a **proportion** of a population between certain values. **Probability** is usually used when talking about the **chance** that the next individual item will have a value between certain values.

The empirical rule is a fairly crude measuring device. With it we are able to find probabilities associated only with whole-number multiples of the standard deviation (within one standard deviation of the mean, and so on). We will often be interested in the probabilities associated with fractional parts of the standard deviation. For example, we might want to know the probability that x is within 1.85 standard deviations of the mean. Therefore, we must refine the empirical rule so that we can deal with more precise measurements. This refinement is discussed in the next section.

6.2 THE STANDARD NORMAL DISTRIBUTION

The key to working with the normal distribution is the **standard score** z. The measures of probabilities associated with the random variable x are determined by the relative position of x with respect to the mean and the standard deviation of the distribution. The empirical rule tells us that approximately 68 percent of the data are within one standard deviation of the mean. The value of the mean and the size of the standard deviation do not change this fact.

Recall that z was defined in Chapter 2 to be

$$\frac{x - (\text{mean of } x)}{(\text{standard deviation of } x)}$$

Therefore, we can write the formula for z as

$$z = \frac{x - \mu}{\sigma} \qquad (6\text{-}3)$$

[Note that if $x = \mu$, then $z = 0$.]

standard normal
distribution

The z score is known as a "standardized" variable because its units are standard deviations. The normal probability distribution associated with this standard score z is called the **standard normal distribution**. Table 5 in Appendix E lists the probabilities associated with the intervals from the mean for specific positive values of z. Other probabilities must be found by addition, subtraction, and so on, based on the concept of symmetry that exists in the normal distribution and the fact that the total area under the curve is 1.0.

area representation
for probability

Let's look at an illustration to see how we read Table 5. The z scores are in the margins: the left margin has the units digit and the tenths digit; the top has the hundredths digit. Let's look up a z score of 1.52. We find 0.4357, as shown in Table 6-1. Now, exactly what is this 0.4357? It is the measure of the area under the standard normal curve between $z = 0$ (which locates the mean) and $z = 1.52$ (a number 1.52 standard deviations larger than the mean). See Figure 6-3. **This area is also the measure of the probability associated with the same interval,** that is,

$$P(0 < z < 1.52) = 0.4357$$

Table 6-1
A portion of Table 5,
Appendix E

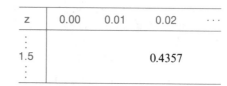

z	0.00	0.01	0.02	$\cdots$
⋮				
1.5			0.4357	
⋮				

Read this result as "the probability that a value picked at random falls between the mean and 1.52 standard deviations above the mean is 0.4357," or, equivalently, "the probability that a z score picked at random will fall between 0 and 1.52 is 0.4357."

Recall that one of the basic properties of probability is that the sum of all probabilities is exactly 1.0. Since the area under the normal curve represents the measure of probability, the total area under the bell-shaped curve is exactly 1 unit.

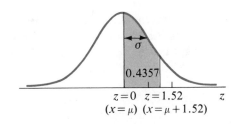

$z=0$ $z=1.52$ z
$(x=\mu)$ $(x=\mu+1.52)$

This distribution is symmetric with respect to the vertical line drawn through $z = 0$, which cuts the area exactly in half at the mean. Can you verify this fact by inspecting formula (6-1)? That is, the area under the curve to the right of the mean is exactly one-half unit, 0.5, and the area to the left is also one-half unit, 0.5. Areas (probabilities) not given directly in the table can be found by relying on these facts.

Now let's look at some illustrations.

Illustration 6-1   Find the area under the normal curve to the right of $z = 1.52$.

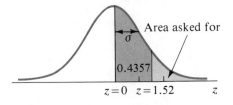

Solution   The area to the right of the mean (all the shading in the figure) is exactly 0.5000. The question asks for the shaded area that is not included in the 0.4357. Therefore, subtract 0.4357 from 0.5000.

$$0.5000 - 0.4357 = \mathbf{0.0643}$$ ■

Suggestion   As we have done here, always draw and label a sketch. It is most helpful.

Illustration 6-2   Find the area to the left of $z = 1.52$.

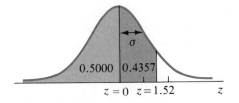

Solution   The total shaded area is made up of the 0.4357 found in the table and the 0.5000 that is to the left of the mean. Therefore, add 0.4357 to 0.5000.

$$0.4357 + 0.5000 = \mathbf{0.9357}$$ ■

Note   The addition and subtraction done in Illustrations 6-1 and 6-2 are correct because the "areas" represent mutually exclusive events (discussed in Section 4-5).

The symmetry of the normal distribution is a key factor in determining probabilities associated with values below (to the left of) the mean. The area between the mean and $z = -1.52$ is exactly the same as the area between the mean and $z = +1.52$. This fact allows us to find values related to the left side of the distribution.

Illustration 6-3   The area between the mean ($z = 0$) and $z = -2.1$ is the same as the area between $z = 0$ and $z = +2.1$.

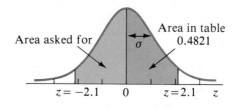

Thus we have

$$P(-2.1 < z < 0) = \mathbf{0.4821}$$ ■

Illustration 6-4   The area to the left of $z = -1.35$ is found by subtracting 0.4115 from 0.5000.

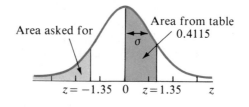

Therefore, we obtain

$$P(z < -1.35) = \mathbf{0.0885}$$ ■

Illustration 6-5   The area between $z = -1.5$ and $z = 2.1$ is found by adding the two areas together.

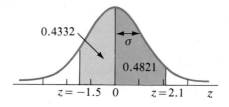

Therefore, we obtain

$$P(-1.5 < z < 0) + P(0 < z < 2.1) = 0.4332 + 0.4821 = \mathbf{0.9153}$$ ■

Illustration 6-6   The area between $z = 0.7$ and $z = 2.1$ is found by subtracting. The area between $z = 0$ and $z = 2.1$ contains all the area between $z = 0$ and $z = 0.7$. The area between $z = 0$ and $z = 0.7$ is therefore subtracted from the area between $z = 0$ and $z = 2.1$

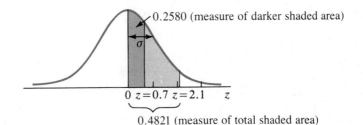

0.2580 (measure of darker shaded area)

$0 \quad z=0.7 \quad z=2.1 \quad z$

0.4821 (measure of total shaded area)

Thus we have

$$P(0.7 < z < 2.1) = 0.4821 - 0.2580 = \mathbf{0.2241}$$ ∎

The normal distribution table can also be used to determine a z score if we are given an area. The next illustration considers this idea.

Illustration 6-7   What is the z score associated with the 75th percentile? (Assume the distribution is normal.) See Figure 6-4.

Figure 6-4
P_{75} and its associated z score

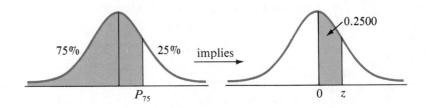

75%   25%   implies   0.2500

P_{75} $0 \quad z$

Solution   To find this z score, look in Table 5, Appendix E, and find the "area" entry that is closest to 0.2500; this area entry is 0.2486. Now read the z score that corresponds to this area. From the table the z score is found to

z	$\cdots$	0.07	0.08	$\cdots$
$\vdots$				
0.6		0.2486	0.2517	
$\vdots$				

be $z = \mathbf{0.67}$. This says that the 75th percentile in a normal distribution is 0.67 (approximately $\frac{2}{3}$) standard deviation above the mean. ∎

Illustration 6-8   What z scores bound the middle 95 percent of a normal distribution? See Figure 6-5.

Solution   The 95 percent is split into two equal parts by the mean; 0.4750 is the area (percentage) between $z = 0$, the mean, and the z score at the right boundary.

Figure 6-5
Middle 95 percent of
distribution and its
associated z score

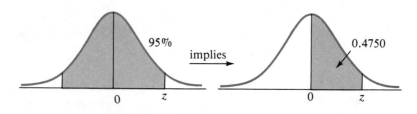

Since we have the area, we look for the entry in Table 5 closest to 0.4750 (it happens to be exactly 0.4750) and read the z score in the margin. We obtain $z = 1.96$. Therefore, $z = -1.96$ **and** $z = 1.96$ bound the middle 95 percent of a normal distribution.

z	$\cdots$	0.06	$\cdots$
$\vdots$			
1.9		0.4750	
$\vdots$			

EXERCISES

6.1 Describe the distribution of the standard normal score z.

6.2 Find the area under the normal curve that lies between the following pairs of z values.

    a. $z = 0$ to $z = 1.30$ **b.** $z = 0$ to $z = 1.28$

    c. $z = 0$ to $z = -3.20$ **d.** $z = 0$ to $z = -1.98$

6.3 Find the probability that a piece of data picked at random from a normal population will have a standard score (z) that lies between the following pairs of z values.

    a. $z = 0$ to $z = 2.10$ **b.** $z = 0$ to $z = 2.57$

    c. $z = 0$ to $z = -1.20$ **d.** $z = 0$ to $z = -1.57$

6.4 Find the area under the standard normal curve that corresponds to the following z values.

    a. between 0 and 1.55   **b.** to the right of 1.55

    c. to the left of 1.55   **d.** between -1.55 and 1.55

6.5 Find the area under the normal curve that lies between the following pairs of z values.

    a. $z = -1.20$ to $z = 1.22$ **b.** $z = -1.75$ to $z = 1.54$

    c. $z = -1.30$ to $z = 2.58$ **d.** $z = -3.5$ to $z = -0.35$

6.6 Find the probability that a piece of data picked at random from a normal population will have a standard score (z) that lies between the following pairs of z values.

    a. $z = -2.75$ to $z = 1.38$ **b.** $z = 0.67$ to $z = 2.95$

    c. $z = -2.95$ to $z = -1.18$

6.7 Find the following areas under the normal curve.

    a. to the right of $z = 0.00$ **b.** to the right of $z = 1.05$

    c. to the right of $z = -2.30$ **d.** to the left of $z = 1.60$

    e. to the left of $z = -1.60$

6.8 Find the probability that a piece of data picked at random from a normally distributed population will have a standard score that is

    a. less than 3.00   **b.** greater than -1.55

    c. less than -0.75 **d.** less than 1.25

    e. greater than -1.25

6.9 Find the following.

    a. $P(0.00 < z < 2.35)$ **b.** $P(-2.10 < z < 2.34)$

    c. $P(z > 0.13)$ **d.** $P(z < 1.48)$

6.10 Find the z score for the standard normal distribution shown on each of the following diagrams.

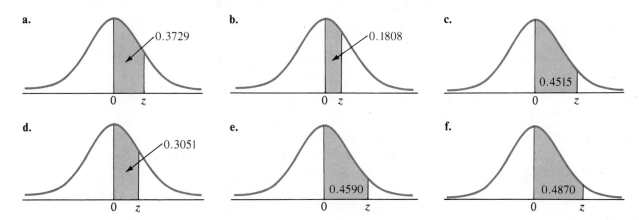

6.11 Find the z score for the standard normal distribution shown in each of the following diagrams.

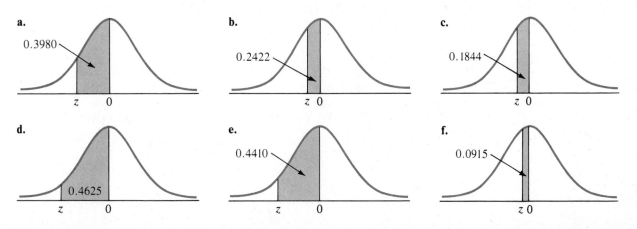

6.12 Find the standard score z shown on each of the following diagrams.

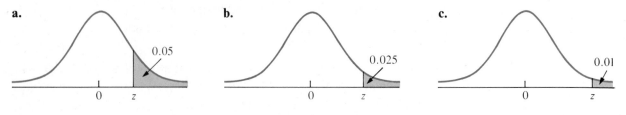

a.   0.05

b.   0.025

c.   0.01

6.13 Find the standard score z shown on each of the following diagrams.

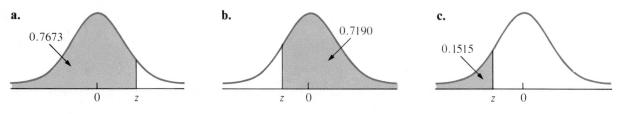

a.   0.7673

b.   0.7190

c.   0.1515

6.14 Find a value of z such that 40 percent of the distribution lies between it and the mean. (There are two possible answers.)

6.15 Find the standard score z such that

    a. 80 percent of the distribution is below (to the left of) this value.

    b. the area to the right of this value is 0.15.

6.16 Find the two z scores that bound the middle 50 percent of a normal distribution.

6.17 Find the two standard scores z such that

    a. the middle 90 percent of a normal distribution is bounded by them.

    b. the middle 98 percent of a normal distribution is bounded by them.

6.18 Assuming a normal distribution, what is the z score associated with the 90th percentile? The 95th percentile? The 99th percentile?

6.3 APPLICATIONS OF NORMAL DISTRIBUTIONS

The probabilities associated with any normal distribution can be found by applying the techniques discussed in Section 6-2. First, however, we must "standardize" the given information. When dealing with a normal distribution, we need to know its mean μ and its standard deviation σ. Once these values are known, any value of the random variable x can be easily converted to the standard score z by using formula (6-3):

$$z = \frac{x - \mu}{\sigma}$$

Illustration 6-9   Consider the intelligence quotient (IQ) scores for people. IQs are normally distributed with a mean of 100 and a standard deviation of 10. If a person is picked at random, what is the probability that his or her IQ is between 100 and 115; that is, what is $P(100 < x < 115)$?

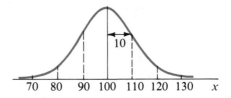

Solution $P(100 < x < 115)$ is represented by the shaded area in the following figure.

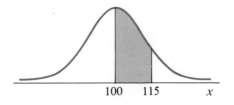

The variable x must be standardized by using formula (6-3). The z values are shown on the next figure.

$$z = \frac{x - \mu}{\sigma}$$

When $x = 100$: $z = \dfrac{100 - 100}{10} = \mathbf{0.0}$

When $x = 115$: $z = \dfrac{115 - 100}{10} = \mathbf{1.5}$

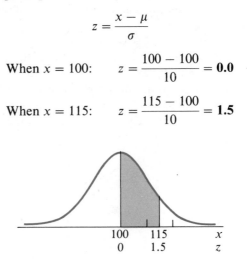

Therefore

$$P(100 < x < 115) = P(0.0 < z < 1.5) = \mathbf{0.4332}$$

(The value 0.4332 is found by using Table 5, Appendix E.) Thus the probability is 0.4332 that a person picked at random has an IQ between 100 and 115.   ∎

Illustration 6-10   Find the probability that a person selected at random will have an IQ greater than 95.

Solution

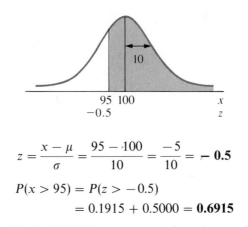

$$z = \frac{x - \mu}{\sigma} = \frac{95 - 100}{10} = \frac{-5}{10} = -\,0.5$$

$$P(x > 95) = P(z > -0.5)$$
$$= 0.1915 + 0.5000 = \mathbf{0.6915}$$

Thus the probability is 0.6915 that a person selected at random will have an IQ greater than 95.   ■

The normal table can be used to answer many kinds of questions that involve a normal distribution. Many times a problem will call for the location of a "cutoff point," that is, a particular value of x such that there is exactly a certain percentage in a specified area. The following illustrations concern some of these problems.

Illustration 6-11   In a large class, suppose that your instructor tells you that you need to obtain a grade in the top 10 percent of your class to get an A on a particular exam. From past experience she is able to estimate that the mean and standard deviation on this exam will be 72 and 13, respectively. What will be the minimum grade needed to obtain an A? (Assume that the grades will be approximately normally distributed.)

Solution

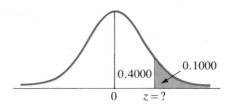

$$10\% = 0.1000 = 0.5000 - 0.4000$$

Look in Table 5 to find the value of z that is closest to 0.4000; it is $z = 1.28$. Thus,

$$P(z > 1.28) = 0.10 \qquad \text{(information given)}$$

Now find the x value that corresponds to $z = 1.28$ by using formula (6-3):

$$z = \frac{x - \mu}{\sigma} \quad \text{or} \quad 1.28 = \frac{x - 72}{13}$$

$$x - 72 = (13)(1.28)$$
$$x = 72 + (13)(1.28) = 72 + 16.64$$
$$= 88.64, \text{ or } \mathbf{89}$$

Thus if you receive an 89 or higher, you can expect to be in the top 10 percent (which means an A). ■

Illustration 6-12   Find the 33d percentile for IQ scores ($\mu = 100$ and $\sigma = 10$ from Illustration 6-9).

Solution

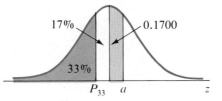

$$P(0 < z < a) = 0.17$$
$$a = 0.44 \quad \text{(cutoff value of } z \text{ from Table 5)}$$
$$\text{33d percentile of } z = -0.44 \quad \text{(below mean)}$$

Now we convert the 33d percentile of the z scores, -0.44, to an x score:

$$z = \frac{x - \mu}{\sigma} \quad \text{[formula (6-3)]}$$

$$-0.44 = \frac{x - 100}{10}$$

$$-4.4 = x - 100$$
$$100 - 4.4 = x$$
$$x = \mathbf{95.6}$$

Thus, 95.6 is the 33d percentile for IQ scores. ■

Illustration 6-13 concerns a situation where you are asked to find the mean μ when given the related information.

Illustration 6-13   The incomes of junior executives in a large corporation are normally distributed with a standard deviation of $1,200. A cutback is pending, at which time those who earn less than $18,000 will be discharged. If such a cut represents 10 percent of the junior executives, what is the current mean salary of the group of junior executives?

Solution   If 10 percent of the salaries are below $18,000, then 40 percent (or 0.4000) are between $18,000 and the mean μ. Table 5 indicates that $z = -1.28$ is the standard score that occurs at $x = \$18,000$. Using formula (6-3) we can find the value of μ:

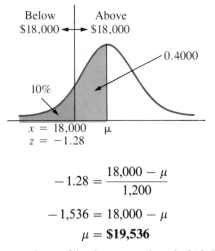

$$-1.28 = \frac{18,000 - \mu}{1,200}$$

$$-1,536 = 18,000 - \mu$$

$$\mu = \mathbf{\$19,536}$$

That is, the current mean salary of junior executives is $19,536.   ■

Referring again to the IQ scores, what is the probability that a person picked at random has an IQ of 125, $P(x = 125)$? (IQ scores are normally distributed with a mean of 100 and a standard deviation of 10.) This situation has two interpretations: (1) theoretical and (2) practical. Let's look at the theoretical interpretation first. Recall that the probability associated with a continuous random variable is represented by the area under the curve. That is, $P(a \leq x \leq b)$ is equal to the area between a and b under the curve. $P(x = 125)$—that is, x is exactly 125—is then $P(125 \leq x \leq 125)$, or the area of the vertical line segment at $x = 125$. This area is zero. However, this is not the *practical* meaning of $x = 125$. It generally means 125 to the nearest integer value. Thus $P(x = 125)$ would most likely be interpreted as

$$P(124.5 < x < 125.5)$$

The interval from 124.5 to 125.5 under the curve has a measurable area and is then nonzero. In situations of this nature, you must be sure of the meaning being used.

EXERCISES

6.19 Given that x is a normally distribution random variable with a mean of 60 and a standard deviation of 10, find the following probabilities.

     a. $P(x > 60)$ **b.** $P(60 < x < 72)$

c. $P(57 < x < 83)$ **d.** $P(65 < x < 82)$

e. $P(38 < x < 78)$ **f.** $P(x < 38)$

6.20 Let h be a normally distributed random variable with a mean of 26.7 and a standard deviation of 3.4. Find the probability that an individual value of h, selected at random, will fall in the following intervals.

a. between 26.7 and 31.7   **b.** between 28.0 and 30.0

c. between 20.0 and 24.0   **d.** between 20.0 and 30.0

6.21 For a particular age group of adult males the distribution of cholesterol readings, in mg/dl, are normally distributed with a mean of 210 and a standard deviation of 15.

a. What percent of this population would have readings exceeding 250?

b. What percent would have readings less than 150?

6.22 Computers are shut down for certain periods of time for routine maintenance, installation of new hardware, and so on. The down times for a particular computer are normally distributed with a mean of 1.5 hours and a standard deviation of 0.4 hours.

a. What percent of the down times exceed 3 hours?

b. What percent are between 1 and 2 hours?

6.23 Final averages are typically approximately normally distributed with a mean of 72 and a standard deviation of 12.5. Your professor says that the top 8 percent of the class will receive A; the next 20 percent, B; the next 42 percent, C; the next 18 percent, D; and the bottom 12 percent, F.

a. What average must you exceed to obtain an A?

b. What average must you exceed to receive a grade better than a C?

c. What average must you obtain to pass the course? (You'll need a D or better.)

6.24 The length of useful life of a fluorescent tube used for indoor gardening is normally distributed. The useful life has a mean of 600 hours and a standard deviation of 40 hours. Determine the probability that

a. a tube chosen at random will last between 620 and 680 hours.

b. such a tube will last more than 740 hours.

6.25 At Pacific Freight Lines, bonuses are given to billing clerks when they complete 300 or more freight bills during an 8-hour day. The number of bills completed per clerk per 8-hour day is approximately normally distributed with a mean of 270 and a standard deviation of 16. What proportion of the time should a randomly selected billing clerk expect to receive a bonus?

6.26 The waiting time x at a certain bank is approximately normally distributed with a mean of 3.7 minutes and a standard deviation of 1.4 minutes.

a. Find the probability that a randomly selected customer has to wait less than 2.0 minutes.

b. Find the probability that a randomly selected customer has to wait more than 6 minutes.

c. Find the value of the 75th percentile for *x*.

6.27 A brewery filling machine is adjusted to fill quart bottles with a mean of 32 ounces of ale, and a variance of 0.003. Periodically, a bottle is checked and the amount of ale is noted. Assuming that the amount of fill is normally distributed, what is the probability that the next randomly checked bottle contains more than 32.02 ounces?

6.28 The weights of ripe watermelons grown at Mr. Smith's farm are normally distributed with a standard deviation of 2.8 pounds. Find the mean weight of Mr. Smith's ripe watermelons if only 3 percent weigh less than 15 pounds.

6.29 A machine fills containers with a mean weight per container of 16.0 ounces. If no more than 5 percent of the containers are to weigh less than 15.8 ounces, what must the standard deviation of the weights equal? (Assume normality.)

6.30 A radar unit is used to measure the speed of automobiles on an expressway during rush-hour traffic. The speeds of individual automobiles are normally distributed with a mean of 62 miles per hour.

a. Find the standard deviation of all speeds if 3 percent of the automobiles travel faster than 72 miles per hour.

b. Using the standard deviation found in (a), find the percentage of these cars that are traveling less than 55 miles per hour.

c. Using the standard deviation found in (a), find the 95th percentile for the variable "speed."

6.31 The MINITAB statistical package is often used to generate simulated samples. Suppose, for example, we want to simulate a sample of adult male heights and that the distribution of these heights is normally distributed with a mean of 68 inches and standard deviation of 2.5 inches. A simulated sample (using MINITAB) follows.

```
MTB > NOTE THIS MINITAB RUN SIMULATES A SAMPLE FROM
MTB > NOTE A NORMAL DISTRIBUTION WITH MEAN = 68
MTB > NOTE AND STANDARD DEVIATION = 2.5
MTB > NRANDOM 25 OBSERVATIONS MU = 68,SIGMA = 2.5,PUT IN C1
      25 NORMAL OBS. WITH MU =      68.000 AND SIGMA =      2.5000
MTB > PRINT C1
C1
   64.9257  66.8683  68.5982  66.3566  69.6508  66.1134  67.3229
   65.1638  64.1846  63.9889  66.1758  65.5503  67.4795  69.7201
   66.9653  70.7060  70.1076  72.3623  66.7146  67.7243  71.1004
   66.0057  70.4767  64.7314  70.1357
MTB > STOP
```

Find the mean and standard deviation of this simulated sample. (Round the numbers to the nearest whole integer first.)

6.4 NOTATION

When working with the standard score z, it is often helpful and necessary to identify z with an area under the normal curve. One standard procedure is to use the area under the curve and to right of the z. We will write this area within parentheses following the z.

Illustration 6-14 $z(0.05)$ is the value of z such that exactly 0.05 of the area under the curve lies to its right, as shown in Figure 6-6.

Figure 6-6
Area associated with z (0.05)

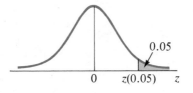

Now let's find the value of $z(0.05)$. We must convert this information into a value that can be read from Table 5; see the areas shown in Figure 6-7.

Figure 6-7
Finding the value of z (0.05)

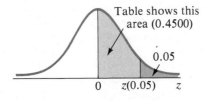

Now we look in Table 5 in Appendix E and find an area as close as possible to 0.4500.

z	$\cdots$	0.04	0.05	$\cdots$
$\vdots$				
1.6	$\cdots$	0.4495	0.4505	
$\vdots$				

Therefore, $z(0.05) = $ **1.65** [*Note*: We will use the z that corresponds to the area closest in value. If the value happens to be exactly halfway between the table entries, always round up to the larger value of z.] ■

$z(0.60)$ is that value of z such that 0.60 of the area lies to its right, as shown in Figure 6-8.

Figure 6-8
Area associated with z (0.60)

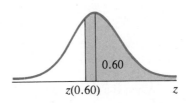

Illustration 6-15   Find the value of $z(0.60)$.

Solution   The value 0.60 is related to Table 5 by use of the area 0.1000, as shown in the following diagram.

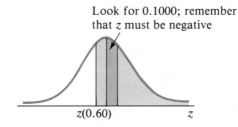

Look for 0.1000; remember that z must be negative

$z(0.60)$ z

The closest values in Table 5 are 0.0987 and 0.1026.

z	$\cdots$	0.05	0.06	$\cdots$
0.2		0.0987	0.1026	

Therefore, $z(0.60)$ is related to 0.25. Since $z(0.60)$ is below the mean, we conclude that $z(0.60) = -0.25$.   ■

In later chapters we will use this notation on a regular basis. The two values of z that will be used regularly come from one of the following situations: (1) the z score such that there is a specified area in one tail of the normal distribution, or (2) the z scores that bound a specified middle proportion of the normal distribution.

Illustration 6-14 showed a commonly used one-tail situation; $z(0.05) = 1.65$ is located so that 0.05 of the area under the normal distribution curve is in the tail to the right.

Illustration 6-16   Find $z(0.95)$.

Solution $z(0.95)$ is located on the left-hand side of the normal distribution since the area to the right is 0.95. The area in the tail to the left then contains the other 0.05, as shown in Figure 6-9. Because of the symmetrical nature of the normal distribution, $z(0.95)$ is $-z(0.05)$; that is, $z(0.05)$ with its sign changed. Thus $z(0.95) = -z(0.05) = -1.65$.   ■

Figure 6-9
Area associated with $z(0.95)$

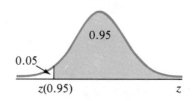

0.95

0.05

$z(0.95)$ z

When the middle proportion of a normal distribution is specified, we can still use the "area to the right" notation to identify the specific z score involved.

Illustration 6-17   Find the z scores that bound the middle 0.95 of the normal distribution.

Solution   Given 0.95 as the area in the middle (Figure 6-10), the two tails must contain a total of 0.05. Therefore each tail contains 0.025, as shown in Figure 6-11. In order to find $z(0.025)$ in Table 5, we must determine the area between the mean and $z(0.025)$. It is 0.4750, as shown in Figure 6-12.

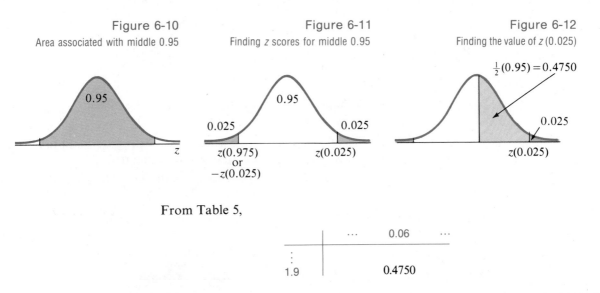

Figure 6-10
Area associated with middle 0.95

Figure 6-11
Finding z scores for middle 0.95

Figure 6-12
Finding the value of z (0.025)

From Table 5,

	⋯	0.06	⋯
⋮			
1.9		0.4750	

Therefore, $z(0.025) = 1.96$ and $z(0.975) = -z(0.025) = -1.96$. The middle 0.95 of the normal distribution is bounded by **−1.96 and 1.96**.   ■

EXERCISES

6.32 Using the $z(\alpha)$ notation (identify the value of α used within the parentheses), name each of the standard normal variable z's shown in the following diagrams.

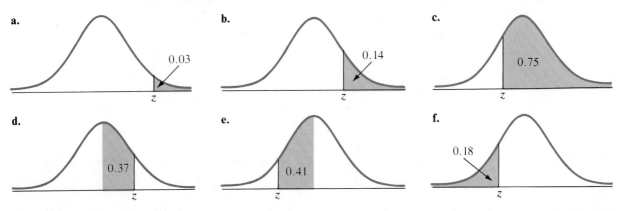

a.   0.03

b.   0.14

c.   0.75

d.   0.37

e.   0.41

f.   0.18

6.33 We are often interested in finding the value of z that bounds a given area in the right-hand tail of the normal distribution, as shown in the accompanying figure. The notation $z(\alpha)$ represents the value of z such that $P(z > z(\alpha)) = \alpha$. Find the following.

    a. $z(0.025)$ **b.** $z(0.05)$ **c.** $z(0.01)$

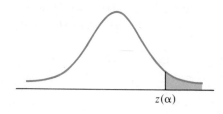

6.34 Use Table 5, Appendix E to find the following values of z.

    a. $z(0.05)$ **b.** $z(0.01)$ **c.** $z(0.025)$ **d.** $z(0.975)$ **e.** $z(0.98)$

6.35 Complete the following charts of z scores. The area A given in the tables is the area to the right under the normal distribution in the figures.

    a. z scores associated with the right-hand tail: Given the area A, find $z(A)$.

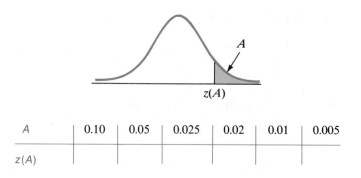

A	0.10	0.05	0.025	0.02	0.01	0.005
$z(A)$						

    b. z scores associated with the left-hand tail: Given the area A, find $z(A)$.

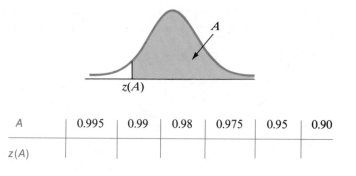

A	0.995	0.99	0.98	0.975	0.95	0.90
$z(A)$						

6.36 a. Find the area under the normal curve for z between $z(0.95)$ and $z(0.025)$.

    b. Find $z(0.025) - z(0.95)$.

6.5   NORMAL APPROXIMATION OF THE BINOMIAL

binomial probability

In Chapter 5 we introduced the binomial distribution. Recall that the binomial distribution is a probability distribution of the discrete random variable x, the number of successes observed in n repeated independent trials. We will now see how **binomial probabilities**, that is, probabilities associated with a binomial distribution, can be reasonably estimated by using the normal probability distribution.

Let's look first at a few specific binomial distributions. Figures 6-13a, 6-13b, and 6-13c show the probabilities of x for 0 to n for three situations: $n = 4$, $n = 8$, and $n = 24$. For each of these distributions, the probability of success for one trial is 0.5. Notice that as n becomes larger, the distribution appears more and more like the normal distribution.

Figure 6-13

Binomial distributions

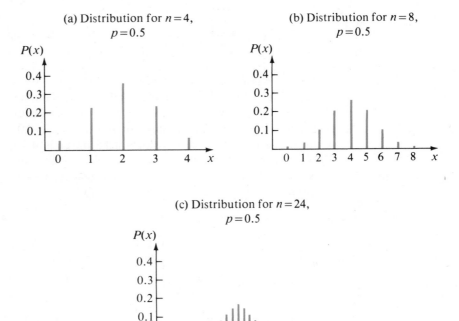

(a) Distribution for $n=4$, $p=0.5$

(b) Distribution for $n=8$, $p=0.5$

(c) Distribution for $n=24$, $p=0.5$

To make the desired approximation, we need to take into account one major difference between the binomial and the normal probability distributions. The binomial random variable is discrete, whereas the normal random variable is continuous. Recall that in Chapter 5 it was demonstrated that the probability assigned to a particular value of x should be shown on a diagram by means of a straight-line segment whose length represents the probability (as in Figure 6-13). It was suggested, however, that we can also use a histogram in which the area of each bar is equal to the probability of x.

Let's look at the distribution of the binomial variable x, where $n = 14$ and $p = 0.5$. The probabilities for each x value can be obtained from Table 4 in Appendix E. This distribution of x is shown in Figure 6-14. In histogram form we see the very same distribution in Figure 6-15.

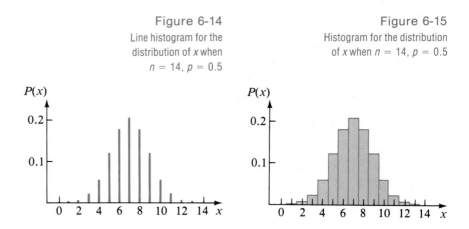

Figure 6-14
Line histogram for the distribution of x when $n = 14, p = 0.5$

Figure 6-15
Histogram for the distribution of x when $n = 14, p = 0.5$

Let's examine $P(x = 4)$ for $n = 14$ and $p = 0.5$ to study the approximation technique. $P(x = 4)$ is equal to 0.061 (*see* Table 4 of Appendix E), the area of the bar above $x = 4$ in Figure 6-16. Area is the product of width and height. In this case the height is 0.061 and the width is 1.0; thus the area is 0.061. Let's take a closer look at the width. For $x = 4$, the bar starts at 3.5 and ends at 4.5, so we are looking at an area bounded by $x = 3.5$ and $x = 4.5$. The addition and subtraction of 0.5 to the x value is commonly called the **continuity correction factor**. It is our method of converting a discrete variable into a continuous variable.

continuity correction factor

Figure 6-16
Area of bar above $x = 4$ is 0.061 when $n = 14$, $p = 0.5$

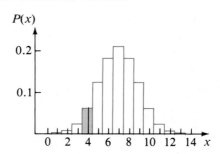

Now let's look at the normal distribution related to this situation. We will first need a normal distribution with a mean and a standard deviation equal to those of the binomial distribution we are discussing. Formulas (5-7) and (5-8) give us these values.

$$\mu = np = (14)(0.5) = \mathbf{7.0}$$
$$\sigma = \sqrt{npq} = \sqrt{(14)(0.5)(0.5)}$$
$$= \sqrt{3.5} = \mathbf{1.87}$$

The probability that $x = 4$ is approximated by the area under the normal curve between $x = 3.5$ and $x = 4.5$, as shown in Figure 6-17. Figure 6-18 shows the entire distribution of the binomial variable x with a normal distribution of the same mean and standard deviation superimposed. Notice that the bars and the interval areas under the curve cover nearly the same area.

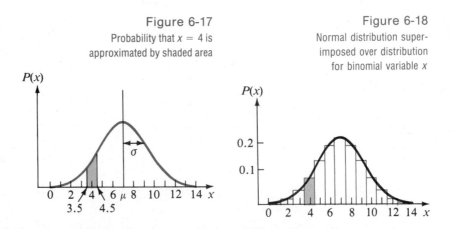

Figure 6-17
Probability that $x = 4$ is approximated by shaded area

Figure 6-18
Normal distribution super-imposed over distribution for binomial variable x

The probability that x is between 3.5 and 4.5 under this normal curve is found by using Table 5 and the methods outlined in Section 6.3.

$$P(3.5 < x < 4.5) = P\left(\frac{3.5 - 7.0}{1.87} < z < \frac{4.5 - 7.0}{1.87}\right)$$

$$= P(-1.87 < z < -1.34)$$

$$= 0.4693 - 0.4099 = \mathbf{0.0594}$$

Since the binomial probability of 0.061 and the normal probability of 0.0594 are reasonably close in value, the normal probability distribution seems to be a reasonable approximation of the binomial distribution.

The normal approximation of the binomial distribution is also useful for values of p that are not close to 0.5. The binomial probability distributions shown in Figures 6-19 and 6-20 suggest that binomial probabilities can be approximated by using the normal distribution. Notice that as n increases in size, the binomial distribution

Figure 6-19
Binomial distributions

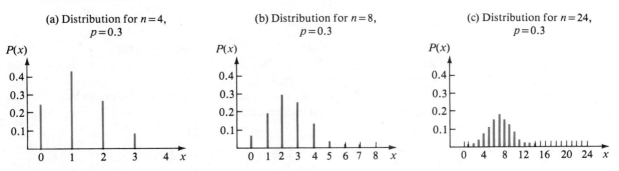

(a) Distribution for $n=4$, $p=0.3$

(b) Distribution for $n=8$, $p=0.3$

(c) Distribution for $n=24$, $p=0.3$

begins to look like the normal distribution. As the value of p moves away from 0.5, a larger n will be needed in order for the normal approximation to be reasonable. The following "rule of thumb" is generally used as a guideline.

Figure 6-20
Binomial distributions

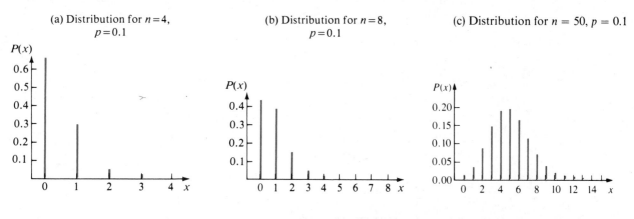

(a) Distribution for $n=4$, $p=0.1$

(b) Distribution for $n=8$, $p=0.1$

(c) Distribution for $n = 50$, $p = 0.1$

Rule

The normal distribution provides a reasonable approximation to a binomial probability distribution whenever the values of np and $n(1 - p)$ both equal or exceed 5.

By now you may be thinking, "So what? I will just use the binomial table and find the probabilities directly and avoid all the extra work." But consider for a moment the situation presented in Illustration 6-18.

Illustration 6-18   An unnoticed mechanical failure has caused $\frac{1}{3}$ of a machine shop's production of 5000 rifle firing pins to be defective. What is the probability that an inspector will find no more than 3 defective firing pins in a random sample of 25?

Solution   In this illustration of a binomial experiment, x is the number of defectives found in the sample, $n = 25$, and $p = P(\text{defective}) = \frac{1}{3}$. To answer the question by using the binomial distribution, we will need to use the binomial probability function [formula (5-6)]:

$$P(x) = \binom{25}{x} \cdot \left(\frac{1}{3}\right)^x \cdot \left(\frac{2}{3}\right)^{25-x} \quad \text{for} \quad x = 0, 1, 2, \ldots, 25$$

We must calculate the values for $P(0)$, $P(1)$, $P(2)$, and $P(3)$, since they do not appear in Table 4. This is a very tedious job because of the size of the exponent. In situations such as this, we can use the normal approximation method.

Now let's find $P(x \leq 3)$ by using the normal approximation method. We first need to find the mean and standard deviation of x [formulas (5-7) and (5-8)]:

$$\mu = np = (25)(\tfrac{1}{3}) = \textbf{8.333}$$
$$\sigma = \sqrt{npq} = \sqrt{(25(\tfrac{1}{3})(\tfrac{2}{3})} = \textbf{2.357}$$

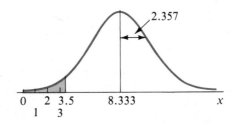

These values are shown in the figure above. The measure of the shaded area $(x < 3.5)$ represents the probability of $x = 0, 1, 2,$ or 3. Remember that $x = 3$, the discrete binominal variable, covers the continuous interval from 2.5 to 3.5.

$$P(x \text{ is no more than } 3) = P(x \leq 3) \quad \text{[for discrete variable } x\text{]}$$
$$= P(x < 3.5) \quad \text{[using a continuous variable } x\text{]}$$

$$P(x < 3.5) = P\left(z < \frac{3.5 - 8.333}{2.357}\right) = P(z < -2.05)$$

$$= 0.5000 - 0.4798 = \mathbf{0.0202}$$

Thus P(no more than 3 defectives) is approximately 0.02. ∎

EXERCISES

6.37 In which of the following binomial distributions does the normal distribution provide a reasonable approximation?

    a. $n = 10, p = 0.3$ **b.** $n = 100, p = 0.005$

    c. $n = 500, p = 0.1$ **d.** $n = 50, p = 0.2$

6.38 In order to see what happens when the normal approximation is improperly used, consider the binomial distribution with $n = 15$ and $p = 0.05$. Since $np = 0.75$, the rule of thumb $(np > 5$ and $nq > 5)$ is not satisfied. Using the binomial tables, find the probability of one or fewer successes and compare this with the normal approximation.

6.39 Find the normal approximation for the binomial probability $P(x = 6)$, where $n = 12$ and $p = 0.6$. Compare this to the value of $P(x = 6)$ obtained from Table 4.

6.40 Find the normal approximation for the binomial probability $P(x = 4, 5)$, where $n = 14$ and $p = 0.5$. Compare this to the value of $P(x = 4, 5)$ obtained from Table 4.

6.41 Find the normal approximation for the binomial probability $P(x \leq 8)$, where $n = 14$ and $p = 0.4$. Compare this to the value of $P(x \leq 8)$ obtained from Table 4.

 6.42 A production run of tee shirts is expected to produce seconds 5 percent of the time. If you randomly select 100 tee shirts from a large run,

    a. what is the mean and the standard deviation of the binomial distribution when samples of 100 are taken?

b. what is the probability that your sample will contain no seconds?

c. what is the probability that your sample will contain no more than two seconds?

6.43 A drug manufacturer states that only 5 percent of the patients using a particular drug will experience side effects. Doctors at a large university hospital use the drug in treating 250 patients. What is the probability that 15 or fewer of the 250 patients experience side effects?

6.44 A company asserts that 80 percent of the customers who purchase its special lawn mower will have no repairs during the first two years of ownership. Your personal study has shown that only 70 of the 100 in your sample lasted the two years without repair expenses. What is the probability of your sample outcome or less, if the actual expenses-free percentage is 80 percent?

6.45 Surveys have shown that 40 percent of the population of the United States have used Crest toothpaste at one time or another. A researcher, seeking to substantiate that 40 percent is correct, took a properly conducted random sample and found that 220 of the 600 individuals contacted had used Crest. If the 40 percent is really correct, what is the probability that a sample of 600 would have fewer than 221 individuals who had used Crest?

6.46 If 30 percent of all students entering a certain university drop out during or at the end of their first year, what is the probability that more than 600 of this year's entering class of 1800 will drop out during or at the end of their first year?

6.47 It is believed that the student body is equally split on a "new pub program" proposal. Assuming this to be the case, what is the probability that the Student Senate's straw poll of 100 student opinions shows at least 60 percent favoring the new proposal?

6.48 Consider a binomial distribution with $n = 500$ and $p = 0.2$.

a. Set up, but do not evaluate, the probability expression for 90 or fewer successes in the 500 trials.

b. Find the normal approximation to (a).

In Retrospect

We now know what a normal distribution is, how to use it, and how it can help us. Let's again consider the news article at the beginning of this chapter, which illustrated a normally distributed random variable. The random variable is the difference between the pre-test and post-test scores and its distribution is mounded and symmetric about the mean value. We have seen many illustrations of normal distributions.

In the next chapter we will examine sampling distributions and learn how to apply the normal distribution to additional applications.

Chapter Exercises

6.49 According to Chebyshev's theorem, there is at least how much area under the standard normal distribution between $z = -2$ and $z = +2$? What is the actual area under the standard normal distribution between $z = -2$ and $z = +2$?

6.50 The news article at the beginning of this chapter includes a histogram of the variable, the difference between pre-test and post-test scores. This difference appears to be approximately normally distributed.

    a. Express this distribution as a frequency distribution.

    b. Using the mean of 4.8 and the standard deviation of 4.3, as reported in the article, find the proportion of this distribution that lies within one standard deviation of the mean. Within two standard deviations of the mean. Within three.

    c. Compare the results found in (b) to the proportions cited by the empirical rule.

    d. Determine what proportion of this distribution lies between the mean and $1\frac{1}{2}$ standard deviations above the mean. What proportion of a normal distribution lies between $z = 0$ and $z = 1.5$?

    e. Do you think it is reasonable to conclude that this distribution is approximately normal?

6.51 The middle 60 percent of a normally distributed population lies between what two standard scores?

6.52 Find the standard score z such that the area above the mean and below z under the normal curve is

    a. 0.3962   **b.** 0.4846   **c.** 0.3712

6.53 Find the standard score z such that the area below the mean and above z under the normal curve is

    a. 0.3212   **b.** 0.4788   **c.** 0.2700

6.54 Find the standard score of a normally distributed variable such that 42 percent of the distribution falls between the mean and this particular value.

6.55 Given that z is the standard normal variable, find the value of k such that

    a. $P(|z| > 1.68) = k$ **b.** $P(|z| < 2.15) = k$

6.56 Given that z is the standard normal variable, find the value of c such that

    a. $P(|z| > c) = 0.0384$ **b.** $P(|z| < c) = 0.8740$

6.57 Find the following values of z.

    a. $z(0.12)$ **b.** $z(0.28)$ **c.** $z(0.85)$ **d.** $z(0.99)$

6.58 Find the area under the normal curve that lies between the following pairs of z values.

    a. $z = -3.00$ and $z = 3.00$ **b.** $z = z(0.975)$ and $z(0.025)$

    c. $z = z(0.10)$ and $z(0.01)$

6.59 The test scores on a computer-science aptitude test are normally distributed with a mean of 15.0 and a standard deviation of 3.0. Find the 90th percentile for this distribution.

6.60 A soft drink vending machine can be regulated so as to dispense an average of μ ounces of soft drink per glass. If the ounces dispensed per glass are normally distributed with a standard deviation of 0.2 ounce, find the setting for μ that will allow a 6-ounce glass to hold (without overflowing) the amount dispensed 99 percent of the time.

6.61 Suppose that a particular normal distribution has a mean of 70 and that the 90th percentile equals 84. Find the standard deviation.

6.62 The length of life of a certain type of refrigerator is approximately normally distributed with a mean of 4.8 years and a standard deviation of 1.3 years.

    a. If this machine is guaranteed for 2 years, what is the probability that the machine you purchased will require replacement under the guarantee?

    b. What period of time should the manufacturer give as a guarantee if he is willing to replace only 0.5 percent of the machines?

6.63 The average length of time required for completing a certain academic achievement test is believed to be 150 minutes, and the standard deviation is 20 minutes. If we wish to allow sufficient time for only 80 percent to complete the test, when should the test be terminated? (Assume that the lengths of time required to complete this test are normally distributed.)

6.64 A machine is programmed to fill 10-ounce containers with a cleanser. However, the variability inherent in any machine causes the actual amounts of fill to vary. The distribution is normal with a standard deviation of 0.02 ounces. What must the mean amount μ be in order that only 5 percent of the containers receive less than 10 ounces?

6.65 In a large industrial complex the maintenance department has been instructed to replace light bulbs before they burn out. It is known that the life of light bulbs is normally distributed with a mean life of 900 hours of use and a standard deviation of 75 hours. When should the light bulbs be replaced so that no more than 10 percent of them will burn out while in use?

6.66 The burn times for a certain solid-fuel rocket are normally distributed with a mean equal to 5.0 minutes and a standard deviation equal to 0.5 minutes. What percent of these rockets have burn times exceeding 6.0 minutes?

6.67 Suppose that x has a binomial distribution with $n = 25$ and $p = 0.3$.

    a. Explain why the normal approximation is reasonable.

    b. Find the mean and standard deviation of the normal distribution that is used in the approximation.

6.68 The MINITAB statistical package was used to compute the binomial probabilities for $n = 50$ and $p = 0.1$. (The output stopped at $k = 14$ because the remaining probabilities are zero to four decimal places.)

```
MTB > BINOMIAL PROBABILITIES FOR N = 50, P = .1
   BINOMIAL PROBABILITIES FOR N = 50 AND P = 0.100000
     K        P(X = K)        P(X LESS OR = K)
     0        0.0052          0.0052
     1        0.0286          0.0338
     2        0.0779          0.1117
     3        0.1386          0.2503
     4        0.1809          0.4312
     5        0.1849          0.6161
     6        0.1541          0.7702
     7        0.1076          0.8779
     8        0.0643          0.9421
```

9	0.0333	0.9755
10	0.0152	0.9906
11	0.0061	0.9968
12	0.0022	0.9990
13	0.0007	0.9997
14	0.0002	0.9999

MTB > STOP

Compute the normal approximation for $x \leq 6$ and compare with this output.

6.69 A test-scoring machine is known to record an incorrect grade on 5 percent of the exams it grades. Find, by the appropriate method, the probability that the machine records

    a. 3 wrong grades in a set of 5 exams.

    b. no more than 3 wrong grades in a set of 5 exams.

    c. no more than 3 wrong grades in a set of 15 exams.

    d. no more than 3 wrong grades in a set of 150 exams.

6.70 It is believed that 58 percent of married couples with children agree on methods of disciplining their children. Assuming this to be the case, what is the probability that in a random survey of 200 married couples, we would find

    a. exactly 110 couples who agree?

    b. less than 110 couples who agree?

    c. more than 100 couples who agree?

6.71 A new drug is supposed to be 85 percent effective in treating a particular illness. (That is, 85 percent of the patients with this illness respond favorably to the drug.) Let x be the number of patients out of every group of 50 who respond favorably. Use the normal approximation method to find the following probabilities.

    a. $P(x > 45)$ **b.** $P(40 < x < 50)$ **c.** $P(x < 35)$

6.72 If 60 percent of the registered voters plan to vote for Ralph Brown for mayor of a large city, what is the probability that less than half of the voters, in a poll of 200 registered voters, plan to vote for Ralph Brown?

6.73 The following triangular distribution provides an approximation to the standard normal distribution. Line segment l_1 has the equation $y = x/9 + 1/3$ and segment l_2 has the equation $y = -x/9 + 1/3$.

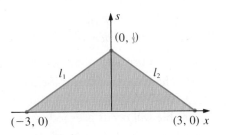

    a. Find the area under the entire triangular distribution.

b. Find the area under the triangular distribution between 0 and 2.

c. Find the area under the standard normal distribution between 0 and 2.

6.74 The grades on an examination whose mean score is 525 and whose standard deviation is 80 are normally distributed.

a. Anyone who scores below 350 will be retested. What percentage does this represent?

b. The top 12 percent are to receive a special commendation. What score must be surpassed to receive this special commendation?

c. The interquartile range of a distribution is the difference between Q_1 and Q_3, $Q_3 - Q_1$. Find the interquartile range for the grades on this examination.

d. Find the grade such that only 1 out of 500 will score above it.

6.75 The designers of the Custom Sport Coupe (Illustration 1-3) want to make the roof in their new model high enough so that there is at least 1 inch of clearance between the headliner (ceiling) and the top of the driver's head for 90 percent of the drivers. If the heights of sports car enthusiasts are normally distributed with a mean of 68 inches and a standard deviation of 2.5 inches, what height must the designers work with?

Vocabulary List

Be able to define each term. In addition, describe, in your own words, and give an example of each term. Your examples should not be ones given in class or in the textbook.

The bracketed numbers indicate the chapter in which the term first appeared, but you should define the terms again to show increased understanding of their meaning.

area representation for probability
bell-shaped curve
binomial distribution [5]
binomial probability
continuity correction factor
discrete random variable [1, 5]
normal approximation of binomial
normal curve

normal distribution
percentage
probability [4]
proportion
random variable [5]
standard normal distribution
standard score [2]
z score [2]

Quiz

Answer "True" if the statement is always true. If the statement is not always true replace the boldface words with words that make the statement always true.

6.1 The normal probability distribution is symmetric about **zero**.

6.2 The total area under the curve of any normal distribution is **one**.

6.3 The theoretical probability that a particular value of a **continuous** random variable will occur is exactly zero.

6.4 The unit of measure of the standard score is **the same as** the unit of measure of the data.

6.5 All **normal distributions** have the same general probability function and distribution.

6.6 When using the notation $z(0.05)$, the number in the parentheses is the measure of the area to the **left** of the z score.

6.7 Standard normal scores have a mean of **one** and a standard deviation of **zero**.

6.8 Probability distributions of **all** continuous random variables are normal.

6.9 We are able to add and subtract the areas under the curve because these areas represent the probabilities of **independent events**.

6.10 The most common distribution of a continuous random variable is the **binomial probability**.

Chapter 7

Sample Variability

HOW POLL WAS CONDUCTED

The findings of this Tribune Poll are based on 250 interviews with registered black voters who live in Chicago Housing Authority family dwellings.

Results can be expected to differ by no more than plus or minus 6 percentage points from a survey of all voters who live in CHA family units, with a 95 percent confidence level.

TRIBUNE POLL

"Candidates for mayor of Chicago in the Democratic primary are Harold Washington and Jane Byrne. If that election were held today; for whom would you vote?"*

1987 primary poll		1983 primary		1983 general election	
Washington	71%	Washington	81%	Washington	89%
Byrne	8%	Byrne	17%	Epton	1%
Undecided	20%	Daley	3%		
Other	1%				

* Figures for the 1987 primary are based on the responses of CHA residents to the Tribune Poll; figures for 1983 are based on a Tribune analysis of Chicago Election Board statistics. Sample size is 250; margin of error ±6%.

Poll research was prepared and analyzed by The Tribune and Market Shares Corp., a marketing and public opinion research firm with offices in Mt. Prospect. Interviews were conducted by Market Shares Jan. 22–25.

"Who do you think cares more about the residents of CHA housing?"

Washington	Byrne	Both the same	Don't know
38%	25%	21%	16%

"All things considered, has the CHA been run better under the Washington adminstration or the Byrne administration?"

Washington	Byrne	Both the same	Dont't know
35%	22%	26%	16%

Chapter
Objectives

In Chapters 1 and 2 we discussed how to obtain and describe a sample. The description of the sample data is accomplished by using three basic concepts: (1) measures of central tendency (the mean is the most popularly used sample statistic), (2) measures of dispersion (the standard deviation is most commonly used), and (3) kind of distribution (normal, skewed, rectangular, and so on). The question that seems to follow is: What can be deduced about the statistical population from which a sample is taken?

To put this query at a more practical level, suppose that we have just taken a sample of 25 rivets made for the construction of airplanes. The rivets were tested for shearing strength, and the force required to break each rivet was the response variable. The various descriptive measures— mean, standard deviation, type of distribution—can be found for this sample. However, it is not the sample itself that we are interested in. The rivets that were tested were destroyed during the test, so they can no longer be used in the construction of airplanes. What we are trying to find out is information about the total population, and we certainly cannot test every rivet that is produced (there would be none left for construction). Therefore, we must somehow deduce information, or make inferences about, the population based on the results observed in the sample.

Suppose that we take another sample of 25 rivets and test them by the same procedure. Do you think that we would obtain the same sample mean from the second sample that we obtained from the first? The same standard deviation?

After considering these questions we might suspect that we would need to investigate the variability in the sample statistics obtained from **repeated sampling**. Thus we need to find (1) measures of central tendency for the sample statistics of importance, (2) measures of dispersion for the sample statistics, and (3) the pattern of variability (distribution) of the sample statistics. Once we have this information, we will be better able to predict the population parameters.

The objective of this chapter is to study the measures and the patterns of variability for the distribution formed by repeatedly observed values of a **sample mean**.

7.1   SAMPLING DISTRIBUTIONS

To make inferences about a population, we need to discuss sample results a little more. A sample mean $\bar{x}$ is obtained from a sample. Do you expect that this value, $\bar{x}$, is exactly equal to the value of the population mean μ? Your answer should be "no." We do not expect that to happen, but we will be satisfied with our sample results if the sample mean is "close" to the value of the population mean. Let's consider a second question: If a second sample is taken, will the second sample have a mean equal to the population mean? Equal to the first sample mean? Again, no, we do not expect it to be equal to the population mean, nor do we expect the second sample mean to repeat the first one. We do, however, again expect the values to be "close." (This argument should hold for any other sample statistic and its corresponding population value.)

The next questions should already have come to mind: What is "close"? How do we determine (and measure) this closeness? Just how would repeated sample statistics be distributed? To answer these questions we must take a look at a *sampling distribution*.

sampling distribution

Sampling Distribution of a Sample Statistic

The distribution of values for that sample statistic obtained from all possible samples of a population. The samples must all be the same size, and the sample statistic could be any descriptive sample statistic.

Illustration 7-1 To illustrate the concept of a sampling distribution, let's consider the mean of each sample of size 2 that can be drawn with replacement from the set of even single-digit integers, (0, 2, 4, 6, 8). There are 25 possible samples of size 2:

$$
\begin{array}{ccccc}
(0, 0) & (2, 0) & (4, 0) & (6, 0) & (8, 0) \\
(0, 2) & (2, 2) & (4, 2) & (6, 2) & (8, 2) \\
(0, 4) & (2, 4) & (4, 4) & (6, 4) & (8, 4) \\
(0, 6) & (2, 6) & (4, 6) & (6, 6) & (8, 6) \\
(0, 8) & (2, 8) & (4, 8) & (6, 8) & (8, 8)
\end{array}
$$

Each of these samples has a mean $\bar{x}$. These means are, respectively,

$$
\begin{array}{ccccc}
0 & 1 & 2 & 3 & 4 \\
1 & 2 & 3 & 4 & 5 \\
2 & 3 & 4 & 5 & 6 \\
3 & 4 & 5 & 6 & 7 \\
4 & 5 & 6 & 7 & 8
\end{array}
$$

Each of these samples is equally likely, and thus each of the 25 sample means can be assigned a probability of $\frac{1}{25} = 0.04$. (Why? See Exercise 7-1c.) The sampling distribution for the sample mean then becomes as shown in Table 7-1. This is a probability distribution of $\bar{x}$ (*see* Figure 7-1).

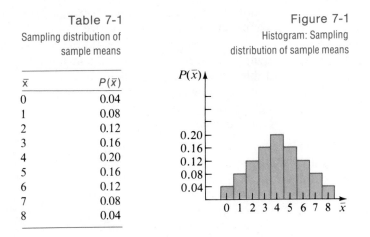

Table 7-1

Sampling distribution of
sample means

$\bar{x}$	$P(\bar{x})$
0	0.04
1	0.08
2	0.12
3	0.16
4	0.20
5	0.16
6	0.12
7	0.08
8	0.04

Figure 7-1

Histogram: Sampling
distribution of sample means

For the same set of all possible samples of size 2, let's find the sampling distribution for sample ranges. Each sample has a range R. These ranges are, respectively,

$$
\begin{array}{ccccc}
0 & 2 & 4 & 6 & 8 \\
2 & 0 & 2 & 4 & 6 \\
4 & 2 & 0 & 2 & 4 \\
6 & 4 & 2 & 0 & 2 \\
8 & 6 & 4 & 2 & 0
\end{array}
$$

Again, each of these 25 sample ranges has a probability of 0.04, and Table 7-2 shows the sampling distribution of sample ranges. This is a probability distribution of R (*see* Figure 7-2).

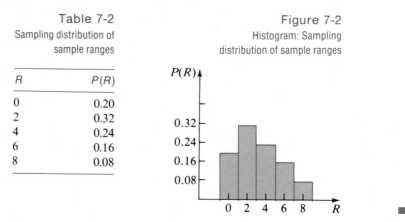

Table 7-2

Sampling distribution of
sample ranges

R	$P(R)$
0	0.20
2	0.32
4	0.24
6	0.16
8	0.08

Figure 7-2

Histogram: Sampling
distribution of sample ranges

Most populations that are sampled are much larger than the one used in Illustration 7-1, and it would be a very tedious job to list all the possible samples. With this in mind, let's investigate a sampling distribution empirically (that is, by experimentation).

Illustration 7-2   Let's consider a portion of the sampling distribution of sample means for samples of size 5 obtained from the rolling of a single die. One sample will consist of 5 rolls of the die, and we will obtain a sample mean $\bar{x}$ from this sample. We repeat the experiment until 30 sample means have been obtained. Table 7-3 shows 30 such samples and their means. The resulting frequency distribution is shown in Figure 7-3. This distribution seems to display characteristics of a normal distribution; it is mounded and nearly symmetric about its mean (approximately 3.5).

Figure 7-3

A portion of a
sampling distribution of
sample means for rolling a die

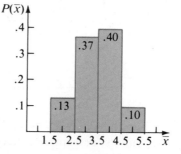

Table 7-3

Sample means for rolling a
single die five times

Trial	Sample	$\bar{x}$	Trial	Sample	$\bar{x}$
1	1, 2, 3, 2, 2	2.0	16	5, 2, 1, 3, 5	3.2
2	4, 5, 5, 4, 5	4.6	17	6, 1, 3, 3, 5	3.6
3	3, 1, 5, 2, 4	3.0	18	6, 5, 5, 2, 6	4.8
4	5, 6, 6, 4, 2	4.6	19	1, 3, 5, 5, 6	4.0
5	5, 4, 1, 6, 4	4.0	20	3, 1, 5, 3, 1	2.6
6	3, 5, 6, 1, 5	4.0	21	5, 1, 1, 4, 3	2.8
7	2, 3, 6, 3, 2	3.2	22	4, 6, 3, 1, 2	3.2
8	5, 3, 4, 6, 2	4.0	23	1, 5, 3, 4, 5	3.6
9	1, 5, 5, 3, 4	3.6	24	3, 4, 1, 3, 3	2.8
10	4, 1, 5, 2, 6	3.6	25	1, 2, 4, 1, 4	2.4
11	5, 1, 3, 3, 2	2.8	26	5, 2, 1, 6, 3	3.4
12	1, 5, 2, 3, 1	2.4	27	4, 2, 5, 6, 3	4.0
13	2, 1, 1, 5, 3	2.4	28	4, 3, 1, 3, 4	3.0
14	5, 1, 4, 4, 6	4.0	29	2, 6, 5, 3, 3	3.8
15	5, 5, 6, 3, 3	4.4	30	6, 3, 5, 1, 1	3.2

The theory involved with sampling distributions that will be described in the remainder of this chapter requires *random sampling.*

random sample

Random Sample

A sample obtained in such a way that each of the possible samples of fixed size has an equal chance of being selected.

Case Study 7-1

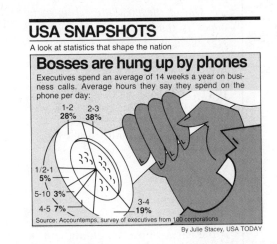

"Bosses are hung up by phones." Copyright 1986, USA TODAY. Reprinted with permission.

BOSSES ARE HUNG UP BY PHONES

A survey of executives from 100 corporations resulted in the information in the *USA SNAPSHOT* "Bosses are hung up by phones." The information is shown in the form of a circle graph. This information could have been represented by a relative frequency distribution or a histogram. This relative frequency distribution might seem to be, but is not part of, the sampling distribution of sample means. Each boss reported his or her average. Why is this distribution of average telephone time not a portion of the sampling distribution of sample means? (*See* Exercise 7.7.)

Case Study 7-2

GALLUPING ATTITUDES

Repeated sampling has been discussed in this chapter as an empirical approach to investigating the sampling distribution of a particular sample statistic. Another use for repeated sampling is demonstrated by the article that forms Case Study 7-2. Public opinion is quite likely to shift from time to time. For example, a recognizable shift has taken place in the American attitude toward a woman's place in life. On the question of a woman president, our abstract of the news article shows a change of opinion from 66 percent no to 76 percent yes. Gallup Polls frequently take samples to estimate the current opinion of the population as a whole. This article does not demonstrate a sampling distribution as such, but it does emphasize the difference between repeated sampling for control or development and repeated sampling for the analysis of sampling distributions.

Since its founding in 1935, the American Institute of Public Opinion, better known as the Gallup Poll, has put approximately 20,000 questions to more than two million people.

One of the most interesting social changes to follow through the years in these pages is that of Americans' attitudes toward women at work and in politics. Herewith, an abbreviated reading of Gallup's progress report.

A WOMAN PRESIDENT

1937

Would you vote for a woman for President, if she qualified in every other respect?

		By Sex	Yes	No
Yes....................	34%	Men...................	27%	73%
No....................	66	Women	41	59

1955

If the party whose candidate you most often support nominated a woman for President of the United States, would you vote for her if she seemed best qualified for the job?

		By Sex	Yes	No
Yes....................	52%	Men...................	47%	48%
No....................	44	Women	57	40

1971

If your party nominated a woman for
President, would you vote for her if
she were qualified for the job?

By Sex		Yes	No
Yes. 66%	Men.	65%	35%
No 29	Women	67	33

1978

If your party nominated a woman for
president, would you vote for her if
she were qualified for the job?

By Sex		Yes	No
Yes. 76%	Men.	76%	19%
No 19	Women	77	18

Copyright by the *Gallup Report*. Reprinted by permission.

EXERCISES

7.1 a. What is the sampling distribution of sample means?

b. A sample of size 3 is taken from a population and the sample mean found. Describe how this sample mean is related to the sampling distribution of sample means.

c. Why is the probability of 0.04 assigned to each of the sample mean values in Illustration 7-1?

7.2 Consider the set of odd single-digit integers $\{1, 3, 5, 7, 9\}$.

a. Make a list of all samples of size 2 that can be drawn from this set of integers. (Sample with replacement; that is, the first number is drawn, observed, then replaced before the next drawing.)

b. Construct the sampling distribution of sample means for samples of size 2 selected from this set.

c. Construct the sampling distributions of sample ranges for samples of size 2.

7.3 Consider the set of even single-digit integers $\{0, 2, 4, 6, 8\}$.

a. Make a list of all the possible samples of size 3 that can be drawn from this set of integers. (Sample with replacement; that is, the first number is drawn, observed, then replaced before the next drawing.)

b. Construct the sampling distribution of the sample medians for samples of size 3.

c. Construct the sampling distribution of the sample means.

7.4 Using the telephone numbers listed in your local directory as your population, obtain randomly 20 samples of size 3. From each number identified as a source, take the fourth, fifth, and sixth digits. (For example, for 245– ⑧②⑥ 8, you would take the

8, the 2, and the 6 as your sample of size 3.)

    a. Calculate the mean of the 20 samples.

    b. Draw a histogram showing the 20 sample means. (Use classes -0.5 to 0.5, 0.5 to 1.5, 1.5 to 2.5, and so on.)

7.5 The following MINITAB program selects 100 samples (each of size 4) from a binomial distribution having $n = 4$ and $p = 0.1$. The means of these 100 samples are placed into column C8 and then a histogram of C8 is formed. Draw a histogram of the binomial distribution and compare it with the histogram of the sample means.

```
MTB > READ VALUE IN C1,PROB IN C2
DATA> 0 .656
DATA> 1 .292
DATA> 2 .049
DATA> 3 .004
DATA> 4 .0001
DATA> END DATA
       5 ROWS READ
MTB > DRANDOM 100 OBS,USING VALUES IN C1,PROB IN C2,PUT IN C3
       100 DISCRETE RANDOM OBSERVATIONS
MTB > DRANDOM 100 OBS,USING VALUES IN C1,PROB IN C2,PUT IN C4
       100 DISCRETE RANDOM OBSERVATIONS
MTB > DRANDOM 100 OBS,USING VALUES IN C1,PROB IN C2,PUT IN C5
       100 DISCRETE RANDOM OBSERVATIONS
MTB > DRANDOM 100 OBS,USING VALUES IN C1,PROB IN C2,PUT IN C6
       100 DISCRETE RANDOM OBSERVATIONS
MTB > ADD C3-C6,PUT IN C7
MTB > DIVIDE C7 BY 4,PUT IN C8
MTB > HISTOGRAM C8

   Histogram of C8 N = 100
   Midpoint Count
       0.0      17 ****************
       0.1       0
       0.2       0
       0.3      26 **************************
       0.4       0
       0.5      39 ***************************************
       0.6       0
       0.7       0
       0.8      10 **********
       0.9       0
       1.0       5 *****
       1.1       0
       1.2       0
       1.3       3 ***
```

7.6 The following MINITAB program selects 100 samples from a uniform probability distribution for integers 1, 2, 3, and 4. Each sample is of size 4 and the means are put into column C6. A histogram of the 100 means is shown. Draw a probability histogram for the uniform distribution and compare it to the histogram of sample means.

```
MTB > IRANDOM 100 OBS BETWEEN 1 AND 4,PUT IN C1
      100 RANDOM INTEGERS BETWEEN      1 AND      4
MTB > IRANDOM 100 OBS BETWEEN 1 AND 4,PUT IN C2
      100 RANDOM INTEGERS BETWEEN      1 AND      4
MTB > IRANDOM 100 OBS BETWEEN 1 AND 4,PUT IN C3
      100 RANDOM INTEGERS BETWEEN      1 AND      4
MTB > IRANDOM 100 OBS BETWEEN 1 AND 4,PUT IN C4
      100 RANDOM INTEGERS BETWEEN      1 AND      4
MTB > ADD C1-C4,PUT IN C5
MTB > DIVIDE C5 BY 4,PUT IN C6
MTB > HISTOGRAM C6
Histogram of C6 N = 100
Midpoint Count
    1.2      1 *
    1.4      0
    1.6      2 **
    1.8      8 ********
    2.0      9 *********
    2.2     15 ***************
    2.4     .0
    2.6     20 ********************
    2.8     16 ****************
    3.0     17 *****************
    3.2      9 *********
    3.4      0
    3.6      3 ***
```

7.7 Refer to the information shown on the circle graph in Case Study 7-1.

a. What variable was reported in this information? (What did each boss report?)

b. Construct a relative frequency histogram picturing this information. Describe the shape of the histogram.

c. Explain why this distribution is not part of the sampling distribution of sample means.

7.2 THE CENTRAL LIMIT THEOREM

On the preceding pages we discussed two types of sampling distributions, for means and for ranges. There are many others that could be discussed; in fact, these two could themselves be discussed further. However, the only sampling distribution of concern to us here is the **sampling distribution of sample means**. The mean is the most commonly used sample statistic and thus is the most important.

The central limit theorem (CLT) tells us about the sampling distribution of sample means of random samples of size n. Recall that there are basically three kinds of information that we want about a distribution: (1) where the center is, (2) how widely it is dispersed, and (3) how it is distributed. The central limit theorem tells us all three.

central limit theorem

Central Limit Theorem

If all possible random samples, each of size n, are taken from any population with a mean μ and standard deviation σ, the sampling distribution of sample means will

1. have a mean $\mu_{\bar{x}}$ equal to μ.
2. have a standard deviation $\sigma_{\bar{x}}$ equal to $\sigma/\sqrt{n}$.
3. be normally distributed when the parent population is normally distributed or will be approximately normally distributed for samples of size 30 or more when the parent population is not normally distributed. The approximation to the normal distribution improves with samples of larger size.

In short, the central limit theorem states the following:

1. $\mu_{\bar{x}} = \mu$; the mean of the $\bar{x}$'s equals the mean of the x's.

2. $\sigma_{\bar{x}} = \sigma/\sqrt{n}$; the standard error of the mean (see the definition that follows) equals the standard deviation of the population divided by the square root of the sample size.

3. The sample means are approximately normally distributed (regardless of the shape of the parent population.)

Note   The n referred to in the central limit theorem is the **size of each sample** in the sampling distribution.

standard error of the mean

Standard Error of the Mean

The standard deviation of the sampling distribution of sample means

We are unable to prove the central limit theorem without using advanced mathematics. However, it is possible to check its validity by examining two illustrations. Let's consider a population for which we can construct the theoretical sampling distribution of all possible samples. For this example let's consider all possible samples of size 2 that could be drawn from a population that contains the three numbers 2, 4, and 6.

First let's look at the population itself. To calculate the mean μ and the standard deviation σ, we must use the formulas from Chapter 5 for discrete probability distributions:

$$\mu = \sum [x \cdot P(x)] \quad \text{and} \quad \sigma = \sqrt{\sum [x^2 \cdot P(x)] - \{\sum [x \cdot P(x)]\}^2}$$

These formulas are necessary because we are not drawing the samples but are discussing the theoretical possibilities (*see* Table 7-4).

Table 7-4

Probability distribution and extensions for $x = 2, 4, 6$

x	$P(x)$	$x \cdot P(x)$	$x^2 \cdot P(x)$
2	$\frac{1}{3}$	$\frac{2}{3}$	$\frac{4}{3}$
4	$\frac{1}{3}$	$\frac{4}{3}$	$\frac{16}{3}$
6	$\frac{1}{3}$	$\frac{6}{3}$	$\frac{36}{3}$
Total	$\frac{3}{3}$	$\frac{12}{3}$	$\frac{56}{3}$

$$\mu = \tfrac{12}{3} = \mathbf{4.0}$$
$$\sigma = \sqrt{56/3 - (12/3)^2} = \sqrt{18.66 - 16.0} = \sqrt{2.66} = \mathbf{1.63}$$

Table 7-5 lists all the possible samples that could be drawn if samples of size 2 were to be drawn from this population. (One number is

Table 7-5

All possible samples of size 2 and their means

Possible Samples	$\bar{x}$
2, 2	2
2, 4	3
2, 6	4
4, 2	3
4, 4	4
4, 6	5
6, 2	4
6, 4	5
6, 6	6

drawn, observed, and then returned to the population before the second number is drawn.) Table 7-5 also lists the means of these samples. The probability distribution for these means and the extensions are given in Table 7-6. Thus we have

$$\mu_{\bar{x}} = \frac{36}{9} = \mathbf{4.0}$$

$$\sigma_{\bar{x}} = \sqrt{\frac{156}{9} - \left(\frac{36}{9}\right)^2} = \sqrt{17.33 - 16} = \sqrt{1.33} = \mathbf{1.15}$$

The histogram for the distribution of possible $\bar{x}$'s is shown in Figure 7-4.
The CLT says that three things will occur in this sampling distribution:

1. It will be approximately normally distributed. The histogram (Figure 7-4) suggests this very strongly.

Table 7-6

Probability distribution for means of all possible samples of size 2 and the extensions

$\bar{x}$	$P(\bar{x})$	$\bar{x} \cdot P(\bar{x})$	$x^2 \cdot P(\bar{x})$
2	1/9	2/9	4/9
3	2/9	6/9	18/9
4	3/9	12/9	48/9
5	2/9	10/9	50/9
6	1/9	6/9	36/9
Total	9/9	36/9	156/9

Figure 7-4

Histogram for distribution of
Table 7-6

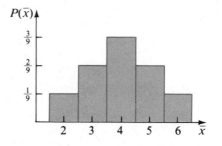

2. The mean $\mu_{\bar{x}}$ of the sampling distribution will equal the mean of the population. They both have the value 4.0.

3. The standard deviation $\sigma_{\bar{x}}$ of the sampling distribution will equal the standard deviation of the population divided by the square root of the sample size ($\sigma/\sqrt{n}$):

$$\sigma_{\bar{x}} = 1.15 \text{ and } \frac{\sigma}{\sqrt{n}} = \frac{1.63}{\sqrt{2}} = \frac{1.63}{1.41} = 1.15$$

This illustration shows that the CLT is true for a **theoretical probability distribution**.

New let's look at the empirical distribution that occurred in Illustration 7-2 and see whether it supports the three claims of the central limit theorem.

First, let's look at the theoretical probability distribution from which these samples were taken (Table 7-7). A histogram showing the probability distribution of the tossing of a die is shown in Figure 7-5. The population mean μ equals **3.5** (*see* Table 7-7). The population standard deviation σ equals $\sqrt{15.17 - (3.5)^2}$, which is $\sqrt{2.92} = \mathbf{1.71}$. (Note that this population has a uniform distribution.)

Table 7-7

Probability distribution and
extensions for rolling a die

x	$P(x)$	$x \cdot P(x)$	$x^2 \cdot P(x)$
1	$\frac{1}{6}$	$\frac{1}{6}$	$\frac{1}{6}$
2	$\frac{1}{6}$	$\frac{2}{6}$	$\frac{4}{6}$
3	$\frac{1}{6}$	$\frac{3}{6}$	$\frac{9}{6}$
4	$\frac{1}{6}$	$\frac{4}{6}$	$\frac{16}{6}$
5	$\frac{1}{6}$	$\frac{5}{6}$	$\frac{25}{6}$
6	$\frac{1}{6}$	$\frac{6}{6}$	$\frac{36}{6}$
Total	$\frac{6}{6} = 1$	$\frac{21}{6} = 3.5$	$\frac{91}{6} = 15.17$

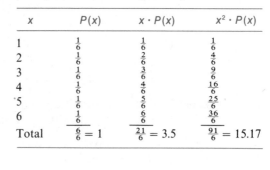

Figure 7-5

Probability distribution for
rolling a die

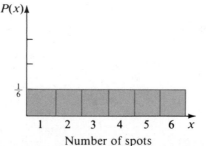

Number of spots

Now let's look at the empirical sampling distribution of the 30 sample means found earlier. Using the 30 values of $\bar{x}$ in Table 7-3, the observed mean of the $\bar{x}$'s turns out to be 3.43 and the observed standard deviations $s_{\bar{x}}$ turns out to be 0.73. The histogram appears in Figure 7-3.

The central limit theorem states that the $\bar{x}$'s should be approximately normally distributed, and the histogram certainly suggests this to be the case. The CLT also says that the mean $\mu_{\bar{x}}$ of the sampling distribution and the mean μ of the population are the same. The mean of the $\bar{x}$'s is 3.43 and $\mu = 3.5$; they seem to be reasonably close. Remember that we have taken only 30 samples, not all possible samples, of size 5.

The theorem says that $\sigma_{\bar{x}}$ should equal $\sigma/\sqrt{n}$. The observed standard deviation of $\bar{x}$'s is $s_{\bar{x}} = 0.73$ and the standard error of the mean is

$$\frac{\sigma}{\sqrt{n}} = \frac{1.71}{\sqrt{5}} = 0.76$$

These two values are very close.

The evidence seen in these two illustrations seems to suggest that the CLT is true, although this does not constitute a proof of the theorem, of course.

Having taken a look at these two specific illustrations, which support the CLT, let's now look at four graphic illustrations that present the same information in slightly different form. In each of these graphic illustrations there are four distributions. The first is a distribution of the parent population, the distribution of the individual x values. Each of the other three graphs show a sampling distribution of sample means, using three different sample sizes. In Figure 7-6 we have a uniform distribution, much like Figure 7-5 for the die illustration, and the resulting distributions of sample means for samples of size 2, 5, and 30. Figure 7-7 shows a U-shaped population and the corresponding sampling distributions. Figure 7-8 shows a J-shaped population and the three corresponding distributions. Figure 7-9 shows a normal distribution population and the three sampling distributions.

Figure 7-6
Uniform distribution

(d) Sampling distribution of $\bar{x}$ when $n = 30$

(c) Sampling distribution of $\bar{x}$ when $n = 5$

(b) Sampling distribution of $\bar{x}$ when $n = 2$

(a) Population

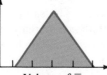

Values of x

Values of $\bar{x}$

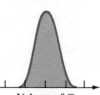

Values of $\bar{x}$

Values of $\bar{x}$

Figure 7-7

U-shaped distribution

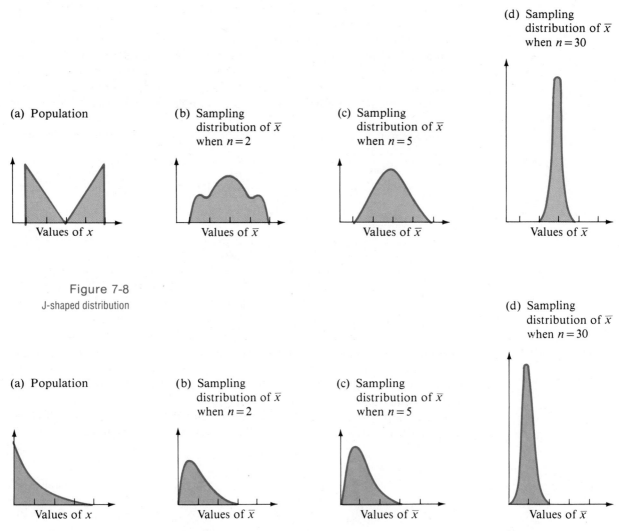

(d) Sampling
distribution of $\bar{x}$
when $n = 30$

(a) Population

(b) Sampling
distribution of $\bar{x}$
when $n = 2$

(c) Sampling
distribution of $\bar{x}$
when $n = 5$

Values of x

Values of $\bar{x}$

Values of $\bar{x}$

Values of $\bar{x}$

Figure 7-8

J-shaped distribution

(d) Sampling
distribution of $\bar{x}$
when $n = 30$

(a) Population

(b) Sampling
distribution of $\bar{x}$
when $n = 2$

(c) Sampling
distribution of $\bar{x}$
when $n = 5$

Values of x

Values of $\bar{x}$

Values of $\bar{x}$

Values of $\bar{x}$

All four illustrations seem to verify the CLT. Note that the sampling distribution of the three nonnormal distributions produced sample means with an approximately normal distribution for samples of size 30. In the normal population (Figure 7-9) the sampling distributions for all sample sizes appear to be normal. Thus you have seen an amazing phenomenon: no matter what the shape of a population, the sampling distribution of the mean becomes approximately normally distributed when n becomes sufficiently large.

You should notice one other point: The sample mean becomes less variable as the sample size increases. Notice that as n increases from 2 to 30, all the distributions become narrower and taller. Can you explain how this implies less variability? How does the CLT state this? (*See* Exercise 7.8.) This point is discussed further in Section 7-3.

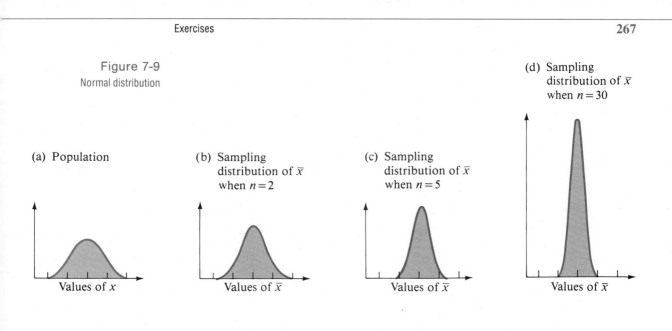

Figure 7-9
Normal distribution

(a) Population

(b) Sampling distribution of $\bar{x}$ when $n=2$

(c) Sampling distribution of $\bar{x}$ when $n=5$

(d) Sampling distribution of $\bar{x}$ when $n=30$

Values of x

Values of $\bar{x}$

Values of $\bar{x}$

Values of $\bar{x}$

EXERCISES

7.8 a. What is the measure of the total area for any probability distribution?

b. How does the CLT state that as n becomes larger, the sample mean becomes less variable?

7.9 A certain population has a mean of 500 and a standard deviation of 30. Many samples of size 36 are randomly selected and their mean calculated.

a. What value would you expect to find for the mean of all these sample means?

b. What value would you expect to find for the standard deviation of all these sample means?

c. What shape distribution would you expect the distribution of all these sample means to have?

7.10 If a normal population has a standard deviation σ of 25 units, what is the standard error of the mean ($\sigma_{\bar{x}}$) if samples of size 16 are used? Of size 25? Of size 50? Of size 100?

7.11 Consider the experiment of taking a standardized mathematics test. The variable x is the raw score received. This exam has a mean of 720 and a standard deviation of 60. A group (sample) of 40 students takes the exam and the sample mean $\bar{x}$ is 725.6. A sampling distribution of means is formed from the means of all such groups of 40 students.

a. Determine the mean of this sampling distribution.

b. Determine the standard deviation for this sampling distribution.

7.12 The following MINITAB program simulates 100 samples, each of size 5, drawn from a normal distribution with mean equal to 5.5 and standard deviation equal to 2.5. The mean for each sample is placed in column C7.

a. From the output, what is the mean of the 100 sample means? What is the standard deviation of the 100 sample means?

b. According to the central limit theorem, if all possible sample means (not just 100 of them) were computed, what would be the mean of the sample means, $\mu_{\bar{x}}$? What would be the standard deviation of the sample means, $\sigma_{\bar{x}}$?

```
MTB > NRANDOM 100 OBS MU = 5.5,SIGMA = 2.5,PUT IN C1
    100 NORMAL OBS. WITH MU =        5.5000 AND SIGMA =     2.5000
MTB > NRANDOM 100 OBS MU = 5.5,SIGMA = 2.5,PUT IN C2
    100 NORMAL OBS. WITH MU =        5.5000 AND SIGMA =     2.5000
MTB > NRANDOM 100 OBS WITH MU = 5.5,SIGMA = 2.5,PUT IN C3
    100 NORMAL OBS. WITH MU =        5.5000 AND SIGMA =     2.5000
MTB > NRANDOM 100 OBS WITH MU = 5.5,SIGMA = 2.5,PUT IN C4
    100 NORMAL OBS. WITH MU =        5.5000 AND SIGMA =     2.5000
MTB > NRANDOM 100 OBS MU = 5.5,SIGMA = 2.5,PUT IN C5
    100 NORMAL OBS. WITH MU =        5.5000 AND SIGMA =     2.5000
MTB > ADD C1-C5,PUT IN C6
MTB > DIVIDE C6 BY 5,PUT IN C7
MTB > AVERAGE C7
   MEAN    =      5.5034
MTB > STANDARD DEVIATION C7
   ST.DEV. =        1.1104
MTB > STOP
```

7.13 A random variable that can take on the values $1, 2, \ldots, n$ (each with probability $1/n$) is called uniform. For such a variable the mean $\mu = (n + 1)/2$ and $\sigma = \sqrt{\frac{n^2-1}{12}}$. [*Hint*: Use these formulas in answering part (b).] The following **MINITAB** program gives 100 simulated samples, each of size 5, from a uniform distribution for integers from 1 to 10. The mean of the 100 samples are found and put into column C7.

a. From the output, what are the mean and standard deviation of the 100 sample means?

b. For all samples of size 5, find $\mu_{\bar{x}}$ and $\sigma_{\bar{x}}$.

```
MTB > IRANDOM 100 OBS BETWEEN 1 AND 10,PUT IN C1
    100 RANDOM INTEGERS BETWEEN     1 AND     10
MTB > IRANDOM 100 OBS BETWEEN 1 AND 10,PUT IN C2
    100 RANDOM INTEGERS BETWEEN     1 AND     10
MTB > IRANDOM 100 OBS BETWEEN 1 AND 10,PUT IN C3
    100 RANDOM INTEGERS BETWEEN     1 AND     10
MTB > IRANDOM 100 OBS BETWEEN 1 AND 10,PUT IN C4
    100 RANDOM INTEGERS BETWEEN     1 AND     10
MTB > IRANDOM 100 OBS BETWEEN 1 AND 10,PUT IN C5
    100 RANDOM INTEGERS BETWEEN     1 AND     10
MTB > ADD C1-C5,PUT IN C6
MTB > DIVIDE C6 BY 5,PUT IN C7
MTB > AVERAGE C7
   MEAN    =      5.6320
MTB > STANDARD DEVIATION C7
   ST.DEV. =        1.2370
MTB > STOP
```

7.3 APPLICATION OF THE CENTRAL LIMIT THEOREM

The central limit theorem tells us about the sampling distribution of sample means by describing the shape of the distribution of all possible sample means. It also specifies the relationship between the mean μ of the population and the mean $\mu_{\bar{x}}$ of the sampling distribution, and the relationship between the standard deviation σ of the population and the standard deviation $\sigma_{\bar{x}}$ of the sampling distribution. Since sample means are approximately normally distributed, we will be able to answer probability questions by using Table 5 of Appendix E.

Illustration 7-3   Consider a normal population with $\mu = 100$ and $\sigma = 20$. If a sample of size 16 is selected at random, what is the probability that this sample will have a mean value between 90 and 110? That is, what is $P(90 < \bar{x} < 110)$?

Solution   The CLT states that the distribution of $\bar{x}$'s is approximately normally distributed. To determine probabilities associated with a normal distribution, we will need to convert the statement $P(90 < \bar{x} < 110)$ to a probability statement concerning z in order to use Table 5, the standard normal distribution table. The sampling distribution is shown in the following figure, with $P(90 < \bar{x} < 110)$ represented by the shaded area.

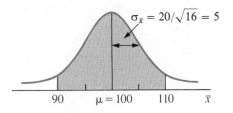

$$\sigma_{\bar{x}} = 20/\sqrt{16} = 5$$

The formula for finding z if a value of $\bar{x}$ is known is

$$z = \frac{\bar{x} - \mu_{\bar{x}}}{\sigma_{\bar{x}}} \qquad (7\text{-}1)$$

However, the CLT tells us that $\mu_{\bar{x}} = \mu$ and $\sigma_{\bar{x}} = \sigma/\sqrt{n}$. Therefore, we will rewrite formula (7-1) in terms of μ and σ:

$$z = \frac{\bar{x} - \mu}{\sigma/\sqrt{n}} \qquad (7\text{-}2)$$

Using formula (7-2), we find that $\bar{x} = 90$ has a standard score of

$$z = \frac{90 - 100}{20/\sqrt{16}} = \frac{-10}{5}$$

$$= -2.0$$

$\bar{x} = 110$ has a standard score of

$$z = \frac{110 - 100}{20/\sqrt{16}} = \frac{10}{5} = \mathbf{2.0}$$

Therefore,

$$P(90 < \bar{x} < 110) = P(-2.0 < z < 2.0)$$
$$= 2(0.4772)$$
$$= \mathbf{0.9544} \qquad \blacksquare$$

   Before we look at more illustrations, let's consider for a moment what is implied by saying that $\sigma_{\bar{x}} = \sigma/\sqrt{n}$. To demonstrate, let's suppose that $\sigma = 20$ and let's use a sampling distribution of samples of size 4. Now $\sigma_{\bar{x}}$ would be $20/\sqrt{4}$, or 10, and approximately 95 percent (0.9544) of all such sample means should be within the interval from 20 below to 20 above the population mean (within two standard deviations of the population mean). However, if the sample size were increased to 16, $\sigma_{\bar{x}}$ would become

$$20/\sqrt{16} = 5$$

and approximately 95 percent of the sampling distribution would be within 10 units of the mean, and so on. As the sample size increases, the size of $\sigma_{\bar{x}}$ becomes smaller, so that the distribution of sample means becomes much narrower. Figure 7-10 illustrates what happens to the distribution of $\bar{x}$'s as the size of the individual samples increases. Recall that the area under the normal curve is always exactly one unit of area. So as the width of the curve narrows, the height will have to increase in order to maintain this area.

Figure 7-10
Distributions of sample
means

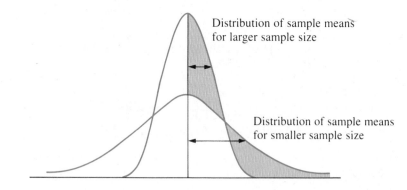

Distribution of sample means
for larger sample size

Distribution of sample means
for smaller sample size

Illustration 7-4   Kindergarten children have heights that are approximately normally distributed about a mean of 39 inches and a standard deviation of 2 inches. A random sample of size 25 is taken and the mean $\bar{x}$ is calculated. What is the probability that this mean value will be between 38.5 and 40 inches?

Solution   We want to find $P(38.5 < \bar{x} < 40.0) = P(? < z < ?)$, where the z scores are as follows:

$$\text{When } \bar{x} = 38.5: z = \frac{38.5 - 39.0}{2/\sqrt{25}} = \frac{-0.5}{0.4}$$

$$= -1.25$$

$$\text{When } \bar{x} = 40.0: z = \frac{40.0 - 39.0}{2/\sqrt{25}} = \frac{10}{0.4}$$

$$= 2.5$$

(See the following figure.)

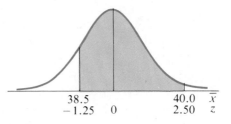

38.5   40.0 $\bar{x}$
−1.25   0   2.50 z

Therefore,

$$P(38.5 < \bar{x} < 40.0) = P(-1.25 < z < 2.50)$$
$$= 0.3944 + 0.4938 = \mathbf{0.8882}$$

■

Illustration 7-5   Referring to Illustration 7-4, within what limits would the middle 90 percent of the sampling distribution of sample means of sample size 100 fall?

Solution   The basic formula is

$$z = \frac{\bar{x} - \mu}{\sigma/\sqrt{n}}$$

and

$$\sigma_{\bar{x}} = \frac{\sigma}{\sqrt{n}} = \frac{2}{\sqrt{100}} = \frac{2}{10} = \mathbf{0.2}$$

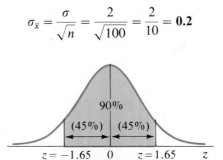

90%

(45%)   (45%)

$z = -1.65$ 0 $z = 1.65$ z

(Recall that the area to the right of the mean, 45 percent (0.4500), is related to the z score of 1.65, according to Table 5.) We want to find the two values for $\bar{x}$.

If $z = -1.65$,

$$-1.65 = \frac{\bar{x} - 39}{0.2}$$

$$(-1.65)(0.2) = \bar{x} - 39$$

$$\bar{x} = 39 - 0.33$$

$$= \mathbf{38.67}$$

If $z = 1.65$,

$$1.65 = \frac{\bar{x} - 39}{0.2}$$

$$(1.65)(0.2) = \bar{x} - 39$$

$$\bar{x} = 39 + 0.33$$

$$= \mathbf{39.33}$$

Thus

$$P(\mathbf{38.67} < \bar{x} < \mathbf{39.33}) = 0.90$$

EXERCISES

7.14 A random sample of size 36 is to be selected from a population that has a mean μ of 50 and a standard deviation σ of 10.

    a. What is the probability that this sample mean will be between 45 and 55?

    b. What is the probability that the sample mean will have a value greater than 48?

    c. What is the probability that the sample mean will be within 3 units of the mean?

7.15 Each of 25 students flips a coin 16 times. Let x represent the number of heads obtained by each student. What is the probability that the mean number of heads for the 25 students is between 7.5 and 8.5?

7.16 Consider the approximately normal population of heights of male students at a college. Assume that the individual heights have a mean of 69 inches and a standard deviation of 4 inches. A random sample of 16 heights is obtained.

    a. Find the mean of this sampling distribution.

    b. Find the standard error of the mean.

    c. Find the shape of this sampling distribution.

d. Find $P(\bar{x} > 70)$.

e. Find $P(\bar{x} < 67)$.

7.17 The heights of the kindergarten children mentioned in Illustration 7-4 are approximately normally distributed with $\mu = 39$ and $\sigma = 2$.

a. If an individual kindergarten child is selected at random, what is the probability that he or she has a height between 38 and 40 inches?

b. A classroom of 30 of these children is used as a sample. What is the probability that the class mean $\bar{x}$ is between 38 and 40 inches?

c. If an individual kindergarten child is selected at random, what is the probability that he or she is taller than 40 inches?

d. A classroom of 30 of these kindergarten children is used as a sample. What is the probability that the class mean $\bar{x}$ is greater than 40 inches?

7.18 A shipment of steel bars will be accepted if the mean breaking strength of a random sample of 10 steel bars is greater than 250 pounds per square inch. In the past, the breaking strength of such bars has had a mean of 235 and a variance of 400.

a. What is the probability, assuming that the breaking strengths are normally distributed, that one randomly selected steel bar will have a breaking strength in the range from 245 to 255?

b. What is the probability that the shipment will be accepted?

7.19 From a sample of 50 employees, taken in a random manner from all of the employees of a large firm, the mean weekly earnings of employed males was $259.30. Given the current wage structure, it has been estimated in labor negotiations that the standard deviation is $34.10. What is the probability of the sample outcome or less if the population mean is really $275.00, as the management negotiator insists?

7.20 The baggage weights for passengers using a particular airline are normally distributed with a mean of 20 pounds and a standard deviation of 4 pounds. If the limit on total luggage weight is 2125 pounds, what is the probability that the limit will be exceeded for 100 such passengers?

7.21 A trucking firm delivers appliances for a large retail operation. The packages (or crates) have a mean weight of 300 pounds and a variance of 2500 pounds.

a. If a truck can carry 4000 pounds and 25 appliances need to be picked up, what is the probability that the 25 appliances will have an aggregate weight greater than the truck's capacity? (Assume that the 25 appliances represent a random sample.)

b. If the truck has a capacity of 8000 pounds, what is the probability that it will be able to carry the entire lot of 25 appliances?

7.22 The following MINITAB program selects 100 samples (each of size 4) from a normal distribution that has a mean of 5 and a standard deviation of 2. The sample means are put into column 6 and printed out.

a. Find $P(4 < \bar{x} < 6)$.

b. How many of the sample means computed fall between 4 and 6?

c. Why do the answers in (a) and (b) differ?

```
MTB > NRANDOM 100 OBS MU = 5,SIGMA = 2,PUT IN C1
     100 NORMAL OBS.WITH MU =       5.0000 AND SIGMA = 2.0000
MTB > NRANDOM 100 OBS MU = 5,SIGMA = 2,PUT IN C2
     100 NORMAL OBS. WITH MU =      5.0000 AND SIGMA = 2.0000
MTB > NRANDOM 100 OBS MU = 5,SIGMA = 2,PUT IN C3
     100 NORMAL OBS. WITH MU =      5.0000 AND SIGMA = 2.0000
MTB > NRANDOM 100 OBS MU = 5,SIGMA = 2,PUT IN C4
     100 NORMAL OBS. WITH MU =      5.0000 AND SIGMA = 2.0000
MTB > ADD C1-C4,PUT IN C5
MTB > DIVIDE C5 BY 4,PUT IN C6
MTB > PRINT C6
C6
  5.64967  4.98275  3.82674  6.04644  3.47323  4.66087  6.46622
  4.49978  5.68478  4.64910  5.38588  5.55961  5.64640  3.72777
  6.64830  5.80450  4.73553  4.32134  5.96149  6.17268  4.05841
  6.46568  4.43453  4.24916  5.68926  4.80518  5.15319  5.07253
  4.89557  5.28228  4.64055  6.54022  4.38253  4.76252  6.20739
  4.84407  4.09733  5.05982  4.48738  6.32345  6.98795  3.86917
  6.10219  5.01370  4.60655  6.16415  5.50360  6.23161  5.55847
  5.08329  5.93220  5.24714  5.15086  3.46194  5.28932  7.39262
  4.46692  3.45167  4.02496  5.66649  4.59504  5.76057  6.08165
  5.27776  5.25800  4.92345  4.45725  4.35168  5.82371  4.29552
  6.22617  5.25266  6.14570  4.91916  5.06789  5.18234  3.34981
  4.99196  5.68083  3.83586  4.86218  4.58404  5.95582  2.07919
  5.02754  6.23805  5.39304  5.35205  4.43152  5.81570  5.18657
  4.95501  4.48976  6.99998  5.37029  5.96970  5.65352  5.15685
  4.83735  4.43587
MTB > STOP
```

In Retrospect

In Chapters 6 and 7 we used the standard normal probability distribution. There are now two formulas for calculating a z score:

$$z = \frac{x - \mu}{\sigma} \quad and \quad z = \frac{\bar{x} - \mu}{\sigma/\sqrt{n}}$$

You must distinguish between these two formulas. The first gives the standard score when dealing with individual values from a normal population (x values). The second uses information provided by the central limit theorem. Sampling distributions of sample means are approximately normally distributed. Therefore, the standard scores and the probabilities in Table 5, Appendix E may be used in connection with sample means ($\bar{x}$ values). The key to distinguishing between the formulas is to decide whether the problem deals with individual values of x from the population or sample means from the sampling distribution. If it deals with the individual values of x, we use the first formula, as presented in Chapter 6. If, on the other hand, the problem deals with the sample means, we use the second formula and proceed as illustrated in this chapter.

The basic purpose for considering what happens under repeated sampling, as discussed in this chapter, is to form sampling distributions. The sampling distribution

is then used to describe the variability that occurs from one sample to the next. Once this pattern of variability is known and understood for a specific sample statistic, we will be able to make accurate predictions about the corresponding population parameters. The central limit theorem describes the distribution for sample means. We will begin to make inferences about population means in Chapter 8.

There are other reasons for repeated sampling. Repeated samples are commonly used in the field of production control, in which samples are taken to determine whether a product is of the proper size or quantity. When the sample is defective, a mechanical adjustment of the machinery is necessary. The adjustment is then followed by another sampling.

The "standard error of the _____" is the name used for the standard deviation of the sampling distribution of whatever statistic is named in the blank. In this chapter we have concerned with the standard error of the mean. However, we could also work with the standard error of the range, median, or whatever.

The article from the *Chicago Tribune* (page 253) discusses the results of a poll taken in Chicago by the *Tribune*. The footnote reports: "margin of error ± 6 percent." This margin of error is related to the standard error of proportion (the standard deviation of the sampling distribution of sample proportion, binomial probability). This relationship between "margin of error" and "standard error" will be discussed in Chapter 9.

You should now be familiar with the concept of a sampling distribution and, in particular, with the sampling distribution of sample means. In Chapter 8 we will begin to make predictions about the values of the various population parameters.

Chapter Exercises

7.23 A sample of size n ($n > 30$) is taken from a population.

    a. What is the probability that $\bar{x}$ will be within 2 standard errors of μ.

    b. What is the probability that x will be within 3 standard errors of μ.

7.24 A population has a normal distribution with an unknown μ and a standard deviation $\sigma = 5$. Find the probability that $\bar{x}$ will be within one unit of μ if n equals the following values.

    a. $n = 25$ **b.** $n = 100$ **c.** $n = 225$

7.25 A population has a normal distribution with an unknown mean μ and a standard deviation $\sigma = 5$. A random sample of size $n = 25$ is taken from this population.

    a. Find the probability that $\bar{x}$ will be within one unit of the mean μ, that is, find $P(-1 < \bar{x} - \mu < 1)$.

    b. If the standard deviation were 2.5 instead of 5, find $P(-1 < \bar{x} - \mu < 1)$.

    c. If the standard deviation were 10 instead of 5, find $P(-1 < \bar{x} - \mu < 1)$.

7.26 Consider a binomial distribution with $n = 400$ and $p = 0.5$. If a sample of size 25 is selected from this binomial distribution, find the probability that the mean of the sample is between 195 and 205.

7.27 Suppose that a box contains three identical blocks numbered 2, 4, and 6. A sample of three numbers is drawn with replacement (that is, the first number is drawn, observed, and returned; then the second number is drawn; then the third). The mean of the sample is determined.

a. Make a list that shows all the possible samples that could result from the sampling. [*Hint*: There should be 27 samples.]

b. Determine the mean of each of these samples and form a sampling distribution of these sample means. (Express as a probability of distribution.)

c. Construct a probability histogram of this probability distribution.

d. Find the mean of this sampling distribution, $\mu_{\bar{x}}$.

e. Find the standard error of the mean $\sigma_{\bar{x}}$ for this sampling distribution.

7.28 Consider the experiment of taking a standardized mathematics test. The variable x is the raw score received. The exam has a mean of 720 and a standard deviation of 60. Assume that the variable x is normally distributed.

a. Consider the experiment of an individual student taking this exam. Describe the distribution (shape of distribution, mean, and standard deviation) for the experiment.

b. A group of 100 students takes the exam; the mean is reported as a result. Describe the sampling distribution (shape of distribution, mean, and standard deviation) for the experiment.

c. What is the probability that the student in (a) scored less than 725.6?

d. What is the probability that the mean of the group in (b) is less than 725.6?

7.29 The diameters of Red Delicious apples in a certain orchard are normally distributed with a mean of 2.63 inches and a standard deviation of 0.25 inch.

a. What percentage of the apples in this orchard have diameters less than 2.25 inches?

b. What percentage of the apples in the orchard are larger than 2.56 inches?

A random sample of 100 apples is gathered and the mean diameter obtained is $\bar{x} = 2.56$.

c. If another sample of size 100 is taken, what is the probability that its sample mean will be greater than 2.56 inches?

d. Why is the z score used in answering parts (a), (b), and (c)?

e. Why is the formula for z used in (c) different from that used in parts (a) and (b)?

7.30 Find a value for e such that 95 percent of the apples in Exercise 7.29 are within e units of the mean 2.63. That is, find e such that $P(2.63 - e < x < 2.63 + e) = 0.95$.

7.31 Find a value for E such that 95 percent of the samples of 100 apples taken from the orchard in Exercise 7.29 will have mean values within E units of the mean 2.63. That is, find E such that $P(2.63 - E < \bar{x} < 2.63 + E) = 0.95$.

7.32 A random sample of 40 part-time employees of fast-food chain outlets in a large metropolitan area showed a mean income of $125.60. The average weekly income for all such employees is $130.60 and the standard deviation is $15.00.

a. What is the probability of a sample mean of $125.60 or less?

b. If the sample size is set at 40, what is the probability that a sample will have a mean between $128.00 and $135.00?

7.33 As a result of several surveys during the last three years, it was concluded that people want pollution controlled and are willing to pay pollution taxes in order to do something about it. The amount that people are willing to pay has a mean of $22.80 per year and a standard deviation of $3.00. Given these values, what is the probability that a random sample of 200 individuals will show a mean differing from $22.80 by more than $0.75?

7.34 Every year, around Halloween, many street signs in a small city are defaced. The average repair cost per sign is $68.00 and the standard deviation is $12.40.

a. If 300 signs are damaged this year, there is a 5 percent chance that the total repair costs for the 300 signs will exceed what value?

b. You are about 68 percent certain that the total repair costs for 300 signs will fall within what interval centered around $20,400?

7.35 After several years of growth, Douglas fir trees being cultivated by a nursery currently have a mean height of 72 inches and a standard deviation of 10 inches. The heights are approximately normally distributed.

a. What proportion of the time will random samples of 100 trees show a mean height between 70 and 75 inches?

b. What mean heights for samples of 100 will fall more than 3 standard errors from the mean of 72 inches?

7.36 A pop-music record firm wants the distribution of lengths of cuts on its records to have an average of 2 minutes and 15 seconds (135 seconds) and a standard deviation of 10 seconds, so that disc jockeys will have plenty of time for commercials within each 5-minute period. The population of times for cuts is approximately normally distributed with only a negligible skew to the right. You have just timed the cuts on a new release and have found that the 10 cuts average 140 seconds.

a. What percent of the time will the average be 140 seconds or longer, if the new release is randomly selected?

b. If the music firm wants 10 cuts to average 140 seconds less than 5 percent of the time, what must the population mean be given that the standard deviation remains at 10 seconds?

7.37 A manufacturer of light bulbs says that its light bulbs have a mean life of 700 hours and a standard deviation of 120 hours. You purchased 144 of these bulbs with the idea that you would purchase more if the mean life of your sample is more than 680 hours. What is the probability that you will not buy again from this manufacturer?

7.38 A tire manufacturer claims (based on years of experience with his tires) that the mean mileage is 35,000 miles and the standard deviation is 5,000 miles. A consumer agency randomly selects 100 of these tires and finds a sample mean of 31,000. Should the consumer agency doubt the manufacturer's claim?

7.39 Take a random sample of 40 single-digit numbers from the random number table (Table 1, Appendix E). Calculate the sample mean $\bar{x}$. Having calculated the sample mean, what do you believe the population mean μ to be?

§ **7.40** For large samples, the sample sum $(\sum x)$ has an approximately normal distribution. The mean of the sample sum is n and the standard deviation is $\sqrt{n} \cdot \sigma$. The distribution of savings per account for a savings and loan institution has a mean equal to $750 and a standard deviation equal to $25. For a sample of 50 such accounts, find the probability that the sum in the 50 accounts exceeds $38,000.

Vocabulary List

Be able to define each term. In addition, describe, in your own words, and give an example of each term. Your examples should not be ones given in class or in the textbook.

The bracketed numbers indicate the chapter in which the term first appeared, but you should define the terms again to show increased understanding of their meaning.

central limit theorem   sampling distribution
frequency distribution [2]   standard error of the mean
probability distribution [5]   theoretical distribution
random sample [2] z score [2, 6]
repeated sampling

Quiz

Answer "True" if the statement is always true. If the statement is not always true, replace the boldface words with words that make the statement always true.

7.1 A sampling distribution **is** a distribution listing all the sample statistics that describe a particular sample.

7.2 The histograms of **all** sampling distributions are symmetrically shaped.

7.3 The mean of the sampling distribution of $\bar{x}$'s is equal to the mean of the **sample**.

7.4 The standard error of the mean is the standard deviation of the population **from which the samples have been taken**.

7.5 The standard error of the mean **increases** as the sample size increases.

7.6 The shape of the distribution of sample means is always that of a **normal** distribution.

7.7 A **probability** distribution of a sample statistic is a distribution of all the values of that statistic that were obtained from all possible samples.

7.8 The central limit theorem provides us with a description of the three characteristics of a sampling distribution of sample **medians**.

7.9 A **frequency** sample is obtained in such a way that all possible samples of a given size have an equal chance of being selected.

7.10 We **do not need** to take repeated samples in order to use the concept of the sampling distribution.

Working with Your Own Data

The central limit theorem is very important to the development of the rest of this course. Its proof, which requires the use of calculus, is beyond the intended level of this course. However, the truth of the CLT can be demonstrated both theoretically and by experimentation. The following series of questions will help to verify the central limit theorem both ways.

A   THE POPULATION

Consider the theoretical population that contains the three numbers 0, 3, and 6 in equal proportions.

  1. a. Construct the theoretical probability distribution for the drawing of a single number, with replacement, from this population.

    b. Draw a histogram of this probability distribution.

    c. Calculate the mean μ and the standard deviation σ for this population.

B   THE SAMPLING DISTRIBUTION, THEORETICALLY

Let's study the theoretical sampling distribution formed by the means of all possible samples of size 3 that can be drawn from the given population.

  2. Construct a list showing all the possible samples of size 3 that could be drawn from this population. (There are 27 possibilities.)

  3. Find the mean for each of the 27 possible samples listed in answer to question 2.

  4. Construct the probability distribution (the theoretical sampling distribution of sample means) for these 27 sample means.

  5. Construct a histogram for this sampling distribution of sample means.

  6. Calculate the mean $\mu_{\bar{x}}$ and the standard error of the mean $\sigma_{\bar{x}}$.

  7. Show that the results found in answers 1c, 5, and 6 support the three claims made by the central limit theorem.

C   THE SAMPLING DISTRIBUTION, EMPIRICALLY

Let's now see whether the central limit theorem can be verified empirically; that is, does it hold when the sampling distribution is formed by the sample means that result from several random samples?

279

8. Draw a random sample of size 3 from the given population. List your sample of three numbers and calculate the mean for this sample.

You may take three identical "tags" numbered 0, 3, and 6, put them in a "hat," and draw your sample using replacement between each drawing. Or you may use dice; let 0 be represented by 1 and 2, let 3 be represented by 3 and 4, and 6 by 5 and 6. You may also use random numbers to simulate the drawing of your samples. Or you may draw your sample from the list of random samples at the bottom of the page. Describe the method you decide to use. (Ask your instructor for guidance.)

9. Repeat question 8 forty-nine (49) more times so that you have a total of fifty (50) sample means that have resulted from samples of size 3.

10. Construct a frequency distribution of the 50 sample means found in answering questions 8 and 9.

11. Construct a histogram of the frequency distribution of observed sample means.

12. Calculate the mean $\bar{x}$ and standard deviation $s_{\bar{x}}$ of the frequency distribution formed by the 50 sample means.

13. Compare the observed values of $\bar{x}$ and $s_{\bar{x}}$ with the values of $\mu_{\bar{x}}$ and $\sigma_{\bar{x}}$. Do they agree?

The following table contains 100 samples of size 3 that were generated randomly by computer.

6 3 0	0 3 0	6 6 0	3 3 6	6 6 3	6 3 3
0 0 3	3 0 6	3 3 0	3 6 6	0 3 0	6 6 3
6 6 6	0 3 0	6 3 6	0 6 3	6 0 3	6 3 3
6 0 0	3 0 6	6 3 3	3 3 0	3 3 0	3 3 3
3 3 3	3 0 0	6 6 6	3 3 6	0 0 6	0 6 3
6 6 6	0 0 6	3 3 0	0 6 6	0 0 3	6 6 3
0 0 6	0 0 6	6 6 6	6 3 6	6 6 0	3 0 0
3 6 6	6 3 0	3 6 3	3 0 0	3 3 6	0 6 0
3 0 0	0 3 6	6 3 3	6 0 6	3 3 6	6 0 3
0 3 6	3 6 3	6 6 3	6 6 0	3 3 3	3 0 0
6 3 0	6 6 0	0 3 0	6 6 0	3 6 6	0 3 6
6 3 3	0 3 0	6 6 0	6 6 3	6 6 0	3 0 3
3 6 3	3 6 0	0 0 6	0 3 3	3 6 6	0 3 6
0 6 0	6 0 0	0 6 0	0 6 6	0 3 3	0 3 6
3 3 6	3 3 3	3 3 6	6 3 6	3 3 3	3 6 6
6 3 3	3 0 0	3 0 6	6 0 3	3 6 6	6 0 3
0 3 3	6 3 0	0 3 6	0 3 6		

The calculations required are accomplished most easily with the assistance of an electronic calculator or a canned program on a computer. Your local computer center can provide a list of the available canned programs and the necessary information and assistance to run your data on the computer.

Part 3

Inferential Statistics

The central limit theorem told us about the sampling distribution of sample means. Specifically, it stated that a distribution of sample means was normally or approximately normally distributed about the mean of the population; and it stated the relationship between the standard deviation of the population and the sampling distribution. With this information we are able to make probability statements about the likelihood of certain sample mean values occurring when samples are drawn from a population with a known mean and a known standard deviation. We are now ready to turn this situation around. We will draw one sample, calculate its

mean value, and then an inference about the value of the population mean will be made based on the sample mean's value.

In this part of the textbook we will learn about making two types of inferences: (1) the decision-making process by use of a hypothesis test procedure, and (2) the procedures for estimating a population parameter. Specifically, we will learn about making these two types of inferences about the population mean μ and the population standard deviation for normally distributed populations, and for the probability parameter p of a binomial population.

Chapter 8

Introduction to Statistical Inferences

EVALUATION OF TEACHING TECHNIQUES FOR INTRODUCTORY ACCOUNTING COURSES

ABSTRACT *This study tests the effect of homework collection and quizzes on exam scores. Expectancy theory as modified by Porter and Lawler [1968] suggests that performance (a student's exam score) is dependent upon effort, abilities and traits, and role perception (regarding the class). An accounting instructor is able to influence a student's effort through assigning different tasks (teaching techniques) which may include homework collection or quizzes....*

The hypothesis for this study is that an instructor can improve a student's performance (exam scores) through influencing the student's perceived effort-reward probability. An instructor accomplishes this by assigning tasks (teaching techniques) which are a part of a student's grade and are perceived by the student as a means of improving his or her grade in the class. The student is motivated to increase effort to complete those tasks which should also improve understanding of course material. The expected final result is improved exam scores.

The null hypothesis for this study is:

H_0: Teaching techniques have no significant effect on student's exam scores....

CONCLUSION

The results of this study were unable to provide evidence that collecting/grading accounting homework assignments and giving quizzes have an effect on students' exam scores. While students may perceive a benefit from homework and quizzes and expect introductory accounting courses to utilize them, their exam performance did not significantly improve in this study when they were rewarded for successfully completing those tasks.

Source: David R. Vruwink and Janon R. Otto, *The Accounting Review*, Vol. LXII, No. 2, April 1987.

Chapter
Objectives

A random sample of 36 pieces of data yields a mean of 4.64. What can we deduce about the population from which the sample was taken? We will be asked to answer two types of questions:

1. Is the sample mean significantly different in value from the hypothesized mean value of 4.5?

2. Based on the sample, what estimate can we make about the value of the population mean?

The first question requires us to make a *decision*, whereas the second question requires us to make an *estimation*.

In this chapter and in Chapters 9 and 10 we will find out how a hypothesis test is used to make a statistical decision about three basic parameters: the mean μ, the standard deviation σ, and the binomial probability of "success" p. We will also learn how an estimation of these three parameters is made. In this chapter we will concentrate on learning about the basic concepts of hypothesis testing and estimation. We will deal with questions about the population mean, using two methods that assume that the population standard deviation is known. This assumption is seldom realized in real-life problems, but it will make our first look at inferences much simpler.

8.1   THE NATURE OF HYPOTHESIS TESTING

Suppose that a certain airline company requires the manufacturer of its aircraft to use rivets, whose mean shearing strength exceeds 120 pounds. Each rivet manufacturer who wants to sell rivets to the aircraft manufacturer must demonstrate that its rivets meet the required specification, namely, that the mean shearing strength of all the manufacturer's rivets, μ, be greater than 120 pounds.

Note 1   Each individual rivet has a shearing strength, which is determined by measuring the force required to shear (break) the rivet. The strength of each rivet is not in question, but rather the mean shearing strength of all such rivets. How can this mean strength be determined? Clearly, not all the rivets can be tested. (If they were, only broken rivets would remain, and there would be none left to build airplanes.) Therefore, a sample of rivets will be tested and a decision about the mean strength of all the untested rivets will be based on the observed mean from those sampled.

Note 2   Throughout Chapter 8 we will treat the standard deviation σ as a known, or given, quantity and concentrate on learning the procedures for making inferences about the population mean μ. We will use $\sigma = 12$ for our rivet illustration. In this illustration the rivet supplier is interested in demonstrating that the mean shearing strength of his rivets is greater than $120\,(\mu > 120)$. The statistical procedure used to make this determination is called a *hypothesis test*.

The statistical hypothesis test is a well-organized four-step procedure. Each step of this procedure will be demonstrated and justified as we investigate the rivet illustration.

Step 1   Formulate the null and alternative hypotheses.

Hypothesis

A statement that something is true.

There are two hypotheses, the *null hypothesis* and the *alternative hypothesis*.

Null Hypothesis, H_0

The hypothesis we will test. Generally this is a statement that a population parameter has a specific value. The null hypothesis is so named because it is the "starting point" for the investigation. The phrase "there is no difference" is often used in its interpretation.

Alternative Hypothesis, H_a

This hypothesis, on which we focus our attention, is a statement about the same population parameter that is used in the null hypothesis. Generally this is a statement that specifies that the population parameter has a value different, in some way, from the value given in the null hypothesis. The rejection of the null hypothesis will imply the acceptance of this alternative hypothesis.

The null hypothesis and the alternative hypothesis are formulated by inspecting the problem or statement to be investigated and first formulating two opposing statements. For our illustration these two opposing statements are: (a) "The mean shearing strength is greater than 120 ($\mu > 120$), the airline company's requirement" versus (b) "The mean shearing strength is not greater than 120 ($\mu \le 120$), the negation of the airline company's requirement."

Note   The Trichotomy Law from algebra states that two numerical values must be related in exactly one of three possible relationships; $<$, $=$, or $>$. All three of these possibilities must be accounted for between the two opposing statements.

Statement (b) becomes the null hypothesis.

$$H_0: \mu = 120 \quad (\le) \quad \text{(the mean is not greater than 120)}$$

The parameter of concern, the population mean μ, is assigned (is equal to) a specific value. Further, 120 is the value on which our attention has been focused. This hypothesis represents the opposition to the rivet supplier's desires and says that his rivets *do not meet* the required standards.

Statement (a) becomes the alternative hypothesis.

$$H_a: \mu > 120 \quad \text{(the mean is greater than 120)}$$

This statement says that the mean is greater than 120. "Greater than 120" is clearly different and separate from "equals 120." This statement represents the rivet supplier's desires and says that his rivets *do meet* the required standards.

From this point on in the hypothesis test procedure, we will work under the assumption that the null hypothesis is a true statement. This situation might be compared to a courtroom trial, where the accused is assumed to be innocent until sufficient evidence has been presented to show otherwise. At the conclusion of the decision   hypothesis test, we will make one of two possible **decisions**. We will decide in agreement with the null hypothesis and say that we "fail to reject H_0" (this corresponds to "fail to convict" or an "acquittal" of the accused in a trial). Or we will decide in opposition to the null hypothesis and say that we "reject H_0" (this corresponds to "conviction" of the accused in a trial).

Before continuing with the illustration we need to look at the different possible situations regarding the truth of the null hypothesis and the correctness of the decision to be reached. There are four possible outcomes that could be reached as a result of the null hypothesis being either true or false and the decision being either "fail to reject" or "reject." Table 8-1 shows these four possible outcomes.

Table 8-1

Four possible outcomes in a hypothesis test

	Null Hypothesis	
Decision	True	False
Fail to reject H_0	Type A Correct decision	Type II Error
Reject H_0	Type I Error	Type B Correct decision

type I and type II errors

A **type A correct decision** occurs when the null hypothesis is true and we decide in its favor. A **type B correct decision** occurs when the null hypothesis is false and our decision is in opposition to the null hypothesis. A **type I error** will be committed when a true null hypothesis is rejected, that is, when the null hypothesis is true but we decided against it. A **type II error** is committed when we decide in favor of a null hypothesis that is actually false.

When we make a decision, it would be nice if it were always a correct decision. This, however, is statistically impossible, since we will be making our decision on the basis of sample information. The best we can hope for is to control the **risk**, or probability, with which an error occurs. The **probability** assigned to the type I error is called **alpha, α** (α is the first letter of the Greek alphabet). The **probability** of the type II error is called **beta, β** (β is the second letter of the Greek alphabet). See Table 8-2. To control these errors we will assign a small probability to them. The most frequently used probability values for α are 0.01 or 0.05. The probability assigned to each error will depend on the seriousness of the error. The more serious the error, the less often we will be willing to allow it to occur, and therefore a smaller probability will be assigned to it. In this text we are going to devote our attention to α, the P(type I error). Further discussion of β, the P(type II error), is beyond the scope of this book.

risk

alpha, beta

Table 8-2

Probability with which error occurs

Error	Type Error	Probability
Rejection of a true hypothesis	I	α
Failure to reject a false null hypothesis	II	β

Let's return to our illustration and proceed with Step 2 of the hypothesis test procedure.

Step 2   Determine the test criteria.

Test Criteria

Consist of (1) specifying a level of significances α, (2) determining a test statistic, (3) determining the critical region(s), and (4) determining the critical value(s).

Level of Significance

The probability of committing the type I error, α.

This can be thought of as a "management decision." Typically, someone in charge determines the amount of risk (probability of) that seems reasonable in view of the seriousness of committing the type I error. If the type I error is a very serious or costly error, alpha should be assigned a very small value. If, on the other hand, the type I error is not that serious, then a larger value can be used.

Note   There is a relationship between the probabilities of the type I and type II errors and the sample size. At this stage of development, α, the probability of the type I error, will be given in the statement of the problem, as will the sample size n. For now we will disregard the role of beta. For our illustration, let $\alpha = 0.05$.

Test Statistic

A random variable used to make the decision "fail to reject H_0" or "Reject H_0."

For our illustration, we will want to compare the value of the sample mean to the hypothesized population mean (the value stated in the null hypothesis). In Chapter 7 we studied the central limit theorem and it told us about the distribution of means from all the possible samples of a given size. You will recall that the distribution of sample means, $\bar{x}$'s, is approximately normally distributed. Therefore, we will use the standard normal variable z as our test statistic for this illustration. Draw a sketch of the standard normal distribution and label it.

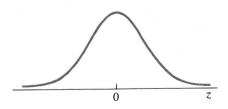

Critical Region

The set of values of the test statistic that will cause us to reject the null hypothesis.

Recall that we are working under the assumption that the null hypothesis is true. Thus, we are assuming that the mean shearing strength of all the rivets is actually 120. If this is the case, then when we select a random sample of 36 rivets, test them, and calculate the mean value for the sample, we can expect this sample mean, $\bar{x}$, to be part of a normal distribution that is centered at 120 and has a standard error of $12/\sqrt{36}$ or 2.0. (Recall that $\mu_{\bar{x}} = \mu$ and $\sigma_{\bar{x}} = \sigma/\sqrt{n}$.) Approximately 95 percent of the sample mean values will be within two standard deviations of 120, or between 116 and 124. Thus, if H_0 is true and $\mu = 120$, then we expect $\bar{x}$ to be between 116 and 124, approximately, 95 percent of the time.

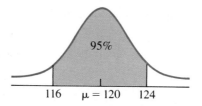

If, however, the value of $\bar{x}$ that we obtain from our sample is larger than 124, say 125, we will have to make a choice. It could be that such an $\bar{x}$-value (125) is either a member of the sampling distribution with mean 120 although it has a very low probability of occurrence (less than 0.05), or $\bar{x} = 125$ is a member of a sampling distribution whose mean is greater than 120, thereby being a value that is more likely to occur.

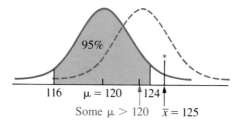

In statistics, we will "bet" on the "more probable to occur" and consider the second choice to be the right one. Thus, the right-hand tail of the z distribution becomes the critical region. And the value of alpha becomes the measure of its area as pictured on the following figure.

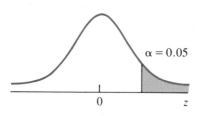

Critical Value

The "first" (or "boundary") value in the critical region.

The critical value for our illustration is $z(0.05)$ and has the value of $+1.65$, as found in Table 5, Appendix E. The notation introduced in Section 6.4 is very handy for identifying critical values.

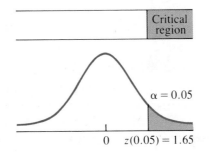

Having completed the ground rules for the test (Steps 1 and 2), we are now ready to obtain and present the sample evidence.

Step 3   Obtain the sample data (evidence) and calculate the value of the test statistic.

At this point the sample data are collected and the value of the sample statistic that corresponds to the parameter being hypothesized about in the null hypothesis is calculated. The value of the test statistic is then determined from this sample statistic.

A sample of 36 rivets is tested and the resulting measurements yield a sample mean of 124.4. This sample statistic must now be converted to a z score using formula (7-2). The resulting z score will be our evidence.

$$z = \frac{\bar{x} - \mu}{\sigma/\sqrt{n}} = \frac{124.4 - 120}{12/\sqrt{36}} = \frac{4.4}{2.0} = \mathbf{2.20}$$

We now have a "calculated value" for the test statistic.

In fact, we now have two values of z. The first value of $z(0.05)$ is 1.65 and the second value is 2.20. Now we must compare these two values. To help keep track of which one is which, we will use an asterisk, *, to identify the calculated value of the test statistic. Thus, $z^* = 2.20$.

Step 4   Make a decision and interpret it.

Make the decision by comparing the value of the calculated test statistic found in Step 3 to the critical value of the test statistic found in Step 2. Graphically this comparison is made by locating an asterisk on the sketch of the z distribution at the location of the value for z^*.

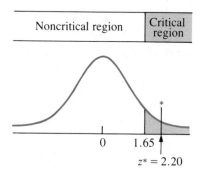

decision rule

Decision Rule

If the test statistic falls within the critical region, we will reject H_0. If the test statistic does not fall within the critical region, we will fail to reject H_0. *Note*: The set of values that are not in the critical region is called the **noncritical region** or, sometimes, the **acceptance region**.

acceptance region

Since the calculated value of z, $z^* = 2.20$, is within the critical region, the decision reached is "reject H_0."

We complete the hypothesis test by interpreting the decision (write the conclusion).

Conclusion Rule

If the decision is "reject H_0," then the conclusion should be worded something like "There is sufficient evidence at the α level of significance to show that...(the meaning of the alternative hypothesis)." If the decision is "fail to reject H_0," then the conclusion should be worded something like "There is *not* sufficient evidence at the α level of significance to show that...(the meaning of the alternative hypothesis)."

For our illustration, the conclusion is: "There is sufficient evidence at the 0.05 level of significance to show that the population of rivets from which the sample was taken does meet the airline company's specifications."

When writing the decision and the conclusion, remember that (1) the decision is about H_0, and (2) the conclusion is a statement about whether or not the contention of H_a was upheld. This is consistent with the "attitude" of the whole hypothesis test procedure. The null hypothesis is the statement that is "on trial" and therefore the decision must be about it. The contention of the alternative hypothesis is the thought (challenge or question) that brought about the need for a decision. Therefore, the question that led to the alternative hypothesis must be answered when the conclusion is written.

Note   Some people prefer to use the phrase "accept H_0" in place of "fail to reject H_0." However, failure to reject the null hypothesis states the decision more accurately, namely, that we failed to find sufficient evidence to convict (reject) the accused (H_0), as in a courtroom trial. We acquitted the defendant; we did not prove the defendant innocent. For these reasons we will use the phrase "fail to reject H_0."

EXERCISES

8.1 As described in this section, the hypothesis-testing procedure has many similarities to a courtroom procedure. The null hypothesis "the accused is innocent" is being tested.

a. Describe the situation involved in each of the four possible outcomes shown in Table 8-1.

b. If the accused is acquitted, does this "prove" him innocent? (Perhaps, in this respect, the phrase "fail to reject H_0" is more accurate in expressing the situation than is "accept H_0.")

c. If the accused is found guilty, does this "prove" his guilt?

8.2 Consider the following nonmathematical situation as a hypothesis test. A medic at the scene of a serious accident tests the null hypothesis "this victim is alive."

a. Carefully state the meaning of the four possible outcomes indicated in Table 8-1.

b. Decide on the seriousness of the two possible errors.

c. If α and β could be controlled statistically, which set of probabilities would you prefer be used if you were the victim?

(1) $\alpha = 0.001$ and $\beta = 0.10$

(2) $\alpha = 0.05$ and $\beta = 0.05$

(3) $\alpha = 0.10$ and $\beta = 0.001$

8.3 Consider the following nonmathematical situation as a hypothesis test. You just received a parachute that was inspected by an inspector whose null hypothesis is "this parachute will open."

a. Carefully state the meaning of the four possible outcomes indicated in Table 8-1.

b. Decide on the seriousness of the two possible errors.

c. If α and β could be controlled statistically, which set of probabilities would you prefer be used if you were going to use the parachute?

(1) $\alpha = 0.001$ and $\beta = 0.10$

(2) $\alpha = 0.05$ and $\beta = 0.05$

(3) $\alpha = 0.10$ and $\beta = 0.001$

 8.4 A paint manufacturer wishes to test the hypothesis that "the addition of an additive will increase the average coverage of the company's paint." The average coverage has been 450 square feet per gallon. Let μ be the average coverage with the additive included. The null hypothesis is "the average coverage will not increase with the addition of the additive," ($\mu = 450$). The alternative hypothesis is "the average coverage will increase with the addition of the additive," ($\mu > 450$). Describe the meaning of the two possible types of errors that can occur in the decision when this test of the hypothesis is conducted.

8.5 a. If the null hypothesis is true, what error in decision could be made?

b. If the null hypothesis is false, what error in decision could be made?

c. If the decision "reject H_0" is made, what error in decision could have been made?

d. If the decision "fail to reject H_0" is made, what error in decision could have been made?

8.6 a. If α is assigned the value 0.001, what are we saying about the type I error?

b. If α is assigned the value 0.05, what are we saying about the type I error?

c. If α is assigned the value 0.10, what are we saying about the type I error?

8.7 a. If β is assigned the value 0.001, what are we saying about the type II error?

b. If β is assigned the value 0.05, what are we saying about the type II error?

c. If β is assigned the value 0.10, what are we saying about the type II error?

8.8 a. If the value of the test statistic falls in the critical region, what decision must we make?

b. If the value of the test statistic does not fall in the critical region (that is, it falls in the noncritical region), what decision must we make?

8.9 Refer to the news article, p. 283. Discuss how the hypothesis for this study, "an instructor can improve a student's performance ...," and the null hypothesis as stated in the news article fit the definitions of the null and alternative hypotheses as defined in this section.

8.10 The director of an advertising agency is concerned with the effectiveness of a television commercial.

a. What null hypothesis is she testing if she commits a type I error when she erroneously says that the commercial is effective?

b. What null hypothesis is she testing if she commits a type II error when she erroneously says that the commercial is effective?

8.11 a. If the null hypothesis is true, the probability of an error in decision is identified by what name?

b. If the null hypothesis is false, the probability of an error in decision is identified by what name?

c. If the test statistic falls in the critical region, what error could be made?

d. If the test statistic falls in the noncritical region, what error might occur?

8.12 a. Suppose that a hypothesis test is to be carried out by using $\alpha = 0.05$. What is the probability of committing a type I error?

b. What proportion of the probability distribution is in the noncritical region provided the null hypothesis is correct?

8.13 a. What is the critical region?

b. What is the critical value?

8.14 Since the size of the type I error can always be made smaller by reducing the size of the critical region, why don't we always choose critical regions that make α extremely small?

8.15 The power of a statistical test is defined to be the probability of rejecting the null hypothesis when the null hypothesis is false. The probability of the type II error is the probability of failing to reject the null hypothesis when it is false. Therefore, the power of a test $= 1 - \beta$, since rejecting a false null hypothesis and failing to reject a false null hypothesis are complementary events. Find the power of a test when the probability of the type II error is

    a. 0.01   **b.** 0.05   **c.** 0.10

8.2 THE HYPOTHESIS TEST (A CLASSICAL APPROACH)

In Section 8.1 we surveyed the steps for and some of the reasoning behind a hypothesis test while looking at a specific illustration. The approach we used is called the classical approach. In this section we are going to continue the study of the classical approach to the hypothesis test procedure as it applies to statements concerning the mean μ of a population. We will continue to impose the restriction that the population standard deviation is known. (This restriction will be removed in Chapter 9.) The first three illustrations deal with additional information about the procedures for formulating the null and alternative hypotheses.

Illustration 8-1   An ecologist would like to show that Rochester has an air pollution problem. Specifically, she would like to show that the mean level of carbon monoxide in downtown Rochester air is higher than 4.9 parts per million. State the null and alternative hypotheses.

Solution   To state the hypotheses we first need to identify the population parameter in question and the value to which it is being compared. The "mean level of carbon monoxide pollution" is the parameter μ and 4.9 parts per million is the specific value. Our ecologist is questioning the value of μ. It could be related to 4.9 in any one of three ways: (1) $\mu < 4.9$, (2) $\mu = 4.9$, or (3) $\mu > 4.9$. These three statements must be arranged to form two statements: one that states what the ecologist is trying to show and one that states the opposite. $\mu > 4.9$ represents the statement "the mean level is higher than 4.9," whereas $\mu < 4.9$ and $\mu = 4.9$ ($\mu \leq 4.9$) represent the opposite, "the mean level is not higher than 4.9." One of these two statements will become the null hypothesis H_0 and the other will become the alternative hypothesis H_a.

Recall   The null hypothesis states that the parameter in question has a specified value.

All this is suggesting that the statement containing the equals sign will become the null hypothesis; the other statement will become the alternative hypothesis.

Thus we have

$$H_0: \mu = 4.9 \quad \text{and} \quad H_a: \mu > 4.9$$

Recall that once the null hypothesis is stated, we proceed with the hypothesis test under the assumption that the null hypothesis is true. Thus $\mu = 4.9$ locates the center of the sampling distribution of sample means. For this reason the null hypothesis will be written with an equals sign only. If $\mu = 4.9$ and $\mu < 4.9$ are stated together as the null hypothesis, the null hypothesis will be expressed as

$$H_0: \mu = 4.9 (\leq)$$

where the $(\leq)$ serves as a reminder of the grouping that took place when the hypothesis was formed.   ■

Note   **The equals sign must be in the null hypothesis**, regardless of the statement of the original problem.

Illustration 8-1 expresses the viewpoint that an ecologist might take. Now let's consider how the Rochester Chamber of Commerce might view the same situation.

Illustration 8-2   In trying to promote the city, the Chamber of Commerce would be more likely to want to conclude that the mean level of carbon monoxide in downtown Rochester is less than 4.9 parts per million. State the null and alternative hypotheses related to this viewpoint.

Solution   Again, the parameter of interest is the mean level μ of carbon monoxide, and 4.9 is the specified value. $\mu < 4.9$ corresponds to "the mean level is less than 4.9," whereas $\mu \geq 4.9$ corresponds to "the mean level is not less than 4.9." Therefore, the hypotheses are

$$H_0: \mu = 4.9 (\geq) \quad \text{and} \quad H_a: \mu < 4.9 \qquad ■$$

A more neutral point of view is suggested in Illustration 8-3.

Illustration 8-3   The "mean level of carbon monoxide in downtown Rochester is not 4.9 parts per million." State the null and alternative hypotheses that correspond to this statement.

Solution   The mean level of carbon monoxide is equal to 4.9 ($\mu = 4.9$) or the mean is not equal to 4.9 ($\mu \neq 4.9$). [*Note:* "Less than or greater than" is customarily expressed as "not equal to."] Therefore,

$$H_0: \mu = 4.9 \quad \text{and} \quad H_a: \mu \neq 4.9 \qquad ■$$

The viewpoint of the experimenter affects the way the hypotheses are formed, as we have seen in these three illustrations. Generally, the experimenter is trying to show that the parameter value is different from the value specified. Thus the experimenter is usually hoping to be able to reject the null hypothesis. Illustrations 8-1, 8-2, and 8-3 represent the three arrangements possible for the $<$, $=$, and $>$ relationships between the parameter μ and the specified value 4.9.

The courtroom trial and the hypothesis test have many similarities, as already emphasized. Let's now look at another similarity. In the courtroom there is a trial only if the prosecutor has shown the court that there is sufficient reason to make the accused stand trial. If the prosecutor did not believe that the accused were guilty, there would be no trial. Much the same is true with the hypothesis test. If the experimenter believes that the null hypothesis is true, he will not challenge its truth and will not be testing it. He proceeds to test the null hypothesis only when he wishes to show that the alternative is correct. Thus the alternative hypothesis is a statement claiming that the value of the parameter is "smaller than" or "larger than" or "different from" the value claimed by the null hypothesis.

Illustration 8-4 demonstrates the complete hypothesis test procedure as it is used in questions dealing with the population mean.

Illustration 8-4   For many semesters an instructor has recorded his students' grades, and the mean μ for all these students' grades is 72. The current class of 36 students seems to be better than average in ability and the instructor wants to show that according to their average "the current class is superior to his previous classes." Does the class mean $\bar{x}$ of 75.2 present sufficient evidence to support the instructor's claim that the current class is superior? Use $\alpha = 0.05$ and $\sigma = 12.0$.

Solution   To be superior, this class must have a mean grade that is higher than the mean of all the previous classes. To be "equal to or less than" would not be superior.

Step 1 $H_0: \mu = 72$ ($\leq$) (class is not superior)

$H_a: \mu > 72$ (class is superior)

Step 2   The level of significance $\alpha = 0.05$ is given in the statement of the problem. The standard score z is used as the test statistic when the null hypothesis is about a population mean and the standard deviation is known. Recall that the central limit theorem tells us that the sampling distribution of sample means is approximately normally distributed. Thus the normal probability distribution will be used to complete the hypothesis test. The critical region—values of the standard score z that will cause a rejection of the null hypothesis—has an area of 0.05 and is located at the extreme right of the distribution. The critical region is on the right because large values of the sample mean suggest "superior," while values near or below 72 support the null hypothesis. See the following figure.

Not superior   Superior ($\mu > 72$)

72 73   90 x

If random samples of size 36 are taken from a population with a mean value equal to 72, then many of the sample means will have values near 72, (71, 72, 71.8, 72.5, 73, and so on). Only sample means that are considerably larger than 72 would cause us to reject the null hypothesis. The critical value, the cutoff between "not superior" and "superior," is determined by α, the probability of the type I error. $\alpha = 0.05$ was given. Thus the critical region (the shaded region of the following figure) has an area of 0.05 and a critical value of $+1.65$. (This value is obtained by using Table 5 of Appendix E.)

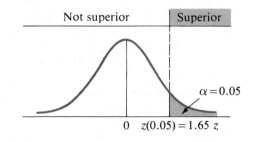

Step 3   The value of the test statistic z will be found by using formula (7-2) and the sample information

$$\bar{x} = 75.2 \quad \text{and} \quad n = 36$$

$$z = \frac{\bar{x} - \mu}{\sigma/\sqrt{n}}$$

Recall   We assumed that μ was equal to 72 in the null hypothesis and that $\sigma = 12.0$ was known.

$$z = \frac{75.2 - 72}{12.0/\sqrt{36}} = \frac{3.2}{2.0} = 1.60$$

$$z^* = \mathbf{1.60}$$

calculated value (*)   We will use an asterisk, *****, to identify the **calculated value** of the test statistic; the asterisk will also be used to locate its value relative to the test criteria (Step 4).

Step 4   We now compare the calculated test statistic, z^*, to the test criteria set up in Step 2 by locating the calculated value on the diagram and placing an asterisk at that value. Since the test statistic (calculated value) falls in the noncritical region (unshaded portion of the diagram), we must reach the following **decision**:

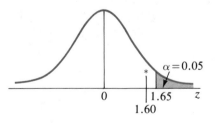

Decision: Fail to reject H_0.

Recall that the critical region was to be shaded in the diagram, and when the test statistic falls in the critical region, we must reject H_0. Step 4 is then completed by

conclusion   stating a **conclusion.**

Conclusion: There is not sufficient evidence at the 0.05 level of significance to show that the current class is superior.   ∎

Does this conclusion seem realistic? 75.2 is obviously larger than 72. Recall that 75.2 is a sample mean, and if a random sample of size 36 is drawn from a population whose mean is 72 and whose standard deviation is 12, then the probability that the sample mean is 75.2 or larger is greater than the risk, α, with which we are willing to make the type I error. (What is $P(\bar{x} > 75.2?)$

Illustration 8-5 It has been claimed that the mean weight of women students at a college is 54.4 kilograms. Professor Schmidt does not believe the statement that the mean is 54.4 kilograms. To test the claim he collects a random sample of 100 weights from among the women students. A sample mean of 53.75 kilograms results. Is this sufficient evidence to reject the null hypothesis? Use $\alpha = 0.05$ and $\sigma = 5.4$ kilograms.

Solution Professor Schmidt's statement suggests that the three possible relationships $(<, =, >)$ between the mean weight μ and the hypothesized 54.4 kilograms be split $(=)$ or $(<, >)$. Therefore, we have the following steps:

Step 1 $H_0: \mu = 54.4$ (mean weight is 54.4 kilograms)

$H_a: \mu \neq 54.4$ (mean weight is not 54.4 kilograms)

The test statistic will be the standard score z, and the normal distribution is used since we are using the sampling distribution of sample means. The sample mean is our estimate for the population mean. Since the alternative hypothesis is "not equal to," a sample mean considerably larger than 54.4, as well as one considerably smaller than 54.4, will be in opposition to the null hypothesis. The values of the sample mean around 54.4 will support the null hypothesis. Therefore, the critical region will be split into two equal parts, one at each extreme of the normal distribution. The area of each region will be $\alpha/2$. Since $\alpha = 0.05$, each part of the critical region will have a probability of 0.025 (see the following figure).

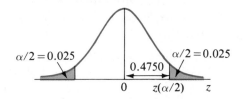

Look in Table 5 of Appendix E for the area 0.4750 (0.4750 = 0.5000 − 0.025). The z score of 1.96 is found and becomes 1.96 on the right of the mean and −1.96 on the left.

Step 2 The test criteria are shown in the following figure.

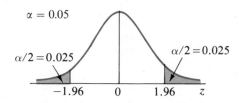

Step 3 Calculate the test statistic z. The sample information is $\bar{x} = 53.75$ and

$n = 100$. σ was given as 5.4, and the null hypothesis assumed that $\mu = 54.4$

$$z = \frac{\bar{x} - \mu}{\sigma/\sqrt{n}} = \frac{53.75 - 54.4}{5.4\sqrt{100}} = \frac{-0.65}{0.54} = -1.204$$

$$z^* = -1.20$$

Locate z^* on the diagram constructed in Step 2.

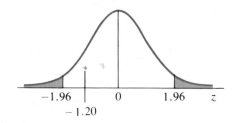

Now we are ready to identify the decision.

Recall   If the test statistic falls in the critical region, we must reject H_0. If the test statistic falls in the noncritical region, we must fail to reject H_0.

Step 4   The calculated value of z, z^*, falls between the critical values cited in Step 2. Therefore, our decision is as follows:

Decision: Fail to reject H_0.

The interpretation of our decision is the only thing left to do. This part of the hypothesis test may very well be the most significant part because the conclusions reached express the results found. We must be very careful to state precisely what is meant by the decision. It is often a temptation to overstate a conclusion. In general, a decision to "fail to reject H_0" may be interpreted to mean that the evidence found does not disagree with the null hypothesis. Note that this does not "prove" the truth of H_0. A decision to "reject H_0" will mean that the evidence found implies the null hypothesis is false and thus indicates the alternative hypothesis to be the case.

Conclusion: The evidence (sample mean) found does not contradict the assumption that the population mean is 54.4 kilograms at the 0.05 level of significance.   ■

Before we look at another illustration, let's summarize briefly some of the details we have seen thus far.

1. The null hypothesis specifies a particular value of a population parameter.

2. The alternative hypothesis can take three forms. Each form dictates a specific location of the critical regions, as shown in the following table.

one-tailed and
two-tailed tests

Sign in the Alternative Hypothesis	$<$	$\neq$	$>$
Critical Region	One region, Left side; **One-tailed test**	Two regions, One on each side; **Two-tailed test**	One region, Right side; **One-tailed test**

3. For many hypothesis tests the sign in the alternative hypothesis "points" in the direction in which the critical region is located. [Think of the not equal to sign ($\neq$) as being both less than ($<$) and greater than ($>$), thus pointing in both directions.]

The value assigned to α is called the significance level of the hypothesis test. Alpha cannot be interpreted to be anything other than the risk (or probability) of rejecting the null hypothesis when it is actually true. We will seldom be able to determine whether the null hypothesis is true or false; we will only decide to reject H_0 or to fail to reject H_0. The relative frequency with which we reject a true hypothesis is α, but we will never know the relative frequency with which we make an error in decision. The two ideas are actually quite different; that is, a type I error and an error in decision are two different things altogether (remember, there are two types of errors).

Let's look at some more illustrations of the hypothesis test.

Illustration 8-6   The student body at many community colleges is considered a "commuter population." The following question was asked of the Student Affairs Office: "How far (one way) does the average community college student commute to college daily?" The office answered: "No more than 9.0 miles." The inquirer was not convinced of the truth of this and decided to test the statement. He took a sample of 50 students and found a mean commuting distance of 10.22 miles. Test the hypothesis stated above at a significance level of $\alpha = 0.05$, using $\sigma = 5$ miles.

Solution

Step 1 $H_0: \mu = 9.0$ ($\leq$)   (no more than 9.0 miles)

$H_a: \mu > 9.0$ (more than 9.0 miles)

The one-way distance traveled by the "average student" would be the same as the mean one-way distance traveled by all students. Therefore, the parameter of concern is μ. The claim "no more than 9.0 miles" implies that the three possible relationships should be grouped "no more than 9.0" ($\leq$) versus "more than 9.0" ($>$).

Step 2   The critical region is on the right (since H_a contains the $>$ sign); H_0 will be rejected if it appears that the $\bar{x}$ observed is significantly greater than the 9.0 claimed. Notice that the sign in $H_a(>)$ points toward the critical region. (*See* the following figure.)

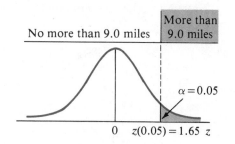

Step 3

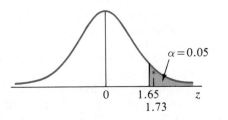

$$z = \frac{\bar{x} - \mu}{\sigma/\sqrt{n}} = \frac{10.22 - 9.0}{5/\sqrt{50}} = \frac{1.22}{0.707} = 1.73$$
$$z^* = \mathbf{1.73}$$

Step 4   *Decision*: Reject H_0. (z^* fell in the critical region)

Conclusion: At the 0.05 level of significance, we conclude that the average community college student probably travels more than 9.0 miles. ■

Illustration 8-7   Draw a sample of 40 single-digit numbers from the random number table (Table 1, Appendix E) and test the null hypothesis $\mu = 4.5$. Use $\alpha = 0.10$ and $\sigma = 2.87$. The standard deviation of the random digits was found in Chapter 5. (*See* Exercise 5.18.)

Solution

Step 1 $H_0: \mu = 4.5$
         $H_a: \mu \neq 4.5$

Step 2 $\alpha = 0.10$ (*See* the following figure.)

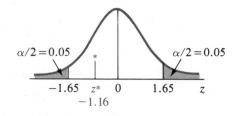

Step 3   The following random sample was drawn from Table 1 of Appendix E:

$$\begin{array}{cccccccccc}
2 & 8 & 2 & 1 & 5 & 5 & 4 & 0 & 9 & 1 \\
0 & 4 & 6 & 1 & 5 & 1 & 1 & 3 & 8 & 0 \\
3 & 6 & 8 & 4 & 8 & 6 & 8 & 9 & 5 & 0 \\
1 & 4 & 1 & 2 & 1 & 7 & 1 & 7 & 9 & 3
\end{array}$$

$$\sum x = 159$$
$$n = 40$$
$$\bar{x} = 3.975$$

$$z = \frac{\bar{x} - \mu}{\sigma/\sqrt{n}} = \frac{3.975 - 4.50}{2.87/\sqrt{40}} = \frac{-0.525}{0.454} = -1.156$$

$$z^* = -1.16$$

z^* falls in the noncritical region, as shown in color on the diagram in Step 2.

Step 4 *Decision*: Fail to reject H_0.

Conclusion: The observed sample mean is not significantly different from 4.5, at the 0.10 level of significance. ∎

Suppose that we were to take another sample of size 40 from the table of random digits. Would we obtain the same results? Suppose that we took a third sample or a fourth? What results might we expect? What is the level of significance α? Yes, its value is 0.10, but what does it measure? Table 8-3 lists the means obtained from 10 different random samples of size 40 that were taken from Table 1. The calculated value of z that corresponds to each $\bar{x}$ and the decision each would dictate are also listed. Each of the 10 calculated z scores is shown in Figure 8-1. The sample number is used to identify each score. Note that one of the samples caused us to reject the null hypothesis, although we know that it is true for this situation. Why did this happen?

Table 8-3

Random samples of size 40 taken from Table 1, Appendix E

Sample Number	Sample Mean ($\bar{x}$)	Calculated z (z^*)	Decision Reached
1	4.62	+0.26	Fail to reject H_0
2	4.55	+0.11	Fail to reject H_0
3	4.08	−0.93	Fail to reject H_0
4	5.00	+1.10	Fail to reject H_0
5	4.30	−0.44	Fail to reject H_0
6	3.65	−1.87	Reject H_0
7	4.60	+0.22	Fail to reject H_0
8	4.15	−0.77	Fail to reject H_0
9	5.05	+1.21	Fail to reject H_0
10	4.80	+0.66	Fail to reject H_0

Remember α is the probability that we "reject H_0" when it is actually a true statement. Therefore, we can anticipate that a type I error will occur α of the time when testing a true null hypothesis.

Figure 8-1

z scores from Table 8-3; $\alpha = 0.10$

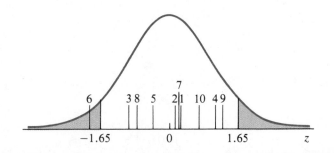

EXERCISES

8.16 State the null hypothesis H_0 and the alternative hypothesis H_a that would be used for a hypothesis test related to each of the following statements.

   a. The mean age of the students enrolled in evening classes at a certain college is greater than 26 years.

   b. The mean weight of packages shipped on Air Express during the last month was less than 36.7 pounds.

   c. The mean life of fluorescent light bulbs is at least 1600 hours.

   d. The mean weight of college football players is no more than 210 pounds.

   e. The mean distance from a home in suburban Chicago to the nearest fire station is less than 4.7 miles.

   f. The mean daily high temperature during February at Daytona Beach, Florida, is not 80 degrees.

   g. The mean grade (measure of steepness) for all ski runs at Greek Peek Ski Center is no less than 20 degrees.

   h. The mean strength of welds by a new process is different from 570 pounds per unit area, the mean strength of welds by the old process.

8.17 Suppose that we want to test the hypothesis that the mean hourly charge for automobile repairs is at least $23 per hour at the repair shops in a nearby city. Explain the conditions that would exist if we make an error in decision by committing a

   a. type I error   **b.** type II error

8.18 Determine the test criteria (critical values and critical region for z) that would be used to test the null hypothesis at the given level of significance, as described in each of the following:

   a. $H_0: \mu = 20$ $(\alpha = 0.10)$
      $H_a: \mu \neq 20$

   b. $H_0: \mu = 24\,(\leq)$ $(\alpha = 0.01)$
      $H_a: \mu > 24$

   c. $H_0: \mu = 10.5\,(\geq)$ $(\alpha = 0.05)$
      $H_a: \mu < 10.5$

   d. $H_0: \mu = 35$ $(\alpha = 0.01)$
      $H_a: \mu \neq 35$

   e. $H_0: \mu = 14.6\,(\geq)$ $(\alpha = 0.02)$
      $H_a: \mu < 14.6$

   f. $H_0: \mu = 6.78\,(\leq)$ $(\alpha = 0.10)$
      $H_a: \mu > 6.78$

   g. $H_0: \mu = 21.50$ $(\alpha = 0.05)$
      $H_a: \mu \neq 21.50$

   h. $H_0: \mu = 0.034$ $(\alpha = 0.01)$
      $H_a: \mu \neq 0.034$

8.19 The calculated value of the test statistic is actually the number of standard errors that the sample mean differs from the hypothesized value of μ in the null hypothesis. Suppose that the null hypothesis is $H_0: \mu = 4.5$, σ is known to be 1.0, and a sample of size 100 results in $\bar{x} = 4.8$.

    a. How many standard errors is $\bar{x}$ above 4.5?

    b. If the alternative hypothesis is $H_a: \mu > 4.5$ and $\alpha = 0.01$, would you reject H_0?

8.20 A machine cuts product parts that have a mean length equal to 15.0 cm with a standard deviation of 0.5 cm. It is known that the standard deviation has not changed, but a change in mean length is possible. A sample of 30 parts is taken with the following results

$$
\begin{array}{cccccccccc}
15.1 & 15.6 & 15.8 & 16.0 & 15.0 & 15.5 & 16.2 & 15.0 & 15.5 & 16.0, \\
15.9 & 14.8 & 16.0 & 15.5 & 15.6 & 15.5 & 16.4 & 15.2 & 15.8 & 16.0 \\
15.5 & 14.9 & 15.9 & 16.2 & 15.9 & 15.8 & 15.0 & 16.1 & 15.7 & 15.6
\end{array}
$$

A MINITAB analysis of the data follows.

    a. Verify the mean, standard deviation, standard error of the mean, and the z value shown.

    b. State H_0 and H_a.

```
MTB > SET DATA IN C1
DATA> 15.1 15.6 15.8 16.0 15.0 15.5 16.2 15.0 15.5 16.0
DATA> 15.9 14.8 16.0 15.5 15.6 15.5 16.4 15.2 15.8 16.0
DATA> 15.5 14.9 15.9 16.2 15.9 15.8 15.0 16.1 15.7 15.6
DATA> END DATA
MTB > ZTEST OF MU = 15.0,SIGMA = .5,DATA IN C1
TEST OF MU = 15.0000 VS MU N.E. 15.0000
THE ASSUMED SIGMA = .0500
                N      MEAN     STDEV    SE MEAN       Z
C1             30    15.6333    0.4270    0.0913    6.94
MTB>STOP
```

8.21 A population has a standard deviation equal to 3.5 and the hypothesis $H_0: \mu = 20.0$ is to be tested against the hypothesis $H_a: \mu > 20.0$. α is specified to be 0.05. For a sample of size 50, how large would the sample mean need to be in order to reject H_0?

8.22 A sample of 36 measurements was taken to test the hypothesis stated in Illustration 8-1. Does the sample mean of 5.3 provide sufficient evidence to reject the null hypothesis in the illustration? Complete the hypothesis test, using $\sigma = 1.8$ and $\alpha = 0.05$.

 8.23 A machine produces ball bearings. The standard deviation remains constant at 0.010 cm. However, the mean diameter can change after the machine has been used for some time. In particular, the mean increases if it changes. A daily sample of 30 bearings is taken to test $H_0: \mu = 3.000$ cm. versus $H_a: \mu > 3.000$ cm. For a given sample we find $\bar{x} = 3.003$. Test the hypothesis and state your conclusion for $\alpha = 0.01$.

8.24 The manager at Air Express feels that the weights of packages shipped recently are less than in the past. Records show that in the past packages have had a mean weight of 36.7 pounds and a standard deviation of 14.2 pounds. A random sample of last month's shipping records yielded a mean weight of 32.1 pounds for 64 packages. Is this sufficient evidence to reject the null hypothesis in favor of the manager's claim? Use $\alpha = 0.01$.

8.25 The manufacturer of the fluorescent light bulbs used by a large industrial complex claims that the bulbs have a mean life of at least 1600 hours. The maintenance foreman had a sample of 100 bulbs randomly identified and kept a record on their use. Does a sample mean of 1562.3 support the foreman's contention that the mean life is less than 1600 hours at the 0.02 level of significance? (Use $\sigma = 150$ hours.)

8.26 A fire insurance company felt that the mean distance from a home to the nearest fire department in a suburb of Chicago was at least 4.7 miles. It set its fire insurance rates accordingly. Members of the community set out to show that the mean distance was less than 4.7 miles. This, they felt, would convince the insurance company to lower its rates. They randomly identified 64 homes and measured the distance to the nearest fire department for each. The resulting sample mean was 4.4. If $\sigma = 2.4$ miles, does the sample show sufficient evidence to support the community's claim, at the $\alpha = 0.05$ level of significance?

8.27 On a popular self-image test the mean score for public assistance recipients is 65 and the standard deviation is 5. A random sample of 42 public assistance recipients in Emerson County are given the test. They achieved a mean score of 60. Do the scores of Emerson County public assistance recipients differ from the average with respect to this variable? Use $\alpha = 0.01$.

8.28 All drugs must be approved by the Food and Drug Administration (FDA) before they can be marketed by a drug manufacturer. The FDA must weigh the error of marketing an ineffective drug, with the usual risks of side effects, against the consequences of not allowing an effective drug to be sold. Suppose, using standard medical treatment, that the mortality rate (r) of a certain disease is known to be a. A manufacturer submits for approval a drug that is supposed to treat this disease. The FDA sets up the hypothesis to test the mortality rate for the drug as (1) $H_0: r = a$, $H_a: r < a$, $\alpha = 0.005$; or (2) $H_0: r = a$, $H_a: r > a$, $\alpha = 0.005$.

  a. If $a = 0.95$, which test do you think the FDA would use?

  b. If $a = 0.05$, which test do you think the FDA would use? Explain.

8.3 THE HYPOTHESIS TEST (A PROBABILITY-VALUE APPROACH)

The classical (or traditional) hypothesis-testing procedure was described in Section 8.2. An alternative approach to the decision-making process in hypothesis testing has gained popularity in recent years, largely as a result of the convenience and the "number crunching" ability of computers. This alternative process is to calculate and report a *probability-value* related to the observed sample statistic. The value reported, p-value, is called the **probability-value**, the **prob-value**, or simply the *p*-**value**. The p-value is then compared to the level of significance α, the probability of the type I error, in order to make the decision to reject H_0 or fail to reject H_0.

prob-value

Prob-value, *p*-value

The prob-value, p-value, of a hypothesis test is the smallest level of significance for which the observed sample information becomes significant, provided the null hypothesis is true. The prob-value is computed by finding the probability that the test statistic could be the value it is or a more extreme value (in the direction of the alternative hypothesis) when the null hypothesis is true. [*Note*: The symbol **P** is often used to represent prob-value, especially in algebraic situations.]

Let's return to Illustration 8-4 and see how its solution would be different if solved using this "prob-value" method.

Illustration 8-8   For many semesters an instructor has recorded his students' grades, and the mean μ for all these students' grades is 72. The current class of 36 students seems to be better than average in ability and the instructor wants to show that, according to their average, "the current class is superior to his previous classes." Does the class mean $\bar{x}$ of 75.2 present sufficient evidence to support the instructor's claim that the current class is superior? Use $\alpha = 0.05$ and $\sigma = 12.0$.

Solution   To be superior, this class must have a mean grade that is higher than the mean of all the previous classes. To be "equal to or less than" would not be superior.

Step 1 $H_0: \mu = 72 \ (\leq)$ (class is not superior)

$H_a: \mu > 72$ (class is superior)

Notice that Step 1 is exactly the same when using either the classical or the prob-value approach to hypothesis testing.

Step 2   Determine α, the probability of the type I error; $\alpha = 0.05$.

Step 3   Using formula (7-2) and the sample information, we will obtain the calculated value of the test statistic z^*.

$$z = \frac{\bar{x} - \mu}{\sigma/\sqrt{n}} \quad \text{and} \quad \bar{x} = 75.2 \quad \text{and} \quad n = 36$$

Recall   We assume that μ was equal to 72 in the null hypothesis and that $\sigma = 12.0$ was known.

$$z = \frac{75.2 - 72.0}{12.0/\sqrt{36}} = \frac{3.2}{2.0} = 1.60$$

$$z^* = 1.60$$

[Reminder: We use an asterisk, *, to identify the calculated value of the test statistic; the asterisk will also be used to locate its value on the probability distribution in the next step.]

Notice that Step 3 in the prob-value approach is exactly the same as Step 3 in the classical approach.

Step 4   Draw a sketch of the probability distribution for the test statistic—z in this case.

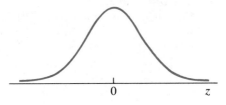

Next locate the value of $z*$ calculated in Step 3 on this distribution (*see* the following figure) and determine what part of the distribution relative to $z*$ represents the prob-value. Then calculate the p-value.

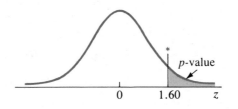

The alternative hypothesis indicates that we are interested in that part of the probability distribution that lies to the right of $z*$, since the "greater than" sign was used.

p-value $= P(z > z*) = P(z > 1.60)$

$\qquad\quad = 0.5000 - 0.4452$ (use Table 5 in Appendix E to find the area under a normal curve)

p-value $= \mathbf{0.0548}$

The results of the statistical analysis are then reported.

Step 5   The prob-value for this hypothesis test is 0.0548. If the decision is made using the 0.05 level of significance (or any α value less than or equal to 0.0548), then the decision will be to "fail to reject the null hypothesis" and the conclusion will be "there is no evidence to show the class as superior." If, however, we had chosen to use a level of significance of 0.10 (or any α value greater than 0.0548), our decision would be to "reject H_0," and our conclusion would be that "the current class is superior" at that level of significance.

Information given to the computer: z-test of $H_o: \mu = 72.0$, $H_a: \mu > 72.0$, $\sigma = 12.0$, and location of sample data

Sample statistics: $n = 36$, $\bar{x} = 75.2$, $s = 11.87$

The calculated z^*

The prob-value

The decision

Computer Solution

MINITAB Printout for Illustration 8-8

```
--ZTEST OF POPULATION MU = 72.0 VS. ALTERNATIVE + 1,
   SIGMA = 12.0, DATA IN C1
   C1    N = 36        MEAN = 75.2      ST.DEV. = 11.87
   TEST OF MU = 72.0   VS. MU G.T.     72.0
   Z = 1.60
   THE TEST IS STATISTICALLY SIGNIFICANT AT ALPHA = 0.0548
   CANNOT REJECT AT ALPHA = 0.05
--STOP
```

Before looking at another illustration, let's summarize the details for this prob-value approach to the hypothesis testing.

1. The null and alternative hypotheses are formulated in the same manner as that used in the classical approach.

2. Determine the level of significance, α, to be used.

3. The value of the test statistic is calculated in Step 3 in exactly the same manner as it was calculated in Step 3 in the classical approach.

4. The prob-value is the area that represents values of z that are more extreme than z^* under the curve of the probability distribution. There are three separate cases: two are one-tailed and one is two-tailed. The direction (or sign) in the alternative hypothesis is the key.

Case I If H_a is one-tailed to the right (that is, contains ">"), then p-value $= P(z > z^*)$, the area to the right of z^*.

Case II If H_a is one-tailed to the left (that is, contains "<"), then p-value $= P(z < z^*)$, the area to the left of z^*.

Case III If H_a is two-tailed (that is, contains "$\neq$"), then

$$p\text{-value} = P(z < -|z^*|) + P(z > |z^*|),$$

the sum of the area to the left of the negative value of z^* and the area to the right of the positive value of z^*. Since both areas are equal, you will probably find one and double it. Thus, p-value $= 2 \times P(z > |z^*|)$.

5. The decision will be made by comparing the p-value to the previously established value of α.

a. If the **calculated prob-value is less than or equal to the desired α**, then the decision must be **reject H_0**.

b. If the **calculated prob-value is greater than the desired α**, then the decision must be **fail to reject H_0**.

6. Conclusions should be worded in the same manner as previously instructed.

Let's look at an illustration involving the two-tailed procedure.

Illustration 8-9   Many of the large companies in a certain city have for years used the Kelley Employment Agency for testing prospective employees. The employment selection test used has historically resulted in scores distributed about a mean of 82 and a standard deviation of 8. The Brown Agency has developed a new test that is quicker and easier to administer and therefore less expensive. Brown claims that their test results are the same as those obtained on the Kelley test. Many of the companies are considering a change from the Kelley Agency to the Brown Agency in order to cut costs. However, they are unwilling to make the change if the Brown test results have a different mean value.

   An independent testing firm tested 36 prospective employees. A sample mean of 80 resulted. Determine the prob-value associated with this hypothesis test.

Solution   The Brown Agency's test results will be different if the mean test score is not equal to 82. They will be the same if the mean is 82. Therefore,

Step 1 $H_0: \mu = 82$ (test results have same mean)
           $H_a: \mu \neq 82$ (test results have different mean)

Step 2   Step 2 is omitted when a question asks for the prob-value and not a decision.

Step 3   The sample information $n = 36$ and $\bar{x} = 80$ and formula (7-2) are used to calculate $z*$:

$$z = \frac{\bar{x} - \mu}{\sigma/\sqrt{n}}$$

$$z = \frac{80 - 82}{8/\sqrt{36}} = \frac{-2}{8/6} = -1.50$$

$$z* = -1.50$$

Step 4   The value of $z*$ is located on the normal distribution. (*See* the accompanying figure.)

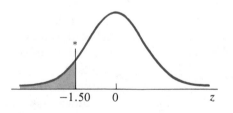

Since the alternative hypothesis indicates a two-tailed test, we must find the probability associated with two areas, namely, $P(z < -|z*|)$ and $P(z > |z*|)$. And since $z* = -1.50$, the value of $|z*| = 1.50$. Thus the p-value $= P(z < -1.50) + P(z > 1.50)$, as shown in Figure 8-2.

$$p\text{-value} = P(z < -1.50) + P(z > 1.50)$$
$$= (0.5000 - 0.4332) + (0.5000 - 0.4332)$$
$$= 0.0668 + 0.0668$$
$$p\text{-value} = \mathbf{0.1336}$$

Figure 8-2

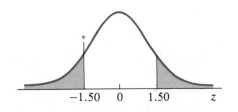

The prob-value for this hypothesis test is 0.1336. Each individual company now will make a decision whether to (a) continue to use Kelley's services or (b) change to the Brown Agency. Each will need to establish the level of significance that best fits its own situation and then make a decision using the decision rule described in this illustration. ■

EXERCISES

8.29 Calculate the prob-value for each of the following.

    a. $H_0: \mu = 10$ $z^* = 1.48$
        $H_a: \mu > 10$

    b. $H_0: \mu = 105$ $z^* = -0.85$
        $H_a: \mu < 105$

    c. $H_0: \mu = 13.4$ $z^* = 1.17$
        $H_a: \mu \neq 13.4$

    d. $H_0: \mu = 8.56$ $z^* = -2.11$
        $H_a: \mu < 8.56$

    e. $H_0: \mu = 110$ $z^* = -0.93$
        $H_a: \mu \neq 110$

    f. $H_0: \mu = 54.2$ $z^* = 0.46$
        $H_a: \mu > 54.2$

8.30 Find the value of z^* for each of the following.

    a. $H_0: \mu = 35$ versus $H_a: \mu > 35$ when p-value $= 0.0582$

    b. $H_0: \mu = 35$ versus $H_a: \mu < 35$ when p-value $= 0.0166$

    c. $H_0: \mu = 35$ versus $H_a: \mu \neq 35$ when p-value $= 0.0042$

8.31 The calculated prob-value for a hypothesis test is p-value $= 0.084$. What decision about the null hypothesis would occur if

    a. the hypothesis test is completed at the 0.05 level of significance?

    b. the hypothesis test is completed at the 0.10 level of significance?

8.32 A paint manufacturer knows from experience that the coverages for its brand of paint are normally distributed with a mean equal to 440 square feet per gallon and a standard deviation equal to 15 square feet. The addition of an additive is known not to effect the variability of coverage, but may increase the mean coverage. Ten randomly selected gallons of the paint (with the additive) are tested and the following coverages obtained.

$$435 \quad 450 \quad 455 \quad 460 \quad 465 \quad 440 \quad 452 \quad 461 \quad 470 \quad 440$$

The following MINITAB output tests the hypothesis $H_0: \mu = 440$ versus $H_a: \mu > 440$.

a. Verify the sample mean, standard deviation, standard error of the mean, calculated z, z^*, and p-value.

b. State H_0 and H_a.

```
MTB > SET DATA IN C1
DATA> 435 450 455 460 465 440 452 461 470 440
DATA> END DATA
MTB > ZTEST OF PAINT WITH MU = 440,ALT = 1,SIGMA = 15,DATA IN C1
TEST OF MU = 440.00 VS MU G.T. 440.00
THE ASSUMED SIGMA = 15.0
                N      MEAN    STDEV    SE MEAN     Z     P VALUE
C1             10     452.80   11.65     4.74     2.70    0.0035
MTB > STOP
```

8.33 An economist claims that when the Dow-Jones average increases, the volume of shares traded on the New York Stock Exchange tends to increase. Over the past two years the daily volume on the exchange has averaged 21.5 million shares and had a standard deviation of 2.5 million. A random sample of 64 days on which the Dow-Jones average increased was selected and the average daily volume was computed. The sample average was 22 million. Calculate the prob-value for this hypothesis test.

8.34 The marketing director of A & B Cola is worried that the product is not attracting enough young consumers. To test this hypothesis, she randomly surveys 100 A & B Cola consumers. The mean age of an individual in the community is 32 years and the standard deviation is 10 years. The surveyed consumers of A & B Cola have a mean age of 35. At the 0.01 level of significance, is this sufficient evidence to conclude that A & B Cola consumers are, on the average, older then the average person living in the community? Complete this hypothesis test using the prob-value approach.

8.4 ESTIMATION

Estimation is the second type of statistical inference. It is the procedure to use when answering a question that asks for the value of a population parameter. For example, "What is the mean one-way distance that Monroe Community College students commute daily?"

If you needed to answer this question, you might take a sample from the population and calculate the sample mean $\bar{x}$. Suppose that you did draw a random sample of 100 one-way distances and a sample mean of 10.22 miles resulted. What is your estimate for the mean value of the population? If you report the sample mean $\bar{x}$ as your estimate, you will be making a point estimate.

point estimate

Point Estimate for a Parameter

The value of the corresponding sample statistic.

That is, the sample mean, $\bar{x} = 10.22$ miles, is the best point estimate for the mean one-way distance for this population. In other words, the mean one-way distance for all students is estimated to be 10.22 miles. We are not really implying that the population mean μ is exactly 10.22 miles. Rather, we intend this point estimate to be interpreted to say "μ is close to 10.22." From our study so far, you probably realize that when we say $\mu = 10.22$, there is very little chance that the statement is true. In previous chapters we drew samples from probability distributions where μ was known, and seldom did we obtain a sample mean exactly equal in value to μ. When, then, should we expect a single sample to yield an $\bar{x}$ equal to μ when μ is unknown? We shouldn't.

What does it mean to say "μ is close to 10.22 miles"? Closeness is a relative term, but perhaps in this case "close" might be arbitrarily defined to be "within 1 mile" of μ. If 1 mile satisfies our intuitive idea of closeness, then to say "μ is close to 10.22" is comparable to saying "μ is between 9.22 ($10.22 - 1.0$) and 11.22 ($10.22 + 1.0$)." This suggests that we might make estimations by using intervals. The confidence interval estimate employs this interval concept and assigns a measure to the interval's reliability in estimating the parameter in question.

interval estimate

Interval Estimate

An interval bounded by two values and used to estimate the value of a population parameter. The values that bound this interval are statistics calculated from the sample that is being used as the basis for the estimation.

level of confidence

Level of Confidence $1 - \alpha$

The probability that the sample to be selected yields boundary values that lie on opposite sides of the parameter being estimated. The level of confidence is sometimes called the **confidence coefficient**.

confidence interval

Confidence Interval

An interval estimate with a specified level of confidence.

The central limit theorem is the source for the information needed to construct confidence interval estimates. Our sample mean ($\bar{x} = 10.22$) is a member of the sampling distribution of sample means. The CLT describes this distribution as being approximately normally distributed with a mean $\mu_{\bar{x}} = \mu$ and a standard deviation $\sigma_{\bar{x}} = \sigma/\sqrt{n}$. The population mean is unknown, however, it exists and is constant. Let's assume that the standard deviation of the population is $\sigma = 6$.

Recall   In this chapter we are studying the inferences about μ under the assumption that σ is known. This assumption is for convenience, and the restriction will be removed in Chapter 9.

Figure 8-3 shows the sampling distribution of sample means to which our sample mean ($\bar{x} = 10.22$ and $n = 100$) belongs.

Figure 8-3
Sampling distribution,
$n = 100$

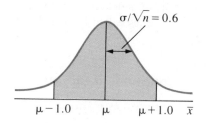

$\sigma/\sqrt{n} = 6.0\,/\sqrt{100} = 0.6$

μ
(unknown)

x

When a sample of size 100 is randomly selected from a population whose mean is μ and whose standard deviation is 6, what is the probability that the sample mean is within one unit of μ? In other words, what is $P(\mu - 1 < \bar{x} < \mu + 1)$? See Figure 8-4. This probability may be found by using the normal probability distribution (Table 5, Appendix E) and formula (7-2):

$$z = \frac{\bar{x} - \mu}{\sigma/\sqrt{n}}$$

If $\bar{x} = \mu - 1$,

$$z = \frac{(\mu - 1) - \mu}{0.6} = \frac{-1.0}{0.6} = -\mathbf{1.67}$$

If $\bar{x} = \mu + 1$,

$$z = \frac{(\mu + 1) - \mu}{0.6} = \frac{+1.0}{0.6} = +\mathbf{1.67}$$

Figure 8-4
Probability that $\bar{x}$ is
within 1 unit of μ

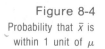

$\sigma/\sqrt{n} = 0.6$

$\mu - 1.0$ μ $\mu + 1.0$ $\bar{x}$

Therefore,

$$P(\mu - 1 < \bar{x} < \mu + 1) = P(-1.67 < z < +1.67)$$
$$= 2 \cdot P(0 < z < 1.67)$$
$$= 2(0.4525) = \mathbf{0.9050}$$

The probability that the mean of a random sample is within one unit of this population mean is 0.9050. Thus the probability that the population mean is within one unit of the mean of a sample is also 0.9050. Therefore, the interval 9.22 to 11.22 is a 0.9050 confidence interval estimate for the mean one-way distance commuted by Monroe students.

maximum error of estimate

Maximum Error of Estimate, E

One-half the width of the confidence interval. In general, E is a multiple of the standard error.

The preceding illustration started with the maximum error being assigned the value of 1 mile. Typically, the maximum error of estimate is determined by the level of confidence that we want our confidence interval to have. That is, $1 - \alpha$ will determine the maximum error. The level of confidence will be split so that half of $1 - \alpha$ is above the mean and half below the mean (Figure 8-5). This will leave $\alpha/2$ as the probability in each of the two tails of the distribution. The z score at the boundary of the confidence interval will be z of $z(\alpha/2)$. [Recall that $\alpha/2$ is the probability (or area) under the curve to the right of this point. *See* Chapter 6.] The standard score z is a number of standard deviations. Therefore, the maximum error of estimate E is

$$E = z(\alpha/2) \cdot \frac{\sigma}{\sqrt{n}} \tag{8-1}$$

Figure 8-5
Each tail contains $\alpha/2$

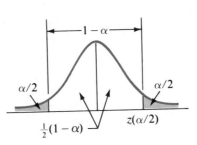

See Figure 8-6.

Figure 8-6
Maximum error of estimate

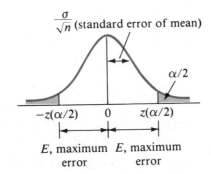

The **1 − α confidence interval** for μ is

$$\bar{x} - z(\alpha/2) \cdot \frac{\sigma}{\sqrt{n}} \quad \text{to} \quad \bar{x} + z(\alpha/2) \cdot \frac{\sigma}{\sqrt{n}} \qquad (8\text{-}2)$$

$\bar{x} - z(\alpha/2) \cdot (\sigma/\sqrt{n})$ is called the **lower confidence limit (LCL)**, and $\bar{x} + z(\alpha/2) \cdot (\sigma/\sqrt{n})$ is called the **upper confidence limit (UCL)** for the confidence interval.

Illustration 8-10   Construct the 0.95 confidence interval for the estimate of the mean one-way distance that the students commute.

Solution

$$1 - \alpha = 0.95$$

$$\alpha = 0.05$$

$$\frac{\alpha}{2} = 0.025 \quad (\textit{see} \text{ the accompanying figure})$$

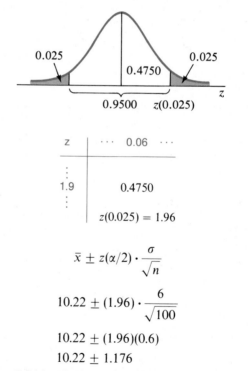

z	$\cdots$	0.06	$\cdots$
$\vdots$			
1.9		0.4750	
$\vdots$			

$$z(0.025) = 1.96$$

$$\bar{x} \pm z(\alpha/2) \cdot \frac{\sigma}{\sqrt{n}}$$

$$10.22 \pm (1.96) \cdot \frac{6}{\sqrt{100}}$$

$$10.22 \pm (1.96)(0.6)$$

$$10.22 \pm 1.176$$

$$10.22 - 1.176 = 9.044 = \textbf{9.04} \quad \text{and} \quad 10.22 + 1.176 = 11.396 = \textbf{11.40}$$

Therefore, with 0.95 confidence we can say that the mean one-way distance is between 9.04 and 11.40, and we will write this as

(9.04, 11.40), the 0.95 confidence interval for μ

Note   Many of the confidence interval formulas that you will be using will have a format similar to formula (8-2), namely, $\bar{x} - E$ to $\bar{x} + E$. However, perhaps the simplest way to handle the arithmetic involved is to think of the formula as $\bar{x} \pm E$, as was done in Illustration 8-10. Using this format you first calculate E. Then you calculate the lower bound by subtracting and you calculate the upper bound by adding.

Illustration 8-11   To estimate the mean score on the first one-hour exam in statistics, we obtain a random sample of 38 exam scores. A sample mean of 74.3 is found. Construct the 0.98 confidence interval estimate for the mean of all first one-hour exam scores. Use $\sigma = 14$.

Solution

$$1 - \alpha = 0.98$$

$$\alpha = 0.02$$

$$\frac{\alpha}{2} = 0.01 \ (\textit{see the accompanying figure})$$

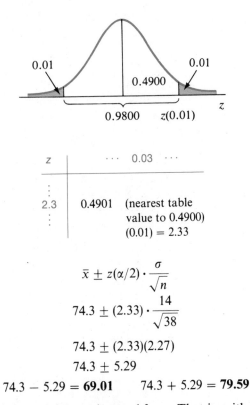

z	$\cdots$ 0.03 $\cdots$
⋮	
2.3	0.4901 (nearest table value to 0.4900) $(0.01) = 2.33$
⋮	

$$\bar{x} \pm z(\alpha/2) \cdot \frac{\sigma}{\sqrt{n}}$$

$$74.3 \pm (2.33) \cdot \frac{14}{\sqrt{38}}$$

$$74.3 \pm (2.33)(2.27)$$

$$74.3 \pm 5.29$$

$$74.3 - 5.29 = \mathbf{69.01} \qquad 74.3 + 5.29 = \mathbf{79.59}$$

(69.01, 79.59) is the 0.98 confidence interval for μ. That is, with 0.98 confidence we can say that the mean exam score for all such exams is between 69.01 and 79.59.

■

Illustration 8-12   Suppose that the sample in Illustration 8-7 had been obtained for the purpose of estimating the population mean μ. The sample results ($n = 40$, $\bar{x} = 3.975$) and $\sigma = 2.87$ are used in formula (8-2) to determine the 0.90 confidence interval.

$$\bar{x} \pm z(\alpha/2) \cdot \frac{\sigma}{\sqrt{n}}$$

$$3.975 \pm 1.65 \cdot \frac{2.87}{\sqrt{40}}$$

$$3.975 \pm (1.65)(0.454)$$

$$3.975 \pm 0.749$$

$$3.975 - 0.749 = 3.226 = \mathbf{3.23} \qquad 3.975 + 0.749 = 4.724 = \mathbf{4.72}$$

(3.23, 4.72) is the 0.90 confidence interval estimate for μ. (*See* the accompanying figure.)

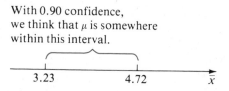

With 0.90 confidence, we think that μ is somewhere within this interval.

Since this sample was taken from a known population—namely, the random digits—we can look back at our answer and say that the confidence interval is correct. That is, the true value of μ, 4.5, does fall within the confidence interval.

◼

Suppose that we were to take another sample of size 40; would we obtain the same results? Suppose we took a third and a fourth. What would happen? What is the level of confidence, $1 - \alpha$? Yes, it has the value 0.90, but what does that mean? Table 8-4 lists the means obtained from 10 different random samples of size 40 taken

Table 8-4

Samples of size 40 taken from Table 1, Appendix E

Sample Number	Sample Mean ($\bar{x}$)	0.90 Confidence Interval Estimate for μ
1	4.64	3.89 to 5.39
2	4.56	3.81 to 5.31
3	3.96	3.21 to 4.71
4	5.12	4.37 to 5.87
5	4.24	3.49 to 4.99
6	3.44	2.69 to 4.19
7	4.60	3.85 to 5.35
8	4.08	3.33 to 4.83
9	5.20	4.45 to 5.95
10	4.88	4.13 to 5.63

from Table 1. The 0.90 confidence interval for the estimate of μ based on each of these samples is also listed in Table 8-4. Since all the interval estimates were for the mean of the random digits ($\mu = 4.5$), we can inspect each confidence interval and see that 9 of the 10 contain μ and would therefore be considered correct estimates. One sample (sample 6) did not yield an interval that contains μ. This is shown in Figure 8-7.

Figure 8-7
Interval estimates from
Table 8-4

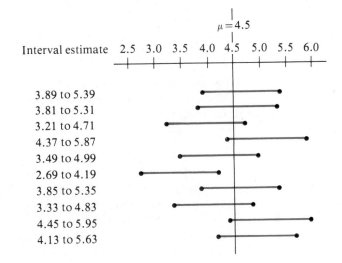

Interval estimate

3.89 to 5.39
3.81 to 5.31
3.21 to 4.71
4.37 to 5.87
3.49 to 4.99
2.69 to 4.19
3.85 to 5.35
3.33 to 4.83
4.45 to 5.95
4.13 to 5.63

Remember $1 - \alpha$ is the probability that we obtain a confidence interval such that μ is contained within it.

The maximum error formula can be used to determine the sample size required to obtain a $1 - \alpha$ confidence interval estimate with a specified maximum error.

Illustration 8-13   Determine the size of the sample that will be necessary to estimate the mean weight of second-grade boys if we want our estimate to be accurate to within 1 pound with 95 percent confidence. Assume that the standard deviation of such weights is 3 pounds.

Solution $1 - \alpha = 0.95$; therefore, $z(0.025) = 1.96$, as found by using Table 5, Appendix E. The maximum error $E = 1$ and $\sigma = 3$. Use formula (8-1).

$$E = z(\alpha/2) \cdot \frac{\sigma}{\sqrt{n}}$$

$$1.0 = 1.96 \cdot \frac{3}{\sqrt{n}}$$

$$1 = \frac{5.88}{\sqrt{n}}$$

$$\sqrt{n} = 5.88$$

$$n = 34.57$$

Therefore,

$$n = 35$$

Note   When solving for the sample size n, all fractional (decimal) values are to be *rounded up* to the next larger integer.   ■

Does a sample size of 35 determine a maximum error of 1 pound in Illustration 8-13?

$$E = z(\alpha/2) \cdot \frac{\sigma}{\sqrt{n}} = 1.96 \cdot \frac{3.0}{\sqrt{35}} = (1.96)(0.51) = \mathbf{0.999}$$

This is a maximum error of just under 1.0.

The use of formula (8-1) can be made a little easier by rewriting the formula in a form that expresses n in terms of the other quantities:

$$n = \left[\frac{z(\alpha/2) \cdot \sigma}{E} \right]^2 \tag{8-3}$$

Illustration 8-14   A sample of what size would be needed to estimate the population mean to within $\frac{1}{5}$ of a standard deviation with 99 percent confidence?

Solution $1 - \alpha = 0.99$; therefore, $\alpha/2 = 0.005$ and $z(0.005) = 2.58$, as found by using Table 5, Appendix E. The maximum error E is to be $\frac{1}{5}$ of σ; that is, $E = \sigma/5$. Using formula (8-3), we have

$$n = \left[\frac{z(\alpha/2) \cdot \sigma}{E} \right]^2 = \left[\frac{(2.58)(\sigma)}{\sigma/5} \right]^2$$

$$= \left[\frac{(2.58\sigma)(5)}{\sigma} \right]^2 = [(2.58)(5)]^2$$

$$= (12.90)^2 = 166.41 = \mathbf{167}$$ ■

EXERCISES

8.35 Discuss the effect that each of the following have on the width of a confidence interval.

    a. level of confidence   **b.** sample size

    c. variability of the characteristic being measured

8.36 Adjustments to a machine sometimes change the mean length of the parts it makes. However, the adjustments do not effect the variability of lengths. The lengths are known to be normally distributed with a standard deviation equal to 0.5 millimeters. After a series of adjustments, the following lengths (in millimeters) were obtained for ten randomly selected parts:

      75.3   76.0   75.0   77.0   75.4   76.3   77.0   74.9   76.5   75.8

The following MINITAB output shows a 95 percent confidence interval estimate for μ. Verify that the given confidence interval is correct.

```
MTB > SET DATA IN C1
DATA> 75.3 76.0 75.0 77.0 75.4 76.3 77.0 74.9 76.5 75.8
DATA> END DATA
MTB > ZINTERVAL 95 PERCENT CONFIDENCE,SIGMA = .5,DATA IN C1
THE ASSUMED SIGMA = 0.500
              N      MEAN    STDEV    SE MEAN    95.0 PERCENT C.I.
C1           10     75.920   0.773     0.158    ( 75.610,   776.230)
MTB > STOP
```

8.37. A certain population of the annual incomes of unskilled laborers has a standard deviation of $1200. A random sample of 36 results in $\bar{x} = \$7280$.

    a. Give a point estimate for the population mean annual income.

    b. Estimate μ with a 95 percent confidence interval.

    c. Estimate μ with a 99 percent confidence interval.

8.38 In order to estimate the mean amount spent for textbooks by students during the fall semester at a large commuter university, a random sample of 75 students were surveyed. The mean of the sample was $85.30. Find the 90 percent confidence interval estimate for the mean cost for all the students. Based on experience with similar studies, it is reasonable to assume $\sigma = \$15.00$.

8.39 A sample of 60 night school students' ages is obtained in order to estimate the mean age of night school students. $\bar{x} = 25.3$ years. The population variance is 16.

    a. Give a point estimate for μ.

    b. Find the 95 percent confidence interval estimate for μ.

    c. Find the 99 percent confidence interval estimate for μ.

8.40 The lengths of 200 fish caught in Cayuga Lake had a mean of 14.3 inches. The population standard deviation is 2.5 inches.

    a. Find the 90 percent confidence interval for the population mean length.

    b. Find the 98 percent confidence interval for the population mean length.

8.41 How large a sample should be taken if the population mean is to be estimated with 99 percent confidence to within $75? The population has a standard deviation of $900.

8.42 A "high tech" company wants to estimate the mean number of years of college education its employees have completed. A good estimate of the standard deviation for the number of years of college is 1.0. How large a sample needs to be taken to estimate μ to within 0.5 years with 99 percent confidence?

8.43 By measuring the amount of time it takes a component of a product to move from one work station to the next, an engineer has estimated that the standard deviation is 5 seconds.

    a. How many measurements should be made in order to be 95 percent certain that the maximum error of estimation will not exceed 1 second?

    b. What sample size is required for a maximum error of 2 seconds?

8.44 A medical products firm has developed a new drug that requires very strict control on the amount of the drug placed in each capsule. It can be hazardous if too much drug is used and ineffective if not enough is used. The maximum error of estimate for the mean is 2 milligrams. The standard deviation is believed to be 7 milligrams. If 99 percent confidence is desired, what size sample should be taken to be certain that the mean amount of the drug per capsule is correct?

In Retrospect

Two forms of inference were studied in this chapter: hypothesis testing and estimation. They may be, and often are, used separately. It would, however, seem quite natural for the rejection of a null hypothesis to be followed by a confidence interval estimate. (If the value claimed is wrong, we will usually want to know just what the true value is.)

These two forms of inference are quite different but they are related. There is a certain amount of crossover between the use of the two inferences. For example, suppose that you had sampled and calculated a 90 percent confidence interval for the mean of a population. The interval was 10.5 to 15.6. Following this, someone claims that the true mean is 15.2. Your confidence interval estimate can be compared to this claim. If the claimed value falls within your interval estimate, you would fail to reject the null hypothesis that $\mu = 15.2$ at a 10 percent level of significance in a two-tailed test. If the claimed value (say 16.0) falls outside the interval, you would then reject the null hypothesis that $\mu = 16.0$ at $\alpha = 0.10$ in a two-tailed test. If a one-tailed test is required, or if you prefer a different value of α, a separate hypothesis test must be used.

Many users of statistics (especially those marketing a product) will claim that their statistical results prove that their product is superior. But remember, the hypothesis test does not prove or disprove anything. The decision reached in a hypothesis test has probabilities associated with the four various situations. If "fail to reject H_0" is the decision, it is possible that an error has occurred. Further, if "reject H_0" is the decision reached, it is possible for this to be an error. Both errors have probabilities greater than zero.

In this chapter we have restricted our discussion of inferences to the mean of a population for which the standard deviation is known. In Chapters 9 and 10, we will discuss inferences about other population parameters and eliminate the restriction about the known value for standard deviation.

Chapter Exercises

8.45 The expected mean of a certain population is 100 and its standard deviation is 12. A sample of 50 measurements gives a sample mean of 96. Using a level of significance of 0.01, a test is to be made to see whether the population mean is really 100 or whether it is different from 100. State or calculate the desired answer in (a) through (j).

    a. H_0 **b.** H_a

    c. α **d.** $z(\alpha/2)$

e. μ (based on H_0)   **f.** $\bar{x}$

g. σ **h.** $\sigma_{\bar{x}}$

i. z score for $\bar{x}$ **j.** decision

k. On the accompanying sketch, locate $\alpha/2$, $z(\alpha/2)$, the rejection region for H_0, and the z score for $\bar{x}$.

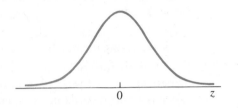

8.46 A sample of 64 measurements is taken from a continuous population and the sample mean is found to be 32.0. The standard deviation of the population is known to be 2.4. An interval estimation is to be made of the mean with a level of confidence of 0.90. State or calculate the following items:

a. $\bar{x}$ **b.** σ

c. n **d.** $1 - \alpha$

e. $z(\alpha/2)$ **f.** $\sigma_{\bar{x}}$

g. E (maximum error of estimate   **h.** upper confidence limit

i. lower confidence limit

8.47 A population has an unknown mean μ and a standard deviation σ equal to 5.0. Compare the concepts of hypothesis testing and confidence interval estimation by working the following:

a. Calculate the value for z^* if the null hypothesis is H_0: $\mu = 40$, given that the sample size is 100 and the sample mean is 40.7. Determine the 95 percent confidence interval estimate for μ.

b. Calculate the value for z^* if the null hypothesis is H_0: $\mu = 40$, given that the sample size is 100 and the sample mean is 42.5. Determine the 95 percent confidence interval estimate for μ.

8.48 The college bookstore tells prospective students that the average cost of its textbooks is $22 per book with a standard deviation of $2.50. The engineering science students think that the average cost of their books is higher than the average for all students. To test the bookstore's claim against their alternative, the engineering students collect a random sample of size 45.

a. If they use $\alpha = 0.05$, what are the cutoff values for $\bar{x}$ that will support their belief?

b. The engineering students' sample data are summarized by $n = 45$ and $\sum x = 1010.25$. Is this sufficient evidence to support their contention?

8.49 A rope manufacturer, after conducting a large number of tests over a long period of time, has found that the rope has a mean breaking strength of 300 pounds and a standard deviation of 24 pounds. Assume that these values are μ and σ. It is believed that by using a recently developed high-speed process, the mean breaking strength has been decreased.

a. Design a null and alternative hypothesis such that rejection of the null hypothesis will imply that the mean breaking strength has decreased.

b. If the decision rule in (a) is used with $\alpha = 0.01$, what is the critical value for the test statistic and what value of $\bar{x}$ corresponds to it if a sample of size 45 is used?

c. Using the decision rule established in (a), what is the prob-value associated with rejecting the null hypothesis when 45 tests result in a sample mean of 295?

8.50 The admissions office at Memorial Hospital recently stated that the mean age of its patients was 42 years. A random sample of 120 ages was obtained from the admissions office records in an attempt to disprove the claim. Is a sample mean of 44.2 years significantly larger than the claimed 42 years, at the $\alpha = 0.05$ level? Use $\sigma = 20$ years.

a. Solve using the classical approach.

b. Solve using the prob-value approach.

8.51 In a large supermarket the customer's waiting time to check out is approximately normally distributed with a standard deviation of 2.5 minutes. A sample of 24 customer waiting times produced a mean of 10.6 minutes. Is this evidence sufficient to reject the supermarket's claim that its customer checkout time averages no more than 9 minutes? Complete this hypothesis test using the 0.02 level of significance.

a. Solve using the classical approach.

b. Solve using the prob-value approach.

8.52 At a very large firm, the clerk-typists were sampled to see whether the salaries differed among departments for workers is similar categories. In a sample of 50 of the firm's accounting clerks, the average annual salary was $16,010. The firm's personnel office insists that the average salary paid to all clerk-typists with the firm is $15,650 and that the standard deviation is $1,800. At the 0.05 level of significance, can we conclude that the accounting clerks receive, on the average, a different salary from that of other clerk-typists?

a. Solve using the classical approach.

b. Solve using the prob-value approach.

8.53 A random sample of 280 rivets is selected from a very large shipment. The rivets are to have a mean diameter of no more than $\frac{5}{16}$ inch. Does a sample mean of 0.3126 show sufficient reason to reject the null hypothesis that the mean diameter is no more than $\frac{5}{16}$ inch, at the $\alpha = 0.01$ level of significance? Use $\sigma = 0.0006$.

a. Solve using the classical approach.

b. Solve using the prob-value approach.

8.54 A manufacturing process produces ball bearings with diameters having a normal distribution and a standard deviation of $\sigma = 0.04$ cm. Ball bearings that have diameters that are too small or too large are undesirable. To test the null hypothesis that $\mu = 0.50$ cm., a sample of 25 is randomly selected and the sample mean is found to be 0.51. Determine the p-value for the test results.

8.55 Suppose that a confidence interval is assigned a level of confidence of $1 - \alpha = 0.95$. How is the 0.95 used in constructing the confidence interval?

8.56 The following MINITAB output shows a simulated sample of size 25 from a normal distribution with $\mu = 130$ and $\sigma = 10$. ZINTERVAL is then used to set a 95 percent confidence interval for μ.

```
MTB > NRANDOM 25 OBS WITH MU = 130,SIGMA = 10,PUT IN C1
        25 NORMAL OBS. WITH MU =    130.0000 AND SIGMA = 10.000
MTB > PRINT C1
C1
    116.187   119.832   121.782   122.320   141.436   129.197   119.172
    120.713   135.765   131.153   122.307   126.155   137.545   141.154
    123.405   143.331   121.767   109.742   140.524   150.600   121.655
    127.992   136.434   139.768   125.594
MTB > ZINTERVAL 95 PERCENT CONFIDENCE,SIGMA = 10,DATA IN C1

THE ASSUMED SIGMA =  10.0

                N      MEAN    STDEV    SE MEAN    95.0 PERCENT C.I.
C1             25    129.02    10.18      2.00    ( 125.10,   132.95)
```

Verify the last line of the MINITAB output; that is, verify the values reported for sample mean, sample standard deviation, standard error of the mean, and the 95 percent confidence interval.

8.57 A random sample of the scores of 100 applicants for clerk-typist positions at a large insurance company showed a mean score of 72.6. The preparer of the test maintained that qualified applicants should average 75.0.

　　a. Determine the 99 percent confidence interval estimate for the mean score of all applicants at the insurance company. Assume that the standard deviation of test scores is 10.5.

　　b. Can the insurance company conclude that it is getting qualified applicants (as measured by this test)?

8.58 Waiting times (in hours) at a popular restaurant are believed to be approximately normally distributed with a variance of 2.25 hours during busy periods.

　　a. A sample of 20 customers revealed a mean waiting time of 1.52 hours. Construct the 95 percent confidence interval for the estimate of the population mean.

　　b. Suppose that the mean of 1.52 hours had resulted from a sample of 32 customers. Find the 95 percent confidence interval.

　　c. What effect does a larger sample size have on the confidence interval?

8.59 The weights of full boxes of a certain kind of cereal are normally distributed with a standard deviation of 0.27 ounce. A sample of 18 randomly selected boxes produced a mean weight of 9.87 ounces.

　　a. Find the 95 percent confidence interval for the true mean weight of a box of this cereal.

b. Find the 99 percent confidence interval for the true mean weight of a box of this cereal.

c. What effect did the increase in the level of confidence have on the width of the confidence interval?

8.60 The standard deviation of a normally distributed population is equal to 10. A sample of size 25 is selected and its mean is found to be 95.

a. Find an 80 percent confidence interval estimate for μ.

b. If the sample size were 100, what would be the 80 percent confidence interval?

c. If the sample size were 25 but the standard deviation were 5 (instead of 10), what would be the 80 percent confidence interval?

8.61 A new paint has recently been developed by a research laboratory. Twenty-five gallons are tested and the mean coverage is found to be 515 square feet per gallon. Assuming the standard deviation to be 50 square feet per gallon, find a 99 percent confidence interval estimate for μ, the mean coverage per gallon of all such paint.

8.62 A large department store has 20,000 customers that use the store's charge account plan. To estimate the amount currently owed to the store, a random sample of 64 accounts was selected. The sample mean is $142, and the population standard deviation is $70.

a. Find the 95 percent confidence interval estimate for the true mean account balance.

b. Find the 95 percent confidence interval estimate for the total amount owed to the store through these charge accounts.

8.63 An automobile manufacturer wants to estimate the mean gasoline mileage that its customers will obtain with its new compact model. How many sample runs must be performed in order that the estimate be accurate to within 0.3 mile per gallon at 95 percent confidence? (Assume that $\sigma = 1.5$.)

8.64 A fish hatchery manager wants to estimate the mean length of her three-year-old hatchery-raised trout. She wants to make a 99 percent confidence interval estimate accurate to within $\frac{1}{3}$ of a standard deviation. How large a sample does she need to take?

8.65 We are interested in estimating the mean life of a new product. How large a sample do we need to take in order to estimate the mean to within $\frac{1}{10}$ of a standard deviation with 0.90 confidence?

Vocabulary List

Be able to define each term. In addition, describe, in your own words, and give an example of each term. Your examples should not be ones given in class or in the textbook.

The bracketed numbers indicate the chapters in which the terms previously appeared, but you should define the terms again to show increased understanding of their meaning.

acceptance region
alpha (α)

alternative hypothesis
beta (β)

calculated value (*)   point estimate
conclusion   prob-value
confidence interval estimate   risk
critical region   sample size
critical value   sample statistic [1, 2]
decision   standard error [7]
estimation   standard error of mean [7]
hypothesis   statistic [1, 2]
hypothesis test   test criteria
level of confidence   test statistic
level of significance   two-tailed test
lower confidence limit   type I error
maximum error of estimate   type II error
null hypothesis   upper confidence limit
one-tailed test $z(\alpha)$ [6]
parameter [1]

Quiz   *Answer "True" if the statement is always true. If the statement is not always true,
replace the boldface words with words that make the statement always true.*

8.1 **Beta** is the probability of a type I error.

8.2 $1 - \alpha$ is known as the level of significance of a hypothesis test.

8.3 The standard error of the mean is the standard deviation of the **sample selected**.

8.4 The maximum error of estimate is controlled by three factors: **level of
confidence, sample size, and standard deviation**.

8.5 Alpha is the measure of the area under the curve of the standard score that lies
in the **rejection region** for H_0.

8.6 The risk of making a **type I error** is directly controlled in a hypothesis test by
establishing a level for α.

8.7 Failing to reject the null hypothesis when it is false is a **correct decision**.

8.8 If the acceptance region in a hypothesis test is made wider (assuming σ and n
remain fixed), **α** becomes larger.

8.9 Rejection of a null hypothesis that is false is a **type II error**.

8.10 To conclude that the mean is higher (or lower) than a claimed value, the value
of the test statistic must fall in the **acceptance region**.

Chapter 9

Inferences Involving One Population

ATTITUDES IN BLACK AND WHITE

Nineteen years after the death of Martin Luther King Jr., black and white Americans alike say the nation is still far from fulfilling his dream of seeing the two races live in harmony. An overwhelming 92% of blacks and 87% of whites polled for TIME last week by Yankelovich Clancy Shulman* agree with the statement "Racial prejudice is still very common in the U.S." Fully 59% of blacks say they have been insulted because of their race at one time or another. But blacks and whites diverge sharply when asked about the reasons and remedies for America's enduring dilemma.

THE ROOTS OF BIGOTRY

Americans may agree that racial prejudice is common, but less than half the whites who were questioned feel that blacks are singled out for discrimination. Only 35% of whites (vs. 51% of blacks) agree that "Most white Americans do not like blacks." Yet 44% of whites (and 45% of blacks) agree that "Most black Americans do not like whites."

Whites are more likely to be afraid of blacks. Although only 26% of the whites say they have ever felt "physically threatened by someone who was black," 64% of the white respondents agree that they would be afraid to be in an all-black neighborhood at night (27% would also feel afraid during the day). Despite news stories about the racial attacks on blacks in Howard Beach, Queens, N. Y., only 30% of the black respondents say they would be afraid in an all-white neighborhood at night; even fewer black respondents (24%) say they have ever felt physically threatened by a white.

Overall, less than half the black respondents report specific instances of discrimination, although a number cite more than one instance: 32% say they have been discriminated against at work, 25% in school and 16% when trying to rent an apartment or buy a house. But they are far less satisfied than whites with the degree of opportunity for blacks.

Do black Americans have the same opportunities as whites?

	Blacks	Whites
In Housing		
Same opportunity	22%	48%
Not the case	75%	47%
In Education		
Same opportunity	38%	73%
Not the case	59%	24%
In Employment		
Same opportunity	26%	59%
Not the case	71%	37%

WHAT IS TO BE DONE?

Blacks and whites alike agree that the "Federal Government should do more to promote equality in housing, education and employment." More than 90% of blacks feel this way, but the figure for whites is just slightly more than 50%. Attitudes vary about specific steps to remedy discrimination:

	Yes	
	Blacks	Whites
Should the Government prosecute landlords who refuse to rent to blacks?	88%	69%
Should colleges admit some black students whose record would not normally qualify them for admission?	33%	15%
Should businesses set a goal of hiring a minimum number of black employees?	62%	32%

---

* The survey of 871 white adults and 93 black adults was conducted by telephone Jan. 19–21. The potential sampling error for the white respondents is plus or minus 3%. For the smaller number of black respondents, the sampling error is larger.

Chapter
Objectives

In Chapter 8 we discussed two forms of statistical inference: (1) hypothesis tests and (2) confidence interval estimation. The study of these two inferences was restricted to the population parameter μ under the restriction that σ was known. However, the population standard deviation is generally not known. Thus the first section of this chapter deals with the inferences about μ when σ is unknown. In addition, we will learn how to perform hypothesis tests and confidence interval estimations about the population parameters σ (standard deviation) and p (binomial probability of success).

9.1 INFERENCES ABOUT THE POPULATION MEAN

In Chapter 8, we learned about making inferences concerning the population mean when the population standard deviation σ was a known quantity. Now we will consider inferences about the population mean when σ is an *unknown* quantity.

The sample standard deviation, s, is a very effective point estimate to use for σ and will be used in place of σ in the various calculations. As a result of this substitution, the test statistic used is also changed. In place of z, the standard normal score, we will use the statistic t. The t statistic is defined as

$$t = \frac{\bar{x} - \mu}{s/\sqrt{n}} \tag{9-1}$$

If the population being sampled is approximately normally distributed, then the calculated t score belongs to the Student's t distribution.

In 1908 W. S. Gosset, an Irish brewery employee, published a paper about this t distribution under the pseudonym "Student." In deriving the t distribution, Gosset assumed that the samples were taken from normal populations. Although this might seem to be quite restrictive, satisfactory results are also obtained when sampling from many nonnormal populations.

The t distribution has the following properties (also *see* Figure 9-1):

Properties of the t Distribution

1. t is distributed with a mean of 0.

2. t is distributed symmetrically about its mean.

3. t is distributed with a variance greater than 1, but as the sample size n increases, the variance approaches 1.

4. t is distributed so as to be less peaked at the mean and thicker at the tails than the normal distribution.

5. t is distributed so as to form a family of distributions, a separate distribution for each sample size. The t distribution approaches the normal distribution as the sample size increases.

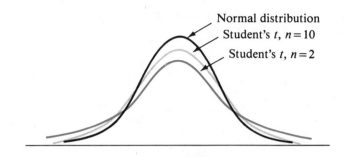

Figure 9-1
Normal and Student's
t distributions

Although there is a separate t distribution for each sample size, $n = 2$, $n = 3, \ldots, n = 20, \ldots, n = 40$, and so on, only certain key critical values of t will be necessary for our work. Consequently, the table for Student's t distribution (Table 6 of Appendix E) is a table of critical values rather than a complete table, such as that for the standard normal distribution for z. As you look at Table 6, you will note that the left side of the table is identified by "df," which means "degrees of freedom." This left-hand column starts at 1 at the top and ranges to 29, then jumps to z at the bottom. As stated previously, as the sample size increases, the distribution approaches that of the normal. By inspecting Table 6, you will note that as you read down any one of the columns, the entry value approaches a familiar z value at the bottom. By now you should realize that as the sample size increases, so does the number of degrees of freedom, df.

degrees of freedom

Degrees of Freedom, df

A parameter in statistics that is very difficult to define completely. Perhaps it is best thought of as an "*index number*" *used for the purpose of identifying the correct distribution to be used.* In the methods presented in this chapter, the value of df for a given situation will be $n - 1$, the sample size minus 1.

As previously stated, as the sample size increases, the t distribution approaches the characteristics of the standard normal z distribution. Once n is larger than 30, the critical values of the t distribution become very close in value to the corresponding critical values of the standard normal distribution. Thus for samples of size n greater than 30 (or degrees of freedom greater than 29), the critical values of the t distribution are approximately the same as z. Thus, statisticians have generally agreed to abbreviate the Student's t distribution table of critical values to include degrees of freedom from 1 to 29, and to use the corresponding critical value for z for all t's with 30 or more degrees of freedom.

Note   It is customary to divide samples into two categories according to size:

small sample: a sample whose size n is 30 or less, or

large sample: a sample whose size n is larger than 30.

The critical value of t to be used either in a hypothesis test or in constructing a confidence interval will be obtained from Table 6, Appendix E. To obtain the value of t, you will need to know two values: (1) df, the number of degrees of freedom, and (2) α, the area under the curve to the right of the right-hand critical value. A notation much like that used with z will be used to identify a critical value. $t(df, \alpha)$, read as "t of df, α," is the symbol for the value of t described in the previous sentence and shown in Figure 9-2.

Figure 9-2
t distribution showing
t (df, α)

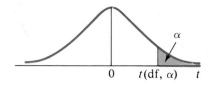

Illustration 9-1   Find the value of $t(10, 0.05)$ (*see* the following diagram).

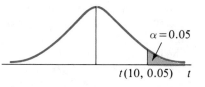

Solution   There are 10 degrees of freedom. So, in Table 6, Appendix E, we look for the column marked $\alpha = 0.05$ and come down to df $= 10$.

	Amount of α in One Tail	
df	⋯ 0.05	⋯
⋮		
10	1.81	
⋮		

From the table we see that $t(10, 0.05) = \mathbf{1.81}$ ◼

For the values of t on the left-hand side of the mean, we can use one of two notations. The t value shown in Figure 9-3 could be $t(\text{df}, 0.95)$, since the area to the right of it is 0.95. Or it could be identified by $-t(\text{df}, 0.05)$, since the t distribution is symmetric about its mean zero.

Figure 9-3
t value on left side of mean

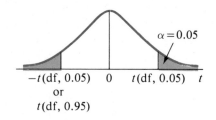

Illustration 9-2   Find the value of $t(15, 0.95)$.

Solution   There are 15 degrees of freedom. In Table 6 we look for the column marked $\alpha = 0.05$ and come down to df $= 15$. The table value is 1.75. Thus

$t(15, 0.95) = -1.75$ (the value is negative because it is to the left of the mean; see the following figure).

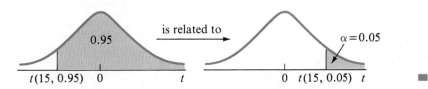

The t statistic is used in problems concerned with μ in much the same manner as z was used in Chapter 8. In hypothesis-testing situations we will use formula (9-1) to calculate the test statistic value in Step 4 of our procedure.

Illustration 9-3   Let's return to the hypothesis of Illustration 8-1 concerning air pollution: "the mean carbon monoxide level of air pollution is no more than 4.9." Does a random sample of 25 readings (sample results: $\bar{x} = 5.1$ and $s = 2.1$) present sufficient evidence to cause us to reject this claim? Use $\alpha = 0.05$.

Solution

Step 1 $H_0: \mu = 4.9 (\leq)$
$H_a: \mu > 4.9$

Step 2 $\alpha = 0.05$, df $= 25 - 1 = 24$, and $t(24, 0.05) = 1.71$, from Table 6, Appendix E. (*See* the accompanying figure.)

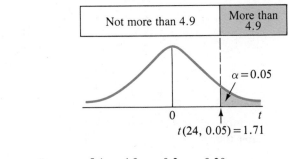

Step 3 $t = \dfrac{\bar{x} - \mu}{s/\sqrt{n}} = \dfrac{5.1 - 4.9}{2.1/\sqrt{25}} = \dfrac{0.2}{2.1/5} = \dfrac{0.20}{0.42} = 0.476$

$t^* = \mathbf{0.48}$

Comparing this value with the test criteria, we have the situation shown in the accompanying figure.

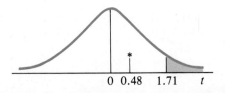

Step 4   *Decision*: Fail to reject H_0 ($t*$ is in the noncritical region).

Conclusion: At the 0.05 level of significance, we do not have enough evidence to reject the claim that the mean carbon monoxide level is no more than 4.9.   ■

Note   **If the value of df** ($df = n - 1$) in an exercise such as Illustration 9-3 **is larger than 29**, then the critical value for $t(df, \alpha)$ actually becomes nearly the same as $z(\alpha)$, the z score listed at the bottom of Table 6. Therefore, $t(df, \alpha)$ is not given when $df > 29$; $z(\alpha)$ is used instead.

Because Table 6 has critical values for the Student's t distribution, the prob-value cannot be calculated for a hypothesis test that involves the use of t. However, the prob-value can be estimated.

Illustration 9-4   Let's return to Illustration 9-3. Note that $t* = 0.48$, $df = 24$, and H_a: $\mu > 4.9$. Thus for Step 4 of the prob-value solution we have

$$P = P(t > 0.48, \text{ knowing df} = 24). \text{ (}see\text{ the following figure)}$$

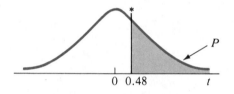

Portion of Table 6

df	0.25
⋮	
24	0.685

By inspecting the $df = 24$ row of Table 6, you can determine that the prob-value is greater than 0.25. The 0.685 entry in the table tells us that the $P(t > 0.685) = 0.25$, as shown in the following figure.

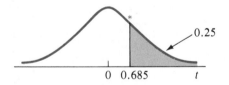

By comparing $t* = 0.48$ (*see* the following figure) we see that the **prob-value p is greater than 0.25**.

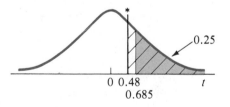

Illustration 9-5   Determine the p-value for the following hypothesis test.

$$H_0: \mu = 55$$
$$H_a: \mu \neq 55$$
$$df = 15 \quad \text{and} \quad t* = -1.84$$

Solution

$$P = P(t < -1.84) + P(t > 1.84) = 2P(t > 1.84)$$

Portion of Table 6

df	0.05	0.025
:		
15	1.75	2.13

By inspecting the df $= 15$ row of Table 6 in Appendix E, we find that $P(t > 1.84)$ is between 0.025 and 0.05. Therefore, **P** is between the double of these amounts. Thus the prob-value is between 0.05 and 0.10, **0.05 < P < 0.10**.   ■

Note   Since Table 6 has only critical values, we can only estimate the p-value when t is the test statistic.

The population mean may be estimated when σ is unknown in a manner similar to that used when σ is known. The difference is the use of Student's t in place of z and the use of s, the sample standard deviation, as an estimate of σ. The formula for the $1 - \alpha$ **confidence interval** of estimation then becomes

$$\bar{x} - t(\text{df}, \alpha/2) \cdot \frac{s}{\sqrt{n}} \quad \text{to} \quad \bar{x} + t(\text{df}, \alpha/2) \cdot \frac{s}{\sqrt{n}} \tag{9-2}$$

where df $= n - 1$.

Illustration 9-6   A random sample of size 20 is taken from the weights of babies born at Northside Hospital during the year 1982. A mean of 6.87 pounds and a standard deviation of 1.76 pounds were found for the sample. Estimate, with 95 percent confidence, the mean weight of all babies born in this hospital in 1982.

Solution   The information we are given is $\bar{x} = 6.87$, $s = 1.76$, and $n = 20$. $1 - \alpha = 0.95$ implies that $\alpha = 0.05$; $n = 20$ implies that df $= 19$. From Table 6, we get $t(19, 0.025) = 2.09$. See the following figure.

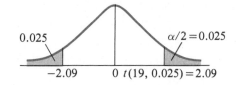

The interval is thus

$$\bar{x} \pm t(19, 0.025) \cdot \frac{s}{\sqrt{n}}$$

$$6.87 \pm 2.09 \cdot \frac{1.76}{\sqrt{20}}$$

$$6.87 \pm \frac{(2.09)(1.76)}{4.472}$$

$$6.87 \pm 0.82$$

$$6.87 - 0.82 = \mathbf{6.05} \quad \text{and} \quad 6.87 + 0.82 = \mathbf{7.69}$$

The 0.95 confidence interval for μ is (6.05, 7.69). That is, with 95 percent confidence we estimate the mean weight to be between 6.05 and 7.69 pounds.   ■

EXERCISES

9.1 Find these critical values using Table 6 in Appendix E.

    a. $t(25, 0.05)$ **b.** $t(10, 0.10)$ **c.** $t(15, 0.01)$ **d.** $t(21, 0.025)$

    e. $t(21, 0.95)$ **f.** $t(26, 0.975)$ **g.** $t(27, 0.99)$ **h.** $t(60, 0.025)$

9.2 Using the notation of Exercise 9.1, name and find the following critical values of t:

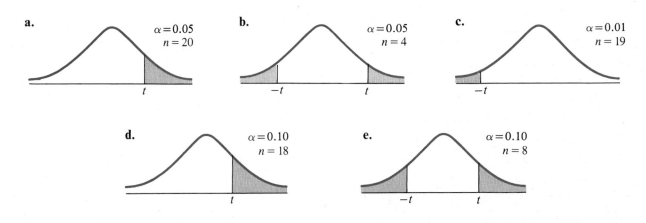

9.3 Ninety percent of Student's t distribution lies between $t = -1.89$ and $t = 1.89$ for how many degrees of freedom?

9.4 a. Find the first percentile of Student's t distribution with 24 degrees of freedom.

    b. Find the 95th percentile of Student's t distribution with 24 degrees of freedom.

    c. Find the first quartile of Student's t distribution with 24 degrees of freedom.

9.5 Find the percent of the Student's t distribution that lies between the following values.

    a. df $= 12$ and t ranges from -1.36 to 2.68

    b. df $= 15$ and t ranges from -1.75 to 2.95

9.6 The variance for a Student's t distribution is equal to df/(df $-$ 2). Find the standard deviation for a Student's t distribution with each of the following degrees of freedom.

    a. 10   **b.** 20   **c.** 30

9.7 a. State two ways in which the normal distribution and the Student's t distribution are alike.

    b. State two ways in which they are different.

9.8 In order to test the null hypothesis "the mean weight for adult males equals 160 pounds" against the alternative "the mean weight for adult males exceeds

160 pounds," the weights of 16 males were determined with the following results.

$$173 \quad 178 \quad 145 \quad 146 \quad 157 \quad 175 \quad 173 \quad 137$$
$$152 \quad 171 \quad 163 \quad 170 \quad 135 \quad 159 \quad 199 \quad 131$$

Verify the results shown in the following MINITAB analysis. [*Note:* ALT $= -1, 0, +1$ represents lower-tail, two-tail, and upper-tail tests, respectively.]

```
MTB > SET THE FOLLOWING MALE WEIGHTS IN C1
DATA> 173 178 145 146 157 175 173 137 199 131 152 171 163 170 135 159
DATA> END DATA
MTB > TTEST OF MU = 160,ALT = 1,DATA IN C1
TEST OF MU = 160.00 VS MU G.T. 160.00
                N      MEAN    STDEV   SE MEAN      T    P VALUE
C1             16     160.25   18.49     4.62     0.05     0.48
MTB > STOP
```

9.9 A student group maintains that the average student must travel for at least 25 minutes in order to reach college each day. The college admissions office obtained a random sample of 36 one-way travel times from students. The sample had a mean of 19.4 minutes and a standard deviation of 9.6 minutes. Does the admissions office have sufficient evidence to reject the students' claim? Use $\alpha = 0.01$.

    a. Solve using the classical approach.

    b. Solve using the prob-value approach.

9.10 Homes in a nearby college town have a mean value of $58,950. It is assumed that homes in the vicinity of the college have a higher value. To test this theory, a random sample of 12 homes are chosen from the college area. Their mean valuation is $62,460 and the standard deviation is $5,200. Complete a hypothesis test using $\alpha = 0.05$.

    a. Solve using the classical approach.

    b. Solve using the prob-value approach.

9.11 The pulse rates for 13 adult women were found to be

$$83 \quad 58 \quad 70 \quad 56 \quad 76 \quad 64 \quad 80 \quad 76 \quad 70 \quad 97 \quad 68 \quad 78 \quad 108$$

Verify the results shown on the last line of this MINITAB output.

```
MTB > SET THE FOLLOWING FEMALE PULSE RATES IN C1
DATA> 83 58 70 56 76 64 80 76 70 97 68 78 108
DATA> END DATA
MTB > TINTERVAL 90 PERCENT CONFIDENCE INTERVAL FOR DATA IN C1
                N      MEAN    STDEV   SE MEAN    90.0 PERCENT C.I.
C1             13     75.69    14.54     4.03    (  68.50,  82.88)
MTB > STOP
```

9.12 Taking a random sample of 25 individuals registering for the draft in a particular town, we find that the mean waiting time in the registration line was 12.6 minutes and the standard deviation was 3.0 minutes. Using a 90 percent confidence interval, estimate the mean waiting time for all individuals registering.

9.13 While doing an article on the high cost of college education, a reporter took a random sample of the cost of textbooks for a semester. The random variable x is the cost of one book. Her sample data can be summarized by $n = 41$, $\sum x = 550.22$, and $\sum (x - \bar{x})^2 = 1617.984$.

    a. Find the sample mean $\bar{x}$.

    b. Find the sample standard deviation s.

    c. Find the 0.90 confidence interval to estimate the true mean textbook cost for the semester, based on this sample.

9.14 Ten randomly selected shut-ins were each asked to list how many hours of television they watched per week. The results are

$$82 \quad 66 \quad 90 \quad 84 \quad 75 \quad 88 \quad 80 \quad 94 \quad 110 \quad 91$$

Determine the 90 percent confidence interval estimate for the mean number of hours of television watched per week by shut-ins.

9.15 It is claimed that the students at a certain university will score an average of 35 on a given test. Is the claim reasonable if a random sample of test scores from this university yields

$$33 \quad 42 \quad 38 \quad 37 \quad 30 \quad 42$$

Complete a hypothesis test using $\alpha = 0.05$.

    a. Solve using the classical approach.

    b. Solve using the prob-value approach.

9.2 INFERENCES ABOUT THE BINOMIAL PROBABILITY OF SUCCESS

Recall that the binomial parameter p was defined to be the theoretical or population probability of success on a single trial in a binomial experiment. Also, the random variable x is the number of successes that occur in a set of n trials. By combining the definition of empirical probability, $P'(A) = n(A)/n$ [formula (4-1)], with the notation of the binomial experiment, we define p', **the observed or sample binomial probability**, to be $p' = x/n$. Also recall that the mean and standard deviation of the binomial random variable x are found by using formulas (5-7) and (5-8): $\mu = np$ and $\sigma = \sqrt{npq}$, where $q = 1 - p$. This distribution of x is considered to be approximately normal if n is larger than 20 and if np and nq are both larger than 5. This commonly accepted rule of thumb allows us to use the normal distribution when making inferences concerning a binomial parameter p.

    Generally it is easier to work with the distribution of p' rather than the distribution of x. Consequently, we will convert formulas (5-7) and (5-8) from the units of x to units of proportions. If we divide formulas (5-7) and (5-8) by n, we should change the units from those of x to those of proportion. The mean of x is np; thus the mean of p', $\mu_{p'}$, should be np divided by n, that is, np/n, or just p. (Does it seem reasonable that the mean of the distribution of observed values of p' should be

p, the true proportion?) Further, the standard error of p' in this sampling distribution is

$$\sigma_{p'} = \sqrt{npq}/n = \sqrt{npq/n^2} = \sqrt{pq/n}$$

We summarize this information as follows:

An observed value of p' belongs to a sampling distribution that

    1. is approximately normal.

    2. has a mean $\mu_{p'}$ equal to p.

    3. has a standard error $\sigma_{p'}$ equal to $\sqrt{pq/n}$.

This approximation to the normal distribution is considered reasonable whenever n is greater than 20 and both np and nq are greater than 5.

Recall   The standard deviation of a sampling distribution is called the standard error.

As a result of these new definitions for μ and σ, the calculated **value of** z in Step 3 of a hypothesis test concerning p is obtained by using the formula

$$z = \frac{p' - p}{\sqrt{pq/n}}, \quad \text{where} \quad p' = \frac{x}{n} \tag{9-3}$$

The value of p to be used in formula (9-3) will be the value stated in the null hypothesis.

Illustration 9-7   While talking about the cars that fellow students drive, Tom made the claim that at least 15 percent of the students drive convertibles. Bill found this hard to believe and decided to check the validity of Tom's claim, so he took a random sample. At a level of significance of 0.10, does Bill have sufficient evidence to reject Tom's claim if there were 17 convertibles in his sample of 200 cars?

Solution

Step 1 H_0: $p = 0.15$ ($\geq$) (at least 15 percent)
             H_a: $p < 0.15$ (less than 15 percent)

Step 2 $\alpha = 0.10$. z is found by using Table 5, Appendix E. See the accompanying figure.

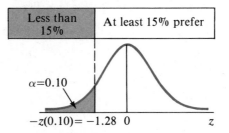

Step 3

$$p' = \frac{17}{200} = 0.085$$

$$z = \frac{p' - p}{\sqrt{pq/n}} = \frac{0.085 - 0.150}{\sqrt{(0.15)(0.85)/200}} = \frac{-0.065}{\sqrt{0.00064}}$$

$$= \frac{-0.065}{0.025} = -2.6$$

$$z^* = -2.6$$

Compare this value with the test criteria, as shown in the accompanying figure.

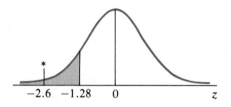

Step 4   *Decision*: Reject H_0 (z^* is in the critical region).

Conclusion: The evidence found contradicts Tom's claim. At the 0.10 level of significance, it appears that less than 15 percent of the students drive convertibles.   ∎

The solution to Illustration 9-7 could have been carried out using the prob-value procedure. This alternative solution follows.

Solution

Step 1 H_0: $p = 0.15$ ($\geq$)   (at least 15 percent)
          H_a: $p < 0.15$ (less than 15 percent)

Step 2 $\alpha = 0.10$

Step 3 $z = \dfrac{p' - p}{\sqrt{pq/n}} = \dfrac{0.085 - 0.150}{\sqrt{(0.15)(0.85)/200}}$

          $z^* = -2.60$

Step 4 $P = P(z < z^*) = P(z < -2.60) = 0.5000 - 0.4953$

          $P = 0.0047$ (*see* the accompanying figure)

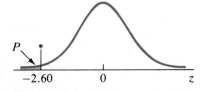

Step 5   At the 0.10 level of significance, the sample information is significant. That is, it appears that less than 15 percent of the students drive convertibles.

■

When we estimate the true population proportion p, we will base our estimations on the observed value p'. The **confidence interval formula** is similar to the previous confidence interval formula.

$$p' - z(\alpha/2) \cdot \sqrt{\frac{p'q'}{n}} \quad \text{to} \quad p' + z(\alpha/2) \cdot \sqrt{\frac{p'q'}{n}} \qquad (9\text{-}4)$$

where $p' = x/n$ and $q' = 1 - p'$.

Notice that the standard error, $\sqrt{pq/n}$, has been replaced by $\sqrt{p'q'/n}$. Since we do not know the value of p, we must use the best replacement available. That replacement is p', the observed value or the point estimate for p. This replacement will cause little change in the width of our confidence interval.

Illustration 9-8   Refer to Illustration 9-7. Suppose that Bill has taken his sample with the intention of estimating the value of p, the proportion of his fellow students who drive convertibles.

a. Find the best point estimate for p that he could use.

b. Determine the 90 percent confidence interval estimate for the true value of p by using formula (9-4).

Solution

a. The best point estimate of p is 0.085, the observed value of p'.
b. The confidence interval is $p' \pm z(\alpha/2) \cdot \sqrt{p'q'/n}$, where $q' = 1 - p'$, and $1 - \alpha = 0.90$; therefore, $z(\alpha/2) = z(0.05) = 1.65$. Thus the confidence interval is

$$0.085 \pm (1.65) \cdot \sqrt{\frac{(0.085)(0.915)}{200}}$$

$$0.085 \pm (1.65)\sqrt{0.000389}$$

$$0.085 \pm (1.65)(0.020)$$

$$0.085 \pm 0.033$$

$$0.085 - 0.033 = \mathbf{0.052} \quad 0.085 + 0.033 = \mathbf{0.118}$$

$$\mathbf{0.052} \quad \text{to} \quad \mathbf{0.118}$$

That is, the true proportion of students who drive convertibles is between 0.052 and 0.118, with 90 percent confidence.

■

By using the maximum-error part of the confidence interval formula, it is possible to determine the size of the sample that must be taken in order to estimate p with a desired accuracy. The **maximum error of estimate for a proportion** is

$$E = z(\alpha/2) \cdot \sqrt{\frac{pq}{n}} \qquad (9\text{-}5)$$

When using this formula, we must decide how accurate our answer must be. (Remember that we are estimating p. Therefore, E will be expressed in hundredths.) We need to establish the level of confidence we wish to work with. If you have any indication of the value of p, use this value for p and $q = 1 - p$. If there is no indication of an approximate value for p, then by assigning p the value 0.5, you will obtain the largest possible sample size that is required.

For ease of use, formula (9-5) can be expressed as

$$n = \frac{[z(\alpha/2)]^2 \cdot p \cdot q}{(E)^2} \tag{9-6}$$

Illustration 9-9   Determine the sample size that is required to estimate the true proportion of blue-eyed community college students if you want your estimate to be within 0.02 with 90 percent confidence.

Solution

Step 1 $1 - \alpha = 0.90$; therefore, $z(\alpha/2) = z(0.05) = 1.65$

Step 2 $E = 0.02$

Step 3   Use $p = 0.5$; therefore, $q = 1 - p = 0.5$.

Step 4   Use formula (9-6) to find n:

$$n = \frac{(1.65)^2 \cdot (0.5) \cdot (0.5)}{(0.02)^2} = \frac{0.680625}{0.0004} = 1701.56$$

$$= \mathbf{1702}$$ ◼

Illustration 9-10   An automobile manufacturer purchases bolts from a supplier who claims his bolts to be approximately 5 percent defective. Determine the sample size that will be required to estimate the true proportion of defective bolts if we want our estimate to be within 0.02 with 90 percent confidence.

Solution

Step 1 $1 - \alpha = 0.90$; therefore, $z(\alpha/2) = z(0.05) = 1.65$

Step 2 $E = 0.02$

Step 3   The supplier's claim is "5 percent defective"; thus $p = 0.05$; therefore, $q = 1 - p = 0.95$.

Step 4   Use formula (9-6) to find n:

$$n = \frac{(1.65)^2(0.05)(0.95)}{(0.02)^2} = \frac{0.12931875}{0.0004} = 323.3$$

$$= \mathbf{324}$$ ◼

Notice the difference in the sample size required in Illustrations 9-9 and 9-10. The only real difference between the problems is the value that was used for p. In

Illustration 9-9 we used $p = 0.5$, and in Illustration 9-10 we used $p = 0.05$. Recall that $p = 0.5$ gives a sample of maximum size. Thus it will be of great advantage to have an indication of the value expected for p if p is much different from 0.5.

Case Study 9-1

POLL FINDS SUSPICION OF REAGAN

The May 1987 New York Times-CBS News Poll showed that only 24 percent of 1343 adults interviewed believed that President Reagan was telling the truth about the Iran-Contra matter. The 24 percent reported is an observed proportion p′ and serves as the point estimate for the binomial parameter p, the probability that a randomly selected adult would answer yes to the question "Do you believe President Reagan is telling the truth, . . ." Note that the closing paragraph mentions "a margin of sampling error of plus or minus three percentage points." This sampling error is the maximum error of estimate. Thus the population percentage is being estimated as 21 to 27 percent (24 ± 3).

NEW YORK   Only one American in four says President Reagan is telling the truth when he denies knowing that money from arms sales to Iran went to help the Nicaraguan rebels, the latest New York Times-CBS News Poll shows.

But the public is evenly divided on whether the congressional hearings on the Iran-Contra matter are producing new information or are merely "for show."

Of 1,343 adults interviewed, 24 percent—down from 31 percent two weeks earlier—said he was telling the truth, and 59 percent said he was lying.

Forty-three percent said the hearings were producing new information and 42 percent said they had "only been for show."

The survey had a margin of sampling error of plus or minus three percentage points.

From *The New York Times*, May 1987. Copyright © 1987 The New York Times Company. Reprinted by permission.

Case Study 9-2

GALLUP REPORT: SAMPLING TOLERANCES

The Gallup Report provided its readers with the accompanying information and table. The table gives the 0.95 confidence level sampling errors (maximum error of estimation) for various percentages and sample sizes. Can you verify the values on this table? (See Exercise 9.25.)

SAMPLING ERROR

In interpreting survey results, it should be borne in mind that all sample surveys are subject to sampling error, that is, the extent to which the results may differ from those that would be obtained if the whole population surveyed had been interviewed. The size of such sampling errors depends largely on the number of interviews.

Table A shows the allowances that should be made for the sampling error of a percentage.

The table should be used as follows: Say a reported percentage is 33 for a group which includes 1500 respondents. Go to the row labeled "percentages near 30" and then to the column headed "1500." The number at this point is 3, which means that the 33 percent obtained in the sample is subject to a sampling error of plus or minus 3 points. Another way of saying it is that very probably (95 times out of 100) the average of repeated samplings would be somewhere between 30 and 36, with the most likely figure the 33 obtained.

TABLE A. Sampling tolerances

	Recommended Allowance for Sampling Error of a Percentage						
	In Percentage Points (at 95 in 100 confidence level) Size of Sample						
	1500	1000	750	600	400	200	100
Percentages near 10	2	2	3	4	4	5	7
Percentages near 20	2	3	4	4	5	7	9
Percentages near 30	3	4	4	5	6	8	10
Percentages near 40	3	4	4	5	6	9	11
Percentages near 50	3	4	4	5	6	9	11
Percentages near 60	3	4	4	5	6	9	11
Percentages near 70	3	4	4	5	6	8	10
Percentages near 80	2	3	4	4	5	7	9
Percentages near 90	2	2	3	4	4	5	7

Source: Copyright 1986 by the Gallup Report. Reprinted by permission.

EXERCISES

9.16 The fairness (balance) of a coin is in question. It is believed that the probability of a head, p, is greater than 0.5. The null and alternative hypotheses are:

$$H_0: p = 0.5 \quad \text{and} \quad H_a: p > 0.5$$

The coin is flipped 14 times and the number of heads is observed. Calculate α for each of the following critical regions.

    a. $x = 10, 11, 12, 13, 14$

    b. $x = 11, 12, 13, 14$

    c. $x = 12, 13, 14$

9.17 We are testing $H_0: p = 0.2$ and decide to reject H_0 if after 15 trials we observe more than 5 successes.

    a. State an appropriate alternative hypothesis.

    b. What is the level of significance of this test?

c. If we observe 5 successes, do we reject H_0?

d. If we observe 6 successes, do we reject H_0?

e. If H_0 is $p = 0.1$ and we use the same decision rule, what happens to the level of significance?

9.18 You are testing the hypothesis $p = \frac{1}{3}$ and have decided to reject this hypothesis if after 15 trials you observe 14 or more successes.

a. If the null hypothesis is true and you observe 13 successes, then which of the following will you do? (1) Correctly fail to reject H_0? (2) Correctly reject H_0? (3) Commit a type I error? (4) Commit a type II error?

b. Find the significance level of your test. (Use normal approximation.)

c. If the true probability of success is $\frac{1}{2}$ and you observe 13 successes, then which of the following will you do? (1) Correctly fail to reject H_0? (2) Correctly reject H_0? (3) Commit a type I error? (4) Commit a type II error?

d. Calculate the prob-value for your hypothesis test after 13 successes are observed.

9.19 An insurance company states that 90 percent of its claims are settled within 30 days. A consumer group selected a random sample of 75 of the company's claims to test this statement. If the consumer group found that 55 of the claims were settled within 30 days, do they have sufficient reason to support their contention that less than 90 percent of the claims are settled within 30 days?

a. Solve using the classical approach.

b. Solve using the prob-value approach.

9.20 A county judge has agreed that he will give up his county judgeship and run for a state judgeship unless there is evidence that more than 25 percent of his party is in opposition. A random sample of 800 party members included 217 who opposed him. Does this sample suggest that he should run for the state judgeship in accordance with his agreement? Carry out this hypothesis test by using $\alpha = 0.10$.

a. Solve using the classical approach.

b. Solve using the prob-value approach.

9.21 A politician claims that she will receive 60 percent of the vote in an upcoming election. The results of a properly designed random sample of 100 voters showed that 50 of those sampled will vote for her. Is it likely that her assertion is correct at the 0.05 level of significance?

a. Solve using the classical approach.

b. Solve using the prob-value approach.

9.22 The full-time student body of a college is composed of 50 percent males and 50 percent females. Does a random sample of students (30 male, 20 female) from an introductory chemistry course show sufficient evidence to reject the hypothesis that the proportion of male and of female students who take this course is the same as that of the whole student body? Use $\alpha = 0.05$.

a. Solve using the classical approach.

b. Solve using the prob-value approach.

9.23 Construct 90 percent confidence intervals for the binomial parameter p for each of the following pairs of values. Write your answers on the chart.

	Observed Proportion $p' = x/n$	Sample Size	Lower Limit	Upper Limit
a.	$p' = 0.3$	$n = 30$		
b.	$p' = 0.7$	$n = 30$		
c.	$p' = 0.5$	$n = 10$		
d.	$p' = 0.5$	$n = 100$		
e.	$p' = 0.5$	$n = 1000$		

f. Compare answers (a) and (b).

g. Compare answers (c), (d), and (e).

9.24 a. Calculate the 0.95 confidence maximum error of estimate for p when p is near 0.25 and $n = 1343$.

b. Compare your answer in (a) to the sampling error described in Case Study 9-1.

c. Most public-opinion polls use a 0.95 confidence level when reporting the sampling error. Explain why the sampling error of the poll in Case Study 9-1 was rounded up to 3 percent instead of 2 percent, the nearest whole percent.

9.25 Calculate the 0.95 confidence maximum error of estimate for p if

a. p is near 0.1 and $n = 1000$.

b. p is near 0.2 and $n = 100$.

c. p is near 0.5 and $n = 50$.

d. Compare these maximum errors to the sampling errors reported in Table A of Case Study 9-2.

9.26 The *Time* poll (news article, page 327) mentions a sampling error of 3 percent for information from white respondents and a larger sampling error for the information from black respondents because of the smaller number. Verify the 3 percent sampling error and find the larger sampling error assuming 95 percent confidence.

9.27 A telephone survey was conducted to estimate the proportion of households with a personal computer. Of the 350 households surveyed, 75 had a personal computer.

a. Give a point estimate for the proportion in the population who have a personal computer.

b. Give the maximum error of estimate with 95 percent confidence.

9.28 A bank randomly selected 250 checking account customers and found that 110 of them also had savings accounts at this same bank. Construct a 0.95 confidence interval estimate for the true proportion of checking account customers who also have savings accounts.

9.29 In a sample of 60 randomly selected students, only 22 favored the amount being budgeted for next year's intermural and interscholastic sports. Construct the 0.99 confidence interval estimate for the proportion of all students who support the proposed budget amount.

9.30 A bank believes that approximately $\frac{2}{5}$ of its checking account customers have used at least one other service provided by the bank within the last six months. How large a sample will be needed to estimate the true proportion to within 5 percent at the 0.98 level of confidence?

9.31 A hospital administrator wants to conduct a telephone survey to determine the proportion of people in a city who have been hospitalized for at least three days in the last five years. How large a sample must she take to be 95 percent confident that the sample proportion will be within 0.03 of the true proportion of the city?

9.32 Members of the student senate are working on a proposal concerning a pub on campus. They feel that if 60 percent of the student body favor the proposal, it should be passed. How large a sampling needs to be taken to enable them to estimate the proportion of students who favor the proposal to within 4 percent with 99 percent confidence?

9.3 INFERENCES ABOUT VARIANCE AND STANDARD DEVIATION

Problems often arise that require us to make inferences about variability. For example, a soft drink bottling company has a machine that fills 32-ounce bottles. It needs to control the variance σ^2 (or standard deviation σ) among the amount x of soft drink put in each bottle. The mean amount placed in each bottle is important, but a correct mean amount does not ensure that the filling machine is working correctly. If the variance is too large, there could be many bottles that are overfilled and that are underfilled. Thus this bottling company will want to maintain as small a variance (or standard deviation) as possible.

We will study two kinds of inferences in this section: (1) the hypothesis test concerning the variance (or standard deviation) of one population and (2) the estimation of the variance (or standard deviation) of one population. In these two inferences it is customary to talk about variance instead of the standard deviation, because the techniques employ the sample variance rather than the standard deviation. However, remember that the standard deviation is the positive square root of the variance; thus to talk about the variance of a population is comparable to talking about the standard deviation.

Let's return to our problem. Suppose the bottling company wishes to detect when the variability in the amount of soft drink placed in each bottle gets out of control. A variance of 0.0004 is considered acceptable, and the company will want to adjust the bottle-filling machine when the variance becomes larger than this value. The decision will be made by using the hypothesis test procedure. The null hypothesis is "the variance is no larger than the specified value 0.0004"; the alternative hypothesis is "the variance is larger than 0.0004."

$$H_0: \sigma^2 = 0.0004 \, (\leq) \quad \text{(variance not out of control)}$$
$$H_a: \sigma^2 > 0.0004 \quad \text{(variance out of control)}$$

chi-square   The test statistic that will be used in making a decision about the null hypothesis is **chi-square**, χ^2 (χ is the Greek lowercase letter chi). The calculated value of chi-square will be obtained by using the formula

$$\chi^2 = \frac{(n-1)s^2}{\sigma^2} \tag{9-7}$$

where s^2 is the sample variance, n is the sample size, and σ^2 is the value specified in the null hypothesis.

When random samples are drawn from a normal population of a known variance σ^2, the quantity $(n-1)s^2/\sigma^2$ possesses a probability distribution that is known as the **chi-square distribution**. The equations that define the chi-square distribution are not given here; they are beyond the level of this book. However, to use the chi-square distribution, we must be aware of the following properties (*see* also Figure 9-4):

Properties of the Chi-square Distribution

1. χ^2 is nonnegative in value; it is zero or positively valued.

2. χ^2 is not symmetrical; it is skewed to the right.

3. There are many χ^2 distributions. As with the t distribution, there is a different χ^2 distribution for each degree-of-freedom value.

Figure 9-4
Various chi-square distributions

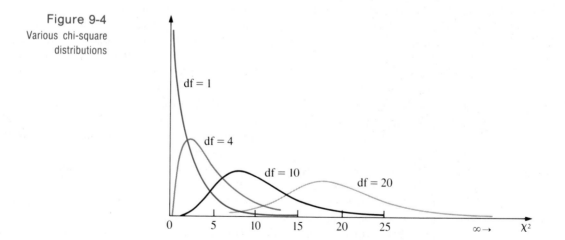

For the inferences discussed in this section, the number of degrees of freedom df is equal to $n-1$.

The critical values for chi-square are obtained from Table 7 in Appendix E. The critical values will be identified by two values: degrees of freedom df and the area under the curve to the right of the critical value being sought. Thus $\chi^2(df, \alpha)$ is the symbol used to identify the critical value of chi-square with df degrees of freedom and with α being the area to the right, as shown in Figure 9-5. Since the chi-square distribution is not symmetrical, the critical values associated with both tails are given in Table 7.

Figure 9-5
Chi-square distribution
showing χ^2 (df, α)

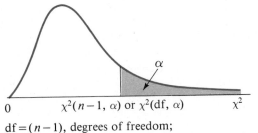

$$0 \qquad \chi^2(n-1, \alpha) \text{ or } \chi^2(df, \alpha) \qquad \chi^2$$

df $=(n-1)$, degrees of freedom;
α is the area under curve to the right of a particular value

Illustration 9-11   Find χ^2 (20, 0.05).

Solution   In Table 7 you will find the value shown in the following table. Therefore, $\chi^2(20, 0.05) = $ **31.4**.

	Area Under Curve to the Right
df	$\cdots$ 0.050 $\cdots$
⋮	
20	31.4
⋮	

■

Illustration 9-12   Find $\chi^2(14, 0.90)$.

Solution   df $= 14$ and the area to the right of the critical value is 0.90, as shown in the following figure. Therefore, $\chi^2(14, 0.90) = $ **7.79**.

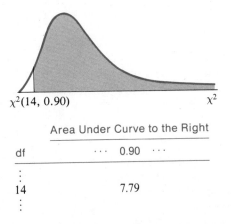

$$\chi^2(14, 0.90) \qquad \qquad \chi^2$$

	Area Under Curve to the Right
df	$\cdots$ 0.90 $\cdots$
⋮	
14	7.79
⋮	

■

Note   When df > 2, the mean value of the chi-square distribution is df. The mean is located to the right of the mode (the value where the curve reaches its high point). By locating the value of df on your sketch of the χ^2 distribution, you will establish an approximate scale so that the values can be located in their respective positions. See Figure 9-6.

Figure 9-6
Location of mean and mode
for χ^2 distribution

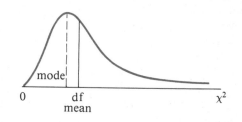

Illustration 9-13   Recall that the soft drink bottling company wanted to control the variance by not allowing the variance to exceed 0.0004. Does a sample of size 28 with a variance of 0.0007 indicate that the bottling process is out of control (with regard to variance) at the 0.05 level?

Solution

Step 1 $H_0: \sigma^2 = 0.0004$ ($\leq$)   (not out of control)
$H_a: \sigma^2 > 0.0004$ (out of control)

Step 2 $\alpha = 0.05$ and $n = 28$; therefore, df $= 27$. The test statistic is χ^2, and the critical region is the right tail, with an area of 0.05. $\chi^2(27, 0.05)$ is found in Table 7. See the following figure.

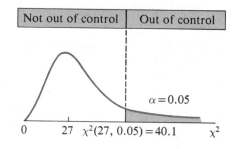

Step 3

$$\chi^2 = \frac{(n-1)s^2}{\sigma^2} = \frac{(28-1)(0.0007)}{0.0004}$$

$$= \frac{(27)(0.0007)}{0.0004} = \frac{0.0189}{0.0004} = 47.25$$

$$\chi^{2*} = \mathbf{47.25}$$

See Figure 9-7.

Figure 9-7

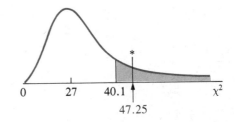

Step 4   *Decision*: Reject H_0 ($\chi^{2}*$ is in the critical region).

Conclusion: At the 0.05 level of significance, we conclude that the bottling process is out of control with regard to the variance.

The prob-value can be estimated for hypothesis tests using the chi-square test statistic in much the same manner as when Student's t was used. ■

Illustration 9-14   Find the prob-value for the hypothesis test in Illustration 9-13.

$$H_0: \sigma^2 = 0.0004$$
$$H_a: \sigma^2 > 0.0004$$
$$\text{df} = 27 \text{ and } \chi^{2}* = 47.25$$

Solution

$$P = P(\chi^2 > 47.25) \text{ (} see \text{ the accompanying figure)}$$

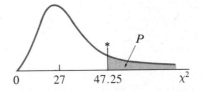

By inspecting the df $= 27$ row of Table 7, we find that 47.25 is between 47.0 and 49.7. Therefore the prob-value is between **0.005** and **0.010**. ■

Illustration 9-15   One of the factors used in determining the usefulness of a particular exam as a measure of students' abilities is the amount of "spread" that occurs in the grades. A set of test results has little value if the range of the grades is very small. However, if the range of grades is quite large, there is a definite difference in the scores achieved by the "better" students and the scores achieved by the "poorer" students.

    On an exam with a total of 100 points, it has been claimed that a standard deviation of 12 points is desirable. To determine whether or not the last one-hour exam a professor gave his class was a good test, he tested this hypothesis at $\alpha = 0.05$ by using the exam scores of the class. There were 28 scores and the standard deviation of those 28 scores was found to be 10.5. Does the professor have evidence, at the 0.05 level of significance, that this exam does not have the specified standard deviation?

Solution   The information given is $n = 28$, $s = 10.5$, and $\alpha = 0.05$.

Step 1 $H_0: \sigma = 12$

$H_a: \sigma \neq 12$

Step 2 $\alpha = 0.05$; the critical values are $\chi^2(27, 0.975) = 14.6$ and $\chi^2(27, 0.025) = 43.2$. See the accompanying figure.

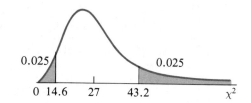

Step 3

$$\chi^2 = \frac{(n - 1)s^2}{\sigma^2} = \frac{(27)(10.5)^2}{(12)^2} = \frac{2976.75}{144} = 20.6719$$

$\chi^{2*} = \mathbf{20.67}$

The accompanying figure shows this value compared to the test criteria.

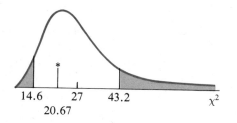

Step 4   *Decision:* Fail to reject H_0.

Conclusion: At the 0.05 level of significance, the professor does not have sufficient evidence to claim that the standard deviation is different from 12. ∎

These tests for variance may be one-tailed or two-tailed tests, in accordance with the statement of the claim being tested.

The formula for chi-square may be reworked to give the values at the extremities of the confidence interval:

$$\chi^2 = \frac{(n - 1) \cdot s^2}{\sigma^2} \quad \text{(solve for } \sigma^2)$$

$$\sigma^2 = \frac{(n - 1) \cdot s^2}{\chi^2} \tag{9-8}$$

When constructing a $1 - \alpha$ confidence interval, the critical values of chi-square are separately substituted into formula (9-8) to obtain the two endpoints of the confidence interval of estimation. Note that $\chi^2(\mathrm{df}, 1 - \alpha/2)$ is less than $\chi^2(\mathrm{df}, \alpha/2)$.

Therefore, after dividing, the numbers will be in the opposite order, yielding the following **confidence interval for variance**:

$$\frac{(n-1)s^2}{\chi^2(df, \alpha/2)} \quad \text{to} \quad \frac{(n-1)s^2}{\chi^2(df, 1-\alpha/2)} \tag{9-9}$$

If the confidence interval for the standard deviation is desired, we need only take the square root of each of the numbers in formula (9-9).

$$s \cdot \sqrt{\frac{(n-1)}{\chi^2(df, \alpha/2)}} \quad \text{to} \quad s \cdot \sqrt{\frac{(n-1)}{\chi^2(df, 1-\alpha/2)}} \tag{9-10}$$

Illustration 9-16   Using the professor's results from Illustration 9-15 ($n = 28$, $s = 10.5$), calculate the 95 percent confidence interval for the estimate of the population variance and standard deviation.

Solution   The information given is $n = 28$ and $s = 10.5$. For a 95 percent confidence interval, $\alpha = 0.05$ and hence $\alpha/2 = 0.025$. The critical values for χ^2 are shown in the accompanying figure. The confidence interval, using formula (9-9), is

$$\frac{(27)(10.5)^2}{43.2} \quad \text{to} \quad \frac{(27)(10.5)^2}{14.6}$$

$$\frac{2976.75}{43.2} \quad \text{to} \quad \frac{2976.75}{14.6}$$

$$\mathbf{68.9} \quad \text{to} \quad \mathbf{203.9}$$

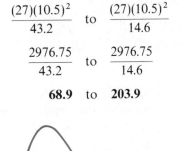

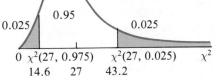

That is, with 95 percent confidence we estimate the population variance to be between 68.9 and 203.9.

The confidence interval for the standard deviation can be found by taking the square root of 68.9 and of 203.9. The 95 percent confidence interval estimate for the standard deviation is **8.3** to **14.3**.   ∎

EXERCISES

9.33 Find these critical values by using Table 7 of Appendix E.

    a. $\chi^2(18, 0.01)$ **b.** $\chi^2(16, 0.025)$ **c.** $\chi^2(8, 0.10)$ **d.** $\chi^2(28, 0.01)$

    e. $\chi^2(22, 0.95)$ **f.** $\chi^2(10, 0.975)$ **g.** $\chi^2(50, 0.90)$ **h.** $\chi^2(24, 0.99)$

9.34 Using the notation of Exercise 9.33, name and find the critical values of χ^2.

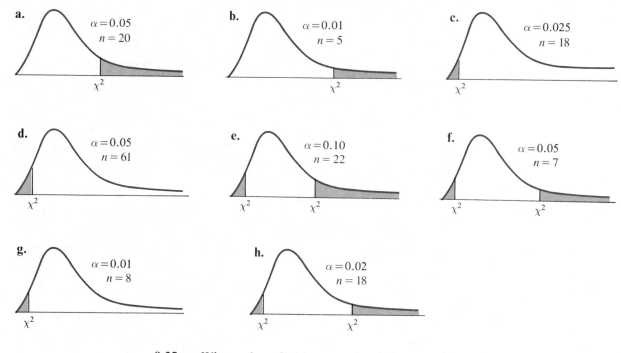

a. $\alpha = 0.05$
    $n = 20$

b. $\alpha = 0.01$
    $n = 5$

c. $\alpha = 0.025$
    $n = 18$

d. $\alpha = 0.05$
    $n = 61$

e. $\alpha = 0.10$
    $n = 22$

f. $\alpha = 0.05$
    $n = 7$

g. $\alpha = 0.01$
    $n = 8$

h. $\alpha = 0.02$
    $n = 18$

9.35 a. What value of chi-square for 5 degrees of freedom subdivides the area under the distribution curve such that 5 percent is to the right and 95 percent is to the left?

b. What is the value of the 95th percentile for the chi-square distribution with 5 degrees of freedom?

c. What is the value of the 90th percentile for the chi-square distribution with 5 degrees of freedom?

9.36 a. The central 90 percent of the chi-square distribution with 11 degrees of freedom lies between what values?

b. The central 95 percent of the chi-square distribution with 11 degrees of freedom lies between what values?

c. The central 99 percent of the chi-square distribution with 11 degrees of freedom lies between what values?

9.37 For a chi-square distribution having 12 degrees of freedom, find the area under the curve for chi-square values ranging from 3.57 to 21.0.

9.38 In the past the standard deviation of weights of certain 32.0-ounce packages filled by a machine was 0.25 ounce. A random sample of 20 packages showed a standard deviation of 0.35 ounce. Is the apparent increase in variability significant at the 0.10 level of significance?

a. Solve using the classical approach.

b. Solve using the prob-value approach.

9.39 A car manufacturer claims that the miles per gallon for a certain model has a mean equal to 40.5 miles with a standard deviation equal to 3.5 miles. Use the following data, obtained from a random sample of 15 such cars, to test the hypothesis that the standard deviation differs from 3.5. Use $\alpha = 0.05$.

$$37.0 \quad 38.0 \quad 42.5 \quad 45.0 \quad 34.0 \quad 32.0 \quad 36.0 \quad 35.5$$
$$38.0 \quad 42.5 \quad 40.0 \quad 42.5 \quad 36.0 \quad 30.0 \quad 37.5$$

   a. Solve using the classical approach.

   b. Solve using the prob-value approach.

9.40 A commercial farmer harvests his entire field of a vegetable crop at one time. Therefore he would like to plant a variety of green beans that mature all at one time (small standard deviation between maturity times of individual plants). A seed company has developed a new hybrid strain of green beans that it believes to be better for the commercial farmer. The maturity time of the standard variety has an average of 50 days and a standard deviation of 2.1 days. A random sample of 30 plants of the new hybrid showed a standard deviation of 1.65 days. Does this sample show a significant lowering of the standard deviation at the 0.05 level of significance?

   a. Solve using the classical approach.

   b. Solve using the prob-value approach.

9.41 The variability of the scores made on the TOEFL exam was of interest. A random sample of ten scores made by foreign-born students on this exam were

$$495 \quad 525 \quad 580 \quad 605 \quad 552 \quad 490 \quad 590 \quad 505 \quad 551 \quad 600$$

Find a 95 percent confidence interval for the standard deviation of TOEFL exam scores.

9.42 x is a random variable with a normal distribution. A sample of size 22 resulted in $\sum x = 397.3$ and $\sum x^2 = 7374.09$.

   a. What would be the point estimate for the population variance?

   b. What would be the 0.90 confidence interval estimate for the population variance?

   b. What would be the 0.90 confidence interval estimate for the population variance?

   c. What would be the 0.90 confidence interval estimate for the population standard deviation?

In Retrospect   We have studied inferences, both hypothesis testing and confidence interval estimation, for three basic population parameters: mean μ, proportion p, and standard deviation σ. When we make inferences about a single population, we are usually concerned with one of these three values. Table 9-1 identifies the formula that is used in each of the inferences for problems involving a single population.

Table 9-1
Formulas to use for
inferences involving
a single population

Situation	Test Statistic	Formula to Be Used	
		Hypothesis Test	Interval Estimate
One mean			
σ known	z	Formula (7-2)	(8-2)
σ unknown	t	Formula (9-1)	(9-2)
One proportion	z	Formula (9-3)	(9-4)
One standard deviation	χ^2	Formula (9-7)	(9-10)
One variance	χ^2	Formula (9-7)	(9-9)

In this chapter we also used the maximum error of estimate term of formula (9-6) to determine the size of sample required to make estimates about the population proportion with the desired accuracy.

Case Study 9-2 reports several observed sample percentages as point estimates of population proportions. The article gives only the point estimates. However, the sampling error explanation that accompanies the article refers to the 95 percent confidence interval estimate. By combining a point estimate with its corresponding maximum error of estimate (sampling tolerance) we can construct an interval estimate. In Exercise 9.25 you did some calculating to verify the 95 percent level of confidence.

In the next chapter we will discuss inferences about two populations whose respective means, proportions, and standard deviations are to be compared.

Chapter Exercises

9.43 The mean for a standardized reading test used in the state of Nebraska is 80. A school district wishes to test the hypothesis that its mean is different from that of the state. Twenty randomly selected students are tested and the results summarized by $\bar{x} = 77.5$ and $s = 2.5$. Complete the hypothesis test using $\alpha = 0.05$.

   a. Solve using the classical approach.

   b. Solve using the prob-value approach.

9.44 A manufacturer of television sets claims that the maintenance expenditures for his product will average no more than $50 during the first year following the expiration of the warranty. A consumer group has asked you to substantiate or discredit the claim. The results of a random sample of 50 owners of such television sets showed that the mean expenditure was $61.60 and the standard deviation was $32.46. At the 0.01 level of significance, should you conclude that the producer's claim is true or not likely to be true?

   a. Solve using the classical approach.

   b. Solve using the prob-value approach.

9.45 It has been suggested that abnormal human males tend to occur more in children born to older-than-average parents. Case histories of 20 abnormal males were obtained and the ages of the 20 mothers were

$$31 \quad 21 \quad 29 \quad 28 \quad 34 \quad 45 \quad 21 \quad 41 \quad 27 \quad 31$$
$$43 \quad 21 \quad 39 \quad 38 \quad 32 \quad 28 \quad 37 \quad 28 \quad 16 \quad 39$$

The mean age at which mothers in the general population give birth is 28.0 years.

    a. Calculate the sample mean and standard deviation.

    b. Does the sample give sufficient evidence to support the claim that abnormal male children have older-than-average mothers? Use $\alpha = 0.05$.

    (1) Solve using the classical approach.

    (2) Solve using the prob-value approach.

9.46 The weights of the drained fruit found in 21 randomly selected cans of peaches packed by Sunny Fruit Cannery were (in ounces)

$$
\begin{array}{ccccccc}
11.0 & 11.6 & 10.9 & 12.0 & 11.5 & 12.0 & 11.2 \\
10.5 & 12.2 & 11.8 & 12.1 & 11.6 & 11.7 & 11.6 \\
11.2 & 12.0 & 11.4 & 10.8 & 11.8 & 10.9 & 11.4
\end{array}
$$

    a. Calculate the sample mean and standard deviation.

    b. Construct the 0.98 confidence interval for the estimate of the mean weight of drained peaches per can.

9.47 In a large cherry orchard the average yield has been 4.35 tons per acre for the last several years. A new fertilizer was tested on 15 randomly selected 1-acre plots. The yields from these plots follow:

$$
\begin{array}{ccccc}
3.56 & 5.00 & 4.88 & 4.93 & 3.92 \\
4.25 & 5.12 & 5.13 & 4.79 & 4.45 \\
5.35 & 4.81 & 3.48 & 4.45 & 4.72
\end{array}
$$

At the 0.05 level of significance, do we have sufficient evidence to claim that there was a significant increase in production?

    a. Solve using the classical approach.

    b. Solve using the prob-value approach.

9.48 Determine the prob-value for the following hypothesis tests involving the Student's t distribution with 10 degrees of freedom.

    a. $H_0: \mu = 15.5$ $H_a: \mu < 15.5$ $t^* = -2.01$

    b. $H_0: \mu = 15.5$ $H_a: \mu > 15.5$ $t^* = 2.01$

    c. $H_0: \mu = 15.5$ $H_a: \mu \neq 15.5$ $t^* = 2.01$

    d. $H_0: \mu = 15.5$ $H_a: \mu \neq 15.5$ $t^* = -2.01$

9.49 One of the objectives of a large medical study was to estimate the mean physician fee for cataract removal. For 25 randomly selected cases the mean fee was found to be $1,550 with a standard deviation of $125. Set a 99 percent confidence interval on μ, the mean fee for all physicians.

9.50 A natural gas utility is considering a contract for purchasing tires for its fleet of service trucks. The decision will be based on expected mileage. For a sample of 100 tires tested, the mean mileage was 36,000 and the standard deviation was 2,000 miles. Estimate the mean mileage that the utility should expect from these tires using a 96 percent confidence interval.

9.51 Oranges are selected at random from a large shipment that just arrived. A sample is taken to estimate the size (circumference, in inches) of the oranges. The sample data are summarized as follows: $n = 100$, $\sum x = 878.2$, and $\sum (x - \bar{x})^2 = 49.91$.

    a. Determine the mean and standard deviation for this sample.

    b. What is the point estimate for μ, the mean circumference of oranges in this shipment?

    c. What is the 0.95 confidence interval estimate for μ?

    d. What is the point estimate for the standard deviation σ of the circumferences?

    e. What is the 0.95 confidence interval for σ?

9.52 A manufacturer wishes to estimate the mean life of a new line of automobile batteries. How large a sample should it take in order to estimate the mean life to within two months at 99 percent confidence, if the lifetimes of batteries of this type typically have a standard deviation of six months?

9.53 A West Coast radio station is promoting a popular music group named Warren Peace and his Atom Bombs. In the past, 60 percent of the listeners of the station have liked the music groups that the station promotes. Out of a randomly selected sample of 200 listeners, 102 of them like the Warren Peace group. At the 0.02 level of significance, test the hypothesis that there is no difference between the attitude of the listeners in this sample and listeners in the past.

    a. Solve using the classical approach.

    b. Solve using the prob-value approach.

9.54 A machine is considered to be operating in an acceptable manner if it produces 0.5 percent or fewer defective parts. It is not performing in an acceptable manner if more than 0.5 percent of its production is defective. The hypothesis $H_0: p = 0.005$ is tested against the hypothesis $H_a: p > 0.005$ by taking a random sample of 50 parts produced by the machine. The null hypothesis is rejected if 2 or more defective parts are found in the sample. Find the probability of the type I error.

9.55 You are interested in comparing the hypothesis $p = 0.8$ against the alternative $p < 0.8$. In 100 trials you observe 73 successes. Calculate the prob-value associated with this result.

9.56 An instructor asks each of the 54 members of his class to write down "at random" one of the numbers 1, 2, 3, ..., 13, 14, 15. Since the instructor believes that students like gambling, he considers that 7 and 11 are lucky numbers. He counts the number of students x who selected 7 or 11. How large must x be before the hypothesis of randomness can be rejected at the 0.05 level?

9.57 A survey concerning the recall of a city's mayor was conducted. In response to the question "Would you vote to recall the mayor?" there were 250 yes, 125 no, and 75 undecided responses. Find a 95 percent confidence interval estimate for the population proportion who would vote to recall the mayor.

9.58 The marketing research department of an instant-coffee firm conducted a survey of married men to determine the proportion of married men who preferred their brand. Twenty of the 100 in the random sample preferred the company's brand.

Use a 95 percent confidence interval to estimate the proportion of all married men that prefer this company's brand of instant coffee. Interpret your answer.

9.59 A company is drafting an advertising campaign that will involve endorsements by noted athletes. In order for the campaign to succeed, the endorser must be both highly respected and easily recognized. A random sample of 100 prospective customers are shown photos of various athletes. If the customer recognizes an athlete, then he or she is asked whether he or she respects the athlete. In the case of a top woman golfer, 16 of the 100 respondents recognized her picture and indicated that they also respected her. At the 95 percent level of confidence, what is the true proportion with which this woman golfer is both recognized and respected?

9.60 A local auto dealership advertises that 90 percent of customers whose autos were serviced by their service department are pleased with the results. As a researcher, you take exception to this statement because you are aware that many people are reluctant to express dissatisfaction even if they are not pleased. A research experiment was set up in which those in the sample had received service by this dealer within the last two weeks. During the interview, the individuals were led to believe that the interviewer was new in town and was considering taking his car to this dealer's service department. Of the 60 sampled, 14 said that they were dissatisfied and would not recommend the department.

    a. Estimate the proportion of dissatisfied customers using a 95 percent confidence interval.

    b. Given your answer to (a), what can be concluded about the dealer's claim?

9.61 In obtaining the sample size to estimate a proportion, the formula $n = \dfrac{[z(\alpha/2)]^2 pq}{E^2}$ is used. If a reasonable estimate of p is not available, then it is suggested that $p = 0.5$ be used because this will give the maximum value for n. Calculate the value of $pq = p(1 - p)$ for $p = 0.1, 0.2, 0.3, \ldots, 0.8, 0.9$ in order to obtain some idea about the behavior of the quantity pq.

9.62 The State Motor Vehicle Department wishes to estimate the proportion of first-year drivers that could be classified as "careless drivers." How large a sample should it take if it wishes to be within 3 percent at 95 percent confidence? (It is expected that approximately $\frac{1}{3}$ of all drivers are careless.)

9.63 A consumer group was interested in determining the proportion of dentists who would accept the patient's insurance payment as the full payment for a routine exam. Determine the sample size needed to estimate the true proportion to within 0.01 with 95 percent confidence.

9.64 In order to test the hypothesis that the standard deviation on a standard test is 12, a sample of 40 randomly selected students were tested. The sample variance was found to be 155. Does this sample provide sufficient evidence to show that the standard deviation differs from 12 at the 0.05 level of significance?

9.65 A production process is considered to be out of control if the produced parts have a mean length different from 27.5 millimeters or a standard deviation that is greater than 0.5 millimeter. A sample of 30 parts yields a sample mean of 27.63

millimeters and a sample standard deviation of 0.87 millimeter.

a. At the 0.05 level of significance, does this sample indicate that the process should be adjusted in order to correct the standard deviation of the product?

b. At the 0.05 level of significance, does this sample indicate that the process should be adjusted in order to correct the mean value of the product?

9.66 Bright-lite claims that its 60-watt light bulb burns with a length of life that is approximately normally distributed with a standard deviation of 81 hours. A sample of 101 bulbs had a variance of 8075. Is this sufficient evidence to reject Bright-lite's claim in favor of the alternative, "the standard deviation is larger than 81 hours," at the 0.05 level of significance?

a. Solve using the classical approach.

b. Solve using the prob-value approach.

9.67 Suppose that a sample of size 30 was used to test

$$H_0: \sigma^2 = 17 \quad \text{versus} \quad H_a: \sigma^2 > 17$$

at $\alpha = 0.05$. How large would s^2 need to be before the null hypothesis would be rejected?

9.68 A drug manufacturer is concerned not only about the mean potency of its 500-mg tablets but is also concerned about the variability in potency. A random sample of 50 tablets was tested and the sample variance was determined to be 0.9 mg. Find a 95 percent confidence interval estimate for the population standard deviation.

9.69 Air pollution is determined by measuring several different elements that can be detected in the air. One of them is carbon monoxide. The sample of daily readings in the following table was obtained from a local newspaper.

Day	1	2	3	4	5	6	7	8	9	10	11	12
Carbon Monoxide	3.5	3.9	2.8	3.1	3.1	3.4	4.8	3.2	2.5	3.5	4.4	3.1

a. Calculate the mean and standard deviation for this sample.

b. Carbon monoxide is measured and interpreted according to the accompanying scale. Does the sample show sufficient evidence to allow us to conclude that the carbon monoxide level is low, that is, $\mu < 4.9$, at $\alpha = 0.05$?

c. Does the sample show sufficient evidence to allow us to reject the claim that variance in the carbon monoxide readings is no more than 0.25 at $\alpha = 0.05$?

9.70 Using the sample given in Exercise 9.69 and the result found in answering part (a), construct the following confidence intervals:

a. 90 percent confidence interval for estimating the mean daily level of carbon monoxide pollution

b. 90 percent confidence interval for estimating the standard deviation of the carbon monoxide pollution

9.71 The case histories for the 20 children in Exercise 9.45 showed their fathers to have ages of

$$
\begin{array}{cccccccccc}
32 & 48 & 33 & 22 & 29 & 30 & 45 & 25 & 26 & 43 \\
36 & 27 & 34 & 20 & 35 & 55 & 52 & 38 & 34 & 37
\end{array}
$$

a. Calculate the sample mean and standard deviation.

b. Estimate the mean age of all fathers of abnormal male children using a 0.95 confidence interval.

c. Estimate the standard deviation of the ages of all fathers for these children using a 0.95 confidence interval.

9.72 Verify algebraically that formula (9-6) is equivalent to formula (9-5).

Vocabulary List

Be able to define each term. In addition, describe, in your own words, and give an example of each term. Your examples should not be ones given in class or in the textbook.

The bracketed numbers indicate the chapters in which the term previously appeared, but you should define the terms again to show increased understanding of their meaning.

acceptance region [8]　　　　　　　parameter [1, 8]
calculated value [8]　　　　　　　　proportion [6]
chi-square　　　　　　　　　　　　random variable [5, 6]
conclusion [8]　　　　　　　　　　response variable [1]
critical region [8]　　　　　　　　sample size [8]
critical value [8]　　　　　　　　　sample statistic [1, 2]
decision [8]　　　　　　　　　　　σ known
degrees of freedom　　　　　　　　σ unknown
inference [8]　　　　　　　　　　　standard error [7, 8]
level of confidence [8]　　　　　　Student's t
level of significance [8]　　　　　　test statistic [8]
maximum error of estimate [8]　　two-tailed test [8]
one-tailed test [8]

Quiz

Answer "True" if the statement is always true. If the statement is not always true, replace the boldface words with words that make the statement always true.

9.1 The Student's t distribution is an approximately normal distribution but is more **dispersed** than the normal distribution.

9.2 The **chi-square** distribution is used for inferences about the mean when σ is unknown.

9.3 The **Student's** t distribution is used for all inferences about a population's variance.

9.4 If the test statistic falls in the critical region, the null hypothesis has **been proved true**.

9.5 When the test statistic is t and the number of degrees of freedom exceeds 30, the critical value of t is very close to that of z.

9.6 When making inferences about one mean when the value of σ is not known, the z **score** is the test statistic.

9.7 The chi-square distribution is a skewed distribution whose mean is always **2**.

9.8 Often the concern with testing the variance (or standard deviation) is to keep its size under control or relatively small. Therefore, many of the hypothesis tests with chi-square will be **one-tailed**.

9.9 $\sqrt{npq}$ is the standard error of proportion.

9.10 The sampling distribution of p' is approximately distributed as **chi-square**.

Chapter 10

Inferences Involving Two Populations

WE LIKE OUR WORK

"Work isn't as bad as we thought it was going to be; we're satisfied"

WHAT WE THINK ABOUT WORK

Living to work or working to live, jobs determine our standard of living and, often, our sense of self-worth. Here's a sampling, from a USA TODAY poll of 802 people of today's attitudes about the work place.

Looking forward to a good day's work

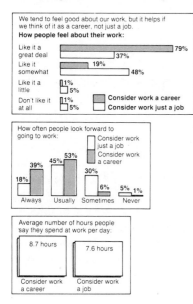

We tend to feel good about our work, but it helps if we think of it as a career, not just a job.

How people feel about their work:

	Consider work a career	Consider work just a job
Like it a great deal	79%	37%
Like it somewhat	19%	48%
Like it a little	1%	5%
Don't like it at all	1%	5%

How often people look forward to going to work:

Consider work just a job / Consider work a career

	Always	Usually	Sometimes	Never
Just a job	18%	45%	30%	5%
A career	39%	53%	6%	1%

Average number of hours people say they spend at work per day:

- 8.7 hours — Consider work a career
- 7.6 hours — Consider work a job

Starting your own business

52% of people have thought about going into business for themselves. Some reasons why people did and why people didn't.

Why people didn't:

Not enough money	51%
Feared risk	12%
Family responsibilities	3%

Why people did:

Make a lot of money	25%
Desire to be own boss	25%
Control life	22%

When you do your job well

The way you're rewarded for good work depends greatly on your field of work. Here's a breakdown of how workers in different fields say their bosses respond to a job well done:

	Professional	Sales	Assembly line	Clerical
Verbal compliment	57%	70%	40%	60%
Nothing	24%	11%	45%	28%
Bonus	10%	21%	12%	7%
Written note	10%	6%	2%	6%

"What We Think About Work," June 15, 1987. Copyright 1987, USA TODAY. Reprinted with permission.

Chapter
Objectives

In Chapters 8 and 9 we introduced the basic concepts of hypothesis testing and confidence interval estimation in connection with inferences about one population and the parameters: mean, proportion, and variance (or standard deviation). In this chapter we continue to investigate the inferences about these same parameters, but we will now use these parameters as a basis for *comparing two populations*.

First we will learn the difference between independent and dependent samples (Section 10.1). Throughout the remainder of this chapter, each section will present the techniques for using two means, or two proportions, or two variances (or two standard deviations) for comparing two populations. We will be studying several difference situations, and the following chart is offered to help you initially organize the relationship of these various inference techniques.

Parameter	Kind of Samples		Section
Difference of Two Means	Independent Samples	σ's known or large samples	10.2
		σ's unknown and small samples	10.4
	Dependent Samples		10.5
Difference of Two Proportions	Independent Samples		10.6
Ratio of Two Variances (standard deviations)	Independent Samples		10.3

10.1   INDEPENDENT AND DEPENDENT SAMPLES

source

dependent and independent samples

In this chapter we are going to study the procedures for making inferences about two populations. When comparing two populations we need two samples, one from each. Two basic kinds of samples can be used: independent and dependent. The dependence or independence of a sample is determined by the sources used for the data. A **source** can be a person, an object, or anything that yields a piece of data. If the same set of sources or related sets are used to obtain the data representing both populations, we have **dependent sampling**. If two unrelated sets of sources are used, one set from each population, we have **independent sampling**. The following illustrations should amplify these ideas.

Illustration 10-1   A test will be conducted to see whether the participants in a physical fitness class actually improve in their level of fitness. It is anticipated that approximately 500 people will sign up for this course, and the instructor decides that she will give 50 of the participants a set of tests before the course actually begins (pre-test) and then will give another set of tests to 50 participants at the end of the course (post-test). The following two sampling procedures are proposed:

Plan A: Randomly select 50 participants and from the list of those enrolled and give them the pre-test. At the end of the course, select another random sample of size 50 and give them the post-test.

Plan B: Randomly select 50 participants and give them the pre-test; give the same set of 50 the post-test upon completion of the course.

Plan A illustrates independent sampling—the sources (class participants) used for each sample (pre-test and post-test) were selected separately. Plan B illustrates dependent sampling—the sources used for both samples (pre-test and post-test) are the same. ∎

Typically, when both pre-test and post-test are used, the same subjects will be used in the study. Thus pre-test versus post-test, (before versus after) studies are usually dependent samples.

Illustration 10-2   A test is being designed to compare the wearing quality of two brands of tires. The automobiles will be selected and equipped with the new tires and then driven under "normal" conditions for one month. Then a measurement will be taken to determine how much wear took place. Two plans are proposed:

Plan C: n cars will be selected randomly and equipped with brand A and driven for the month, and n other cars will be selected and equipped with brand B and driven for the month

Plan D: n cars will be selected randomly, equipped with one tire of brand A and one tire of brand B (the other two tires are not part of the test), and driven for the month.

In this illustration we might suspect that many other factors must be taken into account when testing automobile tires—such as age, weight, and mechanical condition of the car; driving habits of drivers; location of the tire on the car; and where and how much the car is driven. However, at this time we are only trying to illustrate dependent and independent samples. Plan C is independent (unrelated sources) and plan D is dependent (common sources). ∎

Independent and dependent cases each have their advantages; these will be emphasized later. Both methods of sampling are often used.

Case Study 10-1

EXPLORING THE TRAITS OF TWINS

Studies that involve identical twins are a natural for the dependent sampling technique discussed in this section.

A NEW STUDY SHOWS THAT KEY CHARACTERISTICS MAY BE INHERITED

Like many identical twins reared apart, Jim Lewis and Jim Springer found they had been leading eerily similar lives. Separated four weeks after birth in 1940, the Jim twins grew up 45 miles apart in Ohio and were reunited in 1979. Eventually they discovered that both drove the same model blue Chevrolet, chain-smoked Salems, chewed their fingernails and owned dogs named Toy. Each had spent a good deal of time vacationing at the same three-block strip of beach in Florida. More important, when tested for such personality traits as flexibility, self-control and sociability, the twins responded almost exactly alike.

The two Jims were the first of 348 pairs of twins studied at the University of Minnesota, home of the Minnesota Center for Twin and Adoption Research. Much of the investigation concerns the obvious question raised by siblings like Springer and Lewis: How much of any individual's personality is due to heredity?

The project, summed up in a scholarly paper that has been submitted to the *Journal of Personality and Social Psychology*, is considered the most comprehensive of its kind. The Minnesota researchers report the results of six-day tests of their subjects, including 44 pairs of identical twins who were brought up apart. Well-being, alienation, aggression and the shunning of risk or danger were found to owe as much or more to nature as to nurture. Of eleven key traits or clusters of traits analyzed in the study, researchers estimated that a high of 61 percent of what they call "social potency" (a tendency toward leadership or dominance) is inherited, while "social closeness" (the need for intimacy, comfort and help) was lowest, at 33 percent.

All the twins took several personality tests, answering more than 15,000 questions on subjects ranging from personal interests and values to phobias, aesthetic judgment and television and reading habits. Twins reared separately also took medical exams and intelligence tests and were queried on life history and stresses. Not all pairs matched up as well as the two Jims.

Source: Copyright 1987 by Time Inc. All rights reserved. Reprinted by permission of TIME.

EXERCISES

10.1 An insurance company is concerned that garage A charges more for repair work than garage B charges. Thus it sends 25 cars to each garage and obtains separate estimates for the repairs needed for each car. Do these two sets of data represent dependent or independent samples? Explain.

10.2 In trying to estimate the amount of growth that took place in the trees planted by the County Parks Commission recently, 36 trees were randomly selected from the 4000 planted. The heights of these trees were measured and recorded. One year later another set of 42 trees was randomly selected and measured. Do the two sets of data (36 heights, 42 heights) represent dependent or independent samples? Explain.

10.3 Twenty people were selected to participate in a psychology experiment. They answered a short multiple-choice quiz about their attitudes on a particular subject and then viewed a 45-minute film. The following day the same 20 people were asked to answer a follow-up questionnaire about their attitudes. At the completion of the experiment, the experimenter will have two sets of scores. Do these two samples represent dependent or independent samples? Explain.

10.4 An experiment is designed to study the effect diet has on the uric acid level. Twenty white rats are used for the study. Ten rats are randomly selected and given a junk-food diet. The other ten receive a high-fiber, low-fat diet. Uric acid levels of the two groups are determined at the beginning and at the end of the study. Do the resulting sets of data represent dependent or independent samples? Explain.

10.5 Suppose that 400 students in a certain college are taking elementary statistics this semester. Describe how you would obtain two independent samples of size 25 from these 400 students in order to test some precourse skill against the same skill after the students complete the course.

10.6 Describe how you would obtain your samples in Exercise 10.5 if you were to use dependent samples.

10.2 INFERENCES CONCERNING THE DIFFERENCE BETWEEN THE MEANS OF TWO INDEPENDENT SAMPLES (VARIANCES KNOWN OR LARGE SAMPLES)

When comparing the means of two populations, we typically consider the difference between their means, $\mu_1 - \mu_2$. The inferences we will make about $\mu_1 - \mu_2$ will be based on the difference between the observed sample means, $\bar{x}_1 - \bar{x}_2$. This observed difference belongs to a sampling distribution, the characteristics of which are described in the following statement.

independent means

> If independent samples of sizes n_1 and n_2 are drawn randomly from large populations with means μ_1 and μ_2 and variances σ_1^2 and σ_2^2, respectively, **the sampling distribution of $\bar{x}_1 - \bar{x}_2$,** the difference between the means,
>
> **1.** is approximately normally distributed.
> **2.** has a mean of $\mu_{\bar{x}_1 - \bar{x}_2} = \mu_1 - \mu_2$.
> **3.** has a standard error of $\sigma_{\bar{x}_1 - \bar{x}_2} = \sqrt{(\sigma_1^2/n_1) + (\sigma_2^2/n_2)}$.

This normal approximation is good for all sample sizes given that the populations involved are approximately normal and the population variances σ_1^2 and σ_2^2 are known quantities.

Since the sampling distribution is approximately normal, we will use the z **statistic** in our inferences. In the **hypothesis tests**, z will be determined by

$$z = \frac{(\bar{x}_1 - \bar{x}_2) - (\mu_1 - \mu_2)}{\sqrt{(\sigma_1^2/n_1) + (\sigma_2^2/n_2)}} \tag{10-1}$$

if both σ_1 and σ_2 are known quantities.

Illustration 10-3   Suppose that we are interested in comparing the academic success of college students who belong to fraternal organizations with the academic success of those who do not belong to fraternal organizations. These two populations are clearly separate, and we should take independent samples from each of the two populations. The "Greeks" claim that fraternity members achieve academically at a level no lower than that of nonmembers. (Cumulative grade point average is the measure of academic success.) Samples of size 40 are taken from each population. The means obtained are 2.03 for the 40 fraternity members (g) and 2.21 for the 40 nonmembers (n). Assume that the standard deviation of both populations is $\sigma = 0.6$. Complete a hypothesis test of the Greeks' claim, using $\alpha = 0.05$.

Solution   The hypotheses for this test are as follows:

Step 1 $H_0: \mu_g = \mu_n$ $(\geq)$ or $\mu_g - \mu_n = 0$ $(\geq)$ (no lower)
           $H_a: \mu_g < \mu_n$ or $\mu_g - \mu_n < 0$ (lower)

The null hypothesis is usually interpreted as "there is no difference between the means," and, therefore, it is customary to express it by $\mu_1 - \mu_2 = 0$.

Step 2   The test statistic used will be z. The test criteria for $\alpha = 0.05$ will be as shown in the accompanying figure.

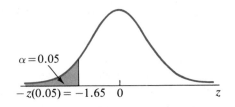

$\alpha = 0.05$

$-z(0.05) = -1.65$ 0 z

Step 3   The formula for the test statistic z is formula (10-1):

$$z = \frac{(\bar{x}_g - \bar{x}_n) - (\mu_g - \mu_n)}{\sqrt{(\sigma_g^2/n_g) + (\sigma_n^2/n_n)}}$$

$$= \frac{(2.03 - 2.21) - 0}{\sqrt{[(0.6)^2/40] + [(0.6)^2/40]}} = \frac{-0.18}{\sqrt{(0.36/40) + (0.36/40)}}$$

$$= \frac{-0.18}{\sqrt{0.009 + 0.009}} = \frac{-0.18}{\sqrt{0.018}} = \frac{-0.180}{0.134} = -1.343$$

$$z^* = -1.34$$

This value is compared to the test criteria in the accompanying figure.

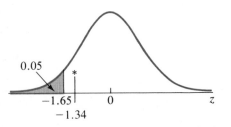

0.05

-1.65 0 z
-1.34

Step 4   *Decision*: Fail to reject H_0 ($z*$ is in the noncritical region).

Conclusion: At the 0.05 level of significance, the claim that the fraternity members achieve at a level no lower than that of nonmembers cannot be rejected.   ■

We often wish to estimate the difference between the means of two different populations. When independent samples are involved, we will use the information about the sampling distribution of $\bar{x}_1 - \bar{x}_2$ and the z statistic to construct our **confidence interval estimate for $\mu_1 - \mu_2$**.

$$(\bar{x}_1 - \bar{x}_2) - z(\alpha/2) \cdot \sqrt{\frac{\sigma_1^2}{n_1} + \frac{\sigma_2^2}{n_2}} \quad \text{to} \quad (\bar{x}_1 - \bar{x}_2) + z(\alpha/2) \cdot \sqrt{\frac{\sigma_1^2}{n_1} + \frac{\sigma_2^2}{n_2}} \quad (10\text{-}2)$$

Illustration 10-4   Construct the 95 percent confidence interval estimate for the difference between the two independent means of Illustration 10-3 (mean grade point average of fraternity members and mean grade point average of nonmembers). Use the sample values found in Illustration 10-3.

Solution   The information given is $\bar{x}_g = 2.03$, $\sigma_g = 0.6$, $n_g = 40$, $\bar{x}_n = 2.21$, $\sigma_n = 0.6$, $n_n = 40$, and $1 - \alpha = 0.95$. The interval is

$$(\bar{x}_g - \bar{x}_n) \pm z(0.025) \cdot \sqrt{\frac{\sigma_g^2}{n_g} + \frac{\sigma_n^2}{n_n}}$$

$$(2.03 - 2.21) \pm (1.96) \cdot \sqrt{\frac{(0.6)^2}{40} + \frac{(0.6)^2}{40}}$$

(*See* Illustration 10-3.)

$$(-0.18) \pm (1.96)(0.134)$$
$$-0.18 \pm 0.26$$
$$\mathbf{-0.44 \quad to \quad 0.08}$$

is the 0.95 confidence interval for $\mu_g - \mu_n$. That is, with 95 percent confidence we estimate the difference between the means to be between -0.44 and $+0.08$.   ■

As we saw in previous chapters, the variance of a population is generally unknown when we wish to make an inference about the mean. Therefore, it is necessary to replace σ_1 and σ_2 with the best estimates available, namely, s_1 and s_2. If both samples have sizes that exceed 30, we may replace σ_1 and σ_2 in formulas (10-1) and (10-2) with s_1 and s_2, respectively, without appreciably affecting our level of significance or confidence. Thus for inferences about the difference between two population means, based on independent samples where the **σ's are unknown** and both $n_1 > 30$ and $n_2 > 30$, we will use

$$z = \frac{(\bar{x}_1 - \bar{x}_2) - (\mu_1 - \mu_2)}{\sqrt{(s_1^2/n_1) + s_2^2/n_2)}} \quad (10\text{-}3)$$

for the calculation of the **test statistic in the hypothesis test**. We will use

$$(\bar{x}_1 - \bar{x}_2) - z(\alpha/2) \cdot \sqrt{\frac{s_1^2}{n_1} + \frac{s_2^2}{n_2}} \quad \text{to} \quad (\bar{x}_1 - \bar{x}_2) + z(\alpha/2) \cdot \sqrt{\frac{s_1^2}{n_1} + \frac{s_2^2}{n_2}} \quad (10\text{-}4)$$

for calculating the endpoints of the $1 - \alpha$ **confidence interval estimate**.

Illustration 10-5   Two independent samples are taken to compare the means of two populations. The sample statistics are given in Table 10-1. Can we conclude that the mean of population A is greater than the mean of population B, at the 0.02 level of significance?

Table 10-1
Sample statistics for
Illustration 10-5

Sample	n	$\bar{x}$	s
A	50	57.5	6.2
B	60	54.4	10.6

Solution   This problem calls for a hypothesis test for the difference of two independent means. Both n's are larger than 30; therefore, formula (10-3) will be used to calculate z.

Step 1 H_0: $\mu_A - \mu_B = 0$ $(\leq)$
            H_a: $\mu_A - \mu_B > 0$

Step 2 $\alpha = 0.02$. The test criteria are shown in the following figure.

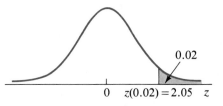

0.02

0     $z(0.02) = 2.05$ z

Step 3   Using formula (10-3),

$$z = \frac{(57.5 - 54.4) - 0}{\sqrt{[(6.2)^2/50] + [(10.6)^2/60]}}$$

$$= \frac{3.1}{\sqrt{0.7688 + 1.8727}}$$

$$= \frac{3.1}{1.625}$$

$$= 1.908$$

$$z* = \mathbf{1.91}$$

Step 4   *Decision*: Fail to reject H_0 (z^* is in the noncritical region).

Conclusion: At the 0.02 level of significance, we do not have sufficient evidence to conclude that $\mu_A > \mu_B$.

Note   The null hypothesis can claim a difference; for example, the mean of A is 10 larger than the mean of B. In this situation the null hypothesis is written $\mu_A - \mu_B = 10$, and the value 10 is used in Step 3 when z is calculated.   ■

Illustration 10-6   Suppose that the samples given in Illustration 10-5 were taken for the purpose of estimating the difference between the two population means. Construct the 0.99 confidence interval for the estimation of this difference.

Solution   This estimation calls for the difference between the means of two independent samples whose sizes are both greater than 30. Therefore, we will use formula (10-4).

$$(57.5 - 54.4) \pm 2.58 \sqrt{\frac{(6.2)^2}{50} + \frac{(10.6)^2}{60}}$$

$$3.1 \pm (2.58)(1.625)$$

$$3.1 \pm 4.19$$

$$\mathbf{-1.09} \quad \text{to} \quad \mathbf{7.29}$$

is the 0.99 confidence interval for $\mu_A - \mu_B$. That is, our 99 percent confidence interval estimate for the difference between the two population means is -1.09 to 7.29.   ■

Case Study 10-2

WHO'S MAKING WHAT?

Traffic Management's *Salary Survey reports average salaries for various subgroups from the population of readers of their magazine. Since only averages are given, and no measures of variation, we can determine only point estimates for the differences. Estimate the difference in average salary for respondents with and without a college education. Does this difference seem large enough to indicate that a college education has an effect on average salary? Explain. (See Exercise 10.9.)*

We haven't hit the $40,000 mark yet, but we're knocking on the door.

The results of *Traffic Management's* second annual salary survey reveal that the average reader now makes $38,350 a year, an increase of $1,800 over last year. If pay increases follow past patterns, next year's average should easily surpass $40,000.

The survey findings tell an interesting story. They depict a mature, experienced, well-educated work force. The average respondent is a 43-year-old male with a college degree who has been working in transportation or distribution for more than 15 years. He's been with his company for 12 years and in his current job for the past six.

RESULTS AT A GLANCE

Average salary is \$38,350. The 4,100 plus readers who responded to the survey earn an average of \$38,350. Men, who make up 88 percent of the total, average \$39,930. Women earn \$25,980.

VPs and GMs make the most. Readers with vice president or general management titles average almost \$62,000. At the other end of the salary spectrum, traffic analysts and specialists earn around \$30,000.

Education pays. Respondents with a college degree average \$42,690. That's 24 percent higher than readers with some college and 33 percent more than high-school-only respondents. MBAs average \$48,310.

Big companies pay best. In companies with sales of \$1 billion or more, six out of 10 readers earn more than \$40,000. By contrast, only 25 percent of readers in \$25-million-and-under companies earn \$40K.

Source: Traffic Management, April 1986.

EXERCISES

10.7 Small metal clips used in a furnace mechanism have a mean weight of 0.6 ounce and a variance of 0.0004. Two random samples, one of 100 observations and the other of 80, are taken on two consecutive days. Assuming that there has been no change in the production process, what is the probability that the two sample means differ by

    a. more than 0.002 ounce?   **b.** less than 0.0015 ounce?

10.8 Determine the p-value for the following hypothesis tests. (Assume that both sample sizes exceed 30.)

    a. $H_0: \mu_1 - \mu_2 = 10,$ $H_a: \mu_1 - \mu_2 > 10,$
       $z^* = 1.85$

    b. $H_0: \mu_1 - \mu_2 = 0,$ $H_a: \mu_1 - \mu_2 \neq 10,$
       $z^* = -2.33$

    c. $H_0: \mu_1 - \mu_2 = 0,$ $H_a: \mu_1 - \mu_2 < 0,$
       $z^* = -2.76$

10.9 Refer to the information about average salaries of respondents to *Traffic Management*'s survey (Case Study 10-2).

    a. Make a point estimate for how much more money respondents make, on the average, when they have a college education.

    b. Does the difference (the answer in (a)), in average salaries seem to suggest that a college education has a significant effect on average salary? Explain. What other information is needed to determine significance?

10.10 Independent samples are taken from normal populations I and II with variances of 200 and 700, respectively. Do the sample data shown in the following table provide sufficient evidence to reject the hypothesis that the populations have equal means, at the 0.10 level of significance?

Sample	n	$\bar{x}$
I	32	104.5
II	40	110.9

    a. Solve using the classical approach.
    b. Solve using the prob-value approach.

10.11 An experiment was conducted to compare the mean absorptions of two drugs in specimens of muscle tissue. Seventy-two tissue specimens were randomly divided into two equal groups. Each group was tested with one of the two drugs. Assume that the variance of absorption is 0.10 for this type of drug. The means found were $\bar{x}_A = 7.9$ and $\bar{x}_B = 8.5$. Construct the 98 percent confidence interval for the difference in the mean absorption rates.

10.12 The purchasing department for a regional supermarket chain is considering two sources from which to purchase 10-pound bags of potatoes. A random sample taken from each source shows the following results:

	Idaho Supers	Idaho Best
Number of Bags Weighed	100	100
Mean Weight	10.2 lbs	10.4 lbs.
Sample Variance	0.36	0.25

At the 0.05 level of significance, is there a difference between the mean weights of the 10-pound bags of potatoes?

    a. Solve using the classical approach.
    b. Solve using the prob-value approach.

10.13 A study was designed to investigate the effect of a calcium-deficient diet on lead consumption in rats. One hundred rats were randomly divided into 2 groups of 50 each. One group served as a control group and the other was the experimental, or calcium-deficient, group. The response recorded was the amount of lead consumed per rat. The results were summarized by:

$$\text{Control group:} \quad n = 50 \quad \bar{x} = 5.2 \quad s = 1.1$$
$$\text{Experimental:} \quad n = 50 \quad \bar{x} = 7.6 \quad s = 1.3$$

Test $H_0: \mu_C - \mu_E = 0$ versus $H_a: \mu_C - \mu_E < 0$ at $\alpha = 0.05$.

    a. Solve using the classical approach.
    b. Solve using the prob-value approach.

10.14 A sample of the heights of female and male students participating in the college intermural program was obtained, with the results as shown in the following table. Do these data provide sufficient evidence to reject the hypothesis that the average male participant is no more than 2.5 inches taller than the average female participant, at the 0.02 level of significance?

Sample	n	$\sum x$	$\sum (x - \bar{x})^2$
Female	30	1952	74.2
Male	40	2757	284.3

a. Solve using the classical approach.

b. Solve using the prob-value approach.

10.15 Two groups of university students were compared with respect to their computer-science aptitude. Group 1 was composed of students who had taken at least one computer-science course in high school, and group 2 was composed of students who had never had a computer-science course. None of the students had had any other computer-related experience. Both groups were given the KSW computer-science aptitude test. The results were

$$\text{group 1:} \quad n = 125 \quad \bar{x} = 15.5 \quad s = 2.7$$
$$\text{group 2:} \quad n = 115 \quad \bar{x} = 14.3 \quad s = 3.0$$

Find a 90 percent confidence interval estimate for $\mu_1 - \mu_2$.

10.16 In a certain species of plant the white-flowered plants appear to have smaller and more abundant flowers than do the red-flowered plants. The following data were collected when the number of flowers on several plants were counted. Use these sample data to construct the 95 percent confidence interval estimate for the difference between the mean number of flowers on the white- and the red-flowered plants ($\mu_W - \mu_R$).

Flower	n	$\sum x$	$\sum (x - \bar{x})^2$
Red	32	2253	32,462.0
White	35	5157	58,600.0

10.3 INFERENCES CONCERNING THE RATIO OF VARIANCES BETWEEN TWO INDEPENDENT SAMPLES

When comparing two populations, it is quite natural that we compare their variances, or standard deviations. The inferences concerning two population variances (or standard deviations) are much like those comparing means. We will study two kinds of inferences about the comparison of the variances of two populations: (1) **the hypothesis test for the equality of the two variances** and (2) **the estimation of the ratio of the two population variances** σ_1^2 / σ_2^2.

The soft drink bottling company discussed in Section 9.3 is trying to decide whether to install a modern high-speed bottling machine. There are, of course, many concerns in making this decision. The variance in the amount of fill per bottle is one of them. In this respect the manufacturer of the new system contends that the variance in fills is no larger with the new machine than it was with the old. A hypothesis test for the equality of the two variances can be used to make a decision in this situation. The null hypothesis will be that the variance of the modern high-speed machine (m) is no larger than the variance of the present machine (p); that is, $\sigma_m^2 \le \sigma_p^2$. The alternative hypothesis will then be $\sigma_m^2 > \sigma_p^2$.

$$H_0: \sigma_m^2 = \sigma_p^2 \quad (\le) \quad \text{or} \quad \frac{\sigma_m^2}{\sigma_p^2} = 1 \quad (\le)$$

$$H_a: \sigma_m^2 > \sigma_p^2 \quad \text{or} \quad \frac{\sigma_m^2}{\sigma_p^2} > 1$$

The test statistic that will be used in making a decision about the null hypothesis is F. The calculated value of F will be obtained by using the following formula:

$$F = \frac{s_1^2}{s_2^2} \tag{10-5}$$

where s_1^2 and s_2^2 are the variances of two independent samples of sizes n_1 and n_2, respectively.

When independent random samples are drawn from normal populations with equal variances, the ratio of the sample variances, s_1^2/s_2^2, will have a probability distribution known as the **F distribution** (*see* Figure 10-1).

F distribution

Figure 10-1
F distribution

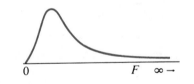

0 F $\infty \rightarrow$

Properties of the F Distribution

1. F is nonnegative in value; it is zero or positively valued.
2. F is nonsymmetrical; it is skewed to the right.
3. There are many F distributions, much like the t and χ^2 distributions.

There is a distribution for each pair of degree-of-freedom values

For the inferences discussed in this section, the degrees of freedom for each of the samples are $df_1 = n_1 - 1$ and $df_2 = n_2 - 1$.

The critical values for the F distribution may be obtained from Tables 8a, 8b, and 8c in Appendix E. Each critical value will be determined by three identification values: df_n, df_d, and the area under the curve to the right of the critical value being sought. Each F distribution has two degrees-of-freedom values: df_n, the degrees of freedom associated with the sample whose variance is in the numerator of the calculated F; and df_d, the degrees of freedom associated with the sample whose variance is in the denominator. Therefore, the symbolic name for a critical value of F will be $F(df_n, df_d, \alpha)$, as shown in Figure 10-2.

Figure 10-2
F distribution showing
$F(df_n, df_d, \alpha)$

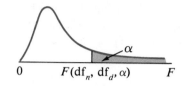

Table 8a in Appendix E shows the critical values for $F(df_n, df_d, \alpha)$, where α is equal to 0.05; Table 8b gives the critical values when $\alpha = 0.025$; Table 8c gives values when $\alpha = 0.01$.

Illustration 10-7   Find $F(5, 8, 0.05)$, the critical F value for samples of size 6 and 9 with 5 percent of the area in the right tail.

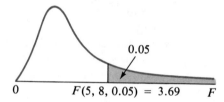

Solution   From Table 8a ($\alpha = 0.05$), we obtain the value shown in the accompanying table. Therefore, $F(5, 8, 0.05) = \textbf{3.69}$

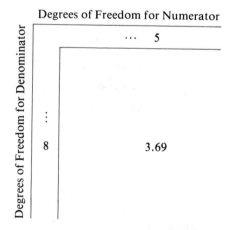

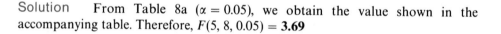

Notice that $F(8, 5, 0.05)$ is 4.82. The degrees of freedom associated with the numerator and with the denominator must be kept in the correct order. (3.69 is quite different from 4.82. Check some other pairs to verify this fact.)

Tables showing the critical values for the F distribution give only the right-hand critical value. If the critical value for the left-hand tail is needed, we obtain it by calculating the reciprocal of the related critical value obtained from the table. Expressed as a formula, this is

$$F(df_1, df_2, 1 - \alpha) = \frac{1}{F(df_2, df_1, \alpha)} \qquad (10\text{-}6)$$

See Figure 10-3. Notice that when the reciprocal is taken, the degrees of freedom for the numerator and the denominator are switched also. (Why is this switch necessary?)

Figure 10-3
Finding the critical value for the left tail of the F distribution

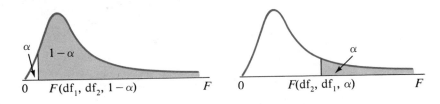

Illustration 10-8   Find the value for $F(10, 15, 0.99)$.

Solution   Using formula (10-6),

$$F(10, 15, 0.99) = \frac{1}{F(15, 10, 0.01)} = \frac{1}{4.56} = 0.219$$

$$= \mathbf{0.22}$$ ∎

Illustration 10-9   Recall that our soft drink bottling company was to make a decision about the equality of the variances of amounts of fill between its present machine and a modern high-speed outfit. Does the sample information in Table 10-2 present sufficient evidence to reject the manufacturer's claim that the modern, high-speed, bottle-filling machine fills bottles with no more variance than the company's present machine? Use $\alpha = 0.01$.

Table 10-2
Sample results for Illustration 10-9

Sample	n	s^2
Present machine (p)	22	0.0008
Modern high-speed machine (m)	25	0.0018

Solution

Step 1 $H_0: \sigma_m^2 = \sigma_p^2$ or $\sigma_m^2/\sigma_p^2 = 1$ $(\leq)$ (no more variance)
$H_a: \sigma_m^2 > \sigma_p^2$ or $\sigma_m^2/\sigma_p^2 > 1$ (more variance)

Step 2 $\alpha = 0.01$. The test statistic to be used is F, since the null hypothesis is about the equality of the variances of two populations. The critical region is one-tailed and on the right because the alternative hypothesis says "greater than."

$F(24, 21, 0.01)$ is the critical value. The number of degrees of freedom for the numerator is $24(25 - 1)$ because the sample from the modern high-speed machine is associated with the numerator, as specified by the null hypothesis. $df_d = 21$ because the sample associated with the denominator has size 22. The critical value is found in Table 8c and is 2.80. See the accompanying figure.

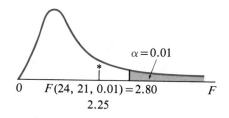

Step 3

$$F = \frac{s_m^2}{s_p^2} = \frac{0.0018}{0.0008} = 2.25$$

$$F* = \mathbf{2.25}$$

This value is shown in the figure.

Step 4   *Decision*: Fail to reject H_0 ($F*$ is in the noncritical region).

Conclusion: At the 0.01 level of significance, the samples do not present sufficient evidence to reject the manufacturer's claim. ■

OPTIONAL TECHNIQUE

When completing a hypothesis test about the equality of two population variances, it would be convenient if we could always use a right-hand critical value for F without having to calculate the left-hand critical value. This can be accomplished by minor adjustments in the null hypothesis and in the calculation of F in Step 3. The two cases we would like to change are (1) the two-tailed test and (2) the one-tailed test where the critical region is on the left. The one-tailed test with the critical region on the right already meets our criterion.

Case 1   When a two-tailed test is to be completed, we will state the hypotheses in the normal way. The calculated value of F, $F*$, will be the larger of s_1^2/s_2^2 or s_2^2/s_1^2. (One of these will have a value between zero and one; the other will be larger than one.) The critical value of F will be $F(df_n, df_d, \alpha/2)$, where df_n is the number of degrees of freedom for the sample whose variance is used in the numerator; df_d represents the degrees of freedom used in the denominator. Only the right-tail critical value will be needed. The test is completed in the usual fashion.

Consider the following hypothesis test situation.

$$H_0: \sigma_1^2/\sigma_2^2 = 1 \quad \text{versus} \quad H_a: \sigma_1^2/\sigma_2^2 \neq 1$$

$$\alpha = 0.05, \; n_1 = 5, \text{ and } n_2 = 11, \text{ with } F* = \frac{5.0}{7.0} = 0.714$$

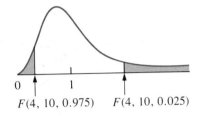

$$F(4, 10, 0.975) \qquad F(4, 10, 0.025)$$

In order to complete the hypothesis test, we will need to determine the left-hand critical value, $F(4, 10, 0.025)$. This will require the use of formula (10-6). Notice that with a few simple changes, we can avoid using formula (10-6.) Let's simply reverse the order of things.

$$H_0: \sigma_2^2/\sigma_1^2 = 1 \quad \text{versus} \quad H_a: \sigma_2^2/\sigma_1^2 = 1$$

$$\alpha = 0.05, n_2 = 11, \text{ and } n_1 = 5, \text{ with } F^* = \frac{7.0}{5.0} = 1.40$$

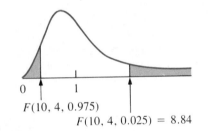

$$F(10, 4, 0.975)$$
$$F(10, 4, 0.025) = 8.84$$

We also need $F(10, 4, 0.025)$ from Table 8b. There were only two changes that were made: the calculated F^* value was inverted and the degrees of freedom for the critical value was reversed. Thus we avoided using formula (10-6).

Case 2   When a one-tailed test where the critical region is on the left is to be completed, we will interchange the position of the two variances in the statement of the hypotheses. This will reverse the direction of the alternative hypothesis and put the critical region on the right. From this point on, the procedure is the same as for the one-tailed test with a critical region on the right.

ratio of variances   Even though the hypothesis test is the most commonly used inference about two variances, you may occasionally be asked to estimate the **ratio of two variances**, σ_A^2/σ_B^2. The best point estimate is s_A^2/s_B^2. The $1 - \alpha$ **confidence interval** for the estimate is constructed by using the following formula:

$$\frac{s_A^2/s_B^2}{F(df_A, df_B, \alpha/2)} \quad \text{to} \quad \frac{s_A^2/s_B^2}{F(df_A, df_B, 1 - \alpha/2)} \tag{10-7}$$

Recall that left-tail values of F are not in the tables. These left-tail values are found by using formula (10-6). Using formula (10-6), we can rewrite formula (10-7) as

$$\frac{s_A^2/s_B^2}{F(df_A, df_B, \alpha/2)} \quad \text{to} \quad \frac{s_A^2}{s_B^2} F(df_B, df_A, \alpha/2) \tag{10-8}$$

(Notice that when the reciprocal is used, the degrees of freedom for the numerator and denominator are switched.)

ratio of standard deviation   If the **ratio of standard deviations** is desired, you need only take the positive square root of each of the bounds of the interval found by using formula (10-8).

Note   The formula for estimating the ratio of population variances, or population standard deviations, requires the use of the ratio of sample variances. If the standard deviation of the sample is given, it must be squared to obtain the variance.

Case Study 10-3

PERSONALITY CHARACTERISTICS OF POLICE APPLICANTS: COMPARISONS ACROSS SUBGROUPS AND WITH OTHER POPULATIONS

Bruce N. Carpenter and Susan M. Raza concluded that "police applicants are somewhat more like each other than are those in the normative population" when the F test of homogeneity of variance resulted in a p-value of less than 0.005. Homogeneity means that the group's scores are less variable than the scores for the normative population. What null and alternative hypotheses did Carpenter and Raza test? What was the critical value for the test? What does "p < .005" mean? (See Exercise 10.21.)

To determine whether police applicants are a more homogeneous group than the normative population, the $\underline{F}$ test of homogeneity of variance was used. With the exception of scales F, K, and 6, where the differences are nonsignificant, the results indicate that the police applicants form a somewhat more homogeneous group than the normative population (mean $\underline{F}(237, 305) = 1.36$, $p < .005$, range of $\underline{F}$ from 1.10 to 1.65). Thus, police applicants are somewhat more like each other than are those in the normative population.

Source: Reproduced from the *Journal of Police Science and Administration*, Vol. 15, No. 1, pp. 10–17, with permission of the International Association of Chiefs of Police, P.O. Box 6010, 13 Firstfield Road, Gaithersburg, Maryland 20878.

EXERCISES

10.17 Using the $F(df_1, df_2, \alpha)$ notation, name each of the critical values shown on the following figures. (For two-tail cases, use $\alpha/2$ in each tail.)

a.

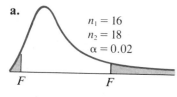

$n_1 = 16$
$n_2 = 18$
$\alpha = 0.02$

F F

b.

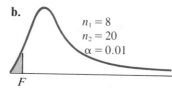

$n_1 = 8$
$n_2 = 20$
$\alpha = 0.01$

F

c.

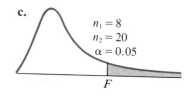

$n_1 = 8$
$n_2 = 20$
$\alpha = 0.05$

F

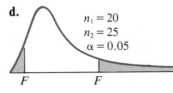

d.
$n_1 = 20$
$n_2 = 25$
$\alpha = 0.05$

F F

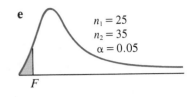

e
$n_1 = 25$
$n_2 = 35$
$\alpha = 0.05$

F

10.18 Find the following critical values for F from Tables 8a, 8b, and 8c in Appendix E.

    a. $F(24, 12, 0.05)$ **b.** $F(30, 40, 0.01)$

    c. $F(12, 10, 0.05)$ **d.** $F(5, 20, 0.01)$

    e. $F(15, 18, 0.025)$ **f.** $(15, 9, 0.025)$

    g. $F(40, 30, 0.05)$ **h.** $F(8, 40, 0.01)$

10.19 Find the following critical values for F. [*Hint*: Use formula (10-6).]

    a. $F(12, 20, 0.95)$ **b.** $F(5, 15, 0.975)$

    c. $F(20, 15, 0.99)$ **d.** $F(17, 20, 0.95)$

10.20 What formula is used to calculate the value of F? Explain it.

10.21 Referring to the article in Case Study 10-3,

    a. what null and alternative hypotheses did Carpenter and Raza test?

    b. what was the critical value for the test?

    c. what does "$p < .005$" mean?

10.22 A bakery is considering buying one of two gas ovens. The bakery requires that the temperature remain constant during a baking operation. A study was conducted to measure the variance in temperature of the ovens during the baking process. The variance in temperature before the thermostat restarted the flame for the Monarch oven was 2.4 for 16 measurements. The variance for the Kraft oven was 3.2 for 12 measurements. Does this information provide sufficient reason to conclude that there is a difference in the variances for the two ovens? Use a 0.01 level of significance.

    a. Solve using the classical approach.

    b. Solve using the prob-value approach.

10.23 A study was conducted to determine whether or not there was equal variability in male and female systolic blood pressures. Random samples of 16 men and 13 women were used to test the experimenters' claim that the variances were unequal. Perform the hypothesis test, at $\alpha = 0.05$, using the following data.

men:	120	120	118	112	120	114	130	114
	124	125	130	100	120	108	112	122
women:	122	102	118	126	108	130	104	
	116	102	122	120	118	130		

10.24 A consumer agency wishes to compare the variability in potency of a comparable drug manufactured by companies 1 and 2. Both drugs are distributed in the form of 250-mg tablets. The potency was determined for 25 tablets from each company, and it was found that $s_1^2 = 1.25$ and $s_2^2 = 1.18$. Find a 95 percent confidence interval for σ_1^2/σ_2^2.

10.25 Assume that the data in Exercise 10.16 were collected for the purpose of estimation.

    a. What would be your point estimate for the ratio of the two variances?

    b. Construct the 95 percent confidence interval for the estimation of the ratio of the two variances.

10.26 Two independent samples, each of size 3, are drawn from a normally distributed population. Find the probability that one of the sample variances is at least 19 times larger than the other one.

10.4 INFERENCES CONCERNING THE DIFFERENCE BETWEEN THE MEANS OF TWO INDEPENDENT SAMPLES (VARIANCES UNKNOWN AND SMALL SAMPLES)

In Section 10.2 we treated the cases for inferences about the difference between the means of two independent samples where the samples were both large. We will now investigate the inference procedures to be used in situations where two independent small samples (that is, one or both samples are of size less than or equal to 30) are taken from approximately normal populations for the purpose of comparing their means. For these inferences we must use Student's t distribution. However, we must distinguish between two possible cases: (1) the variances of the two populations are equal, $\sigma_1^2 = \sigma_2^2$; or (2) the variances of the two populations are unequal, $\sigma_1^2 \neq \sigma_2^2$.

    The two population variances are unknown; therefore, we will use the F test studied in Section 10.3 to determine whether we have case 1 or case 2. The two sample variances will be used in a two-tailed test of the null hypothesis $H_0: \sigma_1^2 = \sigma_2^2$, or $\sigma_1^2/\sigma_2^2 = 1$. If we fail to reject H_0, we will proceed with the methods for case 1. If we reject H_0, we will proceed with case 2. Typically, the same α is used for the F test as is given for the difference between the two means.

Case 1   The procedures here are very similar to those used when the Student's t distribution was employed. The standard error of estimate must be estimated by

$$s_p \sqrt{\frac{1}{n_1} + \frac{1}{n_2}}$$

pooled estimate for the
standard deviation

where s_p symbolizes the **pooled estimate for the standard deviation**. ("Pooled" means that the information from both samples is combined so as to give the best possible estimate.) The formula for s_p is

$$s_p = \sqrt{\frac{(n_1 - 1)s_1^2 + (n_2 - 1)s_2^2}{n_1 + n_2 - 2}} \tag{10-9}$$

The number of degrees of freedom, df, is the sum of the number of degrees of freedom for the two samples, $(n_1 - 1) + (n_2 - 1)$; that is,

$$df = n_1 + n_2 - 2 \tag{10-10}$$

With this information we can now write the formula for the **test statistic** t that will be used in a **hypothesis test**,

$$t = \frac{(\bar{x}_1 - \bar{x}_2) - (\mu_1 - \mu_2)}{s_p\sqrt{(1/n_1) + (1/n_2)}} \tag{10-11}$$

with $n_1 + n_2 - 2$ degrees of freedom.

Illustration 10-10   In studying the nature of the student body at a community college, the following question was raised: Do male and female students tend to drive the same distance (one way) to college daily? To answer this question a random sample of size 25 was taken from each segment of the student body. The results are shown in Table 10-3. Does this evidence contradict the null hypothesis that the two groups travel the same mean distances, at $\alpha = 0.10$?

Table 10-3
Sample results for
Illustration 10-10

Sample	n	$\bar{x}$	s^2
Male (m)	25	10.22	33.95
Female (f)	25	10.55	24.47

Solution   Since the σ's are unknown, we must first test the variances to determine whether we have case 1 or case 2.

Step 1 $H_0: \sigma_m^2 = \sigma_f^2$,   or $\sigma_m^2/\sigma_f^2 = 1$
$H_a: \sigma_m^2 \neq \sigma_f^2$

Step 2 $\alpha = 0.10$. See the following figure for the test criteria.

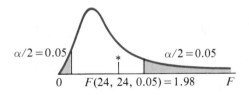

$\alpha/2 = 0.05$ $\alpha/2 = 0.05$

0 $F(24, 24, 0.05) = 1.98$ F

Step 3 $F^* = 33.95/24.47 = \mathbf{1.387}$ (F^* is in the noncritical region).

Step 4   *Decision:* Fail to reject H_0.

Conclusion: Assume $\sigma_m^2 = \sigma_f^2$, at the 0.10 level of significance, and the test for the difference between means will be completed according to case 1 procedures, using t and formula (10-11).

Step 1 $H_0: \mu_m = \mu_f$,   or $\mu_m - \mu_f = 0$
$H_a: \mu_m \neq \mu_f$

Step 2 $\alpha = 0.10; \mathrm{df} = 25 + 25 - 2 = 48; t(48, 0.05) = 1.65.$ See the accompanying figure.

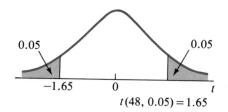

$$t(48, 0.05) = 1.65$$

Step 3

$$t = \frac{(\bar{x}_m - \bar{x}_f) - (\mu_m - \mu_f)}{s_p\sqrt{(1/n_m) + (1/n_f)}}$$

$$= \frac{(10.22 - 10.55) - (0)}{\sqrt{[(24)(33.95) + (24)(24.47)]/48} \cdot \sqrt{(1/25) + (1/25)}}$$

$$= \frac{-0.33}{\sqrt{29.21} \cdot \sqrt{0.08}} = \frac{-0.33}{\sqrt{2.3368}}$$

$$= \frac{-0.33}{1.52866} = -0.2158$$

$$t^* = -0.22$$

Step 4   *Decision*: Fail to reject H_0 (t^* is in the noncritical region).

Conclusion: There appears to be no significant difference between the mean distances traveled to college daily by male and female students.   ∎

Having completed our discussion of case 1, let's continue on to case 2.

Case 2   If we must assume that the two populations have **unequal variances**, then the hypothesis test for the difference between two independent means is completed by using the test statistic t. The **calculated value of t** is obtained by using

$$t = \frac{(\bar{x}_1 - \bar{x}_2) - (\mu_1 - \mu_2)}{\sqrt{(s_1^2/n_1) + (s_2^2/n_2)}} \tag{10-12}$$

with the number of degrees of freedom for the critical value being given by the smaller of $n_1 - 1$ or $n_2 - 1$.

Illustration 10-11   Many students have complained that the soft drink vending machine A (in the student recreation room) dispenses less drink than machine B (in the faculty lounge). To test this belief several samples were taken and weighed carefully, with the results as shown in Table 10-4. Does this evidence support the hypothesis that the mean amount dispensed by A is less than the mean amount dispensed by B, at the 5 percent level of significance?

Table 10-4

Sample results for
Illustration 10-11

Machine	n	$\bar{x}$	s
A	10	5.38	1.59
B	12	5.92	0.83

Solution   First we test for the equality of the variances.

Step 1 $H_0: \sigma_A^2 = \sigma_B^2,$ or $\sigma_A^2/\sigma_B^2 = 1$
$H_a: \sigma_A^2 \neq \sigma_B^2$

Step 2 $\alpha = 0.05$. See the following figure.

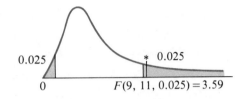

0.025   * 0.025

0          $F(9, 11, 0.025) = 3.59$

Step 3 $F^* = (1.59)^2/(0.83)^2 = \mathbf{3.67}$

Step 4   *Decision*: Reject H_0.

Conclusion: Assume that $\sigma_A^2 \neq \sigma_B^2$, with $\alpha = 0.05$.

Now we are ready to test the difference between the means by use of case 2 procedures.

Step 1 $H_0: \mu_A = \mu_B$ or $\mu_A - \mu_B = 0$ $(\geq)$
$H_a: \mu_A < \mu_B$ or $\mu_A - \mu_B < 0$

Step 2   df $= 10 - 1 = 9$ (smaller sample is of size 10); $\alpha = 0.05$. See the accompanying figure. The critical value is $-t(9, 0.05) = -1.83$.

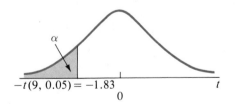

α

$-t(9, 0.05) = -1.83$
0   t

Step 3

$$t = \frac{(5.38 - 5.92) - (0)}{\sqrt{[(1.59)^2/10] + [(0.83)^2/12]}}$$

$$= \frac{-0.54}{\sqrt{0.2528 + 0.0574}} = \frac{-0.54}{0.557} = -0.969$$

$t^* = \mathbf{-0.97}$

Step 4   *Decision*: Fail to reject H_0 (t^* is in the acceptance region).

Conclusion: There is not sufficient evidence shown by these samples to conclude that machine A dispenses less drink than does machine B.   ∎

If confidence interval estimates are to be constructed to estimate the difference between the means of two independent small samples, the *F* test will again be used to determine which of the following two formulas we need to use.

Case 1   **σ's unknown but assumed equal.** The $1 - \alpha$ **confidence interval** is given by

$$(\bar{x}_1 - \bar{x}_2) - t(df, \alpha/2) \cdot s_p \cdot \sqrt{\frac{1}{n_1} + \frac{1}{n_2}} \quad \text{to} \quad (\bar{x}_1 - \bar{x}_2) + t(df, \alpha/2) \cdot s_p \cdot \sqrt{\frac{1}{n_1} + \frac{1}{n_2}}$$

$$(10\text{-}13)$$

where $df = n_1 + n_2 - 2$ and s_p is the pooled estimate for the standard deviation found by using formula (10-9).

Case 2   **σ's unknown but assumed unequal.** The $1 - \alpha$ **confidence interval** is given by

$$(\bar{x}_1 - \bar{x}_2) - t(df, \alpha/2) \cdot \sqrt{\frac{s_1^2}{n_1} + \frac{s_2^2}{n_2}} \quad \text{to} \quad (\bar{x}_1 - \bar{x}_2) + t(df, \alpha/2) \cdot \sqrt{\frac{s_1^2}{n_1} + \frac{s_2^2}{n_2}} \quad (10\text{-}14)$$

where df is the smaller of $n_1 - 1$ or $n_2 - 1$.

Illustration 10-12   The heights of 20 randomly selected women and 30 randomly selected men were obtained from the student body of a certain college in order to estimate the difference in their mean heights. Find (a) a point estimate and (b) a 95 percent confidence interval estimate for the difference between the means, $\mu_m - \mu_f$.

Solution   The sample information is given in Table 10-5.

Table 10-5
Sample results for
Illustration 10-12

Sample	Number	Mean	Standard Deviation
Female (f)	20	63.8	2.18
Male (m)	30	69.8	1.92

a. The point estimate for $\mu_m - \mu_f$ is **6.0**.
b. First we must decide whether we should assume that $\sigma_f^2 = \sigma_m^2$ or $\sigma_f^2 \neq \sigma_m^2$.

$$H_0: \sigma_f^2 = \sigma_m^2 \quad \text{or} \quad \frac{\sigma_f^2}{\sigma_m^2} = 1$$

$$H_a: \sigma_f^2 \neq \sigma_m^2$$

$\alpha = 0.05$; see the following figure for the test criteria. The critical value is $F(19, 29, 0.025) \approx 2.24$.

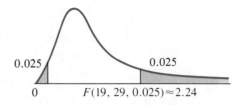

$$F(19, 29, 0.025) \approx 2.24$$

Note When the correct degrees of freedom cannot be found in the table, the critical value is found by using the linear interpolation technique. (*See* Appendix D.)

$$F^* = \frac{(2.18)^2}{(1.92)^2} = 1.289$$

So we fail to reject H_0. Therefore, we assume that $\sigma_f^2 = \sigma_m^2$. The 0.95 confidence interval is then found by using formula (10-13).

$$6.0 \pm t(48, 0.025) \cdot \sqrt{\frac{19(2.18)^2 + 29(1.92)^2}{48}} \cdot \sqrt{\frac{1}{20} + \frac{1}{30}}$$

$$6.0 \pm 1.96\sqrt{4.1084}\sqrt{0.0833}$$

$$6.0 \pm (1.96)(2.0269)(0.2886)$$

$$6.0 \pm 1.1466$$

$$6.0 \pm 1.15$$

4.85 to **7.15**

is the 0.95 confidence interval for $\mu_m - \mu_f$. (No specific illustration is shown for estimation where $\sigma_1^2 \neq \sigma_2^2$. All parts of such a solution can be found in various illustrations in this chapter.) ∎

Case Study 10-4 | **AN EMPIRICAL STUDY OF FACULTY EVALUATION SYSTEMS: BUSINESS FACULTY PERCEPTIONS**

Tong and Bures report a whole series of t-test results from comparing various factors of perceptions between AACSB accredited schools and non-AACSB accredited schools. For each t-test they report a two-tailed probability (p-value). Many of them are 0.000 meaning that the p-value is less than 0.0005. The p-value for "activity in professional societies" is 0.394. What does that mean? (See Exercise 10.31.)

Table 1 shows how respondents from American Assembly of Collegiate Schools of Business (AACSB)-accredited and non-AACSB-accredited schools rated the importance of the ten faculty evaluation factors. Clearly AACSB-accredited schools and non-AACSB-accredited schools are different, with the

AACSB-accredited institutions placing greater weight on research and publication, and with the non-accredited institutions putting more emphasis on classroom teaching, campus committee work, student advising, public service, advisor to student organizations, and consultation (business, government). There is no significant difference between AACSB-accredited and non-AACSB-accredited schools in the importance given to a faculty member's activity in professional societies.

TABLE 1. *t*-test results of importance of faculty evaluation factors, AACSB-accredited vs. non-AACSB-accredited schools

Factor	AACSB-Accredited Schools ($n = 176$)	Non-AACSB-Accredited Schools ($n = 74$)	Two-tailed *t*-test Probability
Articles in professional journals	4.49	2.87	0.000
Classroom teaching	3.34	4.31	0.000
Books as author or editor	3.23	2.45	0.000
Papers at professional meetings	3.09	2.65	0.001
Activity in professional societies	2.55	2.65	0.394
Campus committee work	2.25	3.16	0.000
Student advising	1.80	3.36	0.000
Public service	1.98	2.47	0.000
Advisor to student organizations	1.71	2.30	0.000
Consultation (business, government)	1.73	2.30	0.001

Note: A five-point scale with 1 = "not at all important" to 5 = "extremely important" was used in this analysis.

Source: Hsin-Min Tong and Allen L. Bures in *Journal of Education for Business*, April 1987. Reprinted with permission of the Helen Dwight Reid Educational Foundation. Published by Heldref Publications, 4000 Albemarle St., N.W., Washington, D.C. 20016. Copyright © 1987.

EXERCISES

10.27 Certain "adjustments" are made between the test for the equality of two population means when the variances are unknown and judged to be equal and the test for equality of two population means when variances are unknown and judged to be unequal. (The *t* distribution is used in both cases.) Explain how each of the following adjustments is accomplished.

  a. The test statistic is modified.

  b. The critical region is modified.

  c. The number of degrees of freedom is modified.

10.28 Two independent samples are drawn from normal populations, with the results as shown in the following table. Does this information provide sufficient reason to reject the null hypothesis in favor of the claim that the mean of population

R is significantly larger than the mean of population *S*? Use $\alpha = 0.05$. (Remember to perform the *F* test on the variance first.)

Sample	n	$\sum x$	$\sum (x - \bar{x})^2$
R	10	295	75
S	8	195	90

10.29 Independent samples were taken from each of two normal populations in order to compare the two population means. The sample data are summarized in the following table. Assume that $\sigma_1 \neq \sigma_2$.

a. What effect does the assumption $\sigma_1 \neq \sigma_2$ have on the situation?

b. Determine the number of degrees of freedom to use in comparing μ_1 and μ_2.

c. Is there sufficient reason to reject the hypothesis that $\mu_1 = \mu_2$, at $\alpha = 0.05$?

Sample	n	$\bar{x}$	s^2
1	10	12.3	6.8
2	15	13.9	23.5

10.30 A study was designed to test the hypothesis that men have a higher mean diastolic blood pressure than do women. MINITAB was used to analyze the following sample data.

```
males:   76  76  74  70  80  68  90  70
         90  72  76  80  68  72  96  80

females: 76  70  82  90  68  60  62
         68  80  74  60  62  72
```

```
MTB > SET MALE DIASTOLIC BLOOD PRESSURES IN C1
DATA> 76 76 74 70 80 68 90 70 96 80 90 72 76 80 68 72
DATA> END DATA
MTB > SET FEMALE DIASTOLIC BLOOD PRESSURES IN C2
DATA> 76 70 82 90 68 60 62 60 62 72 68 80 74
DATA> END DATA
MTB > POOLED T,ALT = 1,MALE READINGS IN C1, FEMALE READINGS IN C2

TWOSAMPLE T FOR C1 VS C2
        N       MEAN      STDEV    SE MEAN
C1     16      77.38      8.35      2.09
C2     13      71.08      9.22      2.56

95 PCT CI FOR MU C1 - MU C2: (-0.4064, 13.00)
TTEST MU C1 = MU C2 (VS GT): T = 1.93 P = 0.032 DF = 27.0
```

a. Use the standard deviations to verify that the hypothesis of equal variances is not rejected at $\alpha = 0.05$.

b. Verify the value for t^*.

c. Use the Student's *t* table to bound the *p*-value and compare with the MINITAB value.

10.31 Explain the meaning of the p-value 0.394 reported in Case Study 10-4 for the two-tailed t test comparing the difference between the perception of activities in professional societies.

10.32 If a random sample of 18 homes south of Center Street in Provo, New York, have a mean selling price of $15,000 and a variance of $2,400, and a random sample of 18 homes north of Center Street have a mean selling price of $16,000 and a variance of $4,800, can you conclude that there is a significant difference between the selling price of homes in these two areas of Provo at the 0.05 level?

    a. Solve using the classical approach.

    b. Solve using the prob-value approach.

10.33 To compare the mathematical competency of men and women, the Beckmann-Beal test of mathematical competencies was administered to 15 men and 15 women. The results were:

$$\text{Males:} \quad \bar{x} = 39.5 \quad s = 2.9$$
$$\text{Females:} \quad \bar{x} = 38.4 \quad s = 3.4$$

    a. Test $H_0: \sigma_M = \sigma_F$ versus $H_a: \sigma_M \neq \sigma_F$ at $\alpha = 0.05$.

    b. Test $H_0: \mu_M = \mu_F$ versus $H_a: \mu_M \neq \mu_F$ at $\alpha = 0.05$.

    c. Give the p-value for the test in (b).

10.34 A study was designed to compare the attitudes of two groups of nursing students toward computers. Group 1 had previously taken a statistical methods course that involved significant computer interaction through the use of statistical packages. Group 2 had taken a statistical methods course that did not use computers. The students' attitudes were measured by administering the Computer Anxiety Index (CAIN). The results were:

    group 1 (with computers) $n = 10$ $\bar{x} = 60.3$ $s = 7.5$
    group 2 (without computers): $n = 15$ $\bar{x} = 67.2$ $s = 2.1$

    a. Test for equality of variances at $\alpha = 0.05$.

    b. Do the data show that the mean score for those with computer experience was significantly less than the mean score for those without computer experience? Use $\alpha = 0.05$.

10.35 Two samples of data were obtained to compare the means of two normal populations. The data are summarized in the following table. Is there sufficient evidence to conclude that $\mu_A > \mu_B$ at $\alpha = 0.05$? (*Hint:* Does $\sigma_A = \sigma_B$?)

Sample	n	$\sum x$	$\sum x^2$
A	10	223	6735
B	10	110	1322

10.36 Twenty laboratory mice were randomly divided into two groups of 10. Each group was fed according to a prescribed diet. At the end of three weeks the weight gained by each animal was recorded. Do the data in the following table justify the conclusion that the mean weight gained on diet B was greater than the mean weight gained on diet A, at the $\alpha = 0.05$ level of significance?

Diet A	5	14	7	9	11	7	13	14	12	8
Diet B	5	21	16	23	4	16	13	19	9	21

10.37 Use the sample information in Exercise 10.28 to construct the 0.95 confidence interval estimate for the difference between μ_R and μ_S.

10.38 Construct a 90 percent confidence interval estimate for the difference between μ_1 and μ_2 based on the sample statistics given in Exercise 10.29. (Assume that $\sigma_1 \neq \sigma_2$.)

10.39 The two independent samples shown in the following table were obtained in order to estimate the difference between the two population means. Construct the 0.98 confidence interval estimate.

Sample A	6	7	7	6	6	5	6	8	5	4
Sample B	7	2	4	3	3	5	4	6	4	2

10.40 Show that $s_p \sqrt{\dfrac{1}{n_1} + \dfrac{1}{n_2}}$ reduces to $\sqrt{\dfrac{s_1^2 + s_2^2}{n}}$ when $n_1 = n_2 = n$.

10.5   INFERENCES CONCERNING THE MEAN DIFFERENCE BETWEEN TWO DEPENDENT SAMPLES

The procedure for comparing the means of two dependent samples is quite different from that of comparing the means of two independent samples. When comparing the means of two independent samples we make inferences about the difference between the two population means by using the difference between the two observed sample means. However, because of the relationship between two dependent samples, we will actually compare each data value of the first sample with the data value in the second sample that came from the same source. The two data values, one from each set, that come from the same source are called **paired** data. These pairs of data are compared by using the difference in their numerical values. This difference

paired difference   is called a **paired difference**. The mean difference of the two dependent populations is then tested by using the observed mean of the resulting paired differences. The concept of using paired data for this purpose has a built-in ability to remove many otherwise uncontrollable factors. The tire-wear problem (Illustration 10-2) is an excellent example of such additional factors. The wearing ability of a tire is greatly affected by a multitude of factors: the size and weight of the car, the age and condition of the car, the driving habits of the driver, the number of miles driven, the condition of and types of roads driven on, the quality of the material used to make the tire, and so on. By pairing the tires and considering differences, all of these extraneous causes of variability are removed.

Illustration 10-13   If we were to test the wearing quality of two tire brands by plan D, as described in Illustration 10-2, all the aforementioned factors will have an equal effect on both tire brands. Table 10-6 gives the amount of wear (in thousandths of an inch) that took place in such a test. One tire of each brand was placed on each of the six test cars. The position (left or right side, front or back) was determined with the aid of random number table.

Table 10-6
Sample results for
Illustration 10-13

Tire Brand	Car 1	2	3	4	5	6
A	125	64	94	38	90	106
B	133	65	103	37	102	115

Since the various cars, drivers, and conditions are the same for each paired set of data, it would make sense to introduce a new measure, the paired difference d, that was observed in each pair of related data. Therefore, we add a third row to the data of Table 10-6, d = brand B − brand A.

Car	1	2	3	4	5	6
$d = B - A$	8	1	9	−1	12	9

Do the sample data provide sufficient evidence for us to conclude that the two brands show unequal wear, at the 0.05 level of significance?

Before we can solve the problem posed in this illustration, we need some new formulas. Our two sets of sample data have been combined into one set, a set of n values of d, $d = x_1 - x_2$. The sample statistics that we will use are $\bar{d}$, the observed **mean** value of d,

$$\bar{d} = \frac{\sum d}{n} \tag{10-15}$$

and s_d, the observed **standard deviation** of the d's,

$$S_d = \sqrt{\frac{\sum d^2 - \frac{(\sum d)^2}{n}}{n - 1}} \tag{10-16}$$

The observed values of d in our sample are from the population of *paired differences*, whose distribution is assumed to be approximately normal with a mean value of μ_d and a standard deviation of σ_d. Since σ_d is unknown, it will be estimated by s_d. The inferences about μ_d are completed in the same manner as the inferences about μ in Chapter 8 and in Section 9.1 by using $\bar{d}$ as the point estimate of μ_d. The inferences are completed by using the t distribution and $s_d/\sqrt{n}$ as the approximate measure for the standard error

$$t = \frac{\bar{d} - \mu_d}{s_d/\sqrt{n}} \qquad (10\text{-}17)$$

with $n - 1$ degrees of freedom.

Now let's complete Illustration 10-13.

Solution

mean difference   Step 1 $H_0: \mu_d = 0$ (μ_d is **mean difference**)

$H_a: \mu_d \neq 0$

Step 2 $\alpha = 0.05$. The test criteria are shown in the accompanying figure. The critical values are $\pm t(5, 0.025) = \pm 2.57$.

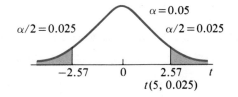

Note n is the number of paired differences (d). df $= n - 1$, and $t(5, 0.025)$ is obtained from Table 6, Appendix E.

Step 3

$$t = \frac{\bar{d} - \mu_d}{s_d/\sqrt{n}}$$

First we must find $\bar{d}$ and s_d; some preliminary calculations are shown in Table 10-7.

Table 10-7
Preliminary calculations

d	d^2
8	64
1	1
9	81
−1	1
12	144
9	81
Total 38	372

$$\bar{d} = \frac{\sum d}{n} = \frac{38}{6} = 6.333 = \mathbf{6.3}$$

$$s_d = \sqrt{\frac{(372) - \dfrac{(38)^2}{6}}{5}} = \sqrt{26.27} = 5.13 = \mathbf{5.1}$$

Therefore,

$$t = \frac{6.3 - 0}{5.1/\sqrt{6}} = \frac{(6.3)\sqrt{6}}{5.1}$$

$$= \frac{(6.3)(2.45)}{5.1} = 3.026$$

$$t^* = \mathbf{3.03}$$

See the following figure.

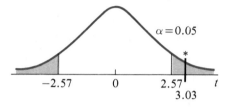

Step 4   *Decision*: Reject H_0 (t^* is in the critical region).

Conclusion: There is a significant difference in the mean amount of wear.

Since the calculated value of t fell in the critical region on the right, we might want to conclude that one brand of these tires is better than the other. This is possible, but we must be extremely careful to interpret the conclusion correctly. Recall that the recorded data concerned the amount of wear. Therefore, the tire that shows less wear should be the better one. Our hypothesis test showed that tire brand B had significantly more wear, thereby implying that brand A is the longer-wearing tire.   ■

It is also possible to estimate the mean difference in paired data. The formula used for this **confidence interval** estimate is

$$\bar{d} - t(\text{df}, \alpha/2) \cdot \frac{s_d}{\sqrt{n}} \quad \text{to} \quad \bar{d} + t(\text{df}, \alpha/2) \cdot \frac{s_d}{\sqrt{n}} \tag{10-18}$$

Illustration 10-14   Construct the 95 percent confidence interval for the estimation of the mean difference in the paired data on tire wear, as found in Illustration 10-13. The given information is $n = 6$ pieces of paired data, $\bar{d} = 6.33$, and $s_d = 5.1$.

Solution

$$\bar{d} \pm t(\text{df}, \alpha/2) \cdot \frac{s_d}{\sqrt{n}}$$

$$6.3 \pm 2.57 \cdot \frac{5.1}{\sqrt{6}}$$

$$6.3 \pm 5.4$$

$$\mathbf{0.9} \quad \text{to} \quad \mathbf{11.7}$$

is the 0.95 confidence interval for $\mu_{d=B-A}$. That is, with 95 percent confidence we can say that the mean difference in the amount of wear is between 0.9 and 11.7.

Note   This confidence interval is quite wide. This is due, in part, to the small sample size. Recall from the central limit theorem that as the sample size increases, the standard error (estimated by $s_d/\sqrt{n}$) decreases.   ∎

EXERCISES

10.41 A group of ten recently diagnosed diabetics were tested to determine whether an educational program was effective in increasing their knowledge of diabetes. They were given a test, before and after the educational program, concerning self-care aspects of diabetes. The scores on the test were as follows.

Patient	1	2	3	4	5	6	7	8	9	10
Before	75	62	67	70	55	59	60	64	72	59
After	77	65	68	72	62	61	60	67	75	68

The following MINITAB output may be used to determine whether the scores inproved as a result of the program. Verify the following as shown on the output: mean difference (MEAN), standard deviation (STDEV), standard error of the difference (SE MEAN), t^* (T), and p-value.

```
MTB > READ BEFORE IN C1, AFTER IN C2
DATA> 75 77
DATA> 62 65
DATA> 67 68
DATA> 70 72
DATA> 55 62
DATA> 59 61
DATA> 60 60
DATA> 64 67
DATA> 72 75
DATA> 59 68
DATA> END DATA
       10 ROWS READ
MTB > SUBTRACT C1 FROM C2, PUT DIFFERENCE IN C3
MTB > TTEST MU = 0, ALT = 1, FOR DIFFERENCES IN C3

TEST OF MU = 0.000 VS MU G.T. 0.000
              N     MEAN    STDEV    SE MEAN      T    P VALUE
C3           10    3.200    2.741     0.867    3.69    0.0025

MTB > STOP
```

10.42 The corrosive effects of various soils on coated and uncoated steel pipe were tested by using a dependent sampling plan. The data collected are summarized by

$$n = 40 \qquad \sum d = 220 \qquad \sum d^2 = 6222$$

where d is the amount of corrosion on the coated portion subtracted from the amount of corrosion on the uncoated portion. Does this sample provide sufficient reason to conclude that the coating is beneficial? Use $\alpha = 0.01$.

a. Solve using the classical approach.

b. Solve using the prob-value approach.

10.43 We want to know which of two types of filters should be used over an oscilloscope to help the operator pick out the image on the screen of the cathode-ray tube. A test was designed in which the strength of a signal could be varied from zero to the point where the operator first detects the image. At this point, the intensity setting is read. The lower the setting, the better the filter. Twenty operators were asked to make one reading for each filter. Do the data in the following table show a significant difference in the filters at the 0.10 level?

	Filter				Filter	
Operator	1	2		Operator	1	2
1	96	92		11	88	88
2	83	84		12	89	89
3	97	92		13	85	86
4	93	90		14	94	91
5	99	93		15	90	89
6	95	91		16	92	90
7	97	92		17	91	90
8	91	90		18	78	80
9	100	93		19	77	80
10	92	90		20	93	90

a. Solve using the classical approach.

b. Solve using the prob-value approach.

10.44 Two men, A and B, who usually commute to work together, decide to conduct an experiment to see whether one route is faster than the other. The men feel that their driving habits are approximately the same, and therefore they decide on the following procedure. Each morning for two weeks A will drive to work on one route and B will use the second route. On the first morning, A will toss a coin. If heads appear, he will use route I; if tails appear, he will use route II. On the second morning, B will toss the coin: heads, route I; tails, route II. The times, recorded to the nearest minute, are shown in the following table. Do these data show that one of the routes is significantly faster than the other, at $\alpha = 0.05$?

	Day									
Route	M	Tu	W	Th	F	M	Tu	W	Th	F
I	29	26	25	25	25	24	26	26	30	31
II	25	26	25	25	24	23	27	25	29	30

a. Solve using the classical approach.

b. Solve using the prob-value approach.

10.45 An experiment was designed to estimate the mean difference in weight gain for pigs fed ration A as compared to those fed ration B. Eight pairs of pigs were used. The pigs within each pair were littermates. The rations were assigned at random to the two animals within each pair. The gains (in pounds) after 45 days are shown in the following table.

Litter	1	2	3	4	5	6	7	8
Ration A	65	37	40	47	49	65	53	59
Ration B	58	39	31	45	47	55	59	51

Find the 95 percent confidence interval estimate for the mean of the differences μ_d, where d = ration A − ration B.

10.46 A sociologist is studying the effects of a certain motion picture film on the attitudes of black men toward white men. Twelve black men were randomly selected and asked to fill out a questionnaire before and after viewing the film. The scores received by the 12 men are shown in the following table. Construct a 95 percent confidence interval estimate for the mean shift in score that takes place when this film is viewed.

Before	10	13	18	12	9	8	14	12	17	20	7	11
After	5	9	13	17	4	5	11	14	13	18	7	12

10.6 INFERENCES CONCERNING THE DIFFERENCE BETWEEN PROPORTIONS OF TWO INDEPENDENT SAMPLES

We are often interested in making statistical comparisons between the proportions, percentages, or probabilities associated with two populations. Such questions as the following are frequently asked: Is the proportion of homeowners who favor a certain tax proposal different from the proportion of renters who favor it? Did a larger percentage of this semester's class than of last semester's class pass statistics? Do students' opinions about the "new code of conduct" differ from those of the faculty? You can probably see the many types of questions involved. In this section we will compare two population proportions by using the difference between the observed proportions, $p'_1 - p'_2$, of two independent samples.

Recall

1. The observed probability is $p' = x/n$, where x is the number of observed successes in n trials.

2. $q' = 1 - p'$.

3. p is the probability of success on an individual trial in a binomial probability experiment of n repeated independent trials (see page 203).

The sampling distribution of $p'_1 - p'_2$ is approximately normally distributed with a mean $\mu_{p'_1 - p'_2} = p_1 - p_2$ and with a standard error of

$$\sqrt{\frac{p_1 q_1}{n_1} + \frac{p_2 q_2}{n_2}}$$

if n_1 and n_2 are sufficiently large. Therefore, when the null hypothesis that there is no difference between two proportions is to be tested, the **test statistic** will be z. The calculated value of z will be determined by formula (10-19):

$$z = \frac{p'_1 - p'_2}{\sqrt{pq[(1/n_1) + (1/n_2)]}} \tag{10-19}$$

Notes

1. The null hypothesis is $p_1 = p_2$, or $p_1 - p_2 = 0$.
2. The numerator of formula (10-19) written in the usual manner is $(p'_1 - p'_2) - (p_1 - p_2)$, but since the null hypothesis is assumed to be true during the test, $p_1 - p_2 = 0$ and the numerator becomes simply $p'_1 - p'_2$.
3. Since the null hypothesis is $p_1 = p_2$, the standard error of $p'_1 - p'_2$ can be written as $\sqrt{pq[(1/n_1) + (1/n_2)]}$, where $p = p_1 = p_2$ and $q = 1 - p$.

Illustration 10-15   A salesman for a new manufacturer of walkie-talkies claims that the percentage of defective walkie-talkies found among his products will be no higher than the percentage of defectives found in a competitor's line. To test his statement, random samples were taken of each manufacturer's product. The sample summaries are given in Table 10-8. Can we reject the salesman's claim, at the 0.05 level of significance?

Table 10-8
Sample results for
Illustration 10-15

Product	Sample Number	Number of Defectives	Number Checked
Salesman's	1	8	100
Competitor's	2	2	100

Solution

Step 1 $H_0: p_1 - p_2 = 0$ ($\leq$)   (no higher than)

$H_a: p_1 - p_2 > 0$ (higher than)

Step 2 $\alpha = 0.05$. The test criteria are shown in the accompanying figure. The critical value is $z(0.05) = 1.65$.

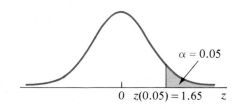

$\alpha = 0.05$

$0 \quad z(0.05) = 1.65 \quad z$

Step 3 $p'_1 = 8/100 = 0.08$, $p'_2 = 2/100 = 0.02$.

Since the values of p_1 and p_2 are unknown but assumed equal, (H_0), the best estimate we have for p $(p = p_1 = p_2)$, is obtained by pooling the two samples to obtain a **pooled observed probability**, p'_p:

$$p'_p = \frac{x_1 + x_2}{n_1 + n_2} \tag{10-20}$$

Our pooled estimate for p, p'_p, then becomes

$$p'_p = \frac{8 + 2}{100 + 100} = \frac{10}{200} = \textbf{0.05} \quad \text{and} \quad q'_p = 1 - p'_p = 1 - 0.05 = \textbf{0.95}$$

The value of the test statistic z is calculated by using formulas (10-19) and (10-20) together. p'_p and q'_p are used as estimates for p and q.

$$z = \frac{p'_1 - p'_2}{\sqrt{p'_p q'_p [(1/n_1) + (1/n_2)]}}$$

$$= \frac{0.08 - 0.02}{\sqrt{(0.05)(0.95)[(1/100) + (1/100)]}}$$

$$= \frac{0.06}{\sqrt{(0.05)(0.95)(0.02)}} = \frac{0.06}{0.031} = 1.9354$$

$$z^* = \textbf{1.94}$$

See the accompanying figure.

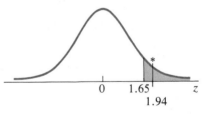

Step 4   *Decision*: Reject H_0 (z^* is in the critical region).

Conclusion: At the 0.05 level of significance, there is sufficient evidence to reject the salesman's claim.   ■

Illustration 10-16   Joe claimed that the probability that a commuting college student has car trouble of some type (car wouldn't start, flat tire, accident) on the way to college in the morning is greater than the probability that the student will have car trouble on the way home after class. He has several theories to explain his claim. The Sports Car Club thinks that the ideas of "before class" and "after class" have nothing to do with whether or not a student has car trouble. The club decides to challenge Joe's claim, and two large samples are gathered in order to test the theory.

Table 10-9

Sample results for
Illustration 10-16

Time of Trouble	Number Sampled (n)	Number Having Trouble
Before (b)	500	30
After (a)	600	28

The resulting sample statistics are given in Table 10-9. Complete the hypothesis test with $\alpha = 0.02$.

Solution

Step 1 $H_0: p_b = p_a$ or $p_b - p_a = 0$ (no difference)
$H_a: p_b > p_a$ or $p_b - p_a > 0$ (Joe's claim)

Step 2 $\alpha = 0.02$. The test criteria are shown in the accompanying figure.

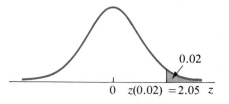

0    $z(0.02) = 2.05$ z

Step 3

$$z = \frac{p'_b - p'_a}{\sqrt{p^* q^* [(1/n_b) + (1/n_a)]}}$$

$$p'_b = \frac{30}{500} = 0.06 \qquad p'_a = \frac{28}{600} = 0.04666 = 0.047$$

The pooled estimate for p is

$$p'_p = \frac{30 + 28}{500 + 600} = \frac{58}{1100} = 0.0527 = \mathbf{0.053}$$

and thus $q'_p = 1 - 0.053 = \mathbf{0.947}$. Therefore,

$$z = \frac{0.06 - 0.047}{\sqrt{(0.053)(0.947)(1/500 + 1/600)}}$$

$$= \frac{0.013}{\sqrt{0.05019(0.002 + 0.0017)}}$$

$$= \frac{0.013}{\sqrt{0.000186}} = \frac{0.013}{0.0136} = 0.956$$

$$z^* = \mathbf{0.96}$$

Step 4   *Decision*: Fail to reject H_0 (z^* is in the acceptance region).

Conclusion: At the 0.02 level of significance, there is no evidence to indicate that the probability of car trouble is greater before class than after class. ■

To estimate the difference between the proportions of two populations, you should use the observed difference $(p_1' - p_2')$ as the point estimate. The $1 - \alpha$ **confidence interval** is given by

$$(p_1' - p_2') - z(\alpha/2) \cdot \sqrt{\frac{p_1' q_1'}{n_1} + \frac{p_2' q_2'}{n_2}} \quad \text{to} \quad (p_1' - p_2') + z(\alpha/2) \cdot \sqrt{\frac{p_1' q_1'}{n_1} + \frac{p_2' q_2'}{n_2}}$$

(10-21)

Note   A pooled sample proportion is not used in the confidence interval because it is not known whether $p_1 = p_2$.

Illustration 10-17   In studying his campaign plans, Mr. Morris wishes to estimate the difference between men's and women's views regarding his appeal as a candidate. He asks his campaign manager to take two samples and find the 99 percent confidence interval estimate of the difference. A sample of 1000 voters was taken from each population, with 388 men and 459 women favoring Mr. Morris.

Solution   The campaign manager calculated the confidence interval estimate by using formula (10-21), as follows.

$$p_m' = \frac{388}{1000} = 0.388 \qquad p_w' = \frac{459}{1000} = 0.459$$

$$(0.459 - 0.388) \pm 2.58 \sqrt{\frac{(0.459)(0.541)}{1000} + \frac{(0.388)(0.612)}{1000}}$$

$$0.071 \pm 2.58 \sqrt{0.000248 + 0.000237}$$

$$0.071 \pm 2.58 \sqrt{0.000485}$$

$$0.071 \pm (2.58)(0.022)$$

$$0.071 \pm 0.057$$

0.014   to   0.128

is the 0.99 confidence interval for $p_w - p_m$. With 99 percent confidence we can say that there is a difference of from 1.4 percent to 12.8 percent in Mr. Morris's voter appeal. That is, a larger proportion of women than men favor Mr. Morris, and the difference in proportion is between 1.47 percent and 12.8 percent.   ■

Case Study 10-5

WHAT WOMEN AND MEN THINK ABOUT JOB BIAS AFFECTING WOMEN

The Gallup Poll results reported in the New York Teacher, *June 8, 1987, show proportions of women and men in both 1975 and 1987 answering "yes" to the question "Do you feel that women in this country have equal job opportunities with men?" We can estimate the difference between the proportion of men and women who answered yes in 1975 and the proportion who answered yes in 1987. We can also estimate the change in proportion from 1975 to 1987 for men and for women. What is the point estimate for each of these differences? (See Exercise 10.51, c and d.)*

DO YOU FEEL THAT WOMEN IN THIS COUNTRY HAVE EQUAL JOB OPPORTUNITIES WITH MEN?

According to a recent Gallup Poll, although there have been improvements in women's employment prospects, fewer women in 1987 than in 1975 believe that they have equal job opportunities with men. In that 12 year period, there has been a sharp decline in the percentage of women who think that their sex has job equality, down from 49 percent to 35 percent. Among the men on the same question, opinions did not change significantly between the 1975 and the 1987 Gallup Polls.

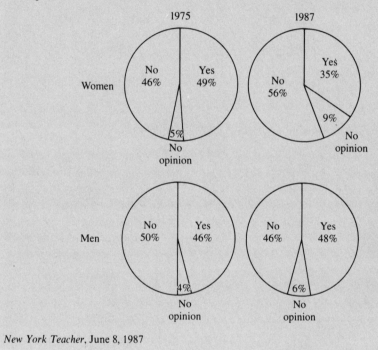

New York Teacher, June 8, 1987

Case Study 10-6

GALLUP REPORT: SAMPLING TOLERANCES

The Gallup Report provides its readers with the following information and tables. The tables give the 0.95 confidence level sampling errors (maximum error of estimation) for various percentages and sample sizes. Can you verify the values in these tables? (See Exercise 10.51, a and b.)

In interpreting survey results, it should be borne in mind that all sample surveys are subject to sampling error, that is, the extent to which the results may differ from those that would be obtained if the whole population surveyed had been interviewed. The size of such sampling errors depends largely on the number of interviews.

In comparing survey results in two sub-samples, such as men and women, the question arises as to how large a difference between them must be before one can be reasonably sure that it reflects a real difference. In Tables 1 and 2, the number of points which must be allowed for in such comparisons is indicated.

For percentages near 20 or 80, use Table 1, for those near 50, Table 2. For percentages in between, the error factor is between that shown in the two tables.

Here is an example of how the tables should be used: Say 50 percent of men and 40 percent of women respond the same way to a question, a difference of 10 percentage points. Can it be said with any assurance that the 10 point difference reflects a real difference between men and women on the question? (Unless otherwise noted, the sample contains approximately 750 men and 750 women.)

Because the percentages are near 50, consult Table 2. Since the two samples are about 750 persons each, look for the place in the table where the column and row labeled "750" converge. The number seven appears there. This means the allowance for error should be seven points, and the conclusion that the percentage among men is somewhere between 3 and 17 points higher than the percentage among women would be wrong only about 5 percent of the time. In other words, there is a considerable likelihood that a difference exists in the direction observed and that it amounts to at least 3 percentage points.

If, in another case, male respondents amount to 22 percent, and female to 24 percent, consult Table 1 because these percentages are near 20. The column and row labeled "750 converge on the number 5. Obviously then, the 2 point difference is inconclusive.

TABLE 1. Sampling tolerances

	Recommended Allowance for Sampling Error of the Difference			
	In Percentage Points (at 95 in 100 confidence level)*			
	Percentages near 20 or percentages near 80			
	750	600	400	200
Size of the Sample				
750	5			
600	6	6		
400	7	7	7	
200	8	8	9	10

TABLE 2. Sampling tolerances

	Percentages near 50			
Size of the Sample				
750	6			
600	8	8		
400	8	8	9	
200	10	11	11	13

* The chances are 95 in 100 that the sampling error is not larger than the figures shown.

Source: *Gallup Reports*, November 1986.

Case Study 10-7

CHEATING: GETTING AN HONEST ANSWER

This Psychology Today *article represents an interesting application for the difference of two proportions to get an honest answer about cheating.*

In the Yankelovich, Skelly and White survey commissioned by the IRS, one taxpayer in five admitted to cheating on their taxes recently. The question that elicited this admission came well into a long interview, presumably after the interviewer had established some rapport. Yet, how accurate was this estimate of tax cheating? How many people denied that they cheated, because they feared getting caught, looking bad or for other reasons?

When people are asked threatening, embarrassing or sensitive questions, many answer with a lie. This problem has received widespread attention from researchers who have proposed ways to combat it.

One method, the randomized-response technique, was used in a 1980 survey commissioned by the IRS to find out about tax cheating. Interviewers presented people with a series of cards, each containing two yes or no statements marked "A" and "B". Statement A might be a sensitive question ("On at least one occasion I have added a dependent that I wasn't entitled to"), and B an innocuous statement ("I have been to a movie within the last year"). The respondent flipped a coin and was told to answer statement A if the coin came up heads and statement B if the coin came up tails. An interviewer, who did not know whether the coin came up heads or tails, recorded only their individual responses, yes or no.

Since the interviewer did not know which statement people were responding to, it's likely that people gave honest answers. Although this method made it impossible for researchers to know exactly how many people answered "yes" to the cheating question, it was possible for them to estimate.

How? Since a toss of a coin determined which statement people responded to, the researchers could assume that approximately half of the people answered statement A (the cheating question) and that half answered statement B (the innocuous question). Since they had asked a similar group of people to answer statement B earlier, the researchers could also estimate the number of people in the second group who answered "yes" to that statement. Combining these two facts, the researchers were then able to estimate how many of the "yeses" that the interviewer tallied earlier applied to question B (the innocuous question) and how many applied to question A (the cheating question).

As an example of how this works, assume that 40 people out of 100 answer "yes." Earlier, when the researchers had asked a similar group of people to answer just question B, assume they found that 3 people out of 10 answered "yes."

Since a coin toss determined which questions the 100 people answered, the researchers could assume that approximately 50 people had answered each question. They also could assume from the earlier test that about 15 people (30 percent of 50) had answered the innocuous statement with a yes. Since the researchers had tallied 40 yeses total, and 15 were for the innocuous question, 25 had to apply to the cheating question.

EXERCISES

10.47 Two manufacturing firms that produce equivalent products claim the same rate of customer preference for their products. A random sample of customers shows that 102 of 300 and 152 of 400 prefer products A and B, respectively. Does this evidence indicate a significant difference between the proportions of preference? Use $\alpha = 0.02$.

    a. Solve using the classical approach.

    b. Solve using the prob-value approach.

10.48 In a survey of working parents (both parents working) one of the questions asked was "Have you refused a job, promotion, or transfer because it would mean less time with your family?" Two hundred men and 200 women were asked this question. Twenty-nine percent of the men and 24 percent of the women responded yes. Based on this survey, can we conclude that there is a difference in the proportion of men and women responding yes at the 0.05 level of significance?

10.49 Two randomly selected groups of citizens were exposed to different media campaigns that dealt with the image of a political candidate. One week later the citizen groups were surveyed to see whether they would vote for the candidate. The results were as follows.

	Exposed to Conservative Image	Exposed to Moderate Image
Number in Sample	100	100
Proportion for the Candidate	0.40	0.50

Is there sufficient evidence to show a difference in the effectiveness of the two image campaigns at the 0.05 level of significance?

    a. Solve using the classical approach.

    b. Solve using the prob-value approach.

10.50 In a survey taken to help understand personality, 22 of 71 persons under the age of 18 expressed a fear of meeting people; 23 of 91 persons 18 years old and over expressed the same fear. Should the null hypothesis "there is no difference in the true proportions" be rejected at the $\alpha = 0.10$ level of significance?

    a. Solve using the classical approach.

    b. Solve using the prob-value approach.

10.51 Calculate the 0.95 confidence maximum error of estimation for the difference between the following proportions and compare your answers to the corresponding values in the tables in Case Study 10-6.

    a. p is near 0.20 and both samples are of size 750.

    b. p is near 0.50 and both samples are of size 750. If each of the four samples reported in Case Study 10-5 had a sample size larger than 750.

c. Would the decrease from 1975 to 1987 in the percentage of women who said yes be significant? Explain.

d. Would the increase from 1975 to 1987 in the percentage of men who said yes be significant? Explain.

10.52 In a random sample of 40 brown-haired individuals, 22 indicated that they used hair coloring. In another random sample of 40 blonde individuals, 26 indicated that they used hair coloring. Use a 0.92 confidence interval to estimate the difference in the proportion of these groups that use hair coloring.

10.53 The proportion of defective parts produced by two machines were compared and the following data were collected.

Machine 1: $n = 150$; number of defective parts $= 10$

Machine 2: $n = 150$; number of defective parts $= 4$

Determine a 90 percent confidence interval estimate for $p_1 - p_2$.

10.54 In a survey of 300 people from city A, 128 preferred New Spring soap to all other brands of deodorant soap. In city B, 149 of 400 people preferred New Spring. Find the 98 percent confidence interval estimate for the difference in the two proportions.

In Retrospect

In this chapter we began the comparisons of two populations by first distinguishing between independent and dependent samples, which are statistically very important and useful sampling procedures. We then proceeded to examine the inferences concerning the comparison of means, variances, and proportions for two populations.

The use of hypothesis testing and interval estimates can sometimes be interchanged; that is, the calculation of a confidence interval can often be used in place of a hypothesis test. For example, Illustration 10-17 called for a confidence interval estimate. Now suppose that Mr. Morris asked: "Is there a difference in my voter appeal to men voters as opposed to women voters?" To answer his question, you would not need to calculate a test statistic if you chose to test at $\alpha = 0.01$ using a two-tailed test. "No difference" would mean a difference of zero, which is not included in the interval from 0.014 to 0.128 (the interval determined in Illustration 10-17). Therefore, a null hypothesis of "no difference" would be rejected, thereby substantiating the conclusion that there is a significant difference in voter appeal between the two groups.

We are always making comparisons between two groups: we compare means and we compare proportions. In this chapter we have learned how to statistically compare two populations by making inferences about their means, proportions, or variances. The news article at the beginning of this chapter talks about two subpopulations in the work force: those who consider their work "just a job" and those who consider their work "a career." Notice the various ways in which these two subgroups may be compared: (1) average number of hours at work per day, (2) proportions concerning how they feel about their job, (3) how their boss says "good job," and so on.

Table 10-10

Formulas to use
for inferences involving
two populations

Situation	Test Statistic	Formula to Be Used	
		Hypothesis Test	Interval Estimate
Two independent means			
σ known	z	Formula (10-1)	(10-2)
σ unknown (large samples)	z	Formula (10-3)	(10-4)
σ unknown (equal and small samples)	t	Formulas (10-9), (10-10), (10-11)	(10-13)
σ unknown (unequal and small samples)	t	Formula (10-12)	(10-14)
Two dependent means	t	Formulas (10-15), (10-16), (10-17)	(10-15), (10-16), (10-18)
Two proportions	z	Formula (10-19)	(10-21)
Two variances	F	Formula (10-5)	(10-8)

For convenience, Table 10-10 identifies the formulas to use when making inferences of comparisons between two populations.

In Chapters 8 through 10 we have introduced and completed hypothesis testing and confidence interval estimation for questions that involve one or two means, proportions, and variances. There is much more to be done. However, this much inferential statistics is sufficient to answer many questions. So in succeeding chapters we are ready to look at some other useful tests for other types of questions that may occur.

Chapter Exercises

10.55 Two semesters ago a sample of 150 math students showed a mean of 41.9 on the departmental final exam. Last semester another sample of 150 math students obtained an average of 44.4 on an equivalent exam. Assume that $\sigma = 12.0$ for both sets of exam grades. Should the null hypothesis "there is no real difference between these two groups" be rejected, at $\alpha = 0.05$?

a. Solve using the classical approach.

b. Solve using the prob-value approach.

10.56 Assuming that the standard deviation of the lifetime for two brands of flashlight batteries is 2.8 hours, do the sample statistics in the following table substantiate the claim that brand S has a longer life than brand K, at the 0.01 level?

Brand	n	$\bar{x}$
S	25	47.6
K	30	45.2

a. Solve using the classical approach.

b. Solve using the prob-value approach.

10.57 A test concerning some of the fundamental facts about AIDS was administered to two groups, one consisting of college graduates and the other consisting of high school graduates. A summary of the test results follows.

College graduates: $n = 75$ $\bar{x} = 77.5$ $s = 6.2$

High school graduates: $n = 75$ $\bar{x} = 50.4$ $s = 9.4$

Do these data show that the college graduates, on the average, score significantly higher on the test? Use $\alpha = 0.05$.

10.58 A test that measures math anxiety was given to 50 male and 50 female students. The results were as follows.

Males: $\bar{x} = 70.5$ $s = 13.2$

Females: $\bar{x} = 75.7$ $s = 13.6$

Find the p-value for the alternative hypothesis $\mu_m \neq \mu_F$.

10.59 The same achievement test is given to soldiers selected at random from two units. The scores they obtained are summarized as follows:

Unit 1: $n_1 = 70$ $\bar{x}_1 = 73.2$ $s_1 = 6.1$

Unit 2: $n_2 = 60$ $\bar{x}_2 = 70.5$ $s_2 = 5.5$

Construct a 90 percent confidence interval estimate for the difference in the mean level of the two units.

10.60 A golfer designed an experiment to compare her consistency on two different golf courses. She played ten rounds on each course and computed the following sample variances for her scores on both courses.

variance on course 1 = 10.5

variance on course 2 = 17.3

Is there sufficient evidence in this information to show a difference in the variances. Use $\alpha = 0.05$

10.61 A manufacturer designed an experiment to compare the difference between men and women with respect to the times they required to assemble a product. Fifteen men and 15 women were tested to determine the time they required, on the average, to assemble the product. The times required by the men had a standard deviation of 4.5 minutes, and the times required by the women had a standard deviation of 2.8 minutes. Do these data show that the amount of time needed by men is more variable than the time needed by women? Use $\alpha = 0.05$.

10.62 A soft-drink distributor is considering two new models of dispensing machines. Both the Harvard Company machine and the Fizzit machine can be adjusted to fill the cups to a certain mean amount. However, the variation in the amount dispensed from cup to cup is a primary concern. Ten cups dispensed from the Harvard machine showed a variance of 0.065, whereas 15 cups dispensed from the Fizzit machine showed a variance of 0.033. The factory representative from the

Harvard Company maintains that his machine had no more variability than the Fizzit machine.

    a. At the 0.05 level of significance, does the sample refute the representative's assertion?

    b. Estimate the variance for the amount of fill dispensed by the Harvard machine using a 90 percent confidence interval.

10.63 To estimate the ratio of the variances among the weights of cans of drained peaches of the leading brand to her own brand, a store manager had samples taken, with the following results:

$$\text{Leading brand:} \quad n = 16 \quad\quad s^2 = 1.968$$
$$\text{Her brand:} \quad\quad n = 25 \quad\quad s^2 = 2.834$$

    a. Give a point estimate for the ratio of "leading brand" variance to "her brand" variance.

    b. Construct the 95 percent confidence interval estimate for the same ratio of variances.

    c. Find the 95 percent confidence interval for the ratio of standard deviations.

10.64 A person might use the variance, or standard deviation, of the daily change in the stock market price as a measure of stability. Suppose that you want to compare the stability of a company's stock this year to its stability last year. You are given the following results of taking random samples from the daily gains and losses over the last two years:

$$\text{This year:} \quad n = 25 \quad\quad s = 2.57$$
$$\text{Last year:} \quad n = 25 \quad\quad s = 1.26$$

    a. Construct the 90 percent confidence interval estimate for the ratio of last year's standard deviation to this year's.

    b. Is the daily gain more, less, or about the same, as far as stability is concerned, for this year as compared to last year?

10.65 The score on a certain psychological test is used as an index of status frustration. The scale ranges from 0 (low frustration) to 10 (high frustration). The test was administered to independent random samples of seven radical rightists and eight Peace Corps volunteers.

Radical Rightists	6	10	3	8	8	7	9	
Peace Corps Volunteers	3	5	2	0	3	1	0	4

Using the 10 percent level of significance, test that the mean score for both groups is the same against the alternative that it is not the same.

    a. Solve using the classical approach.

    b. Solve using the prob-value approach.

10.66 Two diets will be compared. Eighty individuals are selected at random from a population of overweight musicians. Forty-five musicians are assigned diet A and the other 35 are placed on diet B. After one week the weight losses (in pounds) are recorded as shown in the following table.

Diet	Sample Size	Sample Mean (pounds)	Sample Variance
A	45	10.3	7
B	35	7.3	3.25

a. Test the hypothesis that $\sigma_A = \sigma_B$ at $\alpha = 0.10$.

b. Do the data substantiate the conclusion that the expected weight loss μ_A under diet A is greater than the expected weight loss μ_B under diet B? Test at the 0.10 level. Draw the appropriate conclusion.

c. Construct a 90 percent confidence interval for $\mu_A - \mu_B$.

10.67 To compare the merits of two short-range rockets, 8 of the first kind and 10 of the second kind are fired at a target. If the first kind have a mean target error of 36 feet and a standard deviation of 15 feet, while the second kind have a mean target error of 52 feet and a standard deviation of 18 feet, does this indicate that the second kind of rocket is less accurate than the first? Use $\alpha = 0.01$.

10.68 Two methods were used to study the latent heat of ice fusion. Both method A (an electrical method) and method B (a method of mixtures) were conducted with the specimens cooled to $-0.72°C$. The data in the following table represent the change in total heat from $-0.72°C$ to water at $0°C$, in calories per gram of mass.

Method A	Method B
79.98	80.02
80.04	79.94
80.02	79.98
80.04	79.97
80.03	79.97
80.03	80.03
80.04	79.95
79.97	79.97
80.05	
80.03	
80.02	
80.00	
80.02	

Is there a significant difference in the mean values at the 0.05 level?

a. Solve using the classical approach.

b. Solve using the prob-value approach.

10.69 Ten soldiers were selected at random from each of two companies to participate in a rifle-shooting competition. Their scores are shown in the following table. Can you conclude that company B has a higher mean score than company A? Use $\alpha = 0.05$.

Company A	72	29	62	60	68	59	61	73	38	48
Company B	75	43	63	63	61	72	73	82	47	43

10.70 The following data were collected concerning waist sizes of men and women. Do these data present sufficient evidence to conclude that men have larger mean waist sizes than women have at the 0.05 level of significance?

men: 33 33 30 34 34 40 35 35 32
            34 32 35 32 32 34 36 30 38

women: 22 29 27 24 28 28
            27 26 27 26 25

10.71 The performance on an achievement test in a beginning computer science course was administered to two groups. One group had a previous computer science course in high school, the other group did not. The test results were as follows.

group 1 (had high school course): 17 18 27 19 24 36 27 26
                                    35 22 18 29 29 26 33

group 2 (no high school course):   19 25 28 27 21
                           24 18 14 28 21 22 20 21 14
                           29 28 25 17 20 28 31 27

Can we conclude that the mean test score for students who had a high school course in computer science will be higher than the mean score for those students who did not have a high school course? Use $\alpha = 0.05$.

10.72 A group of 17 students participated in an evaluation of a special training session that claimed to improve memory. The students were randomly assigned to two groups: group A, the test group, and group B, the control group. All 17 students were tested for the ability to remember certain material. Group A was given the special training; group B was not. After one month both groups were tested again, with the results as shown in the following table. Do these data support the alternative hypothesis that the special training is effective, at the $\alpha = 0.01$ level of significance?

Group A

Time of Test	\multicolumn{9}{c}{Student}

Time of Test	1	2	3	4	5	6	7	8	9
Before	23	22	20	21	23	18	17	20	23
After	28	29	26	23	31	25	22	26	26

Group B

Time of	Student							
Test	10	11	12	13	14	15	16	17
Before	22	20	23	17	21	19	20	20
After	23	25	26	18	21	17	18	20

10.73 A study was conducted to investigate the effectiveness of two different teaching methods used in physical education: the task method and the command method. Both were used to teach tennis to college students. A group of size 30 was used for each method. The students in both groups were instructed for the same length of time and then tested on the forehand and backhand tennis strokes.

Results	Command Group	Task Group	Calculated t
Forehand	$\bar{x} = 55.6$	$\bar{x} = 59.8$	$t^* = 2.39$
Backhand	$\bar{x} = 35.87$	$\bar{x} = 41.83$	$t^* = 3.59$

a. Is one method more effective than the other in teaching tennis to college students?

b. If so, for which stroke and with what level of significance?

10.74 Twelve automobiles were selected at random to test two new mixtures of unleaded gasolines. Each car was given a measured allotment of the first mixture, x, and driven; then the distance traveled was recorded. The second mixture, y, was immediately tested in the same manner. The order in which the x and y mixtures were tested was also randomly assigned. The results are given in the following table.

Mixture	Car											
	1	2	3	4	5	6	7	8	9	10	11	12
x	7.9	5.6	9.2	6.7	8.1	7.3	8.1	5.4	6.9	6.1	7.1	8.1
y	7.7	6.1	8.9	7.1	7.9	6.7	8.2	5.0	6.2	5.7	6.2	7.5

Can you conclude that there is no real difference in mileage obtained by these two gasoline mixtures, at the 0.10 level of significance?

a. Solve using the classical approach.

b. Solve using the prob-value approach.

10.75 Ten chemical laboratory technicians were asked to determine the weight of a compound on two different analytic scales. The results were as follows:

Technician	1	2	3	4	5	6	7	8	9	10
Scale 1 weight	4.76	4.75	4.70	4.72	4.69	4.75	4.72	4.67	4.73	4.70
Scale 2 weight	4.74	4.70	4.72	4.70	4.69	4.72	4.72	4.73	4.70	4.72

At the 0.05 level of significance, can we conclude that the results are different for the two scales?

10.76 To test the effect of a physical fitness course on one's physical ability, the number of sit-ups that a person could do in 1 minute, both before and after the course, was recorded. Ten randomly selected participants scored as shown in the following table. Can you conclude that a significant amount of improvement took place? Use $\alpha = 0.01$.

Before	29	22	25	29	26	24	31	46	34	28
After	30	26	25	35	33	36	32	54	50	43

    a. Solve using the classical approach.

    b. Solve using the prob-value approach.

10.77 Ten new recruits participated in a rifle-shooting competition at the end of their first day at training camp. The same ten competed again at the end of a full week of training and practice. Their resulting scores are shown in the following table.

Time of Competition	Recruit									
	1	2	3	4	5	6	7	8	9	10
First day	72	29	62	60	68	59	61	73	38	48
One week later	75	43	63	63	61	72	73	82	47	43

    a. Does this set of ten pairs of data show that there was a significant amount of improvement in the recruit's shooting abilities during the week? Use $\alpha = 0.05$.

    b. Notice that the data in Exercises 10.69 and 10.77 are the same set of number values. Compare the types of tests and the results obtained in the solution of Exercise 10.69 and (a) of this exercise.

10.78 Using a 95 percent confidence interval, estimate the difference in IQ between the oldest and the youngest members (brothers and sisters) of a family based on the following random sample of IQs.

Oldest	145	133	116	128	85	100	105	150	97	110	120	130
Youngest	131	119	103	93	108	100	111	130	135	113	108	125

10.79 The diastolic blood pressures for 15 patients were determined using two techniques: the standard method used by medical personnel and a method using an electronic device with a digital readout. The results were as follows.

Patient	1	2	3	4	5	6	7	8	9	10	11	12	13	14	15
Standard method	72	80	88	80	80	75	92	77	80	65	69	96	77	75	60
Digital readout method	70	76	87	77	81	75	90	75	82	64	72	95	80	70	61

Determine the 90 percent confidence interval estimate for the mean difference in the two readings, where d = standard method − digital readout.

10.80 If 250 castings produced on mold A contain 15 defectives, and 300 castings on mold B contain 30 defectives, should the null hypothesis "there is no difference between the true proportions of defectives" be rejected, at the 0.10 level of significance?

    a. Solve using the classical approach.

    b. Solve using the prob-value approach.

10.81 In determining the "goodness" of a test question, a teacher will often compare the percentage of better students who answer it correctly to the percentage of the poorer students who answer it correctly. One expects that the better students will answer the question correctly more frequently than the poorer students. On the last test, 35 of the students with the top 60 grades and 27 with the bottom 60 answered a certain question correctly. Did the students with the top grades do significantly better on this question? Use $\alpha = 0.05$.

    a. Solve using the classical approach.

    b. Solve using the prob-value approach.

10.82 A survey was conducted to determine the proportion of Democrats as well as Republicans who support a "get tough" policy in South America. The results of the survey were as follows.

Democrats:	$n = 250$	number in support = 120
Republicans:	$n = 200$	number in support = 105

Do these data show that there is a significant difference in the proportion who support this policy? Use $\alpha = 0.05$.

10.83 Of a random sample of 100 stocks on the New York Exchange, 32 made a gain today. A random sample of 100 stocks on the American Stock Exchange showed 27 stocks making a gain.

    a. Construct the 99 percent confidence interval, estimating the difference in the proportion of stocks making a gain.

    b. Does the answer to (a) suggest that there is a significant difference between the proportions of stocks making gains on the two different stock exchanges?

10.84 A consumer group compared the reliability of two comparable microcomputers from two different manufacturers. The proportion requiring service within the first year after purchase was determined for samples from each of two manufacturers.

Manufacturer	Sample Size	Proportion Needing Service
1	75	0.15
2	75	0.09

Find a 0.95 confidence interval estimate for $p_1 - p_2$.

10.85 Show that the standard error of $p'_1 - p'_2$, which is $\sqrt{(p_1 q_1/n_1) + (p_2 q_2/n_2)}$, reduces to $\sqrt{pq[(1/n_1) + (1/n_2)]}$ when $p_1 = p_2 = p$.

Vocabulary List
Be able to define each term. In addition, describe, in your own words, and give an example of each term. Your examples should not be ones given in class or in the textbook.

The bracketed numbers indicate the chapters in which the term previously appeared, but you should define the terms again to show increased understanding of their meaning.

confidence interval estimate [8, 9]	large sample
dependent means	mean difference
dependent samples	paired difference
F distribution	pooled estimate for the
hypothesis test [8, 9]	standard deviation
independent means	ratio of standard deviation
independent samples	ratio of variances
	source (of data)

Quiz
Answer "True" if the statement is always true. If the statement is not always true, replace the boldface words with words that make the statement always true.

10.1 When the means of two unrelated samples are used to compare two populations, we are dealing with **two dependent means**.

10.2 The use of **paired data (dependent means)** often allows for the control of unmeasurable or confounding variables because each pair is subjected to these confounding effects equally.

10.3 The **chi-square distribution** is used for making inferences about the ratio of the variances of two populations.

10.4 The **F distribution** is used when two dependent means are to be compared.

10.5 In comparing two independent means when the σ's are unknown, we need to perform an **F test on their variances** to determine the proper formula to use.

10.6 The **standard normal score** is used for all inferences concerning population proportions.

10.7 The *F* distribution is a **symmetric** distribution.

10.8 The number of degrees of freedom for the critical value of *t* is equal to **the smaller of $n_1 - 1$ or $n_2 - 1$** when making inferences about the difference between two independent means for the case where the variances are unknown and assumed to be unequal, and the sample sizes are small.

10.9 A pooled estimate for the standard deviation of two combined populations is calculated in all problems dealing with the difference between two **independent means**.

10.10 A **pooled estimate** for any statistic in a problem dealing with two populations is a value arrived at by combining the two separate sample statistics so as to achieve the best possible point estimate.

Working with Your Own Data

As consumers, we all purchase many bottled, boxed, canned, or packaged products. Seldom, if ever, do we question whether or not the content amount stated on the container is accurate.

Here are a few typical content listings found on various containers:

28 FL OZ (1 PT 12 OZ)

750 ml

5 FL. OZ. (148 ml)

32 FL. OZS. (1 QT.) 0.951

NET WT 10 OZ   283 GRAMS

NET WT. $3\frac{3}{4}$ OZ. 106 g—48 tea bags

140 1-PLY NAPKINS

77 SQ. FT.—92 TWO-PLY SHEETS—11 × 11 IN.

Have you ever thought, "I wonder if I am getting the amount that I am paying for?" And if this thought did cross your mind, did you attempt to check the validity of the content claim? The following article appeared in the *Times Union* of Rochester, New York on February 16, 1972. (This kind of situation occurs frequently.)

Milk Firm Accused of Short Measure

The processing manager of Dairylea Cooperative, Inc., has been named in a warrant charging that the cooperative is distributing cartons of milk in the Rochester area containing less than the quantity represented.

... an investigator found shortages in four quarts of Dairylea milk purchased Friday.

Asst. Dist. Atty. Howard R. Relin, who issued the warrant, said the shortages ranged from $1\frac{1}{8}$ to $1\frac{1}{4}$ ounces per quart. A quart of milk contains 32 fluid ounces.

... the state Agriculture and Markets Law ... provides that a seller of a commodity shall not sell or deliver less of the commodity than the quantity represented to be sold.

... the purpose of the law under which ... the dairy is charged is to ensure honest, accurate, and fair dealing with the public. There is no requirement that intent to violate the law be proved, he said.

From *The Times-Union*, Rochester, N.Y., Feb. 16, 1972.

This situation poses a very interesting legal problem: There is no need to show intent to "short the customer." If caught, the violators are fined automatically and the fines are often quite severe.

A | A HIGH-SPEED FILLING OPERATION

A high-speed piston-type machine used to fill cans with hot tomato juice was sold to a canning company. The guarantee stated that the machine would fill 48-ounce cans with a mean amount of 49.5 ounces, a standard deviation of 0.072 ounce, and a maximum spread of 0.282 ounce while operating at a rate of filling 150 to 170 cans per minute. On August 12, 1986, a sample of 42 cans was gathered and the following weights were recorded. The weights, measured to the nearest $\frac{1}{8}$ oz., are recorded as variations from 49.5 oz.

$-\frac{1}{8}$	0	$-\frac{1}{8}$	0	0	0	$-\frac{1}{8}$	0	0	0
0	0	$-\frac{1}{8}$	0	$\frac{1}{8}$	0	$\frac{1}{8}$	0	$\frac{1}{8}$	0
0	0	$-\frac{1}{8}$	0	0	0	0	$-\frac{1}{8}$	0	0
0	0	0	0	0	0	0	0	0	0
0	0								

1. Calculate the mean $\bar{x}$, the standard deviation s, and the range of the sample data.

2. Construct a histogram picturing the sample data.

3. Does the amount of fill differ from the prescribed 49.5 ounces at the $\alpha = 0.05$ level? Test the hypothesis that $\mu = 49.5$ against an appropriate alternative.

4. Does the amount of variation, as measured by the standard deviation, satisfy the guarantee at the $\alpha = 0.05$ level?

5. Assuming that the filling machine continues to fill cans with an amount of tomato juice that is distributed normally and the mean and standard deviation are equal to the values found in question 1, what is the probability that a randomly selected can will contain less than the 48 ounces claimed on the label?

6. If the amount of fill per can is normally distributed and the standard deviation can be maintained, find the setting for the mean value that would allow only 1 can in every 10,000 to contain less than 48 ounces.

B | YOUR OWN INVESTIGATION

Select a packaged product that has a quantity of fill per package that you can and would like to investigate.

1. Describe your selected product, including the quantity per package, and describe how you plan to obtain your data.

2. Collect your sample of data. (Consult your instructor for advice on size of sample.)

3. Calculate the mean $\bar{x}$ and standard deviation s for your sample data.

4. Construct a histogram or stem-and-leaf diagram picturing the sample data.

5. Does the mean amount of fill meet the amount prescribed on the label? Test using $\alpha = 0.05$.

6. Assume that the item you selected is filled continually. The amount of fill is normally distributed and the mean and standard deviation are equal to the values found in question 3. What is the probability that one randomly selected package contains less than the prescribed amount?

Part 4

More Inferential Statistics

In Part 3 we studied the two inferential statistical techniques: hypothesis testing and confidence interval estimating, for one and two populations, and the three population parameters: mean, μ, standard deviation, σ, and proportion, p. In Part 4 we will learn how to use the hypothesis test and the interval estimation techniques in other statistical situations. Some of these situations will be extensions of previous methods. (For example, analysis of variance will be used to deal with more than two populations when the question involves the mean, and the binomial experiment will be expanded to a multinomial experiment. Other situations will be alternatives to methods previously studied (such as the nonparametric techniques), and others will deal with new inferences (such as the nonparametric methods and the regression and correlation inferential techniques).

Chapter 11

Additional Applications of Chi-Square

TRIAL BY NUMBER

Karl Pearson's chi-square test measured the fit between theory and reality, ushering in a new sort of decision making.

Our world is inundated with statistics. Every medical fear or triumph is charted by a complex analysis of chances. Think of cancer, heart disease, AIDS: The less we know, the more we hear of probabilities. This daily barrage is not a matter of mere counting but of inference and decision in the face of uncertainty. No committee changes our schools or our prisons without studies on the effects of busing or early parole. Money markets, drunken driving, family life, high energy physics, and deviant human cells are all subject to tests of significance and data analysis.

This all began in 1900 when Karl Pearson published his chi-square test of goodness of fit, a formula for measuring how well a theoretical hypothesis fits some observations. The basic idea is simple enough. Suppose that you think a die will fall equally often on each of its six faces. You roll it 600 times. It seems to come up six all too frequently. Could this simply be chance? How well does the hypothesis— that the die is fair—fit the data? The result that would best fit your theory would be that each face came up just 100 times in 600. In practice the ratios are almost always different, even with a fair die, because even for many throws there will always be the factor of chance. How different should they be to make you suspect a poor fit between your theory and your 600 observations? Pearson's chi-square test gives one measure of how well theory and data correspond.

The chi-square test can be used for hypotheses and data where observations naturally fall into discrete categories that statisticians call cells. If, for example, you are testing to find whether a certain treatment for cholera is worthless, then the patients divide among four cells: treated and recovered, treated and died, untreated and recovered, and untreated and died. If the treatment is worthless, you expect no difference in recovery rate between treated and untreated patients. But chance and uncontrollable variables dictate that there will almost always be some difference. Pearson's test takes this into account, telling you how well your hypothesis—that the treatment is worthless—fits your observations, ...

The chi-square test was a tiny event in itself, but it was the signal for a sweeping transformation in the ways we interpret our numerical world. Now there could be a standard way in which ideas were presented to policy makers and to the general public....

Statistical talk has now invaded many aspects of daily life, creating sometimes spurious impressions of objectivity. Policy makers demand numbers to measure every risk and hope. Thus the 1975 *Reactor Safety Study* of the Nuclear Regulatory Commission attached probabilities to various kinds of danger. Some critics suggest that this is just a way of escaping the responsibility of admitting ignorance. Other critics point to the increasing use of complex models generated to "fit" reams of data in some statistical sense—models that can become a fantasyland without any connection to the real world.

For better or worse, statistical inference has provided an entirely new style of reasoning. The quiet statisticians have changed our world—not by discovering new facts or technical developments but by changing the ways we reason, experiment, and form our opinions about it.

Source: Ian Hacking. Reprinted by permission from the November issue of SCIENCE '84.

Chapter
Objectives

Previously we discussed and used the chi-square distribution both to test and to estimate the value of the variance (or standard deviation) of a single population. The chi-square distribution may also be used for tests in other types of situations. In this chapter we are going to investigate some of these uses. Specifically, we will look at two tests: a multinomial experiment and the contingency table. These two types of tests will be used to compare experimental results with expected results in order to determine (1) preferences, (2) independence, and (3) homogeneity.

The information that we will be using in these techniques will be **enumerative**, that is, the data will be placed in categories and counted. These counts become the enumerative information used.

11.1 CHI-SQUARE STATISTIC

There are many problems for which the data are categorized and the results shown by way of counts. For example, a set of final exam scores can be displayed as a frequency distribution. These frequency numbers are counts, the number of data that fall in each cell. A survey asks voters whether they are registered as Republican, Democrat, or other, and whether or not they support a particular candidate. The results are usually displayed on a chart that shows the number of voters for each possible category. There were numerous illustrations throughout the previous ten chapters.

cell

Suppose that we have a number of **cells** into which n observations have been sorted. (The term *cell* is synonymous with the term *class*; the terms *class* and *frequency* were defined and first used in earlier chapters. Before continuing, a brief review of Sections 2.2, 2.3, and 2.4 might be beneficial.) The **observed frequencies** in each cell are denoted by $O_1, O_2, O_3, \ldots, O_k$ (*see* Table 11-1). Note that the sum of all observed frequencies is equal to $O_1 + O_2 + \cdots + O_k = n$, where n is the sample size. What we would like to do is to compare the observed frequencies with some

expected frequency

expected or theoretical frequencies, denoted by $E_1, E_2, E_3, \ldots, E_k$, for each of these cells. Again, the sum of these expected frequencies must be exactly

$$E_1 + E_2 + \cdots + E_k = n$$

Table 11-1
Observed frequencies

	k Categories					Total
	1st	2d	3d	$\cdots$	kth	
Observed Frequency	O_1	O_2	O_3	$\cdots$	O_k	n

We will then decide whether the observed frequencies seem to agree or seem to disagree with the expected frequencies. This will be accomplished by a hypothesis test using **chi-square**, χ^2.

The calculated value of the test statistic will be

$$\chi^2 = \sum_{\text{all cells}} \frac{(O - E)^2}{E} \qquad (11\text{-}1)$$

This calculated value for chi-square will be the sum of several nonnegative numbers, one from each cell (or category). The numerator of each term in the formula for χ^2 is the square of the difference between the values of the observed and the expected frequencies. The closer together these values are, the smaller the value of $(O - E)^2$; the farther apart, the larger the value of $(O - E)^2$. The denominator for each cell puts the size of the numerator into perspective. That is, a difference $(0 - E)$ of 10 resulting from frequencies of 110 (O) and 100 (E) seems quite different from a difference of 10 resulting from 15 (O) and 5 (E).

These ideas suggest that small values of chi-square indicate agreement, whereas larger values indicate disagreement, between the two sets of frequencies. Therefore, it is customary for these tests to be one-tailed, with the critical region on the right.

In repeated sampling, the calculated value of χ^2 [formula (11-1)] will have a sampling distribution that can be approximated by the chi-square probability distribution when n is large. This approximation is generally considered adequate when all the expected frequencies are equal to or greater than 5. Recall that there is a separate chi-square distribution for each degree of freedom, df. The appropriate value of df will be described with each specific test. The critical value of chi-square, $\chi^2(\text{df}, \alpha)$, can be found in Table 7 of Appendix E.

In this chapter we will permit a certain amount of "liberalization" with respect to the null hypothesis and its testing. In previous chapters the null hypothesis has always been a statement about a population parameter (μ, σ, or p). However, there are other types of hypotheses that can be tested, such as "this die is fair," or "height and weight of individuals are independent." Notice that these hypotheses are not claims about a parameter, although sometimes they could be stated with parameter values specified.

Suppose that I claim that "this die is fair" $p = P(\text{any one number}) = \frac{1}{6}$, and you want to test it. What would you do? Was your answer something like "Roll this die many times, recording the results"? Suppose that you decide to roll the die 60 times. If the die is fair, what do you expect will happen? Each number $(1, 2, \ldots, 6)$ should appear approximately $\frac{1}{6}$ of the time (that is, 10 times). If it happens that approximately 10 of each number occur, you will certainly accept the claim of fairness ($p = \frac{1}{6}$). If it happens that the die seems to favor some numbers, you will reject the claim. (The test statistic χ^2 will have a large value in this case, as we will soon see.)

11.2 INFERENCES CONCERNING MULTINOMIAL EXPERIMENTS

multinomial experiment

The preceding die problem is a good illustration of a **multinomial experiment**. Let's consider this problem again. Suppose that we want to test this die (at $\alpha = 0.05$) and decide whether to fail to reject or reject the claim "this die is fair." (The probability of each number is $\frac{1}{6}$.) The die is rolled from a cup onto a smooth flat surface 60 times, with the frequencies shown in the following table.

Number	1	2	3	4	5	6
Occurrences	7	12	10	12	8	11

The null hypothesis that the die is fair is assumed to be true. This allows us to calculate the expected frequencies. If the die was fair, we certainly would expect 10 occurrences for each number.

Now let's calculate an observed value of χ^2. These calculations are shown in Table 11-2. The calculated value is $\chi^{2*} = 2.2$.

Table 11-2
Computations for calculating χ^2

Number	Observed (O)	Expected (E)	O − E	(O − E)²	(O − E)²/E
1	7	10	−3	9	0.9
2	12	10	2	4	0.4
3	10	10	0	1	0.0
4	12	10	2	4	0.4
5	8	10	−2	4	0.4
6	11	10	1	1	0.1
Total	60	60	0		2.2

Note $\sum(O - E)$ must equal zero since $\sum O = \sum E = n$. You can use this fact as a check, as shown in Table 11-2.

Before continuing, let's set up the hypothesis-testing format.

Step 1 H_0: The die is fair (each $p = \frac{1}{6}$).

H_a: The die is not fair (at least one p is different).

Step 2 $\alpha = 0.05$. See the accompanying figure for the test criteria. In the multinomial test, df $= k - 1$, where k is the number of cells. The critical value is $\chi^2(5, 0.05) = 11.1$.

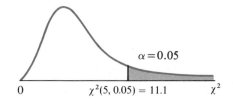

Step 3   The test statistic was calculated in Table 11-2 and was found to be $\chi^{2*} = 2.2$

Step 4   *Decision*: Fail to reject H_0 (χ^{2*} is not in the critical region).

Conclusion: At the 0.05 level of significance, the observed frequencies are not significantly different from those expected of a fair die.

Before we look at other illustrations, we must define the term *multinomial experiment* and we must state the guidelines for completing the chi-square test for it.

Multinomial Experiment

An experiment with the following characteristics:

1. It consists of n identical independent trials.

2. The outcome of each trial fits into exactly one of k possible cells.

3. There is a probability associated with each particular cell, and these individual probabilities remain constant during the experiment. (It must be true that $p_1 + p_2 + \cdots + p_k = 1$.)

4. The experiment will result in a set of observed frequencies, $O_1, O_2, \ldots, O_k$, where each O_i is the number of times a trial outcome falls into that particular cell. (It must be the case that

$$O_1 + O_2 + \cdots + O_k = n.)$$

The die illustration meets the definition of a multinomial experiment because it has all four of the characteristics described in the definition.

1. The die was rolled n (60) times in an identical fashion, and these trials were independent of each other. (The result of each trial was unaffected by the results of other trials.)

2. Each time the die was rolled, one of six numbers resulted, and each number was associated with a cell.

3. The probability associated with each cell was $\frac{1}{6}$, and this was constant from trial to trial. (Six values of $\frac{1}{6}$ sum to 1.0.)

4. When the experiment was complete, we had a list of frequencies (7, 12, 10, 12, 8, and 11) that summed to 60, indicating that each of the outcomes was taken into account.

The **testing procedure** for multinomial experiments is very much as it was in previous chapters. The biggest change comes with the statement of the null hypothesis. It may be a verbal statement, such as in the die illustration: "This die is fair." Often the alternative to the null hypothesis is not stated. However, in this book the alternative hypothesis will be shown, since it seems to aid in organizing and understanding the problem. However, it will not be used to determine the location of the critical region, as was the case in previous chapters. **For multinomial experiments we will always use a one-tailed critical region, and it will be the right-hand tail of the χ^2 distribution, because large deviations from the expected values lead to a large calculated χ^2 value.**

The critical value will be determined by the level of significance assigned (α) and the number of degrees of freedom. The number of degrees of freedom (df) will be 1 less than the number of cells (k) into which the data are divided:

$$df = k - 1 \qquad (11\text{-}2)$$

Each expected frequency, E_i, will be determined by multiplying the corresponding probability (p_i) for that cell by the total number of trials n. That is,

$$E_i = n \cdot p_i \qquad (11\text{-}3)$$

One guideline should be met to ensure a good approximation to the chi-square distribution: Each expected frequency should be at least 5 (that is, each $E_i \geq 5$). Sometimes it is possible to combine "smaller" cells to meet this guideline. If this guideline cannot be met, corrective measures to ensure a good approximation should be used. These corrective measures are not covered in this book but are discussed in many other sources.

Illustration 11-1   College students have regularly insisted on freedom of choice when registering for courses. This semester there were seven sections of a particular mathematics course. They were scheduled to meet at various times with a variety of instructors. Table 11-3 shows the number of students who selected each of the seven sections. Do the data indicate that the students had a preference for certain sections or do they indicate that each section was equally likely to be chosen?

Table 11-3
Data for Illustration 11-1

	Section							
	1	2	3	4	5	6	7	Total
Number of Students	18	12	25	23	8	19	14	119

Solution   If no preference was shown in the selection of sections, we would expect the 119 students to be equally distributed among the seven classes. Thus if no preference was the case, then we would expect 17 students to register for each section. The test is completed as shown in Steps 1–4, at the 5 percent level of significance.

Step 1 H_0: There was no preference shown (equally distributed).

H_a: There was a preference shown (not equally distributed).

Step 2 $\alpha = 0.05$. See the accompanying figure for the test criteria.

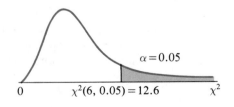

$\alpha = 0.05$

$0 \qquad \chi^2(6, 0.05) = 12.6 \qquad \chi^2$

Step 3 We use formula (11-1) to calculate the test statistic.

$$\chi^2 = \frac{(18-17)^2}{17} + \frac{(12-17)^2}{17} + \frac{(25-17)^2}{17}$$

$$+ \frac{(23-17)^2}{17} + \frac{(8-17)^2}{17} + \frac{(19-17)^2}{17} + \frac{(14-17)^2}{17}$$

$$= \frac{(1)^2 + (-5)^2 + (8)^2 + (6)^2 + (-9)^2 + (2)^2 + (-3)^2}{17}$$

$$= \frac{1 + 25 + 64 + 36 + 81 + 4 + 9}{17} = \frac{220}{17}$$

$$= 12.9411$$

$$\chi^{2*} = \mathbf{12.94}$$

Step 4 *Decision*: Reject H_0 (χ^{2*} falls in the critical region).

Conclusion: At the 0.05 level of significance, there does seem to be a preference shown.

We cannot determine, from the given information, what the preference is. It could be teacher preference, time preference, or a case of schedule conflict. Conclusions must be worded carefully to avoid suggesting conclusions that we cannot support. ■

Not all multinomial experiments result in equal expected frequencies, as we will see in Illustration 11-2.

Illustration 11-2 The Mendelian theory of inheritance claims that the frequencies of round and yellow, wrinkled and yellow, round and green, and wrinkled and green will occur in the ratio 9:3:3:1 when two specific varieties of peas are crossed. In testing this theory, Mendel obtained frequencies of 315, 101, 108, and 32, respectively. Do these sample data provide sufficient evidence to reject this theory, at the 0.05 level of significance?

Solution

Step 1 H_0: 9:3:3:1 is the ratio of inheritance.

H_a: 9:3:3:1 is not the ratio of inheritance.

Step 2 $\alpha = 0.05$, $k = 4$, and df = 3. See the accompanying figure.

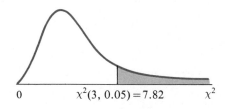

$0 \qquad \chi^2(3, 0.05) = 7.82 \qquad \chi^2$

Step 3   The ratio 9:3:3:1 indicates probabilities of $\frac{9}{16}$, $\frac{3}{16}$, $\frac{3}{16}$, and $\frac{1}{16}$. Therefore, the expected frequencies are $9n/16$, $3n/16$, $3n/16$, and $n/16$, where

$$n = \sum O_i = 315 + 101 + 108 + 32 = 556$$

The computations necessary for calculating χ^2 are given in Table 11-4. Thus $\chi^{2*} = \mathbf{0.47}$.

Table 11-4

Computations needed to calculate χ^2

O	E	$O - E$	$(O - E)^2/E$
315	312.75	2.25	0.0162
101	104.25	-3.25	0.1013
108	104.25	3.75	0.1349
32	34.75	-2.75	0.2176
556	556.00	0	0.4700

Step 4   *Decision*: Fail to reject H_0

Conclusion: At the 0.05 level of significance, there is not sufficient evidence to reject Mendel's theory. ■

Case Study 11-1

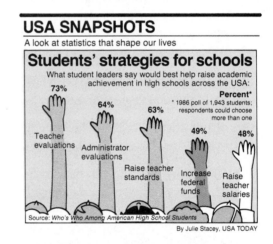

STUDENTS' STRATEGIES FOR SCHOOLS

The *USA SNAPSHOTS*' "Students' strategies for schools" reports the results of asking 1,943 high school leaders what they think would best raise academic achievement in high schools across the United States. The results reported are the percentages of students that chose each of the various answers. This might appear to be a multinomial experiment, but it is not. Explain why. (*See* Exercise 11.6.)

Source: "Students' Strategies for Schools," September 15, 1986. Copyright 1986, USA TODAY. Reprinted with permission.

EXERCISES

11.1 A manufacturer of floor polish conducted a consumer-preference experiment to determine which of five different floor polishes was the most appealing in appearance. A sample of 100 consumers viewed five patches of flooring that had each received one of the five polishes. Each consumer indicated the patch he or she preferred. The lighting and background were approximately the same for all patches. The results were as follows.

Polish	A	B	C	D	E	Total
Frequency	27	17	15	22	19	100

  a. State the hypothesis for "no preference" in statistical terminology.
  b. What test statistic will be used in testing this null hypothesis?
  c. Complete the hypothesis test using $\alpha = 0.10$.
    (1) Solve using the classical approach.
    (2) Solve using the prob-value approach.

11.2 A certain type of flower seed will produce magenta, chartreuse, and ochre flowers in the ratio 6:3:1 (one flower per seed). A total of 100 seeds are planted and all germinate, yielding the following results.

Magenta	Chartreuse	Ochre
52	36	12

  a. If the null hypothesis (6:3:1) is true, what is the expected number of magenta flowers?
  b. How many degrees of freedom are associated with chi-square?
  c. Complete the hypothesis test using $\alpha = 0.10$.
    (1) Solve using the classical approach.
    (2) Solve using the prob-value approach.

11.3 A standard diet is recommended to reduce cholesterol in individuals who have elevated cholesterol levels. The expected response to the standard diet after six months may be categorized as follows:

Category	Percentage
1. Marked decrease in cholesterol	60
2. Moderate decrease in cholesterol	20
3. Slight decrease in cholesterol	10
4. No change in cholesterol	10

An experimental diet is to be compared to the standard diet. Two hundred individuals were placed on the experimental diet and the following results were obtained.

Category	1	2	3	4
Observed Cell Counts	110	30	10	50

Do the results show sufficient reason to reject the hypothesis that "the responses to the two diets are the same"? Use $\alpha = 0.05$.

    a. Solve using the classical approach.

    b. Solve using the prob-value approach.

11.4 Previous sales records show that Inboard Custom Boats sold 20 percent of its boats in its Northeast sales district, and 28 percent, 8 percent, 12 percent, and 32 percent, respectively, in its Southeast, North Central, South Central, and West Coast sales districts. Of the first 500 boats sold this year, 120, 128, 43, 66, and 143 were sold in each of the five districts, respectively. Does the sales distribution so far this year seem to be the same as in previous years? Use $\alpha = 0.05$.

    a. Solve using the classical approach.

    b. Solve using the prob-value approach.

11.5 A program for generating random numbers on a computer is to be tested. The program is instructed to generate 100 single-digit integers between 0 and 9. The frequencies of the observed integers were as follows.

Integer	0	1	2	3	4	5	6	7	8	9
Frequency	11	8	7	7	10	10	8	11	14	14

At the 0.05 level of significance, is there sufficient reason to believe that the integers are not being generated uniformly?

11.6 The *USA SNAPSHOT* in Case Study 11-1 shows the responses of 1,943 student leaders when asked what they think would best raise academic achievement in high schools across the United States. Explain why this survey does not fit the definition of a multinomial experiment.

11.7 A sampling from the import distribution center for Volkswagen shows that the proportions of color choices for the Jetta are as shown in the following table. Take a sample of $n = 100$ or 200 Jettas and compare the color preferences of local Jetta owners to the data shown in the table.

Color	Yellow	Orange	Light Blue	Beige	Green	Red	Dark Blue
Percentage	25	17	15	12	11	10	10

    a. What null hypothesis will you test?

    b. What is the alternative hypothesis?

    c. Complete the test at $\alpha = 0.10$.

11.3 INFERENCES CONCERNING CONTINGENCY TABLES

contingency table   A **contingency table** is an arrangement of data into a two-way classification. The data are sorted into cells and the number of data in each cell is reported. The contingency table involves two factors (or variables), and the usual question concerning such tables is whether the data indicate that the two variables are independent or dependent.

TEST OF INDEPENDENCE

To illustrate the use and analysis of a contingency table, let's consider the sex classification of liberal arts college students and their favorite academic area.

Illustration 11-3   Each person in a group of 300 students was identified as male or female and then asked whether he or she preferred taking liberal arts courses in the area of math-science, social science, or humanities. Table 11-5 is a contingency table that shows the frequencies found for these categories. Does this sample present sufficient evidence to reject the null hypothesis "preference for math-science, social science, or humanities is independent of the sex of a college student," at the 0.05 level of significance?

Table 11-5
Contingency table showing sample results for Illustration 11-3

Sex of Student	Favorite Subject Area			
	Math-Science (M-S)	Social Science (SS)	Humanities (H)	Total
Male	37	41	44	122
Female	35	72	71	178
Total	72	113	115	300

Solution

Step 1 H_0: Preference for math-science, social science, or humanities is independent of the sex of a college student.

H_a: Subject preference is not independent of the sex of the student.

Step 2   To determine the critical value of chi-square, we need to know df, the number of degrees of freedom, involved. In the case of contingency tables, the number of degrees of freedom is exactly the same number as the number of cells in the table that may be filled in freely when you are given the totals. The totals in this illustration are shown in the following table.

			122
			178
72	113	115	300

Given these totals, you can fill in only two cells before the others are all determined. (The totals must, of course, be the same.) For example, once we pick two arbitrary values (say 50 and 60) for the first two cells of the first row (see the following table), the other four cell values are fixed.

50	60	C	122
D	E	F	178
72	113	115	300

They have to be $C = 12$, $D = 22$, $E = 53$, and $F = 103$. Otherwise the totals will not be correct. Therefore, for this problem there are two free choices. Each free choice corresponds to one degree of freedom. Hence the number of degrees of freedom for our example is 2 (df = 2). Thus, if $\alpha = 0.05$ is used, the critical value is $\chi^2(2, 0.05) = 6.00$. See the accompanying figure. (Following the conclusion of this illustration, a formula for finding df will be discussed.)

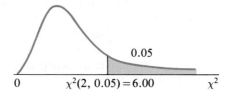

$$0.05$$

$$0 \qquad \chi^2(2, 0.05) = 6.00 \qquad \chi^2$$

Step 3   Before the calculated value of chi-square can be found, we need to determine the expected values E for each cell. To do this we must recall the null hypothesis, which asserts that these factors are independent. Therefore, we would expect the values to be distributed in proportion to the **marginal totals**. There are 122 males; we would expect them to be distributed among M-S, SS, and H proportionally to the 72, 113, and 115 totals. Thus the expected cell counts for males are

margin: marginal totals

$$\frac{72}{300} \cdot 122 \qquad \frac{113}{300} \cdot 122 \qquad \frac{115}{300} \cdot 122$$

Similarly, we would expect

$$\frac{72}{300} \cdot 178 \qquad \frac{113}{300} \cdot 178 \qquad \frac{115}{300} \cdot 178$$

for the females.

Thus the expected values are as shown in Table 11-6. (Always check the new totals against the old totals.)

Table 11-6
Expected values

	M-S	SS	H	Total
	29.28	45.95	46.77	122.00
	42.72	67.05	68.23	178.00
Total	72.00	113.00	115.00	300.00

Note   We can think of the computation of the expected values in a second way. Recall that we assume the null hypothesis to be true until there is evidence to reject it. Having made this assumption in our example, in effect we are saying that the event that a student picked at random is male and the event that a student picked at random prefers math-science courses are independent. Our point estimate for the probability that a student is male is $122/300$, and the point estimate for the probability that the student prefers the math-science courses is $72/300$. Therefore, the probability that both events occur is the product of the probabilities. [Refer to formula (4-8a).] Thus $(122/300) \cdot (72/300)$ is the probability of a selected student being male and preferring math-science. Therefore, the number of students out of 300 that are expected to be male and prefer math-science is found by multiplying the probability (or proportion) by the total number of students (300). Thus the expected number of males who prefer math-science is $(122/300)\ (72/300)\ (300) = (122/300)(72) = 29.28$. The other expected values can be determined in the same manner.

Typically the contingency table is written so that it contains all this information (*see* Table 11-7).

Table 11-7
Contingency table showing
sample results and expected
values

Sex of Student	Favorite Subject Area			Total
	Math-Science (M-S)	Social Science (SS)	Humanities (H)	
Male	37 (29.28)	41 (45.95)	44 (46.77)	122
Female	35 (42.72)	72 (67.05)	71 (68.23)	178
Total	72	113	155	300

The calculated chi-square is

$$\chi^2 = \sum \frac{(O - E)^2}{E}$$

$$= \frac{(37 - 29.28)^2}{29.28} + \frac{(41 - 45.95)^2}{45.95} + \frac{(44 - 46.77)^2}{46.77}$$

$$+ \frac{(35 - 42.72)^2}{42.72} + \frac{(72 - 67.05)^2}{67.05} + \frac{(71 - 68.23)^2}{68.23}$$

$$= 2.035 + 0.533 + 0.164 + 1.395 + 0.365 + 0.112$$

$$= 4.604$$

$$\chi^{2*} = \mathbf{4.604}$$

Step 4   *Decision*: Fail to reject H_0 (χ^{2*} did not fall in the critical region.)

Conclusion: At the 0.05 level of significance, the evidence does not allow us to reject the idea of independence between the sex of a student and the student's preferred academic subject area. ∎

In general, the **$r \times c$ contingency table** (r is the number of rows; c is the number of columns) will be used to test the independence of the row factor and the column factor. The number of degrees of freedom will be determined by

$$df = (r - 1) \cdot (c - 1) \qquad (11\text{-}4)$$

where r and c are both greater than 1.

(This value for df should agree with the number of cells counted according to the general description on pages 431 and 432).

Note   If either r or c is 1, then $df = k - 1$, where k represents the number of cells [formula (11-2)].

The expected values for an $r \times c$ contingency table will be found by means of the formulas found in each cell in Table 11-8, where $n =$ grand total. In general, the **expected value** at the intersection of the **ith row and the jth column** is given by

$$E_{i,j} = \frac{R_i \times C_j}{n} \qquad (11\text{-}5)$$

We should again observe the previously mentioned guideline: each $E_{i,j}$ should be at least 5.

Note   The notation used in Table 11-8 and formula (11-5) may be unfamiliar to you. For convenience in referring to cells or entries in a table, $E_{i,j}$ (or E_{ij}) can be used to denote the entry in the ith row and the jth column. That is, the first letter in the subscript corresponds to the row number and the second letter corresponds to the column number. Thus $E_{1,2}$ is the entry in the first row, second column, and $E_{2,1}$ is the entry in the second row, first column. Referring to Table 11-6, $E_{1,2}$ for that table is 45.95 and $E_{2,1}$ is 42.72. The notation used in Table 11-8 is interpreted in a similar manner, that is, R_1 corresponds to the total from row 1, and C_1 corresponds to the total from column 1.

<table>
<tr><td colspan="8" style="text-align:center">Columns</td></tr>
<tr><td>Rows</td><td>1</td><td>2</td><td>$\cdots$</td><td>j</td><td>$\cdots$</td><td>c</td><td>Total</td></tr>
<tr><td>1</td><td>$\dfrac{R_1 \times C_1}{n}$</td><td>$\dfrac{R_1 \times C_2}{n}$</td><td></td><td>$\dfrac{R_1 \times C_j}{n}$</td><td></td><td>$\dfrac{R_1 \times C_c}{n}$</td><td>$R_1$</td></tr>
<tr><td>2</td><td>$\dfrac{R_2 \times C_1}{n}$</td><td></td><td></td><td>$\vdots$</td><td></td><td></td><td>R_2</td></tr>
<tr><td>$\vdots$</td><td>$\vdots$</td><td>$\vdots$</td><td></td><td></td><td>$\vdots$</td><td></td><td>$\vdots$</td></tr>
<tr><td>i</td><td>$\dfrac{R_i \times C_1}{n}$</td><td></td><td></td><td>$\dfrac{R_i \times C_j}{n}$</td><td></td><td></td><td>R_i</td></tr>
<tr><td>$\vdots$</td><td>$\vdots$</td><td></td><td></td><td>$\vdots$</td><td></td><td></td><td>$\vdots$</td></tr>
<tr><td>r</td><td>$\dfrac{R_r \times C_1}{n}$</td><td></td><td></td><td></td><td></td><td>$\dfrac{R_r \times C_c}{n}$</td><td>$R_r$</td></tr>
<tr><td>Total</td><td>C_1</td><td>C_2</td><td>$\cdots$</td><td>C_j</td><td>$\cdots$</td><td>C_c</td><td>n</td></tr>
</table>

Table 11-8
Expected values for an $r \times c$ contingency table

TEST OF HOMOGENEITY

homogeneity   There is another type of contingency table problem. It is called a **test of homogeneity**. This test is used when one of the two variables is controlled by the experimenter so that the row (or column) totals are predetermined.

For example, suppose that we were to poll registered voters in reference to a piece of legislation proposed by the governor. In the poll, 200 urban, 200 suburban, and 100 rural residents will be randomly selected and asked whether they favor or oppose the governor's proposal. That is, a simple random sample is taken for each of these three groups. A total of 500 voters are to be polled. But notice that it has been predetermined (before the sample is taken) just how many are to fall within each row category, as shown in Table 11-9, and each category is sampled separately.

Table 11-9
Registered voter poll with predetermined row totals

| Type of Residence | Governor's Proposal | | |
	Favor	Oppose	Total
Urban			200
Suburban			200
Rural			100
Total			500

In a test of this nature, we are actually testing the hypothesis "the distribution of proportions within the rows is the same for all rows." That is, the distribution of proportions in row 1 is the same as in row 2, is the same as in row 3, and so on. The alternative to this is "the distribution of proportions within the rows is not the same for all rows." This type of example may be thought of as a comparison of several multinomial experiments.

Beyond this conceptual difference, the actual testing for independence and homogeneity with contingency tables is the same. Let's demonstrate this by completing the polling illustration.

Illustration 11-4   Each person in a random sample of 500 registered voters (200 urban, 200 suburban, and 100 rural residents) was asked his or her opinion about the governor's proposed legislation. Does the sample evidence shown in Table 11-10 support the hypothesis that" voters within the different residence groups have different opinions about the governor's proposal"? Use $\alpha = 0.05$.

Table 11-10
Sample results for Illustration 11-4

| Type of Residence | Governor's Proposal | | |
	Favor	Oppose	Total
Urban	143	57	200
Suburban	98	102	200
Rural	13	87	100
Total	254	246	500

Solution

Step 1 H_0: The proportion of voters favoring the proposed legislation is the same in all three groups.

H_a: The proportion of voters favoring the proposed legislation is not the same in all three groups. (That is, in at least one group the proportions are different from the others.)

Step 2 $\alpha = 0.05$ and df $= (3 - 1)(2 - 1) = 2$. See the accompanying figure.

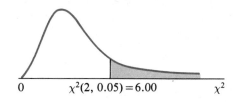

$$0 \qquad \chi^2(2, 0.05) = 6.00 \qquad \chi^2$$

Step 3   The expected values are found by using formula (11-5) and are given in Table 11-11.

Table 11-11
Sample results and
expected values

| Type of Residence | Governor's Proposal | | Total |
	Favor	Oppose	
Urban	143	57	200
	(101.6)	(98.4)	
Suburban	98	102	200
	(101.6)	(98.4)	
Rural	13	87	100
	(50.8)	(49.2)	
Total	254	246	500

Note   Each expected value is used twice in the calculation of $\chi^{2}*$; therefore, it is a good idea to keep extra decimal places while doing the calculations.

$$\chi^2 = \frac{(143 - 101.6)^2}{101.6} + \frac{(57 - 98.4)^2}{98.4} + \frac{(98 - 101.6)^2}{101.6}$$

$$+ \frac{(102 - 98.4)^2}{98.4} + \frac{(13 - 50.8)^2}{50.8} + \frac{(87 - 49.2)^2}{49.2}$$

$$= 16.87 + 17.42 + 0.13 + 0.13 + 28.13 + 29.04$$

$$= 91.72$$

$$\chi^2* = \textbf{91.72}$$

Step 4   *Decision*: Reject H_0.

Conclusion: The three groups of voters do not all have the same proportions favoring the proposed legislation.   ■

Computer Solution

MINITAB Printout for Illustration 11-4

Information given to computer

```
--READ TABLE INTO C1-C2
   COLUMN C1           C2
   COUNT   3            3
ROW
   1          143.              57.
   2           98.             102.
   3           13.              87.
```

Contingency table with expected values, compare to Table 11-11

```
--CHISQUARE FOR DATA IN C1-C2
EXPECTED FREQUENCIES ARE PRINTED BELOW OBSERVED
FREQUENCIES
           |   C1    |   C2   |  TOTALS
-----------+---------+--------+---------
      1    |  143    |   57   |   200
           | 101.6   | 98.4   |
-----------+---------+--------+---------
      2    |   98    |  102   |   200
           | 101.6   | 98.4   |
-----------+---------+--------+---------
      3    |   13    |   87   |   100
           |  50.8   | 49.2   |
-----------+---------+--------+---------
 TOTALS    |  254    |  246   |   500
```

The calculated value of chi-square, cell by cell, compare to solution above

The calculated chi-square value, χ^{2*}
The number of degrees of freedom

```
TOTAL CHI SQUARE =
     16.87 + 17.42 +
      0.13 +  0.13 +
     28.13 + 29.04 +
                       ———————————→  = 91.72
DEGREES OF FREEDOM = (3 - 1) × (2 - 1) = 2
--STOP
```

Case Study 11-2

WANTED: MORE ACTION

Newsweek's "Wanted: More Action" shows the percentage of people who answered "too much power," "right amount of power," and "should use powers more vigorously" when asked which of these statements comes closest to his or her view about government power. This poll was taken in 1978, 1982, 1984, and 1986. Does this information form a contingency table? Explain. (*See* Exercise 11.10.)

Despite President Reagan's anti-big-government line, the desire for a more active federal role is growing.

Powerful Government
Which of these statements comes closest to your view about government power?

	1978	1982	1984	1986
Federal government has too much power	38%	38%	35%	28%
Federal government is using about the right amount of power	18%	18%	25%	24%
Federal government should use its powers more vigorously	36%	30%	34%	41%

Source: Gallup for the Commission on Intergovernmental Relations, in *Newsweek*, November 17, 1986.

Case Study 11-3

50% OPPOSED TO BUILDING NUCLEAR PLANTS

The results of a poll by the National Center for Telephone Research of New York for Gannett News Service asked registered voters "Do you favor or oppose building more nuclear power plants in New York State?" Percentages of voters who answered yes, no, or not sure for each of four geographical regions of the state as well as for the whole state are shown in the table. If the actual number of voters in each category were given, we would have a contingency table and we would be able to complete a hypothesis test about the homogeneity of the four regions. Explain why the following question could be tested using the chi-square statistic: "Is the support for building more nuclear power plants the same in all four regions of New York State?" (See Exercise 11.11.)

NEW YORK STATE POLL

The New York Poll was conducted Dec. 15–18 by the National Center for Telephone Research of New York for Gannett News Service. Trained interviewers surveyed 1,000 registered voters. The calls were placed at random, but in proportion to past voter turnouts around the state.

Half of all New Yorkers oppose construction of more nuclear power plants in the state, the New York State Poll shows.

And women, far more than men, account for the opposition.

Exactly 50 percent of New Yorkers surveyed said they opposed more nuclear plants in the state. That matches the percentage of voters who said they were against more plants two years ago in a New York State Poll.

The latest poll, however, shows a drop since 1977 in the percentage that favors nuclear plants—down to 32 percent from 38 percent. There is a corresponding increase of 6 percent among those who say they're not sure.

Resistance to nuclear plants, regionally, ran highest in western New York, and lowest in the New York City north suburbs.

There is a great deal of uncertainty about the safety of nuclear plants, the poll found. But slightly more people seem to think they're safe than think they are unsafe.

Thus, it seems that most New Yorkers agree with Gov. Carey's decision last year during the campaign, to not allow more nuclear plants to be built until the radioactive waste problem is solved.

Here are the figures on the main question. "Do you favor or oppose building more nuclear power plants in New York State?":

	Yes	No	Not Sure
Total	32	50	18
Upstate	35	48	17
NY City	30	54	16
No. suburbs	32	45	23
Western NY	28	56	16

Source: John Omicinski, Chief, GNS Albany Bureau. Copyright 1979 by the Gannett News Service, Albany Bureau. Reprinted by permission.

EXERCISES

11.8 The manager of an assembly process wants to determine whether the number of defective articles manufactured is depend on the day of the week the articles are produced. She collected the following information.

Day of Week	Mon.	Tues.	Wed.	Thurs.	Fri.
Nondefective	85	90	95	95	90
Defective	15	10	5	5	10

Is there sufficient evidence to reject the hypothesis that the number of defective articles is independent of the day of the week on which they are produced? Use $\alpha = 0.05$.

11.9 Eighty-five college sophomores were classified according to their high school academic performance (HSAP) as well as their current college academic performance (CCAP). Excellent performance was coded 1, above average was coded 2, and average or below was coded 3.

		CCAP		
		1	2	3
	1	10	10	5
HSAP	2	6	25	6
	3	5	8	10

The following **MINITAB** analysis shows the expected values as well as the calculated values for chi-square.

a. Verify the expected numbers, the chi-square values.

b. Find the *p*-value.

```
MTB > READ DATA IN C1-C3
DATA> 10 10 5
DATA> 6 25 6
DATA> 5 8 10
DATA> END DATA
       3 ROWS READ
MTB > CHISQUARE ANALYSIS ON DATA IN C1-C3
Expected counts are printed below observed counts
              C1        C2        C3   Total
    1         10        10         5      25
             6.2      12.6       6.2

    2          6        25         6      37
             9.1      18.7       9.1

    3          5         8        10      23
             5.7      11.6       5.7

 Total        21        43        21      85

 ChiSq =   2.37 +    0.55 +    0.22 +
           1.08 +    2.11 +    1.08 +
           0.08 +    1.14 +    3.28 = 11.91
 df = 4
MTB > STOP
```

11.10 Refer to the table of information given in Case Study 11-2.

a. How is this information related to a contingency table?

b. What information would you need to make a contingency table for this same information?

c. Would the resulting contingency table show independence or homogeneity?

11.11 Refer to Case Study 11-3.

a. Explain why the following question could be tested using the chi-square statistic: "Is the support for building more nuclear power plants the same in all four regions of New York State?"

b. Explain why this is a test of homogeneity.

11.12 A psychologist is investigating how a person reacts to a certain situation. He feels the reaction may be influenced by how ethnically pure the person's neighborhood is. He collects data on 500 people and finds the results shown in the following table.

Ethically Pure Neighborhood	Reaction			
	Mild	Medium	Strong	Total
Yes	170	100	30	300
No	70	100	30	200
Total	240	200	60	500

Does there appear to be a relationship between neighborhood and reaction at the 0.10 level of significance?

a. Solve using the classical approach.

b. Solve using the prob-value approach.

11.13 A survey of travelers who visited the service station restrooms of a large U.S. petroleum distributor showed the following results.

Sex of Respondent	Quality of Restroom Facilities			
	Above Average	Average	Below Average	Total
Female	7	24	28	59
Male	8	26	7	41
Total	15	50	35	100

Using $\alpha = 0.05$, test to see whether the null hypothesis "quality of responses are independent of the sex of the respondent" can be rejected.

a. Solve using the classical approach.

b. Solve using the prob-value approach.

11.14 Given the following four instructors' grade distributions, determine whether or not it can be concluded that the grade distributions are different for these instructors at the $\alpha = 0.05$ level of significance.

			Grades		
Instructor	A	B	C	D or Below	Totals
Smith	33	89	66	22	210
Jones	13	50	66	9	138
Johnson	18	38	31	8	95
Anderson	16	24	13	0	53
Totals	80	201	176	39	496

a. State the null hypothesis.

b. Solve using the classical approach.

c. Solve using the prob-value approach.

11.15 Fear of darkness is a common emotion. The following data were obtained by asking 200 individuals in each age group whether they had serious fears of darkness. At $\alpha = 0.01$, do we have sufficient evidence to reject the hypothesis that "the same proportion of each age group has serious fears of darkness"? [*Hint*: The contingency table must account for all 1000 people.]

Age Group	Elementary	Jr. high	Sr. high	College	Adult
No. Who Fear Darkness	83	72	49	36	114

a. Solve using the classical approach.

b. Solve using the prob-value approach.

In
Retrospect

In this chapter we have been concerned with tests of hypotheses using chi-square, with the cell probabilities associated with the multinomial experiment, and with the simple contingency table. In each case the basic assumptions are that a large number of observations have been made and that the resulting test statistic, $\sum [(O - E)^2/E]$, is approximately distributed as chi-square. In general, if n is large and the minimum allowable expected cell size is 5, this assumption is satisfied.

The article in Case Study 11-3 shows a contingency table with the results listed as relative frequencies. This is a common practice when reporting the results of public opinion polls. However, to statistically interpret the results, the actual frequencies must be used. (*See* Exercise 11.11.)

The contingency table can also be used to test homogeneity. The test for homogeneity and the test for independence look very similar and, in fact, are carried out in exactly the same way. The concepts being tested, however, equal distributions and independence are quite different. The two tests are easily distinguished from one another, for the test of homogeneity has predetermined marginal totals in one direction in the table. That is, before the data are collected, the experimenter determines how many subjects will be observed in each category. The only predetermined number in the test of independence is the grand total.

A few words of caution: The correct number of degrees of freedom is critical if the test results are to be meaningful. The degrees of freedom determine, in part, the critical region, and its size is important. Like other tests of hypothesis, failure to reject H_0 does not mean outright acceptance of the null hypothesis.

Chapter Exercises

11.16 The psychology department at a certain college claims that the grades in its introductory course are distributed as follows: 10 percent A's, 20 percent B's, 40 percent C's, 20 percent D's, and 10 percent F's. In a poll of 200 randomly selected students who had completed this course, it was found that 16 had received A's, 43 B's, 65 C's, 48 D's, and 28 F's. Does this sample contradict the department's claim, at the 0.05 level?

    a. Solve using the classical approach.

    b. Solve using the prob-value approach.

11.17 When interbreeding two strains of roses, we expect the hybrid to appear in three genetic classes in the ratio 1:3:4. If the results of an experiment yield 80 hybrids of the first type, 340 of the second type, and 380 of the third type, do we have sufficient evidence to reject the hypothesized genetic ratio at the 0.05 level of significance?

    a. Solve using the classical approach.

    b. Solve using the prob-value approach.

11.18 A sample of 200 individuals are tested for their blood type and the results are used to test the hypothesized distribution of blood types:

Blood Type	A	B	O	AB
Percent	0.41	0.09	0.46	0.04

The observed results were as follows.

Blood Type	A	B	O	AB
Number	75	20	95	10

At the 0.05 level of significance, is there sufficient evidence to show that the stated distribution is incorrect?

11.19 The weights (x) of 300 adult males were determined and used to test the hypothesis that the weights were normally distributed with a mean of 160 pounds and a standard deviation of 15 pounds. The data were grouped into the following classes:

Weight (x)	Observed Frequency
$x \leq 130$	7
$130 \leq x < 145$	38
$145 \leq x < 160$	100
$160 \leq x < 175$	102
$175 \leq x < 190$	40
190 and over	13

Using the normal tables, the percentages for the classes are 2.28 percent, 13.59 percent, 34.13 percent, 34.13 percent, 13.59 percent, and 2.28 percent, respectively. Do the observed data show significant reason to discredit the hypothesis that the weights are normally distributed with a mean of 160 pounds and a standard deviation of 15 pounds? Use the prob-value approach.

11.20 a. Case Study 11-3 (p. 438) reports the findings of a sample of 1000 people. Suppose that the percentages reported in this news article had come from the following contingency table. Can the null hypothesis "the distribution of yes, no, and not sure opinions is the same for all four geographic areas of New York State" be true? Use $\alpha = 0.05$.

	Yes	No	Not Sure
Upstate	96	150	54
NY City	105	144	51
No. suburbs	64	90	46
Western NY	56	112	32

b. Suppose that there had been only 400 people in the sample, 100 from each geographic location. Calculate the observed value of chi-square, χ^{2*}. Compare this value to that found in (a). Explain the difference.

11.21 A survey of employees at an insurance firm was concerned with worker-supervisor relationships. One statement for evaluation was "I am not sure what my supervisor expects." The results of the survey are found in the following contingency table.

Years of Employment	I Am Not Sure What My Supervisor Expects		
	True	Not True	Totals
Less than 1 year	18	13	31
1–3 years	20	8	28
3–10 years	28	9	37
10 years or more	26	8	34
Total	92	38	130

Can we reject the null hypothesis that "the responses to the statement and years of employment are independent" at the 0.10 level of significance?

a. Solve using the classical approach.

b. Solve using the prob-value approach.

 11.22 Based on the results of a survey questionnaire, 400 individuals are classified as either politically conservative, moderate, or liberal. In addition, each is classified by age, as shown in the following table.

	Age Group			
	20–35	36–50	Over 50	Total
Conservative	20	40	20	80
Moderate	80	85	45	210
Liberal	40	25	45	110
Total	140	150	110	400

Is there sufficient evidence to reject the hypothesis that "political preference is independent of age"? Use the prob-value approach.

11.23 A random sample of 500 married men from all over the country was taken. Each person was cross-classified as to the size of community that he was presently residing in and the size of community that he was reared in. The results are shown in the following table. Does this sample contradict the claim of independence, at the 0.01 level of significance?

a. Solve using the classical approach.

b. Solve using the prob-value approach.

Size of Community Reared In	Size of Community Residing In			
	Under 10,000	10,000 to 49,999	50,000 or Over	Total
Under 10,000	24	45	45	114
10,000 to 49,999	18	64	70	152
50,000 or over	21	54	159	234
Total	63	163	274	500

11.24 Four brands of popcorn were tested for popping. One hundred kernels of each brand were popped and the number of kernels not popped was recorded in each test (*see* the following table). Can we reject the null hypothesis that all four brands pop equally? Test at $\alpha = 0.05$.

Brand	A	B	C	D
No. Not Popped	14	8	11	15

a. Solve using the classical approach.

b. Solve using the prob-value approach.

11.25 The tumor-producing potential of a new drug was tested. One hundred rats were used as a control group, 100 were exposed to a low dose of the drug, and 100 were exposed to a high dose. The results were as follows.

| | Number of Tumors | | |
Rat Group	None	One or More	Total
Control	93	7	100
Low dose	89	11	100
High dose	86	14	100
Total	268	32	300

Is there sufficient evidence to conclude that the dosage does, in fact, affect the occurrence of tumors? Use $\alpha = 0.05$.

11.26 In order to determine what improvements are needed in a shopping mall, the patrons were asked a variety of questions regarding the conditions at the mall. An appraisal of the aesthetics of the mall resulted in the following contingency table.

| Sex of the Respondent | How Are the Aesthetics of the Mall? | | | | |
	Excellent	Fair	Good	Poor	Total
Female	6	16	21	7	50
Male	8	11	10	21	50
Totals	14	27	31	28	100

It had been decided in advance to structure the sample so that exactly 50 men and 50 women responded to the survey. Are the attitudes toward the aesthetics of the mall equally distributed for both sexes? Use $\alpha = 0.05$.

a. Solve using the classical approach.

b. Solve using the prob-value approach.

11.27 Last year's work record for the incidence of absenteeism in each of four categories for 100 randomly selected employees is compiled in the following table. Do these data provide sufficient evidence to reject the hypothesis that the rate of absenteeism is the same for all categories of employees? Use $\alpha = 0.01$ and 240 work days for the year.

	Married Male	Single Male	Married Female	Single Female
Number of Employees	40	14	16	30
Days Absent	180	110	75	135

a. Solve using the classical approach.

b. Solve using the prob-value approach.

11.28 The following table shows the number of reported crimes committed last year in the inner part of a large city. The crimes were classified according to type of crime and district of the inner city where it occurred. Do these data show sufficient evidence to reject the hypothesis that the type of crime and the district in which it occurred are independent? Use $\alpha = 0.01$.

| | Crime | | | | |
District	Robbery	Assault	Burglary	Larceny	Stolen Vehicle
1	54	331	227	1090	41
2	42	274	220	488	71
3	50	306	206	422	83
4	48	184	148	480	42
5	31	102	94	596	56
6	10	53	92	236	45

11.29 Consider the following set of data.

| | Response | | |
	Yes	No	Total
Group 1	75	25	100
Group 2	70	30	100
Total	145	55	200

a. Compute the value of the test statistic $z*$ that would be used to test the null hypothesis that $p_1 = p_2$, where p_1 and p_2 are the proportions of yes responses in the respective groups.

b. Compute the value of the test statistic χ^{2*} that would be used to test the hypothesis that "response is independent of group."

c. Show that $\chi^2 = (z*)^2$.

Vocabulary List

Be able to define each term. In addition, describe, in your own words, and give an example of each term. Your examples should not be ones given in class or in the textbook.

The bracketed numbers indicate the chapters in which the term previously appeared, but you should define the terms again to show increased understanding of their meaning.

cell
chi-square [9]
column
contingency table
degrees of freedom [9, 10]
enumerative data
expected frequency
frequency distribution [2]

homogeneity
independence [4]
marginal totals
multinomial experiment
observed frequency [2, 4]
rows
statistic [1, 2, 8]

Quiz   *Answer "True" if the statement is always true. If the statement is not always true, replace the boldface words with words that make the statement always true.*

11.1 The number of degrees of freedom for a test of a multinomial experiment is **equal to** the number of cells in the experimental data.

11.2 The **expected frequency** in a chi-square test is found by multiplying the hypothesized probability of a cell by the number of pieces of data in the sample.

11.3 The **observed** frequency of a cell is not allowed to be smaller than five when a chi-square test is being conducted.

11.4 In the **multinomial experiment** we have $(r - 1)$ times $(c - 1)$ degrees of freedom (r is the number of rows and c is the number of columns).

11.5 A multinomial experiment consists of n **identical, independent trials**.

11.6 A **multinomial experiment** arranges the data into a two-way classification such that the totals in one direction are predetermined.

11.7 The charts for both the multinomial experiment and the contingency table **must** be set up in such a way that each piece of data will fall into exactly one of the categories.

11.8 The test statistic $\sum [(O - E)^2/E]$ has a distribution that is **approximately normal**.

11.9 The data used in a chi-square multinomial test are always **enumerative** in nature.

11.10 The null hypothesis being tested by a test of **homogeneity** is that the distribution of proportions is the same for each of the subpopulations.

Chapter 12

Analysis of Variance

TILLAGE TEST PLOTS REVISITED

All tillage tools are not created equal although the weather can certainly make them look that way.

That's our analysis after putting six of your favorite primary tillage tools through the paces last season. As we told you in our 1978 Planting Issue, we conducted those tests because we felt conservation tillage had a place on the fine textured, poorly drained soils long sacred to the moldboard plow. Despite a photo finish among yields, our test plots reinforced that opinion. Here's a wrap-up of the things that shape that optimism.

We staked out 18 plots on a gently sloping field with a Chalmers silty clay loam soil type. The Chalmers is similar to the Drummer, Brookston and Webster soils common to the Corn Belt. With last fall's soils wet underfoot we used the six primary tillage tools....

By the second day of May we were ready for the planter and the temperature and growth measurements that would tell us if our theories about heavy-duty conservation tillage held water.

How good is a tie. The yields from individual tillage treatments showed less variability than we had hoped, probably due to the late planting. That's why university researchers conduct these studies over many years. The only yield differences worth noting are the 7.4 bu/A spreads between the standard chisel and light disk at 115.2 bu. and the moldboard plow at 122.6 bu. The coulter-chisel and heavy disk, and surprisingly enough the V-chisel, yielded virtually the same as the moldboard plow. However, we think there are other factors that add even more merit to heavy-duty conservation tillage.

Tillage plot yields—1978
Bu/A @ 15.5% moisture

Tillage	Rep I	Rep II	Rep III	Tillage Average
Plow	118.3	125.6	123.8	122.6
V chisel	115.8	122.5	118.9	119.1
Coulter chisel	124.1	118.5	113.3	118.6
Std. chisel	109.2	114.0	122.5	115.2
Hvy. disk	118.1	117.5	121.4	119.0
Lt. disk	118.3	113.7	113.7	115.2

Source: Larry Reichenberger, Associate Machine Editor. Reprinted with permission from *Successful Farming*, February 1979. Copyright 1979 by Meredith Corporation. All rights reserved.

Chapter
Objectives

Previously we have tested hypotheses about two means. In this chapter we are concerned with testing a hypothesis about several means. The **analysis of variance technique (ANOVA),** which we are about to explore, will be used to test a hypothesis about several means, for example,

$$H_0: \mu_1 = \mu_2 = \mu_3 = \mu_4 = \mu_5$$

By using our former technique for hypotheses about two means, we could test several hypotheses if each stated a comparison of two means. For example, we could test

$$H_1: \mu_1 = \mu_2 \qquad H_2: \mu_1 = \mu_3$$
$$H_3: \mu_1 = \mu_4 \qquad H_4: \mu_1 = \mu_5$$
$$H_5: \mu_2 = \mu_3 \qquad H_6: \mu_2 = \mu_4$$
$$H_7: \mu_2 = \mu_5 \qquad H_8: \mu_3 = \mu_4$$
$$H_9: \mu_3 = \mu_5 \qquad H_{10}: \mu_4 = \mu_5$$

In order to test the null hypothesis, H_0, that all five means are equal, we would have to test each of these ten hypotheses using our former technique for two means. Rejection of any one of the ten hypotheses about two means would cause us to reject the null hypotheses that all five means are equal. If we failed to reject all ten hypotheses about the means, we would fail to reject the main null hypothesis. Suppose we tested a null hypothesis that dealt with several means by testing all the possible pairs of two means; the overall type-I error rate would become much larger than the value of α associated with a single test. The ANOVA techniques allow us to test the null hypothesis (all means are equal) against the alternative hypothesis (at least one mean value is different) with a specified value of α.

In this chapter we will introduce ANOVA. ANOVA experiments can be very complex, depending on the situation. We will restrict our discussion to the most basic experimental design, the single-factor ANOVA.

12.1 INTRODUCTION TO THE ANALYSIS OF VARIANCE TECHNIQUE

We will begin our discussion of the analysis of variance technique by looking at an illustration.

Illustration 12-1   The temperature at which a plant is maintained is believed to affect the rate of production in the plant. The data in Table 12-1 are the number, x, of units produced in one hour for randomly selected one-hour periods when the production process in the plant was operating at each of three temperature **levels**. The data values from repeated samplings are called **replicates**. Four replicates, or data values, were obtained for two of the temperatures and five were obtained for the third temperature. Do these data suggest that temperature has a significant effect on the production level at the 0.05 level?

level

replicate

Table 12-1
Sample results for
Illustration 12-1

	Temperature Levels		
	Sample from 68°F ($i = 1$)	Sample from 72°F ($i = 2$)	Sample from 76°F ($i = 3$)
	10	7	3
	12	6	3
	10	7	5
	9	8	4
		7	
Column Totals	$C_1 = 41$ $\bar{x}_1 = 10.25$	$C_2 = 35$ $\bar{x}_2 = 7.0$	$C_3 = 15$ $\bar{x}_3 = 3.75$

The level of production is measured by the mean value; $\bar{x}_i$ indicates the observed production mean at level i, where $i = 1, 2,$ and 3 corresponds to temperatures of 68°F, 72°F, and 76°F, respectively. There is a certain amount of variation among these means. Since sample means do not necessarily repeat when repeated samples are taken from a population, some variation can be expected, even if all three population means are equal. We will next pursue the question "Is this variation among the $\bar{x}$'s due to chance or is it due to the effect that temperature has on the production rate?"

Solution   The null hypothesis that we will test is

$$H_0: \mu_{68} = \mu_{72} = \mu_{76}$$

That is, the true production mean is the same at each temperature level tested. In other words, the temperature does not have a significant effect on the production rate. The alternative to the null hypothesis is

$$H_a: \text{not all temperature level means are equal}$$

Thus we will want to reject the null hypothesis if the data show that one or more of the means are significantly different from the others.

We will make the decision to reject H_0 or fail to reject H_0 by using the F distribution and the F test statistic. Recall from Chapter 10 that the calculated value of F is the ratio of two variances. The analysis of variance procedure will separate the variation among the entire set of data into two categories. To accomplish this separation, we first work with the numerator of the fraction used to define sample variance, formula (2-7):

$$s^2 = \frac{\sum (x - \bar{x})^2}{n - 1}$$

sum of squares   The numerator of this fraction is called the **sum of squares**:

$$\text{sum of squares} = \sum (x - \bar{x})^2 \qquad (12\text{-}1)$$

We calculate the **total sum of squares, SS(total)**, for the total set of data by using a formula that is equivalent to formula (12-1) but does not require the use of $\bar{x}$. This equivalent formula is

$$SS(total) = \sum (x^2) - \frac{(\sum x)^2}{n} \qquad (12\text{-}2)$$

Now we can find the SS(total) for our illustration by using formula (12-2):

$$\sum (x^2) = 10^2 + 12^2 + 10^2 + 9^2 + 7^2 + 6^2 + 7^2$$
$$+ 8^2 + 7^2 + 3^2 + 3^2 + 5^2 + 4^2$$
$$= 731$$
$$\sum x = 10 + 12 + 10 + 9 + 7 + 6 + 7$$
$$+ 8 + 7 + 3 + 3 + 5 + 4$$
$$= 91$$

$$SS(total) = 731 - \frac{(91)^2}{13}$$
$$= 731 - 637 = \mathbf{94}$$

Next, 94, the SS(total), must be separated into two parts: the sum of squares, SS(temperature), due to temperature levels; and the sum of squares, SS(error), due to experimental error of replication. This splitting is often called **partitioning**, since SS(temperature) + SS(error) = SS(total); that is, in our illustration SS(temperature) + SS(error) = 94.

partitioning

The sum of squares **SS(factor)**, [SS(temperature) for our illustration] that measures the **variation between the factor levels** (temperatures) is found by using formula (12-3):

variation between the factor levels

$$SS(factor) = \left(\frac{C_1^2}{k_1} + \frac{C_2^2}{k_2} + \frac{C_3^2}{k_3} + \cdots\right) - \frac{(\sum x)^2}{n} \qquad (12\text{-}3)$$

where C_i represents the column total, k_i represents the number of replicates at each level of the factor, and n represents the total sample size ($n = \sum k_i$).

Note　The data have been arranged so that each column represents a different level of the factor being tested.

Now we can find the SS(temperature) for our illustration by using formula (12-3):

$$SS(temperature) = \left(\frac{41^2}{4} + \frac{35^2}{5} + \frac{15^2}{4}\right) - \frac{(91)^2}{13}$$
$$= (420.25 + 245.00 + 56.25) - 637.0$$
$$= 721.5 - 637.0 = \mathbf{84.5}$$

The sum of squares **SS(error)** that measures the **variation within the rows** is found by using formula (12-4):

variation within rows

$$SS(error) = \sum (x^2) - \left(\frac{C_1^2}{k_1} + \frac{C_2^2}{k_2} + \frac{C_3^2}{k_3} + \cdots\right) \qquad (12\text{-}4)$$

The SS(error) for our illustration can now be found by using formula (12-4):

$$\sum(x^2) = 731 \text{ (found previously)}$$

$$\left(\frac{C_1^2}{k_1} + \frac{C_2^2}{k_2} + \frac{C_3^2}{k_3}\right) = 721.5 \text{ (found previously)}$$

$$\text{SS(error)} = 731.0 - 721.5 = \mathbf{9.5}$$

Note SS(total) = SS(factor) + SS(error). Inspection of formulas (12-2), (12-3), and (12-4) will verify this.

For convenience we will use an **ANOVA table** to record the sums of squares and to organize the rest of the calculations. The format of an ANOVA table is shown in Table 12-2.

Table 12-2
Format for ANOVA table

Source	SS	df	MS
Factor			
Error			
Total			

We have calculated the three sums of squares for our illustration. The degrees of freedom, df, associated with each of the three sources are determined as follows:

1. df(factor) is one less than the number of levels (columns) for which the factor is tested:

$$\text{df(factor)} = c - 1 \qquad (12\text{-}5)$$

where c represents the number of levels for which the factor is being tested (number of columns on the data table).

2. df(total) is one less than the total number of data:

$$\text{df(total)} = n - 1 \qquad (12\text{-}6)$$

where n represents the number of data in the total sample (that is, $n = k_1 + k_2 + k_3 + \cdots$, where k_i is the number of replicates at each level tested).

3. df(error) is the sum of the degrees of freedom for all the levels tested (columns in the data table). Each column has $k_i - 1$ degrees of freedom; therefore,

$$\text{df(error)} = (k_1 - 1) + (k_2 - 1) + (k_3 - 1) + \cdots$$

or

$$\text{df(error)} = n - c \qquad (12\text{-}7)$$

The degrees of freedom for our illustration are

$$\text{df(temperature)} = c - 1 = 3 - 1 = \mathbf{2}$$
$$\text{df(total)} = n - 1 = 13 - 1 = \mathbf{12}$$
$$\text{df(error)} = n - c = 13 - 3 = \mathbf{10}$$

The sums of squares and the degrees of freedom must check. That is,

$$SS(factor) + SS(error) = SS(total) \qquad (12\text{-}8)$$

$$df(factor) + df(error) = df(total) \qquad (12\text{-}9)$$

mean square   The **mean square** for the factor being tested, the **MS(factor)**, and for error, the **MS(error)**, will be obtained by dividing the sum-of-square value by the corresponding number of degrees of freedom. That is,

$$MS(factor) = \frac{SS(factor)}{df(factor)} \qquad (12\text{-}10)$$

$$MS(error) = \frac{SS(error)}{df(error)} \qquad (12\text{-}11)$$

The mean squares for our illustration are

$$MS(temperature) = \frac{SS(temperature)}{df(temperature)} = \frac{84.5}{2}$$

$$= \mathbf{42.25}$$

$$MS(error) = \frac{SS(error)}{df(error)} = \frac{9.5}{10}$$

$$= \mathbf{0.95}$$

The completed ANOVA table appears as shown in Table 12-3. The

Table 12-3
ANOVA table for
Illustration 12-1

Source	SS	df	MS
Temperature	84.5	2	42.25
Error	9.5	10	0.95
Total	94.0	12	

hypothesis test is now completed using the two mean squares as measures of variance. The calculated value of the **test statistic, F^*,** is found by dividing the MS(factor) by the MS(error):

$$F = \frac{MS(factor)}{MS(error)} \qquad (12\text{-}12)$$

The calculated value of F for our illustration is found by using formula (12-12):

$$F = \frac{MS(temperature)}{MS(error)} = \frac{42.25}{0.95}$$

$$F^* = \mathbf{44.47}$$

The decision to reject H_0 or fail to reject H_0 will be made by comparing the calculated value of F, $F^* = 44.47$, to a one-tailed critical value of F obtained from

Table 8a of Appendix E. See the accompanying figure. The critical value is $F(2, 10, 0.05) = 4.10$.

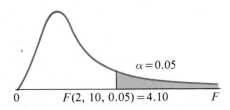

Note   Since the calculated value of F, F^*, is found by dividing MS(temperature) by MS(error), the number of degrees of freedom for the numerator is df(temperature) = 2 and the number of degrees of freedom for the denominator is df(error) = 10.

We rejected H_0 because the value F^* fell in the critical region. Therefore we conclude that at least one of the room temperatures does have a significant effect on the production rate. The differences in the mean production rates at the tested temperature levels were found to be significant. The mean at 68°F is certainly different from the mean at 76°F, since the sample means for these levels are the largest and smallest, respectively. Whether any other pairs of means are significantly different cannot be determined from the ANOVA procedure alone. ■

In this section we saw how the ANOVA technique separated the variance among the sample data into two measures of variance: (1) MS(factor), the measure of variance between the levels tested, and (2) MS(error), the measure of variance within the levels being tested. Then these measures of variance can be compared. For our illustration, the between-level variance was found to be significantly larger than the within-level variance (experimental error). This led us to the conclusion that temperature did have a significant effect on the variable x, the number of units of production completed per hour.

In the next section we will use several illustrations to demonstrate the logic of the analysis of variance technique.

12.2 THE LOGIC BEHIND ANOVA

Many experiments are conducted to determine the effect that different levels of some test factor has on a response variable. The test factor may be temperature (as in Illustration 12-1), the manufacturer of a product, the day of the week, or any number of other things. In this chapter we are investigating the **single-factor analysis of variance**. Basically, the design for the single-factor ANOVA is to obtain independent random samples at each of the several levels of the factor being tested. We will then make a statistical decision concerning the effect that the levels of the test factor have on the response (observed) variable.

Illustrations 12-2 and 12-3 demonstrate the logic of the analysis of variance technique. Briefly, the reasoning behind the technique proceeds like this: In order to compare the means of the levels of the test factor, a measure of the **variation between the levels** (between columns on the data table), the **MS(factor)**, will be compared to a measure of the **variation within the levels** (within the columns on the data table), the **MS(error)**. If the MS(factor) is significantly larger than the MS(error), then we will conclude that the means for each of the factor levels being tested are not all the same. This implies that the factor being tested does have a significant effect on the response variable. If, however, the MS(factor) is not significantly larger than the MS(error), we will not be able to reject the null hypothesis that all means are equal.

variation between levels

variation within levels

Illustration 12-2   Do the data in Table 12-4 show sufficient evidence to conclude that there is a difference in the three population means μ_F, μ_G, and μ_H?

Table 12-4
Sample results for
Illustration 12-2

	Factor Levels		
	Sample from Level F	Sample from Level G	Sample from Level H
	3	5	8
	2	6	7
	3	5	7
	4	5	8
Column Totals	$C_F = 12$ $\bar{x} = 3.00$	$C_G = 21$ $\bar{x} = 5.25$	$C_H = 30$ $\bar{x} = 7.50$

Figure 12-1 shows the relative relationship among the three samples. A quick look at the figure suggests that the three sample means are different from each other, implying that the sampled populations have different mean values.

Figure 12-1

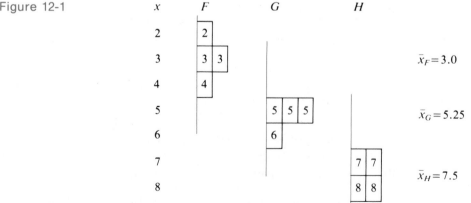

These three samples demonstrate relatively little **within-sample variation**, although there is a relatively large amount of **between-sample variation**. Let's look at another illustration.

Illustration 12-3   Do the data in Table 12-5 show sufficient evidence to conclude that there is a difference in the three population means μ_J, μ_K, and μ_L?

Table 12-5

Sample results for
Illustration 12-3

	Factor Levels		
	Sample from Level J	Sample from Level K	Sample from Level L
	3	5	6
	8	4	2
	6	3	7
	4	7	5
Column Totals	$C_J = 21$ $\bar{x}_J = 5.25$	$C_K = 19$ $\bar{x}_K = 4.75$	$C_L = 20$ $\bar{x}_L = 5.00$

Figure 12-2 shows the relative relationship among the three samples. A quick look
at the figure **does not suggest** that the three sample means are different from each
other.

Figure 12-2

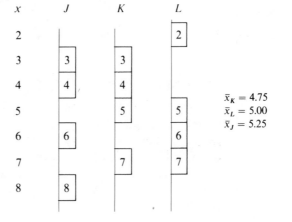

$\bar{x}_K = 4.75$
$\bar{x}_L = 5.00$
$\bar{x}_J = 5.25$

■

There is little between-sample variation for these three samples (that is, the
sample means are relatively close in value), whereas the within-sample variation is
relatively large (that is, the data values within each sample cover a relatively wide
range of values).

To complete a hypothesis test for analysis of variance, we must agree on some
ground rules, or assumptions. In this chapter we will use the following three basic
assumptions:

1. Our goal is to investigate the effect that various levels of the factor under test
have on the response variable. Typically we want to find the level that yields the
most advantageous values of the response variable. This, of course, means that
we will probably want to reject the null hypothesis in favor of the alternative.
Then a follow-up study could determine the "best" level of the factor.

2. We must assume that the effects due to chance and due to untested factors are normally distributed and that the variance caused by these effects is constant throughout the experiment.

3. We must assume independence among all observations of the experiment. (Recall that independence means that the results of one observation of the experiment do not affect the results of any other observation.) We will usually conduct the tests in a randomly assigned order to ensure independence. This technique also helps to avoid data contamination.

Case Study 12-1 |

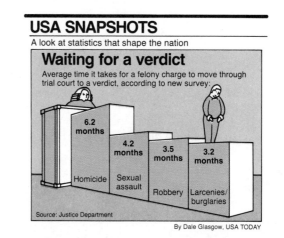

USA SNAPSHOTS

A look at statistics that shape the nation

Waiting for a verdict

Average time it takes for a felony charge to move through trial court to a verdict, according to new survey:

6.2 months — Homicide
4.2 months — Sexual assault
3.5 months — Robbery
3.2 months — Larcenies/burglaries

Source: Justice Department

By Dale Glasgow, USA TODAY

WAITING FOR A VERDICT

This *USA SNAPSHOT* reports the average amount of time it takes for each of several types of felony charges to move through trial and to a verdict. Does the type of felony appear to have an effect on the average amount of time required? What additional information would be needed in order to determine whether the type of felony has a significant effect on the waiting time? (*See* Exercise 12.4.)

Source: "Waiting for a verdict," August 19, 1986. Copyright 1986, USA TODAY. Reprinted with permission.

12.3 APPLICATIONS OF SINGLE-FACTOR ANOVA

Before continuing our ANOVA discussion, let's identify the notation, particularly the subscripts that are used (*see* Table 12-6). Notice that each piece of data has two subscripts; the first subscript indicates the column number (test factor level) and the second subscript identifies the replicate (row) number. The column totals, C_i, are listed across the bottom of the table. The grand total, T, is equal to the sum of all x's and is found by adding the column totals. Row totals can be used as a cross-check but serve no other purpose.

Replication	Sample from Level 1	Sample from Level 2	Sample from Level 3	$\cdots$	Sample from Level C
		Factor Levels			
$k = 1$	$x_{1,1}$	$x_{2,1}$	$x_{3,1}$		$x_{c,1}$
$k = 2$	$x_{1,2}$	$x_{2,2}$	$x_{3,2}$		$x_{c,2}$
$k = 3$	$x_{1,3}$	$x_{2,3}$	$x_{3,3}$		$x_{c,3}$
$\vdots$					
Column Totals	C_1	C_2	C_3	$\cdots$	C_c	T
	$T = \text{grand total} = \text{sum of all } x\text{'s} = \sum x = \sum C_i$				

A mathematical model (equation) is often used to express a particular situation. In Chapter 3 we used a mathematical model to help explain the relationship between the values of bivariate data. The equation $\hat{y} = b_0 + b_1 x$ served as the model when we believed that a straight-line relationship existed. The probability functions studied in Chapter 5 are also examples of mathematical models. For the single-factor ANOVA, the **mathematical model**, formula (12-13), is an expression of the composition of each piece of data entered in our data table.

mathematical model

$$x_{c,k} = \mu + F_c + \varepsilon_{k(c)} \tag{12-13}$$

We interpret each term of this model as follows:

1. μ is the mean value for all the data without respect to the test factor.

2. F_c is the effect that the factor being tested has on the response variable at each different level c.

experimental error

3. $\varepsilon_{k(c)}$ (ε is the lowercase Greek letter epsilon) is the **experimental error** that occurs among the k replicates in each of the c columns.

Let's look at another hypothesis test concerning an analysis of variance.

Illustration 12-4   A rifle club performed an experiment on a randomly selected group of beginning shooters. The purpose of the experiment was to determine whether shooting accuracy is affected by the method of sighting used: only the right eye open, only the left eye open, or both eyes open. Fifteen beginning shooters were selected and split into three groups. Each group experienced the same training and practicing procedures with one exception: the method of sighting used. After completing training, each student was given the same number of rounds and asked to shoot at a target. Their scores appear in Table 12-7.

Table 12-7
Sample results for
Illustration 12-4

Method of Sighting		
Right Eye	Left Eye	Both Eyes
12	10	16
10	17	14
18	16	16
12	13	11
14		20
		21

At the 0.05 level of significance, is there sufficient evidence to reject the claim that these methods of sighting are equally effective?

Solution   In this experiment the factor is "method of sighting" and the levels are the three different methods of sighting (right eye, left eye, and both eyes open). The replicates are the scores received by the students in each group. The null hypothesis to be tested is "the three methods of sighting are equally effective", or "the mean scores attained using each of the three methods are the same."

Step 1 $H_0: \mu_R = \mu_L = \mu_B$

H_a: The means are not all equal (that is, at least one mean is different).

Step 2 $\alpha = 0.05$. The test criteria are shown in the accompanying figure.

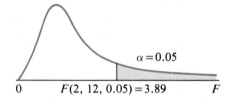

df(numerator) = df(method) = 3 − 1 = 2   [using formula (12-5)]
df(denominator) = df(error) = 15 − 3 = 12   [using formula (12-7)]

The critical value is $F(2, 12, 0.05) = 3.89$.

Step 3   Calculate the test statistic F^*. Table 12-8 is used to find column totals.

Table 12-8
Sample results for
Illustration 12-4

| Replicates | Factor Levels: Method of Sighting | | |
	Right Eye	Left Eye	Both Eyes
$k = 1$	12	10	16
$k = 2$	10	17	14
$k = 3$	18	16	16
$k = 4$	12	13	11
$k = 5$	14		20
$k = 6$			21
Totals	$C_R = 66$	$C_L = 56$	$C_B = 98$

First, the summations $\sum x$ and $\sum x^2$ need to be calculated:

$$\sum x = 12 + 10 + 18 + 12 + 14 + 10 + 17 + \cdots + 21 = \mathbf{220}$$
$$\sum x^2 = 12^2 + 10^2 + 18^2 + 12^2 + 14^2 + 10^2 + \cdots + 21^2 = \mathbf{3392}$$

Using formula (12-2), we find

$$SS(total) = 3392 - \frac{(220)^2}{15} = 3392 - 3226.67 = \mathbf{165.33}$$

Using formula (12-3), we find

$$SS(method) = \left(\frac{66^2}{5} + \frac{56^2}{4} + \frac{98^2}{6}\right) - 3226.67$$

$$= 3255.87 - 3226.67 = \textbf{29.20}$$

Using formula (12-4), we find

$$SS(error) = 3392 - 3255.87 = \textbf{136.13}$$

Use formula (12-8) to check the sum of squares.

$$SS(method) + SS(error) = SS(total)$$
$$29.20 + 136.13 = 165.33$$

The number of degrees of freedom are found using formulas (12-5), (12-6), and (12-7):

$$df(method) = 3 - 1 = \textbf{2}$$
$$df(total) = 15 - 1 = \textbf{14}$$
$$df(error) = 15 - 3 = \textbf{12}$$

Using formulas (12-10) and (12-11), we find

$$MS(method) = \frac{29.20}{2} = \textbf{14.60}$$

$$MS(error) = \frac{136.13}{12} = \textbf{11.34}$$

The results of these computations are combined in the ANOVA table shown in Table 12-9.

Table 12-9
ANOVA table for
Illustration 12-4

Source	SS	df	MS
Method	29.20	2	14.60
Error	136.13	12	11.34
Total	165.33	14	

The calculated value of the test statistic is then found using formula (12-12).

$$F^* = \frac{14.60}{11.34} = \textbf{1.287}$$

Step 4 The decision to fail to reject or reject the null hypothesis is made by comparing F^* to the critical value shown in the test criteria (Step 2).

Decision: Fail to reject H_0.

Conclusion: The data show no evidence that would give reason to reject the null hypothesis that the three methods are equally effective.

Computer Solution
MINITAB printout for Illustration 12-4

Information given to computer

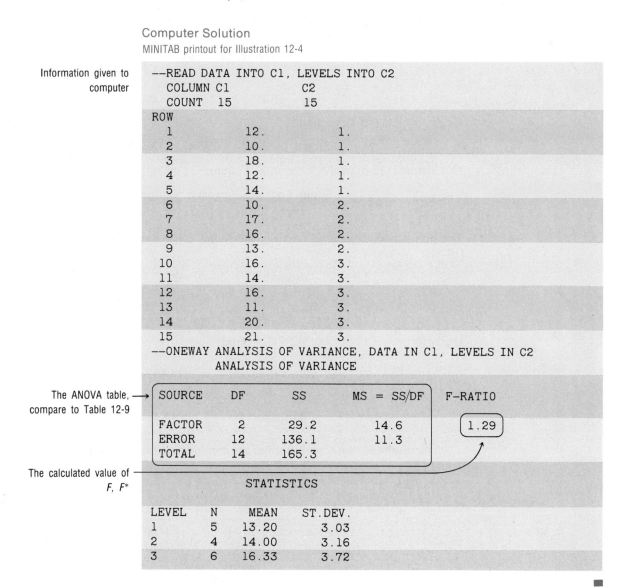

The ANOVA table, compare to Table 12-9

The calculated value of F, F*

```
--READ DATA INTO C1, LEVELS INTO C2
   COLUMN C1                C2
   COUNT  15                15
ROW
    1            12.            1.
    2            10.            1.
    3            18.            1.
    4            12.            1.
    5            14.            1.
    6            10.            2.
    7            17.            2.
    8            16.            2.
    9            13.            2.
   10            16.            3.
   11            14.            3.
   12            16.            3.
   13            11.            3.
   14            20.            3.
   15            21.            3.
--ONEWAY ANALYSIS OF VARIANCE, DATA IN C1, LEVELS IN C2
            ANALYSIS OF VARIANCE
```

SOURCE	DF	SS	MS = SS/DF	F-RATIO
FACTOR	2	29.2	14.6	1.29
ERROR	12	136.1	11.3	
TOTAL	14	165.3		

```
                  STATISTICS
```

LEVEL	N	MEAN	ST.DEV.
1	5	13.20	3.03
2	4	14.00	3.16
3	6	16.33	3.72

Recall the null hypothesis, that there is no difference between the levels of the factor being tested. A "fail to reject H_0" decision must be interpreted as the conclusion that there is no evidence of a difference due to the levels of the tested factor, whereas the rejection of H_0 implies that there is a difference between the levels. That is, at least one level is different from the others. If there is a difference, the next problem is to locate the level or levels that are different. Locating this difference may be the main object of the analysis. In order to find the difference, the only method that is appropriate at this stage is to inspect the data. It may be obvious which level(s) caused the rejection of H_0. In Illustration 12-1 it seems quite obvious that at least one of the levels (level 1 (68°F) or level 3 (76°F), because they have the largest and smallest sample means) is different from the other two. If the higher values are more desirable for finding the "best" level to use, we would choose that corresponding level of the factor.

Thus far we have discussed analysis of variance for data dealing with one factor. It is not unusual for problems to have several factors of concern. The ANOVA techniques presented in this chapter can be developed further and applied to more complex cases.

| Case Study 12-2 | **THE EFFECT OF THE CLASS EVALUATION METHOD ON LEARNING IN CERTAIN MATHEMATICS COURSES** |

In the following article, Larry Stephens discusses how the statistical techniques of one-way analysis of variance can be used to evaluate three teaching methods. The percentages in Table 1 are the students' test scores. Verify the results shown. (See Exercise 12.8.)

Much has been written concerning the merits of different teaching methods and their effectiveness in different disciplines. The purpose of this paper is to investigate the effects of different evaluation methods on the learning process....

To determine whether different methods of evaluation influence the learning in such a course, three different methods were used. Method I can be described as follows. Homework was collected weekly and mid-term and comprehensive final exams were administered. Method II consisted of four tests administered at the end of each four weeks covering the material of the past four weeks. Method III consisted of 30-minute weekly tests and a comprehensive final. Homework was assigned but not collected in methods II and III....

The same text and instructor were used in all three methods, and the three groups were representative of the junior and senior students from the school....

The response variable measured was a percentage of test points obtained by each student.....

TABLE 1. Percentages, means and standard deviations for the three methods

Method I	91, 62, 77, 84, 52, 67, 58, 78, 88, 72 87, 75, 93, 62, 63, 56, 83, 97, 72, 68 72, 85, 66, 94, 63, 60, 75, 79, 87	$n_1 = 29$ $\bar{x} = 74.7, s = 12.5$
Method II	85, 91, 79, 84, 73, 47, 92, 91, 91, 62 64, 73, 67, 75, 87, 77, 92, 83	$n_2 = 18$ $\bar{x} = 78.5, s = 12.6$
Method III	79, 85, 74, 89, 79, 80, 94, 91, 79, 71 73, 86, 70, 71, 71	$n_3 = 15$ $\bar{x} = 79.5, s = 8.0$

Table 1 shows that the weekly test with homework not taken up produced the largest average of the three methods. To answer the question of whether these sample means are statistically different requires an analysis of variance. In other words, if μ_1, μ_2 and μ_3 are the average scores made by all students taking such a course and being evaluated by methods I, II and III, respectively, what can be inferred about μ_1, μ_2 and μ_3? To test the hypothesis that $\mu_1 = \mu_2 = \mu_3$, i.e., that there is no difference in the three methods, the sample results given in Table 1 are subjected to an analysis of variance.

TABLE 2. One-way analysis of variance for the three methods

Source	df.	SS	MS	F
Total	61	8257		
Between methods	2	286.4	143.2	
Within methods	59	7970.6	135.1	1.06

Table 2 shows the results of a one-way analysis of variance for the three methods. The F-value is very nonsignificant, indicating that, statistically speaking $\mu_1 = \mu_2 = \mu_3$.

The results of this experiment indicate that collecting and grading homework does not produce significantly higher test scores. Also, the frequency of testing does not seem to produce any different results.

Source: Larry J. Stephens, *International Journal of Math, Education, Science, and Technology*, 1977, Vol. 8., No. 4., pp. 477–479.

EXERCISES

12.1 Suppose that an F test (as described in this chapter) has a critical value of 2.2, as shown in the following figure.

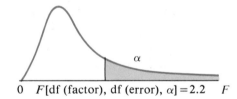

$$0 \quad F[\text{df (factor), df (error)}, \alpha] = 2.2 \quad F$$

    a. What is the interpretation of a calculated value of F larger than 2.2?

    b. What is the interpretation of a calculated value of F smaller than 2.2?

    c. What is the interpretation if the calculated F were 0.1? 0.01?

12.2 Why does df(factor), the number of degrees of freedom, always appear first in the critical value notation F [df(factor), df(error), α]?

12.3 Consider the following table for a single-factor ANOVA. Find the following:

    a. $x_{1,2}$ **b.** $x_{2,4}$ **c.** C_1 **d.** $\sum x$ **e.** $\sum (C_i)^2$

	Level of Factor		
Replicates	1	2	3
1	3	2	7
2	0	5	4
3	1	4	5
4	5	3	6
5	7	6	2

12.4 Refer to Case Study 12-1.

a. Does the type of felony charge appear to have an effect on the average amount of time it takes to move through the trial to a verdict?

b. What additional information would be needed in order to determine whether the type of felony has a significant effect on the waiting time?

12.5 a. State the null and alternative hypotheses that would be under test in Exercise 12.3.

b. How would a "reject the null hypothesis" decision be interpreted?

c. How would a "fail to reject the null hypothesis" decision be interpreted?

d. How is such a decision reached?

12.6 For the following data, find SS(error) and show that

$$SS(error) = [(k_1 - 1)s_1^2 + (k_2 - 1)s_2^2 + (k_3 - 1)s_3^2]$$

where s_i is the variance for the ith factor level.

Factor Level		
1	2	3
8	6	10
4	6	12
2	4	14

12.7 Students with three different high school academic backgrounds were compared with respect to their aptitude in computer science. The students were classified as having excellent, above average, or average or below average high school academic backgrounds. Each student was given the KSW computer-science aptitude test and the score was recorded. The results were as follows.

High School Academic Performance		
Excellent	Above Average	Average or Below
16	21	4
22	19	20
15	16	13
20	17	18
23	5	8
16	20	6
21	18	11
	19	
	14	
	22	
	13	

Refer to the following MINITAB analysis of the experiment and verify the analysis of variance table and test for the null hypothesis $\mu_E = \mu_{AA} = \mu_{AB}$, at $\alpha = 0.05$.

```
MTB > READ DATA IN C1,LEVELS IN C2
DATA> 16 1
DATA> 22 1
DATA> 15 1
DATA> 20 1
DATA> 23 1
DATA> 16 1
DATA> 21 1
DATA> 21 2
DATA> 19 2
DATA> 16 2
DATA> 17 2
DATA> 5 2
DATA> 20 2
DATA> 18 2
DATA> 19 2
DATA> 14 2
DATA> 22 2
DATA> 13 2
DATA> 4 3
DATA> 20 3
DATA> 13 3
DATA> 18 3
DATA> 8 3
DATA> 6 3
DATA> 11 3
DATA> END DATA
       25 ROWS READ
MTB > ONEWAY ANOVA,DATA IN C1,LEVELS IN C2

ANALYSIS OF VARIANCE ON C1
SOURCE     DF      SS       MS       F
C2          2     214.7    107.4    4.65
ERROR      22     507.9     23.1
TOTAL      24     722.6
```

12.8 Refer to Case Study 12-2.

    a. State the null hypothesis being tested.

    b. Verify the n, $\bar{x}$, and s reported for method I.

    c. What was the value of the calculated F?

    d. Approximate the critical value of F if $\alpha = 0.05$.

    e. Why did Mr. Stephens say "The F-value is very nonsignificant"?

12.9 A new operator was recently assigned to a crew of workers who perform a certain job. From the records that show the number of units of work completed by each worker each day last month, a sample of size five was randomly selected for each of the two experienced workers and the new worker. At the 0.05 level of significance, does the evidence provide sufficient reason to reject the claim that there is no difference in the amount of work done by the three workers?

Units of work (replicates)

	Workers		
	New	A	B
	8	11	10
	10	12	13
	9	10	9
	11	12	12
	8	13	13

12.10 A cookie salesman interested in increasing the sales volume of cookies in a large supermarket arranged with the manager to display his product in three different locations in the market over a 15-week period. Each week he moved the display to a different location. The following sales figures resulted.

Amount of cookie
sales per week

Location of Display		
By Meat Counter	By Checkout	In Cookie Section
32	45	30
44	63	28
37	44	16
39	59	21
23	34	22

At the 0.05 level of significance, do the sample data provide sufficient reason to reject the null hypothesis that all locations are equally desirable for selling cookies?

In
Retrospect

In this chapter we have presented an introduction to the statistical techniques known as analysis of variance. The techniques studied here were restricted to the test of a hypothesis that dealt with questions about means from several populations. We were restricted to normal populations and populations with homogeneous (equal) variances. The test of multiple means is accomplished by partitioning the sum of squares into two segments: (1) the sum of squares due to variation between the levels of the factor being tested, and (2) the sum of squares due to the variation between the replicates within each level. The null hypothesis about means is then tested by using the appropriate variance measurements.

Note that we restricted our development to one-factor experiments. This one-factor technique represents only a beginning to the study of analysis of variance techniques.

Refer back to the news article at the beginning of this chapter and you will see that the data given use the method of tillage (plow, and so on) as the factor and the yield from each field plot as the replicate of a one-factor analysis of variance experiment. (*See* Exercise 12.15.)

Chapter
Exercises

12.11 An experiment compared typing speeds for clerk-typists on a standard electric typewriter and on a Teletype Model 43 computer terminal. Twelve typists were randomly assigned to type on the two machines. The scores are shown in the following table.

Standard Electric	Terminal
62	52
78	60
48	47
63	48
55	52
51	40

Is there sufficient evidence to conclude that there is a difference in the population means for the two types of machines? Use $\alpha = 0.05$.

12.12 An employment agency wants to see which of three types of ads in the help-wanted section of local newspapers is the most effective. Three types of ads (big headline, straightforward, and bold print) were randomly alternated over a period of weeks and the number of people responding to the ads were noted each week. Do these data support the null hypothesis that there is no difference in the effectiveness of the ads, as measured by the mean number responding, at the 0.01 level of significance?

Number of responses (replicates)

Type of Advertisement		
Big Headline	Straightforward	Bold Print
23	19	28
42	31	33
36	18	46
48	24	29
33	26	34
26		34

12.13 Fifteen patients were randomly divided into three groups for the purpose of testing three different methods of controlling their hypertension. One method used exercise, one method used dietary techniques, and another method used a standard drug treatment. The reduction in diastolic blood pressure was recorded after a six-month period. The results were as follows:

Method		
Diet	Exercise	Drug
10	8	8
12	12	6
10	10	6
14	10	10
10	6	6

Is there significant variation in the data to reject the null hypothesis $H_0: \mu_{Diet} = \mu_{Exer.} = \mu_{Drug}$, at $\alpha = 0.05$?

12.14 A farm consisting of 15 plots was randomly treated with three concentrations of fertilizer, each on 5 plots. The yield of corn (in bushels) was recorded for each plot. The results were as follows:

	Concentration	
1	2	3
40	50	30
45	52	36
42	50	36
46	54	38
44	57	40

Do these data show sufficient evidence to reject the null hypothesis "there is no difference in yield due to concentration differences" at $\alpha = 0.05$?

12.15 Do the sample results given in the news article at the beginning of this chapter show sufficient evidence to support the claim that the tillage tool used to prepare the ground for planting affects the mean yield? Complete the hypothesis test at the 0.05 level of significance.

12.16 The following table shows the number of arrests made last year for violations of the narcotic drug laws in 24 communities. The data given are rates of arrest per 10,000 inhabitants. At $\alpha = 0.05$, is there sufficient evidence to reject the hypothesis that the mean rates of arrests are the same in all four sizes of communities?

Cities (over 250,000)	Cities (under 250,000)	Suburban Communities	Rural Communities
45	23	25	8
34	18	17	16
41	27	19	14
42	21	28	17
37	26	31	10
28	34	37	23

12.17 The distance required to stop a vehicle on wet pavement was measured to compare the stopping power of four major brands of tires. A tire of each brand was tested on the same vehicle on a controlled wet pavement. The resulting distances are shown in the following table. At $\alpha = 0.05$, is there sufficient evidence to conclude that there is a difference in the mean stopping distance?

Distance (replicate)

	Brand of Tire		
A	B	C	D
37	37	33	41
34	40	34	41
38	37	38	40
36	42	35	39
40	38	42	41
32		34	43

12.18 To compare the effectiveness of three different methods of teaching reading, 26 children of equal reading aptitude were divided into three groups. Each group was instructed for a given period of time using one of the three methods. After completing the instruction period, all students were tested. The test results are shown in the following table. Is the evidence sufficient to reject the hypothesis that all three instruction methods are equally effective? Use $\alpha = 0.05$.

Test scores (replicates)

Method I	Method II	Method III
45	45	44
51	44	50
48	46	45
50	44	55
46	41	51
48	43	51
45	46	45
48	49	47
47	44	

12.19 An experiment was designed to compare the lengths of time that four different drugs provided pain relief following surgery. The results (in hours) are shown in the following table.

Drug			
A	B	C	D
8	6	8	4
6	6	10	4
4	4	10	2
2	4	10	
		12	

Is there enough evidence to reject the null hypothesis that there is no significant difference in the length of pain relief for the four drugs at $\alpha = 0.05$?

12.20 A certain vending company's soft-drink dispensing machines are supposed to serve six ounces of beverage. Various machines were sampled and the resulting amounts of dispensed drink were recorded, as shown in the following table. Does this sample evidence provide sufficient reason to reject the null hypothesis that all five machines dispense the same average amount of soft drink? Use $\alpha = 0.01$.

Amounts of soft drink dispensed

Machines				
A	B	C	D	E
3.8	6.8	4.4	6.5	6.2
4.2	7.1	4.1	6.4	4.5
4.1	6.7	3.9	6.2	5.3
4.4		4.5		5.8

12.21 Seven golf balls from each of six manufacturers were randomly selected and tested for durability. Each ball was hit 300 times or until failure occurred, whichever came first. Do these sample data show sufficient reason to reject the null hypothesis that the six different brands tested withstood the durability test equally well? Use $\alpha = 0.05$.

Number of hits (replicate)

		Manufacturer Brand			
A	B	C	D	E	F
300	190	228	276	162	264
300	164	300	296	175	168
300	238	268	62	157	254
260	200	280	300	262	216
300	221	300	230	200	257
261	132	300	175	256	183
300	156	300	211	92	93

12.22 For the following data, show that

$$SS(factor) = k_1(\bar{x}_1 - \bar{x})^2 + k_2(\bar{x}_2 - \bar{x})^2 + k_3(\bar{x}_3 - \bar{x})^2$$

where $\bar{x}_1, \bar{x}_2, \bar{x}_3$ are the means for the three factor levels and $\bar{x}$ is the overall mean.

Factor Level		
1	2	3
6	13	9
8	12	11
10	14	7

Vocabulary List

Be able to define each term. In addition, describe in your own words and give an example of each term. Your examples should not be ones given in class or in the textbook.

The bracketed numbers indicate the chapters in which the term previously appeared, but you should define the terms again to show increased understanding of their meaning.

analysis of variance
ANOVA
degrees of freedom [9, 10, 11]
experimental error
level
mathematical model
mean square
partition

randomize [2]
replicate
response variable [1]
sum of squares
variance [2, 9, 10]
variation between levels
variation within a level
variation within rows

Quiz

Answer "True" if the statement is always true. If the statement is not always true, replace the boldface words with words that make the statement always true.

12.1 To partition the sum of squares for the total is to separate the numerical value of SS(total) into two values such that the **sum** of these two values is equal to SS(total).

12.2 A **sum of squares** is actually a measure of variance.

12.3 **Experimental error** is the name given to the variability that takes place between the levels of the test factor.

12.4 **Experimental error** is the name given to the variability that takes place among the replicates of an experiment as it is repeated under constant conditions.

12.5 **Fail to reject** H_0 is the desired decision when the means for the levels of the factor being tested are all different.

12.6 The **mathematical model** for a particular problem is an equational statement showing the anticipated makeup of an individual piece of data.

12.7 The degrees of freedom for the factor are equal to **the number** of factors tested.

12.8 The measure of a specific level of a factor being tested in an ANOVA is the **variance** of that factor level.

12.9 We **need not** assume that the observations are independent to do analysis of variance.

12.10 The rejection of H_0 **indicates** that you have identified the level(s) of the factor that is (are) different from the others.

Chapter 13

Linear Correlation and Regression Analysis

NUTRITIONAL OUTCOME OF 207 VERY LOW-BIRTH-WEIGHT INFANTS IN AN INTENSIVE CARE UNIT

Feeding and growth outcome for 207 critically ill, very low-birth-weight infants were highly correlated with severity of illness.

During the past decade, intensive care centers for newborns throughout the country have witnessed a dramatic increase in the number of admissions and in the survival rate of very low-birth-weight (VLBW) infants. From 1974 to 1977, the number of babies admitted to the James Whitcomb Riley Hospital for Children who weighed from 501 to 1,000 gm. doubled, and the number who weighed from 1,001 to 1,500 gm. tripled, while total admissions remained unchanged. Over the same period, survival of the 1,001 to 1,500 gm. infants reached 80 percent, while that of the 501 to 1,000 gm. babies approached 50 percent.

Provision of adequate nutrition to the VLBW infant is a major challenge because of the extreme physiological immaturity and severe medical complications often encountered. A nutrition care protocol and arbitrary guidelines for growth were developed (1) to provide a more uniform approach to the nutrition support of these extremely high-risk infants. Data for the present investigation were collected in a prospective fashion over a two-year period to: (a) evaluate the early feeding and growth outcome of a large population of VLBW infants managed with the same general nutrition guidelines and (b) assess the impact of several factors, including sex, appropriateness of size for gestational age, and degree of illness, on feeding and growth outcome during the first weeks of life.

PROCEDURES AND METHODS

Population The population consisted of 207 VLBW infants less than 48 hours old who were sequentially admitted to the tertiary (level III) intensive care unit for newborns at the James Whitcomb Riley Hospital for Children during 1978 and 1979. All infants were 35 weeks' gestation or less, with birth weights of less than 1,500 gm. Eighty-two percent of the infants were outborn, requiring transport to the James Whitcomb Riley Hospital for Children. Eighty-two percent of the babies required ventilator support, reflecting the intensity of care required by this group of infants.

Complete data were obtained prospectively for each infant. During the initial hospitalization, 167 babies survived. The babies who survived and their respective weights and gestational ages are indicated in Figure 1.

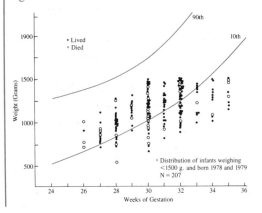

FIG. 1. The weights and gestational ages of very low-birth-weight infants who survived (•) or died (○) during 1978 and 1979. The reference weight percentiles by gestational age are those of Lubchenco *et al* (14).

Source: Rickard *et al.*: "Nutritional Outcome of 207 Very Low-Birth-Weight Infants in an Intensive Care Unit." Copyright The American Dietetic Association. Reprinted by permission from *Journal of the American Dietetic Association*, Vol. 81: 674, 1982.

Chapter
Objectives

In Chapter 3 the basic ideas of regression and linear correlation analysis were introduced. (If these concepts are not fresh in your mind, review Chapter 3 before beginning this chapter.) Chapter 3 was only a first look—a presentation of the basic graphic and descriptive statistical aspects of linear correlation and regression analysis. In this chapter we will take a second, more detailed look at linear correlation and regression analysis.

Previously we used the linear correlation coefficient to measure the strength of the linear relationship between two variables. Now we will determine whether there is a linear relationship by using a hypothesis test in which the probability of a type I error is fixed by the value assigned to α. In Chapter 3 we introduced a set of formulas for finding the equation of the straight line of best fit. Now we wish to ask "Is the linear equation of any real use?" Previously we used the equation of the line of best fit to make point predictions. Now we will make confidence interval estimations. In short, this second look at linear correlation and regression analysis will be a much more complete presentation than that in Chapter 3.

Recall that **bivariate data** are ordered pairs of numerical values. The two values are each response variables. They are paired with each other as a result of a common bond (see page 203).

13.1 LINEAR CORRELATION ANALYSIS

In Chapter 3 the linear correlation coefficient was presented as a quantity that measures the strength of a linear relationship (dependency). Now let's take a second look at this concept and see how r, the coefficient of linear correlation, works. Intuitively, we want to think about how to measure the mathematical linear dependency of one variable on another. As x increases, does y tend to increase (decrease)? How strong (consistent) is this tendency? We are going to use two measures of dependence—covariance and the coefficient of linear correlation—to measure the relationship between two variables. We'll begin our discussion by examining a set of bivariate data and identifying some related facts as we prepare to define covariance.

Illustration 13-1   Let's consider the following sample of six pieces of bivariate data: (2, 1), (3, 5), (6, 3), (8, 2), (11, 6), (12, 1). (*See* Figure 13-1.) The mean of the six x values (2, 3, 6, 8, 11, 12) is $\bar{x} = 7$. The mean of the six y values (1, 5, 3, 2, 6, 1) is $\bar{y} = 3$.

Figure 13-1
Graph of data for
Illustration 13-1

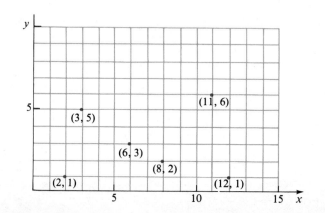

The point $(\bar{x}, \bar{y})$, which is (7, 3), is located as shown on the graph of the sample

centroid   points in Figure 13-2. The point $(\bar{x}, \bar{y})$ is called the **centroid** of the data. If a vertical and a horizontal line are drawn through the centroid, the graph is divided into four sections, as shown in Figure 13-2. Each point (x, y) lies a certain distance from each of these two lines. $(x - \bar{x})$ is the horizontal distance from (x, y) to the vertical line passing through the centroid. $(y - \bar{y})$ is the vertical distance from (x, y) to the horizontal line passing through the centroid. Both the horizontal and vertical distances of each data point from the centroid can be measured, as shown in Figure 13-3. The distances may be positive, negative, or zero, depending on the position of the point (x, y) in reference to $(\bar{x}, \bar{y})$. [$(x - \bar{x})$ and $(y - \bar{y})$ are represented by means of braces, with positive or negative signs, as shown in Figure 13-3.]

Figure 13-2

The point (7, 3) is
the centroid

Figure 13-3

Measuring the distance of each
data point from the centroid

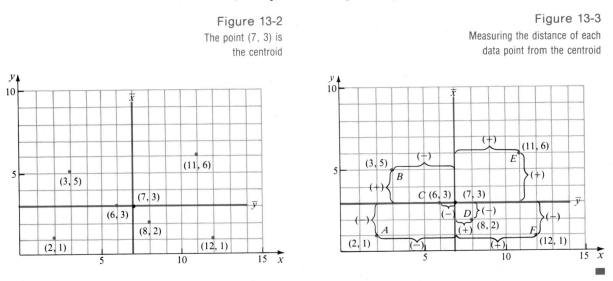

covariance   One measure of linear dependency is the covariance. The **covariance of x and y is** defined as the sum of the products of the distances of all values of x and y from the centroid, $\sum[(x - \bar{x})(y - \bar{y})]$, divided by $n - 1$:

$$\text{covar}(x, y) = \frac{\sum_{i=1}^{n}(x_i - \bar{x})(y_i - \bar{y})}{n - 1} \qquad (13\text{-}1)$$

The covariance for the data given in Illustration 13-1 is calculated in Table 13-1. The covariance, written as covar (x, y), of the data is $+\frac{3}{5} = 0.6$.

Table 13-1

Calculations for finding
covar (x, y) for the data of
Illustration 13-1

Points	$x - \bar{x}$	$y - \bar{y}$	$(x - \bar{x})(y - \bar{y})$
A(2, 1)	-5	-2	10
B(3, 5)	-4	2	-8
C(6, 3)	-1	0	0
D(8, 2)	1	-1	-1
E(11, 6)	4	3	12
F(12, 1)	5	-2	-10
Total	0	0	3

Note $\sum (x - \bar{x}) = 0$ and $\sum (y - \bar{y}) = 0$. This will always happen. Why? (See page 59).

The covariance is positive if the graph is dominated by points to the upper right and to the lower left of the centroid. The products of $(x - \bar{x})$ and $(y - \bar{y})$ are positive in these two sections. If the majority of the points are in the upper-left and lower-right sections relative to the centroid, the sum of the products is negative. Figure 13-4 shows data that represent a positive dependency (a), a negative dependency (b), and little or no dependency (c). The covariances for these three situations would definitely be positive in part (a), negative in (b), and near zero in (c). (The sign of the covariance is always the same as the sign of the slope of the regression line.)

Figure 13-4

Data and covariance

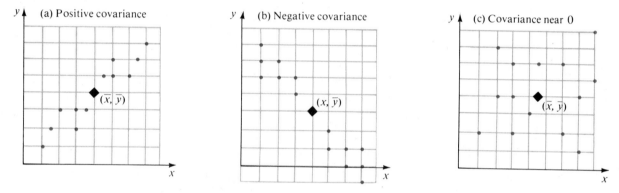

The biggest disadvantage of covariance as a measure of linear dependency is that it does not have a standardized unit of measure. One reason for this is that the spread of the data is a strong factor in the size of the covariance. For example, if we were to multiply each data point in Illustration 13-1 by 10, we would have (20, 10), (30, 50), (60, 30), (80, 20), (110, 60), and (120, 10). The relationship of the points to each other would be changed only in that they would be much more spread out. However, the covariance for this new set of data is 60. (This calculation is assigned in Exercise 13.4.) Does this mean that the amount of dependency between the x and y variables is stronger than in the original case? No, it does not; the relationship is the same, even though each data point was multiplied by 10. This is the trouble with covariance as a measure. We must find some way to eliminate the effect of the spread of the data when we measure dependency.

If we standardize x and y by dividing the distance of each from the respective mean by the respective standard deviation,

$$x' = \frac{x - \bar{x}}{s_x} \quad \text{and} \quad y' = \frac{y - \bar{y}}{s_y}$$

and then compute the covariance of x' and y', we will have a covariance that is *not* affected by the spread of the data. This is exactly what is accomplished by the linear correlation coefficient. It divides the covariance of x and y by a measure of the spread of x and by a measure of the spread of y (the standard deviations of x and of y are used as measures of spread). Therefore, by definition, the **coefficient of linear correlation** is

$$r = \text{covar}(x', y') = \frac{\text{covar}(x, y)}{s_x \cdot s_y} \qquad (13\text{-}2)$$

Pearson's product
moment

The coefficient of linear correlation standardizes the measure of dependency and allows us to compare the relative strengths of dependency of different sets of data. (Formula (13-2) for linear correlation is also commonly referred to as **Pearson's product moment, r.**)

The value of r, the coefficient of linear correlation, for the data in Illustration 13-1 can be found by calculating the two standard deviations and then dividing:

$$s_x = 4.099 \qquad \text{and} \qquad s_y = 2.098$$

$$r = \frac{0.6}{(4.099)(2.098)} = \textbf{0.07}$$

Finding the correlation coefficient by using formula (13-2) can be a very tedious arithmetic process. The formula can be written in a more workable form, as it was given in Chapter 3:

$$r = \frac{\text{covar}(x, y)}{s_x \cdot s_y} = \frac{\sum [(x - \bar{x}) \cdot (y - \bar{y})]/(n - 1)}{s_x \cdot s_y}$$

$$= \frac{\text{SS}(xy)}{\sqrt{\text{SS}(x) \cdot \text{SS}(y)}} \qquad (13\text{-}3)$$

Formula (13-3) avoids the separate calculations of $\bar{x}$, $\bar{y}$, s_x, and s_y, and, more importantly, the calculations of the deviations from the means. Therefore, formula (13-3) is much easier to use. (Refer to Chapter 3 for an illustration of the use of this formula.)

EXERCISES

13.1 Explain why $\sum (x - \bar{x}) = 0$ and $\sum (y - \bar{y}) = 0$.

13.2 The weight, x, and the waist size, y, were determined for 11 women. The data were as follows:

x	110	143	120	127	143	111	137	154	123	104	128
y	22	29	27	26	27	24	28	28	26	25	27

Verify the correlation coefficient in the following MINITAB output.

```
MTB > READ X IN C1,Y IN C2
DATA> 110 22
DATA> 143 29
DATA> 120 27
DATA> 127 26
DATA> 143 27
DATA> 111 24
DATA> 137 28
DATA> 154 28
DATA> 123 26
DATA> 104 25
DATA> 128 27
DATA> END DATA
        11 ROWS READ
MTB > CORRELATION C1,C2

Correlation of C1 and C2 = 0.805

MTB > STOP
```

13.3 a. Construct a scatter diagram of the following bivariate data.

	Point									
	A	B	C	D	E	F	G	H	I	J
x	1	1	3	3	5	5	7	7	9	9
y	1	2	2	3	3	4	4	5	5	6

  b. Calculate the covariance.

  c. Calculate s_x and s_y.

  d. Calculate r using formula (13-2).

  e. Calculate r using formula (13-3).

13.4 a. Calculate the covariance of the set of data (20, 10), (30, 50), (60, 30), (80, 20), (110, 60), and (120, 10).

  b. Calculate the standard deviation of the six x values and the standard deviation of the six y values.

  c. Calculate r, the coefficient of linear correlation, for the data in part (a).

  d. Compare these results to those found in the text for Illustration 13-1.

13.5 Consider the accompanying bivariate data.

	Point									
	A	B	C	D	E	F	G	H	I	J
x	0	1	1	2	3	4	5	6	6	7
y	6	6	7	4	5	2	3	0	1	1

  a. Draw a scatter diagram for the data.

  b. Calculate the covariance.

  c. Calculate s_x and s_y.

    d. Calculate r by formula (13-2).

    e. Calculate r by formula (13-3).

13.6 A formula that is sometimes given for computing the correlation coefficient is

$$r = \frac{n(\sum xy) - (\sum x)(\sum y)}{\sqrt{n(\sum x^2) - (\sum x)^2}\sqrt{n(\sum y^2) - (\sum y)^2}}$$

Use this expression as well as the formula

$$r = \frac{SS(xy)}{\sqrt{SS(x) \cdot SS(y)}}$$

to compute r for the data in the following table.

x	2	4	3	4	0
y	6	7	5	6	3

13.2 INFERENCES ABOUT THE LINEAR CORRELATION COEFFICIENT

After the linear correlation coefficient r has been calculated for the sample data, it seems necessary to ask this question: Does the value of r indicate that there is a linear dependency between the two variables in the population from which the sample was drawn? To answer this question we can perform a hypothesis test. The null hypothesis is "the two variables are linearly unrelated" ($\rho = 0$), where ρ (the lowercase Greek letter rho) is the **linear correlation coefficient for the population**. The alternative hypothesis may be either one-tailed or two-tailed. Most frequently it is two-tailed. However, when we suspect that there is only a positive or only a negative correlation, we should use a one-tailed test. The alternative hypothesis of a one-tailed test is $\rho > 0$ or $\rho < 0$.

    The critical region for the test is on the right when a positive correlation is expected and on the left when a negative correlation is expected. The test statistic used to test the null hypothesis is the calculated value of r from the sample. Critical values for r are found in Table 9 of Appendix E at the intersection of the column identified by the appropriate value of α and the row identified by the degrees of freedom. The number of degrees of freedom for the r statistic is 2 less than the sample size, **df = $n - 2$**.

    The rejection of the null hypothesis means that there is evidence of a linear dependency between the two variables in the population.

Caution

The sample evidence may say only that the pattern of behavior of the two variables is related in that one can be used effectively to predict the other. **This does not mean that you have established a cause-and-effect relationship.**

Failure to reject the null hypothesis is interpreted as meaning that linear dependency between the two variables in the population has not been shown.

Now let's look at such a hypothesis test.

Illustration 13-2   For Illustration 13-1, where $n = 6$, we found $r = 0.07$. Is this significantly different from zero, at the 0.02 level of significance?

Solution

Step 1 $H_0: \rho = 0$.

$H_a: \rho \neq 0$.

Step 2 $\alpha = 0.02$, df $= n - 2 = 6 - 2 = 4$. See the accompanying figure.

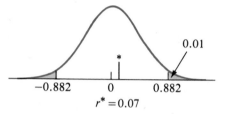

The critical values (-0.882 and 0.882) were obtained from Table 9 of Appendix E.

Step 3   The calculated value of r, r^*, was found earlier. It is $r^* = 0.07$.

Step 4   *Decision*: Fail to reject H_0.

Conclusion: At the 0.02 level of significance, we have failed to show that x and y are correlated.   ■

As in other problems, a confidence interval estimate of the population correlation coefficient is sometimes required. It is possible to estimate the value of ρ, the linear correlation coefficient of the population. Usually this is accomplished by using a table showing **confidence belts**. Table 10 in Appendix E gives confidence belts for 95 percent confidence interval estimates. This table is a bit tricky to read, so be extra careful when you use it. The next illustration demonstrates the procedure for estimating ρ.

confidence belts

Illustration 13-3   A sample of 15 ordered pairs of data have a calculated r value of 0.35. Find the 95 percent confidence interval estimate for ρ, the population linear correlation coefficient.

Solution   Find $r = 0.35$ at the bottom of Table 10. (*See* the arrow on Figure 13-5.) Visualize a vertical line through that point. Find the two points where the belts marked for the correct sample size cross the vertical line. The sample size is 15. These two points are circled in Figure 13-5. Now look horizontally from the two circled points to the vertical scale on the left and read the confidence interval. The values are 0.72 and -0.20. Thus the 95 percent confidence interval estimate for ρ, the population coefficient of linear correlation, is **-0.20 to 0.72**.

Figure 13-5

Using Table 10 of Appendix E,
confidence belts for the
correlation coefficient

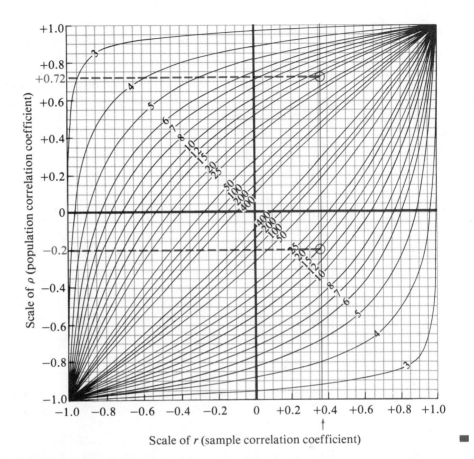

Scale of r (sample correlation coefficient)

Case Study 13-1

PERSONALITY CHARACTERISTICS OF
POLICE APPLICANTS

*Pearson's linear correlation coefficient is often used in the study of bivariate data.
Bruce N. Carpenter and Susan M. Raza report correlation coefficients for several
age-related variables (and their p-values) in their study of personality characteristics of police applicants. The following excerpt also contains the conclusions
Carpenter and Raza reached as a result of the correlation analysis.*

The Minnesota Multiphasic Personality Inventory (MMPI) continues to be
a very widely used measure in the selection of police officers and other law
enforcement officials (for example, Parisher, Rios, and Reilley 1979). The
characterization of police applicants with the MMPI is therefore important both
for better understanding how the measure works as a selection device and for
gaining increased understanding of applicant characteristics, the latter of which
is examined in this article.

The age of the male applicants was correlated with their MMPI scale scores
to reveal any age-related differences. Older male applicants obtained significantly
lower K scores, $\underline{r}(237) - .11$ p < .05. However, they obtained significantly higher
scores on scale 1, $\underline{r}(237) = .27$ p < .001; scale 2, $\underline{r}(237) = .22$ p < .001; scale 3,

$\underline{r}(237) = .15$ $\underline{p} < .01$; and scale 0, $\underline{r}(237) = .14$ $\underline{p} < .05$ than younger male applicants. This suggests that older applicants tend to be less satisfied, to have more physical complaints, to be more likely to develop physical symptoms in response to stress, and to be more conservative than their younger counterparts. With the exception of developing physical symptoms in response to stress, these personality characteristics have been found to be related to age in a random sample of United States males (Colligan *et al.* 1984).

Source: Bruce N. Carpenter and Susan M. Raza, reproduced from the *Journal of Police Science and Administration*, Vol. 15, No. 1, pp. 10–17, with permission of the International Association of Chiefs of Police, P.O. Box 6010, 13 Firstfield Road, Gaithersburg, Maryland 20878.

Case Study 13-2

DETROIT HANGS TOUGH

Two graphs appear in Financial World's *article "Detroit Hangs Tough." One shows the cost of a new car for each year from 1968 through 1986 and the other shows the annual median family income for the same years. The graphs are similar in that both show a consistently increasing pattern. This suggests that the two variables are linearly related. Construct a scatter diagram using x as the annual median income (the input variable) and y as the annual cost and expenditures per new car (the output variable). You will see a strong linear relationship. (See Exercise 13.14.)*

At the same time, personal incomes have been rising slowly but steadily, so that cars now cost less to buy and run as a percentage of disposable income. Two decades ago, the last time Detroit had it really palmy, a new car cost the equivalent of 21.1 weeks of average income. Today, a vastly improved, safer, more fuel-conscious car costs an American only 17.2 weeks of income. Meanwhile, the average auto loan maturity has increased from less than three years to almost five, so that auto payments have grown relatively slowly in comparison with other finance costs. Put it all together and cars are now almost bargains compared to other major purchases. That helps explain why Americans are renewing their long-standing love affair with bigger American cars.

New car prices and consumer expenditures per new car

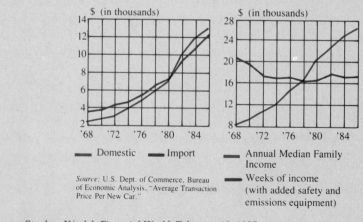

Source: U.S. Dept. of Commerce. Bureau of Economic Analysis. "Average Transaction Price Per New Car."

Source: Stephen Kindel, *Financial World*, February 10, 1987.

EXERCISES

13.7 Using graphs to illustrate, explain the meaning of a correlation coefficient whose value is, (a) -1, (b) 0, (c) $+1$.

13.8 A sample of 20 pieces of bivariate data has a linear correlation coefficient of $r = 0.43$. Does this provide sufficient evidence to reject the null hypothesis that $\rho = 0$ in favor of a two-sided alternative? Use $\alpha = 0.10$.

13.9 If a sample of size 18 has a linear correlation coefficient of -0.50, is there significant reason to conclude that the linear correlation coefficient of the population is negative? Use $\alpha = 0.01$.

13.10 A sample of size 10 produced $r = -0.67$. Is this sufficient evidence to conclude that ρ is different from zero, at the 0.05 level of significance?

13.11 Is a value of $r = +0.24$ significant in trying to show that ρ is greater than zero for a sample of 62 data, at the 0.05 level of significance?

13.12 Use Table 10 of Appendix E to determine a 0.95 confidence interval estimate for the true population linear correlation coefficient based on the following sample statistics:

    a. $n = 8, r = 0.20$ **b.** $n = 100, r = -0.40$

    c. $n = 25, r = +0.65$ **d.** $n = 15, r = -0.23$

13.13 "The ages of mated pairs of California gulls," the news article for Chapter 3, reported a linear coefficient of $r = 0.77$ and an associated p-value of $p < 0.001$.

    a. What two variables were being studied?

    b. Explain the meaning of "$r = 0.77$ with $p < 0.001$."

13.14 a. Construct a scatter diagram using the information shown in Case Study 13-2. Let x represent the annual median family income and y represent the annual cost and expenditures for a new car for the corresponding year.

    b. Estimate the linear correlation coefficient for the variables shown on the scatter diagram in (a). Do these variables appear to be linearly correlated? Explain.

13.15 The population (in millions) and the violent crime rate (per 1000) were recorded for ten metropolitan areas. The data are shown in the following table.

Population	10.0	1.3	2.1	7.0	4.4	0.3	0.3	0.2	0.2	0.4
Crime Rate	12.0	9.5	9.2	8.4	8.2	7.3	7.1	7.0	6.9	6.9

Do these data provide evidence to reject the null hypothesis that $\rho = 0$ in favor of $\rho \neq 0$, at $\alpha = 0.05$.

13.16 The Test-Retest Method is one way of establishing the reliability of a test. The test is administered and then, at a later date, the same test is readministered to the same individuals. The correlation coefficient is computed between the two sets of scores. The following test scores were obtained in a Test-Retest situation.

First Score	75	87	60	75	98	80	68	84	47	72
Second Score	72	90	52	75	94	78	72	80	53	70

Find r and set a 95 percent confidence interval for ρ.

13.3 LINEAR REGRESSION ANALYSIS

Recall that the line of best fit results from an analysis of two or more related variables. (We will restrict our work to two variables. However, on occasion more than two will be mentioned to clarify the analysis.) When two variables are studied jointly, we often would like to control one variable by means of controlling the other. Or we might want to predict the value of a variable based on knowledge about another variable. In both cases we want to find the line of best fit, provided one exists, that will best predict the value of the dependent, or output, variable. Recall that the variable we know or can control is called the **independent**, or **input, variable**; the variable resulting from using the equation of the line of best fit is called the **dependent**, or **predicted, variable**.

Recall that in Chapter 3 the *method of least squares* was developed. From this concept formulas (3-6) and (3-7) were obtained and are used to calculate β_0 (the y-intercept) and b_1 (the slope of the line of best fit).

$$b_0 = \frac{1}{n}\left(\sum y - b_1 \cdot \sum x\right) \tag{3-6}$$

$$b_1 = \frac{SS(xy)}{SS(x)} \tag{3-7}$$

Then these two coefficients are used to write the equation for the line of best fit in the form

$$\hat{y} = b_0 + b_1 x$$

When the line of best fit is plotted, it does more than just show us a pictorial representation. It tells us two things: (1) whether or not there really is a functional (equational) relationship between the two variables, and (2) the quantitative relationship between the two variables. Recall that the line of best fit is of no use when a change in the input variable does not seem to have a definite effect on the output variable. When there is no relationship between the variables, a horizontal line of best fit will result. A horizontal line has a slope of zero, which implies that the value of the input variable has no effect on the output variable. (This idea will be amplified later in this chapter.)

The result of regression analysis is the mathematical equation that is the equation of the line of best fit. We will, as mentioned before, restrict our work to the simple linear case, that is, one input variable and one output variable, where the line of best fit is straight. However, you should be aware that not all relationships are of

curvilinear regression

this nature. If the scatter diagram suggests something other than a straight line, we have **curvilinear regression**. In cases of this type we must introduce terms to higher powers, x^2, x^3, and so on, or other functions, e^x, log x, and so on; or we must introduce other input variables. Maybe two or three input variables would improve the usefulness of our regression equation. These possibilities are examples of curvilinear regression and multiple regression.

The **linear model** used to explain the behavior of linear bivariate data **in the population** is

$$y = \beta_0 + \beta_1 x + \varepsilon \qquad (13\text{-}4)$$

This equation represents the linear relationship between the two variables in a population. β_0 **is the y-intercept and** β_1 **is the slope.** ε (lowercase Greek letter epsilon) **is the random experimental error** in the observed value of y at a given value of x.

experimental error
(ε or e)

The regression line from the sample data gives us b_0, which is **our estimate of** β_0, and b_1, which is **our estimate of** β_1. The **error** ε **is approximated by** $e = y - \hat{y}$, the difference between the observed value of y and the predicted value of y, $\hat{y}$, at a given value of x.

$$e = y - \hat{y} \qquad (13\text{-}5)$$

The random variable e is positive when the observed value of y is larger than the predicted value $\hat{y}$; e is negative when y is smaller than $\hat{y}$. The sum of the errors for the different values of y for a given value of x is exactly zero. (This is part of the least squares criteria.) Thus the mean value of the experimental error is zero; its variance is σ_ε^2. Our next goal is to estimate this variance of the experimental error.

Before we estimate the variance of ε, let's try to understand exactly what the error represents. ε is the amount of error in our observed value of y. That is, it is the difference between the observed value of y and the mean value of y at that particular value of x. Since we do not know the mean value of y, we will use the regression equation and estimate it with $\hat{y}$, the predicted value of y at this same value of x. Thus the best estimate that we have for ε is $e = (y - \hat{y})$, as shown in Figure 13-6.

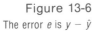

Figure 13-6
The error e is $y - \hat{y}$

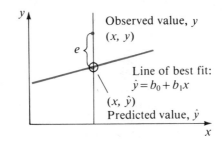

Note e is the observed error in measuring y at a specified value of x.

If we were to observe several values of y at a given value of x, we could plot a distribution of y values about the line of best fit (about $\hat{y}$, in particular). Figure 13-7 shows a sample of bivariate values for which the value of x is the same. Figure 13-8 shows the theoretical distribution of all possible y values at a given x value. A similar

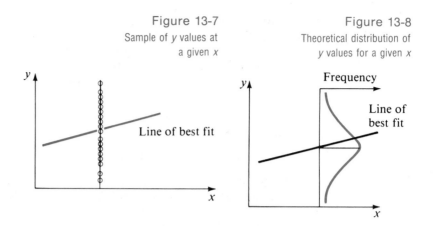

Figure 13-7

Sample of y values at a given x

Line of best fit

Figure 13-8

Theoretical distribution of y values for a given x

Frequency

Line of best fit

distribution occurs at each different value of x. The mean of the observed y's at a given value of x varies, but it can be estimated by $\hat{y}$.

Before we can make any inferences about a regression line, we must assume that the distribution of y's is approximately normal and that the variances of the distributions of y at all values of x are the same. That is, the standard deviation of the distribution of y about $\hat{y}$ is the same for all values of x, as shown in Figure 13-9.

Figure 13-9

Standard deviation of the distribution of y values for all x is the same

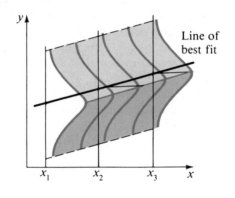

Line of best fit

Before looking at the variance of e, let's review the definition of sample variance. The sample variance s^2 is defined as $\sum(x - \bar{x})^2/(n - 1)$, the sum of the squares of each deviation divided by the number of degrees of freedom, $n - 1$, associated with a sample of size n. The variance of y involves an additional complication: There is a different mean for y at each value of x. (Notice the many distributions in Figure 13-9.) However, each of these "means" is actually the predicted value, $\hat{y}$, that corresponds to the x that fixes the distribution. So the **variance of the error e** is estimated by the formula

$$s_e^2 = \frac{\sum(y - \hat{y})^2}{n - 2}$$

(13-6)

where $n - 2$ is the number of degrees of freedom.

Note   The variance of y about the line of best fit is the same as the variance of the error e. Recall that $e = y - \hat{y}$.

Formula (13-6) can be rewritten by substituting $b_0 + b_1 x$ for $\hat{y}$. Since $\hat{y} = b_0 + b_1 x$, then s_e^2 becomes

$$s_e^2 = \frac{\sum(y - b_0 - b_1 x)^2}{n - 2} \qquad (13\text{-}7)$$

With some algebra and some patience, this formula can be rewritten once again into a more workable form. The form we will use is

$$s_e^2 = \frac{(\sum y^2) - (b_0)(\sum y) - (b_1)(\sum xy)}{n - 2} \qquad (13\text{-}8)$$

For ease of discussion let's agree to call the numerator of formulas (13-6), (13-7), and (13-8) the **sum of squares for error (SSE)**.

Now let's see how all this information can be used.

Illustration 13-4   Suppose that you move into a new city and take a job. You will, of course, be concerned about the problems you will face commuting to and from work. For example, you would like to know how long (in minutes) it will take you to drive to work each morning. Let's use "one-way distance to work" as a measure of where you live. You live x miles away from work and want to know how long it will take you to commute each day. Your new employer, foreseeing this question, has already collected a random sample of data to be used in answering your question. Fifteen of your co-workers were asked to give their one-way travel time and distance to work. The resulting data are shown in Table 13-2. (For convenience the data have been arranged so that the x values are in numerical order.) Find the line of best fit and the variance of y about the line of best fit, s_e^2.

Table 13-2
Data for Illustration 13-4

Co-worker	Miles (x)	Minutes (y)	x^2	xy	y^2
1	3	7	9	21	49
2	5	20	25	100	400
3	7	20	49	140	400
4	8	15	64	120	225
5	10	25	100	250	625
6	11	17	121	187	289
7	12	20	144	240	400
8	12	35	144	420	1,225
9	13	26	169	338	676
10	15	25	225	375	625
11	15	35	225	525	1,225
12	16	32	256	512	1,024
13	18	44	324	792	1,936
14	19	37	361	703	1,369
15	20	45	400	900	2,025
Total	184	403	2,616	5,623	12,493

Solution   The extensions and summations needed for this problem are shown in Table 13-2. The line of best fit, using formulas (2-9), (3-4), (3-7), and (3-6), can now be calculated.

Using formula (2-9):

$$SS(x) = 2{,}616 - \frac{(184)^2}{15} = 358.9333$$

Using formula (3-4):

$$SS(xy) = 5{,}623 - \frac{(184)(403)}{15} = 679.5333$$

Using formula (3-7):

$$b_1 = \frac{679.5333}{358.9333} = 1.893202 = 1.89$$

Using formula (3-6):

$$b_0 = \frac{1}{15}[403 - (1.893202)(184)] = 3.643387$$

$$= 3.64$$

Therefore, the equation for the line of best fit is:

$$\hat{y} = 3.64 + 1.89x$$

The variance of y about the regression line is calculated by using formula (13-8).

$$s_e^2 = \frac{SSE}{n-2} = \frac{(\sum y^2) - (b_0)(\sum y) - (b_1)(\sum xy)}{n-2}$$

$$= \frac{12{,}493 - (3.643387)(403) - (1.893202)(5{,}623)}{15-2}$$

$$= \frac{379.2402}{13} = 29.1723$$

$$= 29.17$$

Note   Extra decimal places are often needed for this type of calculation. Notice that b_1 (1.893202) was multiplied by 5,623. If 1.89 had been used instead, that one product would have changed the numerator by approximately 18. That, in turn, would have changed the final answer by almost 1.4, and that is a sizable round-off error.

$s_e^2 = 29.17$ is the variance of the 15 e's. In Figure 13-10 the 15 e's are shown as vertical line segments.   ■

In the sections that follow, the variance of e will be used in much the same way as the variance of x (as calculated in Chapter 2) was used in Chapters 8, 9, and 10 to complete the statistical inferences studied there.

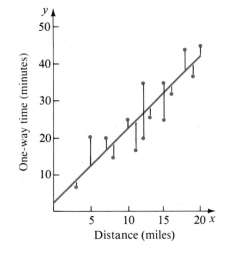

Figure 13-10
The 15 random errors
as line segments

EXERCISES

$ **13.17** Ten salespeople were surveyed and the average number of client contacts per month, x, and the sales volume, y (in thousands), were recorded for each, as shown in the following table.

x	12	14	16	20	23	46	50	48	50	55
y	15	25	30	30	30	80	90	95	110	130

Refer to the following MINITAB output and verify that the equation of the line of best fit is $\hat{y} = -13.4 + 2.3x$ and that $s_e = 10.17$.

```
          READ C1, C2
DATA> 12 15
DATA> 14 25
DATA> 16 30
DATA> 20 30
DATA> 23 30
DATA> 46 80
DATA> 50 90
DATA> 48 95
DATA> 50 110
DATA> 55 130
DATA> END DATA
       10 ROWS READ
MTB > REGRESS Y IN C2 ON 1 PRED IN C1
```

The regression equation is
$$C2 = -13.4 + 2.30 \ C1$$

Predictor	Coef
Constant	-13.414
C1	2.3028

$$s = 10.17$$

13.18 The computer-science aptitude score, x, and the achievement score, y (measured by a comprehensive final), were measured for 20 students in a beginning computer science course. The results were as follows.

x	4	16	20	13	22	21	15	20	19	16	18	17	8	6	5	20	18	11	19	14
y	19	19	24	36	27	26	25	28	17	27	21	24	18	18	14	28	21	22	20	21

Find the equation of the line of best fit and s_e^2.

13.19 In a regression problem where $x =$ the number of days on a diet, and $y =$ the number of pounds lost, we obtain the following:

$$n = 4 \qquad \sum xy = 132$$
$$\sum x = 20 \qquad \sum x^2 = 110$$
$$\sum y = 24 \qquad \sum y^2 = 162$$

a. Find the equation of the line of best fit.

b. How much is the average weight loss in pounds for each additional day on the diet?

13.20 a. Using the ten points shown in the following table, find the equation of the line of best fit, $\hat{y} = b_0 + b_1 x$, and graph it on a scatter diagram.

				Point						
	A	B	C	D	E	F	G	H	I	J
x	1	1	3	3	5	5	7	7	9	9
y	1	2	2	3	3	4	4	5	5	6

b. Find the ordinates $\hat{y}$ for the points on the line of best fit whose abscissas are $x = 1, 3, 5, 7,$ and 9.

c. Find the value of e for each of the points in the given data
$$(e = y - \hat{y})$$

d. Find the variance s_e^2 of those points about the line of best fit by using formula (13-6).

e. Find the variance s_e^2 by using formula (13-8). (Answers to (d) and (e) should be the same.)

13.21 The following data show the number of hours studied for an exam, x, and the grade received on the exam, y, (y is measured in 10s; that is, $y = 8$ means that the grade, rounded to the nearest 10 points, is 80.)

x	2	3	3	4	4	5	5	6	6	6	7	7	7	8	8
y	5	5	7	5	7	7	8	6	9	8	7	9	10	8	9

a. Draw a scatter diagram of the data.

b. Find the equation of the line of best fit and graph it on the scatter diagram.

c. Find the ordinates $\hat{y}$ that correspond to $x = 2, 3, 4, 5, 6, 7,$ and 8.

d. Find the five values of e that are associated with the points where $x = 3$ and $x = 6$.

e. Find the variance s_e^2 of all the points about the line of best fit.

13.4 INFERENCES CONCERNING THE SLOPE OF THE REGRESSION LINE

Now that the equation of the line of best fit has been determined and the linear model has been verified (by inspection of the scatter diagram), we are ready to determine whether we can use the equation to predict y. We will test the null hypothesis "the equation of the line of best fit is of no value in predicting y given x." That is, the null hypothesis to be tested is β_1 (the slope of the relationship in the population) is zero. If $\beta_1 = 0$, then the linear equation will be of no real use in predicting y. To test this hypothesis we will use a t test.

Before we look at the hypothesis test, let's discuss the sampling distribution of the slope. If random samples of size n are repeatedly taken from a bivariate population, the calculated slopes, the b_1's, would form a sampling distribution that is approximately normally distributed with a mean of β_1, the population value of the slope, and with a variance of $\sigma_{b_1}^2$, where

$$\sigma_{b_1}^2 = \frac{\sigma_\varepsilon^2}{\sum (x - \bar{x})^2} \tag{13-9}$$

provided there is no lack of fit. An appropriate estimator for $\sigma_{b_1}^2$ is obtained by replacing σ_ε^2 by s_e^2, the estimate of the variance of the error about the regression line:

$$s_{b_1}^2 = \frac{s_e^2}{\sum (x - \bar{x})^2} \tag{13-10}$$

This formula may be rewritten in the following, more manageable form:

$$s_{b_1}^2 = \frac{s_e^2}{\text{SS}(x)} = \frac{s_e^2}{\sum x^2 - [(\sum x)^2/n]} \tag{13-11}$$

Note   The "standard error of __" is the standard deviation of the sampling distribution of __. Therefore, the standard error of regression (slope) is σ_{b_1} and is estimated by s_{b_1}.

We are now ready to **test the hypothesis** $\beta_1 = 0$. Let's use the line of best fit determined in Illustration 13-4: $\hat{y} = 3.64 + 1.89x$. That is, we want to determine whether this equation is of any use in predicting travel time y. In this type of hypothesis test, the null hypothesis is always $H_0: \beta_1 = 0$.

Step 1 $H_0: \beta_1 = 0$. (This implies that x is of no use in predicting y; that is, that $\hat{y} = \bar{y}$ would be as effective.)

The alternative hypothesis can be either one-tailed or two-tailed. If we suspect that the slope is positive, as in Illustration 13-4 (we would expect travel time (y) to increase as the distance (x) increased), a one-tailed test is appropriate:

$$H_a: \beta_1 > 0.$$

Step 2   The test statistic is t. The number of degrees of freedom for this test is $n - 2$, **df = $n - 2$**. Thus for our example, df $= 15 - 2 = 13$. If we use $\alpha = 0.05$, the critical value of t is $t(13, 0.05) = 1.77$, as found in Table 6 of Appendix E (*see* the accompanying figure).

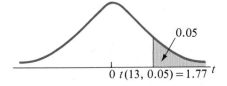

The formula used to calculate the value of the **test statistic t** for inferences about the slope is

$$t = \frac{b_1 - \beta_1}{s_{b_1}} \qquad (13\text{-}12)$$

Step 3   In our illustration of travel times and distances, the variance among the b_1's is estimated by use of formula (13-11):

$$s_{b_1}^2 = \frac{29.1723}{358.9333} = 0.081275 = 0.0813$$

Using formula (13-12), the observed value of t becomes

$$t = \frac{b_1 - \beta_1}{s_{b_1}} = \frac{1.89 - 0}{\sqrt{0.0813}} = 6.629$$

$$t^* = \mathbf{6.63}$$

Step 4   *Decision*: Reject H_0 (t^* is in the critical region; see the figure that accompanies Step 2).

Conclusion: At the 0.05 level of significance, we conclude that the slope of the line of best fit in the population is greater than zero. The evidence indicates that there is a linear relationship and that the one-way distance (x) is useful in predicting the travel time to work (y).

The slope β_1 of the regression line of the population can be estimated by means of a confidence interval. The **confidence interval** is given by

$$b_1 \pm t(n - 2, \alpha/2) \cdot s_{b_1} \qquad (13\text{-}13)$$

The 95 percent confidence interval for the estimate of the population's slope, β_1, for Illustration 13-4 is

$$1.89 \pm (2.16)(\sqrt{0.0813})$$
$$1.89 \pm 0.6159$$
$$1.89 \pm 0.62$$
$$\mathbf{1.27} \quad \text{to} \quad \mathbf{2.51}$$

Thus 1.27 to 2.51 is the 0.95 confidence interval for β_1. That is, we can say that the slope of the line of best fit of the population from which the sample was drawn is between 1.27 and 2.51, with 95 percent confidence.

EXERCISES

13.22 The undergraduate grade point average (GPA) and the composite score on the graduate record exam (GRE) were recorded for 15 graduates in a given department at a university. The data are shown in the following table.

GPA (x)	2.30	3.65	3.00	2.75	3.10	2.55	2.50	2.30	2.90	3.15	3.25	2.00	2.75	2.65	3.13
GRE (y)	925	1300	1150	1400	900	825	950	1050	1200	1200	1100	700	850	990	1000

Verify that the value for testing $\beta_1 = 0$ versus $\beta_1 \neq 0$ is $t^* = 2.70$, as shown on the following MINITAB printout. Complete the test using $\alpha = 0.05$.

```
            READ X IN C1, Y IN C2
     DATA> 2.30 925
     DATA> 3.65 1300
     DATA> 3.00 1150
     DATA> 2.75 1400
     DATA> 3.10 900
     DATA> 2.55 825
     DATA> 2.50 950
     DATA> 2.30 1050
     DATA> 2.90 1200
     DATA> 3.15 1200
     DATA> 3.25 1100
     DATA> 2.00 700
     DATA> 2.75 850
     DATA> 2.65 990
     DATA> 3.13 1000
     DATA> END DATA
            15 ROWS READ
     MTB > REGRESS Y IN C2 ON 1 PRED IN C1
```

```
The regression equation is
C2 = 297 + 264 C1

Predictor        Coef     Stdev    t-ratio
Constant        296.9     276.4      1.07
C1              264.09     97.66      2.70

s = 158.0
```

13.23 A sample of 10 students were asked for the distance and the time required to commute to college yesterday. The data collected are shown in the following table.

Distance (x)	1	3	5	5	7	7	8	10	10	12
Time (y)	5	10	15	20	15	25	20	25	35	35

a. Draw a scatter diagram of these data.

b. Find the equation that describes the regression line for these data.

c. Does the value of b_1 show sufficient strength to conclude that β_1 is greater than zero, at the $\alpha = 0.05$ level?

d. Find the 0.98 confidence interval for the estimation of β_1. (Retain these answers for use in Exercise 13.26.)

13.24 Interest rates are alleged to have an effect on the level of employment. The data in the following table show, by quarters, the bank interest rates on short-term loans and the unemployment rate.

Interest Rate	12.27	12.34	12.31	15.81	15.67	17.75	11.56	15.71	19.91	19.99	21.11
Unemployment Rate	5.9	5.6	5.9	5.9	6.2	7.6	7.5	7.3	7.6	7.2	8.3

a. Calculate the equation of the line of best fit.

b. Does this sample present sufficient evidence to reject the null hypothesis (slope is zero) in favor of the alternative hypothesis that the slope is positive, at the 0.05 level of significance?

13.5 CONFIDENCE INTERVAL ESTIMATES FOR REGRESSION

Once the equation for the line of best fit has been obtained and determined usable, we are ready to use the equation to make predictions. There are two different quantities that we can estimate: (1) the mean of the population y values at a given value of x, written $\mu_{y|x_0}$, and (2) the individual y value selected at random that will occur at a given value of x, written y_{x_0}. The best point estimate, or prediction, for

predicted value of y $(\hat{y})$ | both $\mu_{y|x_0}$ and y_{x_0} is $\hat{y}$. This is the y value obtained when an x value is substituted into the equation of the line of best fit. Like other point estimates it is seldom correct. The actual values for both $\mu_{y|x_0}$ and y_{x_0} will vary above and below the calculated value of $\hat{y}$.

Before developing confidence intervals for $\mu_{y|x_0}$ and y_{x_0}, recall the development of confidence intervals for the population mean μ in Chapter 8 when the variance was known, and in Chapter 9 when the variance was estimated. The sample mean $\bar{x}$ was the best point estimate of μ. We used the fact that $\bar{x}$ is normally distributed with a standard deviation of $\sigma/\sqrt{n}$ to construct formula (8-2) for the confidence interval for μ. When σ had to be estimated, we used formula (9-2) for the confidence interval.

The confidence intervals for $\mu_{y|x_0}$ and y_{x_0} are constructed in a similar fashion. $\hat{y}$ replaces $\bar{x}$ as our point estimate. If we were to take random samples from the population, construct the line of best fit for each sample, calculate $\hat{y}$ for a given x using each regression line, and plot the various $\hat{y}$ values (they would vary since each sample would yield a slightly different regression line), we would find that the $\hat{y}$ values form a normal distribution. That is, the sampling distribution of $\hat{y}$ is normal, just as the sampling distribution of $\bar{x}$ is normal. What about the appropriate standard deviation of $\hat{y}$? The standard deviation in both cases ($\mu_{y|x_0}$ and y_{x_0}) is calculated by multiplying the square root of the variance of the error by an appropriate correction factor. Recall that the variance of the error, s_e^2, is calculated by means of formula (13-8).

Before looking at the correction factors for the two cases, let's see why they are necessary. Recall that the line of best fit passes through the point $(\bar{x}, \bar{y})$, the centroid. In Section 13.3 we formed a confidence interval estimate for the slope β_1 (*see* Illustration 13-4) by using formula (13-13). If we draw lines with slopes equal to the extremes of that confidence interval, 1.27 to 2.51, through the point $(\bar{x}, \bar{y})$ [which is (12.3, 26.9)] on the scatter diagram, we will see that the value for $\hat{y}$ fluctuates considerably for different values of x (Figure 13-11). Therefore, we should suspect a need for a wider confidence interval as we select values of x that are further away from $\bar{x}$. Hence we need a correction factor to adjust for the distance between x_0 and $\bar{x}$. This factor must also adjust for the variation of the y values about $\hat{y}$.

Figure 13-11

Lines representing the confidence interval for slope

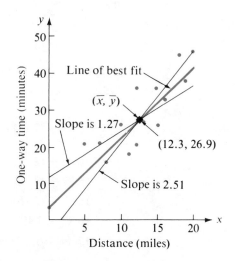

First, let's estimate the **mean value of y** at a given value of x, $\mu_{y|x_0}$. The confidence interval estimate formula is

$$\hat{y} \pm t(n-2, \alpha/2) \cdot s_e \cdot \sqrt{\frac{1}{n} + \frac{(x_0 - \bar{x})^2}{\sum(x - \bar{x})^2}} \qquad (13\text{-}14)$$

Note   The numerator of the second term under the radical sign is the square of the distance of x_0 from $\bar{x}$. The denominator is closely related to the variance of x and has a "standardizing" effect on this term.

Formula (13-14) can be modified to avoid having $\bar{x}$ in the denominator. The new form is

$$\hat{y} \pm t(n-2, \alpha/2) \cdot s_e \cdot \sqrt{\frac{1}{n} + \frac{(x_0 - \bar{x})^2}{SS(x)}} \qquad (13\text{-}15)$$

Let's compare formula (13-14) with formula (9-2). $\hat{y}$ replaces $\bar{x}$, and

$$s_e \sqrt{\frac{1}{n} + \frac{(x_0 - \bar{x})^2}{\sum(x - \bar{x})^2}} \qquad \text{(the standard error of } \hat{y}\text{)}$$

the estimated standard deviation of $\hat{y}$ in predicting $\mu_{y|x_0}$, replaces $s/\sqrt{n}$, the standard deviation of $\bar{x}$. The degrees of freedom are now $n - 2$ instead of $n - 1$ as before. These ideas are explored in the next illustration.

Illustration 13-5   Construct a 95 percent confidence interval estimate for the mean travel time for the co-workers who travel 7 miles to work (refer to Illustration 13-4).

Solution

Step 1   Find $\hat{y}_{x_0}$ when $x_0 = 7$:

$$\hat{y} = 3.64 + 1.89x = 3.64 + (1.89)(7) = \mathbf{16.87}$$

Step 2   Find s_e:

$$s_e^2 = 29.17 \qquad \text{(found in Illustration 13-4)}$$
$$s_e = \sqrt{29.17} = \mathbf{5.40}$$

Step 3   Find $t(13, 0.025) = 2.16$ (from Table 6 in Appendix E).

Step 4   Use formula (13-15):

$$16.87 \pm (2.16) \cdot (5.40) \cdot \sqrt{\frac{1}{15} + \frac{(7 - 12.27)^2}{358.9333}}$$

$$16.87 \pm (2.16)(5.40)\sqrt{0.06667 + 0.07738}$$
$$16.87 \pm (2.16)(5.40)\sqrt{0.14405}$$
$$16.87 \pm (2.16)(5.40)(0.38)$$
$$16.87 \pm 4.43$$
$$\mathbf{12.44} \quad \text{to} \quad \mathbf{21.30}$$

is the 0.95 confidence interval for $\mu_{y|x_0 = 7}$. This confidence interval estimate is shown in Figure 13-12 by the heavy vertical line. The **confidence interval belt** showing the upper and lower boundaries of all interval estimates at 95 percent confidence is also shown. Notice that the boundary lines for x values far away from $\bar{x}$ become close to the two lines that represent the equations having slopes equal to the extreme values of the 95 percent confidence interval estimate for the slope (*see* Figure 13-11).

Figure 13-12

Confidence belts for μ_{yx_0}

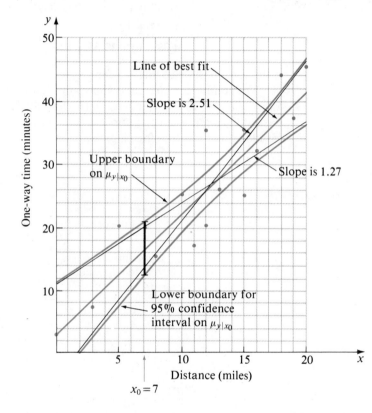

Often when making a prediction, we want to predict the value of an individual y. For example, you live 7 miles from your place of business and you are interested in an estimate of how long it will take you to get to work. You are somewhat less interested in the average time for all those who live 7 miles away. The formula for the interval estimate of the value of a **single randomly selected y** is

$$\hat{y} \pm t(n - 2, \alpha/2) \cdot s_e \cdot \sqrt{1 + \frac{1}{n} + \frac{(x_0 - \bar{x})^2}{SS(x)}} \qquad (13\text{-}16)$$

Illustration 13-6   What is the 95 percent confidence interval prediction for the time it will take you to commute to work if you live 7 miles away?

Solution

Step 1 $\hat{y}$ at $x_0 = 7$ is 16.87, as found in Illustration 13-5.

Step 2 $s_e = 5.40$, as found in Illustration 13-5.

Step 3 $t(13, 0.025) = 2.16$.

Step 4   Use formula (13-16).

$$16.87 \pm (2.16)(5.40)\sqrt{1 + 0.06667 + 0.07738}$$
$$16.87 \pm (2.16)(5.40)(1.0696)$$
$$16.87 \pm 12.48$$
$$\mathbf{4.39} \quad \text{to} \quad \mathbf{29.35}$$

is the 0.95 confidence interval for $y_{x_0 = 7}$.

The confidence interval is shown in Figure 13-13 as the vertical line segment at $x_0 = 7$. Notice that it is much longer than the confidence interval for $\mu_{y|x_0 = 7}$. The dashed lines represent the upper and lower boundaries of the confidence intervals for individual y values for all given x values.

Figure 13-13
Confidence belts for y_{x_0}

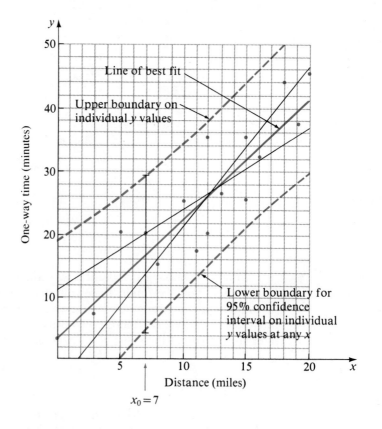

Can you justify the fact that the confidence interval for individual values of y is wider than the confidence interval for the mean values? Think about "individual values" and "mean values" and study Figure 13-14.

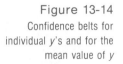

Figure 13-14

Confidence belts for
individual *y*'s and for the
mean value of *y*

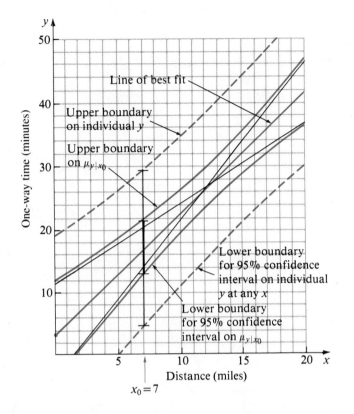

The techniques of coding (for large numbers) and the techniques of frequency distributions (for large sets of numbers) are adaptable to the calculations presented in this chapter. Information about these techniques and about models other than linear models may be found in many other textbooks. It is worth noting here that if the data suggest a logarithmic or an exponential functional relationship, a simple change of variable will allow for the use of linear analysis.

There are three basic precautions that you need to be aware of as you work with regression analysis.

1. Remember that the regression equation is meaningful only in the domain of the *x* variable studied. Prediction outside this domain is extremely dangerous; it requires that we know or assume that the relationship between *x* and *y* remains the same outside the domain of the sample data. For example, Joe says that he lives 75 miles from work, and he wants to know how long it will take him to commute. We certainly can use $x = 75$ in all the formulas, but we do not expect the answers to carry the confidence or validity of the values of *x* between 3 and 20, which were in the sample. The 75 miles may represent a distance to the heart of a nearby major city. Do you think the estimated times, which were based on local distances of 3 to 20 miles, would be good predictors in this situation? Also, at $x = 0$ the equation has no real meaning. However, although projections outside the interval may be somewhat dangerous, they may be the best predictors available.

Computer Solution
MINITAB Printout for parts of Illustrations 13-4, 13-5, and 13-6

Information given to computer

```
--READ DATA, X-VALUES INTO C1, Y-VALUES INTO C2
     COLUMN C1          C2
     COUNT  15          15
ROW
  1        3.            7.
  2        5.           20.
  3        7.           20.
  4        8.           15.
  5       10.           25.
  6       11.           17.
  7       12.           20.
  8       12.           35.
  9       13.           26.
 10       15.           25.
 11       15.           35.
 12       16.           32.
 13       18.           44.
 14       19.           37.
 15       20.           45.
```

Scatter diagram of given data

```
--PLOT Y-VALUES IN C2 VS X-VALUES IN C1
      C2
   50.0+
      -
      -
      -                                              *
      -                                           *
   40.0+
      -
      -                                   *
      -                             *        *
      -                                        *
   30.0+
      -
      -
      -                      *        *        *
      -
   20.0+
      -              *        *           *
      -                              *
      -                   *
      -
   10.0+
      -        *
      -
      -
      -
    0.0+
      +-----+-----+-----+-----+-----+C1
      0.0    4.0   8.0   12.0   16.0   20.0
```

Calculated correlation coefficient, $r = +0.879$

```
--CORRELATION COEFFICIENT BETWEEN DATA IN C1 AND C2
   CORRELATION OF C1 AND C2 = 0.879
```

Equation of line of best fit, $\hat{y} = 3.64 + 1.89x$; see page 490, solution of Illustration 13-4

```
--REGRESS Y-VARIABLE IN C2 ON 1 X-VARIABLE IN C1
THE REGRESSION EQUATION IS
Y = 3.64 + (1.89     ) × 1
```

Given data →

Values of $\hat{y}$ for each given x-value using $\hat{y} = 3.6434 + 1.8932x$

	X IN COL.1	Y IN COL.2	PRED Y VALUES	STD DEV PRED Y	RES.	STD. RES.
ROW						
1	3.000	7.000	9.323	2.987	-2.323	-0.516
2	5.000	20.000	13.109	2.497	6.891	1.439
3	7.000	20.000	16.896	2.049	3.104	0.621
4	8.000	15.000	18.789	1.851	-3.789	-0.747
5	10.000	25.000	22.575	1.537	2.425	0.468
6	11.000	17.000	24.469	1.441	-7.469	-1.435
7	12.000	20.000	26.362	1.397	-6.362	-1.219
8	12.000	35.000	26.362	1.397	8.638	1.656
9	13.000	26.000	28.255	1.410	-2.255	-0.433
10	15.000	25.000	32.041	1.598	-7.041	-1.365
11	15.000	35.000	32.041	1.598	2.959	0.573
12	16.000	32.000	33.935	1.754	-1.935	-0.379
13	18.000	44.000	37.721	2.149	6.279	1.267
14	19.000	37.000	39.614	2.373	-2.614	-0.539
15	20.000	45.000	41.507	2.609	3.493	0.738

Calculated values of b_0 and b_1

Calculated value of S_{b_1}, $S_{b_1} = 0.285 (\sqrt{0.0813})$; compare to $S_{b_1}^2 = 0.0813$ as found in Step 3 on page 494

```
X  COLUMN      ESTIMATES  FROM LEAST SQUARES FIT
0   **        COEFFICIENTS   S.D. OF COEFF.   RATIO
1    1          3.6434          3.76           0.97
                1.8932          0.285          6.64
```

THE ESTIMATED ST. DEV. OF Y ABOUT THE REGRESSION LINE IS 5.4011 WITH (15 − 2) = 13 DEGREES OF FREEDOM

Calculated value of S_e, $S_e = 5.4011 (\sqrt{29.1723})$; compare to $S_e^2 = 29.1723$ as found in solution of Illustration 13-4, page 489

2. Don't get caught by the common fallacy of applying the regression results inappropriately. For example, this fallacy would include applying the results of Illustration 13-4 to another company. But suppose that the second company had a city location, whereas the first company had a rural location, or vice versa. Do you think the results for a rural location would also be valid for a city location? Basically, the results of one sample should not be used to make inferences about a population other than the one from which the sample was drawn.

3. Don't jump to the conclusion that the results of the regression prove that x causes y to change. (This is perhaps the most common fallacy.) Regressions only measure movement between x and y; they **never prove causation.** A judgment of causation can be made only when it is based on theory or knowledge of the relationship separate from the regression results. The most common difficulty in this regard occurs because of what is called the missing variable, or third-variable, effect. That is, we observe a relationship between x and y because a third variable, one that is not in the regression, affects both x and y.

EXERCISES

13.25 An experiment was conducted to study the effect of a new drug in lowering the heart rate in adults. The data collected are shown in the following table.

Drug Dose in mg (x)	0.5	0.75	1.00	1.25	1.50	1.75	2.00	2.25	2.50	2.75
Heart-rate Reduction (y)	10	7	15	12	15	14	20	20	18	21

a. Find the 95 percent confidence interval estimate for the mean heart-rate reduction for a dose of 2.00 mg.

b. Find the 95 percent confidence prediction interval for the heart-rate reduction expected for an individual receiving a dose of 2.00 mg.

13.26 Use the data and the answers found in Exercise 13.23 to make the following estimates.

a. Give a point estimate for the mean time required to commute 4 miles.

b. Give a 0.90 confidence interval estimate for the mean travel time required to commute 4 miles.

c. Give a 0.90 confidence interval estimate for the travel time required for one person to commute the 4 miles.

d. Answer (a), (b), and (c) for $x = 9$.

13.27 Explain why a 95 percent confidence interval estimate for the mean value of y at a particular x is much narrower than a 95 percent confidence interval for an individual y value at the same value of x.

13.28 When $x_0 = \bar{x}$, is the formula for the standard error of $\hat{y}_{x_0}$ what you might have expected it to be, $s \cdot \sqrt{1/n}$?

13.6   UNDERSTANDING THE RELATIONSHIP BETWEEN CORRELATION AND REGRESSION

Now that we have taken a closer look at both correlation and regression analysis, it is necessary to decide when to use them. Do you see any duplication of work?

The primary use of the linear correlation coefficient is in answering the question "Are these two variables linearly related?" There are other words that may be used to ask this basic question. For example "Is there a linear correlation between the annual consumption of alcoholic beverages and the salary paid to firemen?"

The linear correlation coefficient can be used to indicate the usefulness of x as a predictor of y in the case where the linear model is appropriate. The test concerning the slope of the regression line ($H_0: \beta_1 = 0$) also tests this same basic concept. Either one of the two is sufficient to determine the answer to this query.

Although the "lack-of-fit" test can be statistically determined, it is beyond the scope of this text. However, we are very likely to carry out this test at a subjective level when we view the scatter diagram. If the scatter diagram is carefully constructed, this subjective decision will determine the mathematical model for the regression line that you believe will fit the data.

The concepts of linear correlation and regression are quite different, because each measures different characteristics. It is possible to have data that yield a strong linear correlation coefficient and have the wrong model. For example, the straight line can be used to approximate almost any curved line if the interval of domain is restricted sufficiently. In such a case the linear correlation coefficient can become quite high but the curve will still not be a straight line. Figure 13-15 suggests one such interval where r could be significant, but the scatter diagram does not suggest a straight line.

Figure 13-5
The value of r is high but the
relationship is not linear

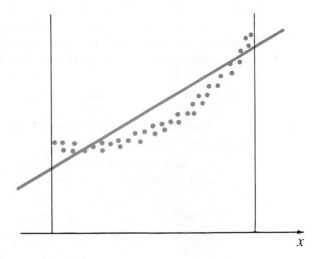

x

Regression analysis should be used to answer questions about the relationship between two variables. Such questions as "What is the relationship?" How are two variables related?" and so on, require this regression analysis.

In
Retrospect

In this chapter we have made a more thorough inspection of the linear relationship between two variables. Although the curvilinear and multiple regression situations were only mentioned in passing, the basic techniques and concepts have been explored. We would only have to modify our mathematical model and our formulas if we wanted to deal with these other relationships.

In this chapter we have drawn scatter diagrams and constructed confidence belts for estimations. But both were done for one population at a time. The news article at the beginning of this chapter shows a unique graphic presentation. The graph in the article shows a set of confidence belts for one population (the reference weight percentiles of Lubchenco *et al.*) and then plots a scatter diagram of the very low-birth-weight infants on the same graph. One definitely gets the feeling that there is a difference between the weights of the infants from these two different populations.

Although it was not directly emphasized, we have applied many of the topics of earlier chapters in this chapter. The ideas of hypothesis testing and confidence interval estimates were applied to the regression problem. Reference was made to the sampling distribution of the sample slope b_1. This allowed us to make inferences about β_1, the slope of the population from which the sample was drawn. We estimated the mean value of y at a fixed value of x by pooling the variance for the slope with the variance of the y's. This was allowable since they are independent. You should recall that in Chapter 10 we presented formulas for combining the variances of independent samples. The idea here is much the same. Finally, we added a measure of variance for individual values of y and made estimates for these individual values of y at fixed values of x.

As this chapter ends you should be aware of the basic concepts of regression analysis and correlation analysis. You should now be able to collect the data for and do a complete analysis on any two-variable linear relationship.

Chapter Exercises

13.29 Answer the following as "sometimes," "always," or "never." Explain each "never" and "sometimes" response.

    a. The correlation coefficient has the same sign as the slope of the least squares line fitted to the same data.

    b. A correlation coefficient of 0.99 indicates a strong causal relationship between the variables under consideration.

    c. An r value greater than zero indicates that ordered pairs with high x-values will have low y-values.

    d. The two coefficients for the line of best fit have the same sign.

    e. If x and y are independent, then the population correlation coefficient equals zero.

13.30 Suppose that you calculated the correlation coefficient between two variables and found it to be -1.50. What conclusion would you reach?

13.31 The correlation coefficient, r, is related to the slope of the line of best fit, b_1, by the equation

$$r = b_1 \sqrt{\frac{SS(x)}{SS(y)}}$$

Verify this equation using the following data.

x	1	2	3	4	6
y	4	6	7	9	12

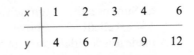

13.32 One would usually expect that the number of hours studied, x, in preparation for a particular examination would have direct (positive) correlation with the grade, y, attained on that exam. The hours studied and the grades obtained by ten students selected at random from a large class are shown in the following table.

	Student									
	1	2	3	4	5	6	7	8	9	10
Hours Studied (x)	10	6	15	11	7	19	17	3	13	17
Exam Grade (y)	51	36	67	63	44	89	80	26	50	85

a. Draw the scatter diagram and estimate r.

b. Calculate r. (How close was your estimate?)

c. Is there significant positive correlation shown by these data? Test at $\alpha = 0.01$.

d. Find the 95 percent confidence interval estimate for the true population value of ρ.

13.33 When buying most items, it is advantageous to buy in as large a quantity as possible. The unit price is usually less for the larger quantities. The following data were obtained to test this theory.

Number of Units (x)	1	3	5	10	12	15	24
Cost per Unit (y)	55	52	48	36	32	30	25

a. Calculate r.

b. Does the number of units ordered have an effect on the unit cost? Test at $\alpha = 0.05$.

c. Find the 95 percent confidence interval estimate for the true value of ρ.

13.34 When two dice are rolled simultaneously, it is expected that the result on each roll will be independent of the result on the other. To test this, roll a pair of dice 12 times. Identify one die as x and the other as y. Record the number observed on each of the two dice.

a. Draw the scatter diagram of x versus y.

b. If the two do behave independently, what value of r can be expected?

c. Calculate r.

d. Test for independence of these dice at $\alpha = 0.10$.

13.35 The coding scheme used for a variable can determine whether a negative or a positive correlation coefficient is obtained between two variables.

a. Suppose that high school academic performance (HSAP) is coded as: 1 = excellent, 2 = above average, and 3 = average or below. The score made on a final exam in a college algebra course is also recorded. The resulting data are shown in the following table.

HSAP	1	2	3	1	3	2	1	3	3	2
Final Exam Score	79	80	65	89	70	75	90	65	70	80

Compute r for these data.

b. Now suppose that HSAP is coded as: 3 = excellent, 2 = above average, and 1 = average or below. The data would now appear as follow.

HSAP	3	2	1	3	1	2	3	1	1	2
Final Exam Score	79	80	65	89	70	75	90	65	70	80

Compute r for these data.

c. What connection do you see between the two resulting correlation coefficients?

13.36 The following data show the gain in reading speed, y, and the number of weeks in a speed-reading program, x, for eight students.

x	3	5	2	8	6	9	3	4
y	80	110	50	190	160	230	70	110

For these data, find

a. the equation of the line of best fit.

b. s_e.

c. the standard error of b_1.

13.37 The following data resulted from an experiment performed for the purpose of regression analysis. The input variable x was set at five different levels, and observations were made at each level.

x	0.5	1.0	2.0	3.0	4.0
y	3.8	3.2	2.9	2.4	2.3
	3.5	3.4	2.6	2.5	2.2
	3.8	3.3	2.7	2.7	2.3
		3.6	3.2	2.3	

a. Draw a scatter diagram of the data.

b. Draw the regression line by eye.

c. Place an asterisk, *, at each level, approximately where the mean of the observed y values is located. Does your regression line look like the line of best fit for these five mean values?

d. Calculate the equation of the regression line.

e. Find the standard deviation of y about the regression line.

f. Construct a 95 percent confidence interval estimate for the true value of β_1.

g. Construct a 95 percent confidence interval estimate for the mean value of y at $x = 3.0$. At $x = 3.5$.

h. Construct a 95 percent confidence interval prediction for an individual value of y at $x = 3.0$. At $x = 3.5$.

13.38 The following set of 25 scores was randomly selected from a teacher's class list. Let x be the prefinal average and y the final examination score. (The final examination had a maximum of 75 points.)

Student	x	y	Student	x	y
1	75	64	14	73	62
2	86	65	15	78	66
3	68	57	16	71	62
4	83	59	17	86	71
5	57	63	18	71	55
6	66	61	19	96	72
7	55	48	20	96	75
8	84	67	21	59	49
9	61	59	22	81	71
10	68	56	23	58	58
11	64	52	24	90	67
12	76	63	25	92	75
13	71	66			

a. Draw a scatter diagram for these data.

b. Draw the regression line (by eye) and estimate its equation.

c. Estimate the value of the coefficient of linear correlation.

d. Calculate the equation of the line of best fit.

e. Draw the line of best fit on your graph. How does it compare with your estimate?

f. Calculate the linear correlation coefficient. How does it compare with your estimate?

g. Test the significance of r at $\alpha = 0.10$.

h. Find the 95 percent confidence interval estimate for the true value of ρ.

i. Find the standard deviation of the y values about the regression line.

j. Calculate a 95 percent confidence interval estimate for the true value of the slope, β_1.

k. Test the significance of the slope at $\alpha = 0.05$.

l. Estimate the mean final exam grade that all students with an 85 prefinal average will obtain (95 percent confidence interval).

m. Using the 95 percent confidence interval, predict the grade that John Henry will receive on his final, knowing that his prefinal average is 78.

13.39 Twenty-one mature flowers of a particular species were dissected and the number of stamens and carpels present in each flower were counted. See the following table.

a. Is there sufficient evidence to claim a linear relationship between these two variables, at $\alpha = 0.05$?

b. What is the relationship between the number of stamens and carpels in this variety of flower?

c. Is the slope of the regression line significant, at $\alpha = 0.05$?

Stamens (x)	Carpels (y)	Stamens (x)	Carpels (y)
52	20	74	29
68	31	38	28
70	28	35	25
38	20	45	27
61	19	72	21
51	29	59	35
56	30	60	27
65	30	73	33
43	19	76	35
37	25	68	34
36	22		

d. Give the 0.95 confidence interval prediction for the number of carpels that one would expect to find in a mature flower of this variety if the number of stamens were 64.

13.40 It is believed that the amount of nitrogen fertilizer used per acre has a direct effect on the amount of wheat produced. The following data show the amount of nitrogen fertilizer used per test plot and the amount of wheat harvested per test plot.

Pounds of Fertilizer (x)	100 Pounds of Wheat (y)	Pounds of Fertilizer (x)	100 Pounds of Wheat (y)
30	5	70	19
30	9	70	23
30	14	70	31
40	6	80	24
40	14	80	32
40	18	80	35
50	12	90	27
50	14	90	32
50	23	90	38
60	18	100	34
60	24	100	35
60	28	100	39

a. Is there sufficient reason to conclude that the use of more fertilizer results in a higher yield? Use $\alpha = 0.05$.

b. Estimate, with a 0.98 confidence interval, the mean yield that could be expected if 50 pounds of fertilizer were used per plot.

c. Estimate, with a 0.98 confidence interval, the mean yield that could be expected if 75 pounds of fertilizer were used per plot.

13.41 The following equation is known to be true for any set of data.

$$\sum (y - \bar{y})^2 = \sum (y - \hat{y})^2 + \sum (\hat{y} - \bar{y})^2$$

Verify this equation with the following data.

x	0	1	2
y	1	3	2

Vocabulary List   *Be able to define each term. In addition, describe in your own words and give an example of each term. Your examples should not be ones given in class or in the textbook.*

   The bracketed numbers indicate the chapters in which the term previously appeared, but you should define the terms again to show increased understanding of their meaning.

bivariate data [3]
coefficient of linear correlation [3]
confidence belts
confidence interval estimate
   [8, 9, 10]
covariance
curvilinear regression
experimental error (ε or e)
intercept (b_0 or β_0) [3]
linear regression [3]
line of best fit [3]

multiple regression
Pearson's product moment r
predicted value of y ($\hat{y}$)
regression line [3]
rho (ρ)
sampling distribution [7, 10]
scatter diagram [3]
slope (b_1 or β_1) [3]
standard error [7, 8, 9]
sum of squares for error [1, 2]
variance (s^2 or σ^2) [2, 8, 9, 10, 12]

Quiz   *Answer "True" if the statement is always true. If the statement is not always true, replace the boldface words with words that make the statement always true.*

13.1 The error **must be** normally distributed if inferences are to be made.

13.2 x and y **must both be** normally distributed.

13.3 A high correlation between x and y **proves** that x causes y.

13.4 The values of the input variable **must be** randomly selected to achieve valid results.

13.5 The output variable must be **normally distributed** about the regression line for each value of x.

13.6 Covariance measures the strength of the linear relationship and is a standardized measure.

13.7 The **sum of squares for error** is the name given to the numerator portion of the formula for the calculation of the variance of y about the line of regression.

13.8 Correlation analysis attempts to find the equation of the line of best fit for two variables.

13.9 There are $n - 3$ degrees of freedom involved with the inferences about the regression line.

13.10 $\hat{y}$ serves as the **point estimate** for both $\mu_{y|x_0}$ and y_{x_0}.

Chapter 14

Elements of Nonparametric Statistics

STUDY FINDS THAT MONEY CAN'T BUY JOB SATISFACTION

When it comes to getting workers to produce—do their level best—money is less than everything. Feeling appreciated—having a sense of being recognized—is more important.

That is the thesis of Kenneth A. Kovach, of George Mason University....

Kovach reproduces a consequential survey of 30 years ago in which workers and supervisors gave their opinions of "what workers want from their jobs and what management thinks they want." See the accompanying table.

Managements today are more "behavioral" in their approach to managing. Nevertheless, Kovach contends that a wide gap still exists, a gap that is suggested by what workers rank as No. 1 in importance to them ("appreciation") and what supervisors consider No. 1 to workers ("good wages").

	Worker Ranking	Boss Ranking
Full appreciation of work done	1	8
Feeling of being in on things	2	10
Sympathetic help on personal problems	3	9
Job security	4	2
Good wages	5	1
Interesting work	6	5
Promotion and growth in the organization	7	3
Personal loyalty to employees	8	6
Good working conditions	9	4
Tactful disciplining	10	7

Source: Reprinted by permission of the *Philadelphia Inquirer*, 29 December 1976.

Nonparametric methods of statistics dominate the success story of statistics in recent years. Unlike their parametric counterparts, many of the best-known nonparametric tests, also known as distribution-free tests, are founded on a basis of elementary probability theory. The derivation of most of these tests is well within the grasp of the student who is competent in high school algebra and understands binomial probability. Thus the nonmathematical statistics user is much more at ease with the nonparametric techniques.

This chapter is intended to give you a feeling for basic concepts involved in nonparametric techniques and to show you that nonparametric methods are extremely versatile and easy to use, once a table of critical values is developed for a particular application. The selection of the nonparametric methods presented here includes only a few of the common tests and applications. You will learn about the sign test, the Mann-Whitney U test, the runs test, and Spearman's rank correlation test. These will be used to make inferences corresponding to both one- and two-sample situations.

14.1 NONPARAMETRIC STATISTICS

parametric method

Most of the statistical procedures that we have studied in this book are known as **parametric methods**. For a statistical procedure to be parametric, we either assume that the parent population is at least approximately normally distributed or we rely on the central limit theorem to give us a normal approximation. This is particularly true of the statistical methods studied in Chapters 8, 9, and 10.

nonparametric, or
distribution-free, methods

The **nonparametric methods**, or **distribution-free methods**, as they are also known, do not depend on the distribution of the population being sampled. The nonparametric statistics are usually subject to much less confining restrictions than are their parametric counterparts. Some, for example, require only that the parent population be continuous.

The recent popularity of nonparametric statistics can be attributed to the following characteristics:

1. Nonparametric methods require few assumptions about the parent population.

2. Nonparametric methods are generally easier to apply than their parametric counterparts.

3. Nonparametric methods are relatively easy to understand.

4. Nonparametric methods can be used in situations where the normality assumptions cannot be made.

5. Nonparametric methods appear to be wasteful of information in that they sacrifice the value of the variable for only a sign or a rank number. However, nonparametric statistics are generally only slightly less efficient than their parametric counterparts.

14.2 THE SIGN TEST

sign test   The ordinary **sign test** is a versatile and exceptionally easy-to-apply nonparametric method that **uses only plus and minus signs**. (There are several specific techniques.) The sign test is useful in two situations: (1) a hypothesis test concerning the value of the median for one population and (2) a hypothesis test concerning the median difference (paired difference) for two dependent samples. Both tests are carried out using the same basic procedures and are nonparametric alternatives to the t tests used with one mean (Section 9.1) and the difference between two dependent means (Section 10.5).

SINGLE-SAMPLE HYPOTHESIS TEST PROCEDURE

The sign test can be used when a random sample is drawn from a population with an unknown **median, M**, and the population is assumed to be continuous in the vicinity of M. The null hypothesis to be tested concerns the value of the population median M. The test may be either one- or two-tailed. This test procedure is presented in the following illustration.

Illustration 14-1   A random sample of 75 students was selected and each student was asked to carefully measure the amount of time required to commute from his or her front door to the college parking lot. The data collected were used to test the hypothesis "the median time required for students to commute is 15 minutes" against the alternative that the median is unequal to 15 minutes. The 75 pieces of data were summarized as follows:

$$
\begin{array}{ll}
\text{Under 15:} & 18 \\
\text{15:} & 15 \\
\text{Over 15:} & 42
\end{array}
$$

Use the sign test to test the null hypothesis against the alternative hypothesis.

Solution   The data are converted to $(+)$ and $(-)$ signs. A plus sign will be assigned to each piece of data larger than 15, a minus sign to each piece of data smaller than 15, and a zero to those data equal to 15. The sign test uses only the plus and minus signs; therefore, the zeros are discarded and the usable sample size becomes 60. That is, $n(+) = 42, n(-) = 18$, and $n = n(+) + n(-) = 42 + 18 = 60$.

Step 1 $H_0: M = 15$
          $H_a: M \neq 15$

Step 2 $\alpha = 0.05$ for a two-tailed test. The **test statistic** that will be used is the **number of the less frequent sign,** the smaller of $n(+)$ and $n(-)$, which is $n(-)$ for our illustration. We will want to reject the null hypothesis whenever the number of the less-frequent sign is extremely small. Table 11 of Appendix E gives the maximum allowable number of the less-frequent sign, k, that will allow us to reject the null hypothesis. That is, if the number of the less-frequent sign is less than or equal to the critical value in the table, we will reject H_0. If the observed value of the less-frequent sign is larger than the table value, we will fail to reject H_0. In the table, n is the total number of signs, not including zeros.

For our illustration, $n = 60$ and the critical value from the table is 21. See the accompanying figure.

Number of less frequent sign

Step 3   The observed statistic is $x = n(-) = 18$, and it falls in the critical region.

Step 4   *Decision*: Reject H_0.

Conclusion: The sample shows sufficient evidence at the 0.05 level to reject the claim that the median is 15 minutes.   ■

TWO-SAMPLE HYPOTHESIS TEST PROCEDURE

The sign test may also be applied to a hypothesis test dealing with the **median difference** between paired data that result from **two dependent samples**. A familiar application is the use of before-and-after testing to determine the effectiveness of some activity. In a test of this nature, the signs of the differences are used to carry out the test. Again, zeros are disregarded. The following illustration shows this procedure.

Illustration 14-2   A new no-exercise, no-starve weight-reducing plan has been developed and advertised. To test the claim that "you will lose weight within 2 weeks or ...," a local statistician obtained the before-and-after weights of 18 people who had used this plan. Table 14-1 lists the people, their weights, a minus ($-$) for those who lost weight during the 2 weeks, a 0 for those who remained the same, and a plus ($+$) for those who actually gained weight.

The claim being tested is that people are able to lose weight. The null hypothesis that will be tested is that there is no weight loss (or the median weight loss is zero), meaning that only a rejection of the null hypothesis will allow us to conclude in favor of the advertised claim. Actually we will be testing to see whether there are significantly more minus signs than plus signs. If the weight-reducing plan is of absolutely no value, we would expect to find an equal number of plus and minus signs. If it works, there should be significantly more minus signs than plus signs. Thus the test performed here will be a one-tailed test. (We will want to reject the null hypothesis in favor of the advertised claim if there are "many" minus signs.)

Solution

Step 1 $H_0: M = 0$ (no weight loss).
             $H_a: M < 0$ (weight loss).

Table 14-1

Sample results for
Illustration 14-2

| Person | Weight | | Sign of Difference, Before to After |
	Before	After	
Mrs. Smith	146	142	−
Mr. Brown	175	178	+
Mrs. White	150	147	−
Mr. Collins	190	187	−
Mr. Gray	220	212	−
Miss Collins	157	160	+
Mrs. Allen	136	135	−
Mrs. Noss	146	138	−
Miss Wagner	128	132	+
Mr. Carroll	187	187	0
Mrs. Black	172	171	−
Mrs. McDonald	138	135	−
Miss Henry	150	151	+
Miss Greene	124	126	+
Mr. Tyler	210	208	−
Mrs. Williams	148	148	0
Mrs. Moore	141	138	−
Mrs. Sweeney	164	159	−

Step 2   Use $\alpha = 0.05$. $n(+) = 5$, $n(-) = 11$, and $n = 16$. The critical value from Table 11 shows $k = 4$ as the maximum allowable number. (You must use the $\alpha = 0.10$ column, because the table is set up for two-tailed tests.)

Reject H_0					Fail to reject H_0
0	1	2	3	4	5

Number of less frequent sign

Step 3 $x = n(+) = 5$

Step 4   *Decision*: Fail to reject H_0 (we have too many plus signs).

Conclusion: The evidence observed is not sufficient to allow us to reject the no-weight-loss null hypothesis at the 0.05 level of significance.   ■

The sign test may be carried out by means of a **normal approximation** using the standard normal variable z. The normal approximation will be used if Table 11 does not show the particular levels of significance desired or if n is large. z will be calculated by using the formula

$$z = \frac{x' - n/2}{\left(\frac{1}{2}\right)\sqrt{n}}$$ (14-1)

(*See* Note 3 with regard to x'.)

Notes

1. x may be the number of the less-frequent sign or the most-frequent sign. You will have to determine this in such a way that the direction is consistent with the interpretation of the situation.

2. x is really a binomial random variable, where $p = \frac{1}{2}$. The sign test statistic satisfies the properties of a binomial experiment (*see* page 203). Each sign is the result of an independent trial. There are n trials, and each trial has two possible outcomes ($+$ or $-$). Since the median is used, the probabilities for each outcome are both $\frac{1}{2}$. Therefore, the mean, μ_x, is equal to

$$n/2 \quad (\mu = np = n \cdot \tfrac{1}{2} = n/2)$$

and the standard deviation, σ_x, is equal to

$$(\tfrac{1}{2})\sqrt{n} \quad (\sigma = \sqrt{npq} = \sqrt{n \cdot \tfrac{1}{2} \cdot \tfrac{1}{2}} = \tfrac{1}{2}\sqrt{n})$$

3. x is a discrete variable. But recall that the normal distribution must be used only with continuous variables. However, although the binomial random variable is discrete, it does become approximately normally distributed for large n. Nevertheless, when using the normal distribution for testing, we should make an adjustment in the variable so that the approximation is more accurate. (*See* Section 6.5 on the normal approximation.) This adjustment is illustrated in Figure 14-1 and is called a **continuity correction**. For this discrete variable the area that represents the probability is a rectangular bar. Its width is 1 unit wide, from $\frac{1}{2}$ unit below to $\frac{1}{2}$ unit above the value of interest. Therefore, when z is to be used, we will need to make a $\frac{1}{2}$-unit adjustment before calculating the observed value of z. x' will be adjusted value for x. If x is larger than $n/2$, $x' = x - \frac{1}{2}$. If x is smaller than $n/2$, $x' = x + \frac{1}{2}$. The test is then completed by the usual procedure.

Figure 14-1
Continuity correction

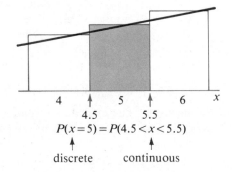

$$P(x=5) = P(4.5 < x < 5.5)$$

discrete   continuous

Illustration 14-3   Use the sign test to test the hypothesis that the median number of hours, M, worked by students of a certain college is at least 15 hours per week. A survey of 120 students was taken; a plus sign was recorded if the number of hours the student worked last week was equal to or greater than 15, and a minus sign was recorded if the number of hours was less than 15. Totals showed 80 minus signs and 40 plus signs.

Solution

Step 1 $H_0: M = 15$ ($\geq$)   (at least as many plus signs as minus signs).

$\quad\quad\quad\quad H_a: M < 15$ (fewer plus signs than minus signs).

Step 2 $\alpha = 0.05$; x is the number of plus signs. The critical value is shown in the accompanying figure.

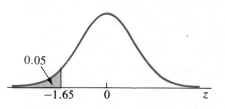

0.05

-1.65 0 z

Step 3

$$z = \frac{x' - n/2}{(\frac{1}{2})\sqrt{n}} = \frac{40.5 - 60}{(\frac{1}{2})\sqrt{120}} = \frac{-19.5}{(\frac{1}{2})(10.95)}$$

$$= \frac{-19.5}{5.475} = -3.562$$

$$z^* = -3.56$$

Step 4   *Decision*: Reject H_0.

Conclusion: At the 0.05 level, there are significantly more minus signs than plus signs, thereby implying that the median is less than the claimed 15 hours. ∎

CONFIDENCE INTERVAL PROCEDURE

The sign test techniques can be applied to obtain a confidence interval estimate for the unknown population median M. To accomplish this we will need to arrange the data in ascending order (smallest to largest). The data are identified as x_1 (smallest), $x_2, x_3, \ldots, x_n$ (largest). The critical value k for the maximum allowable number of signs, obtained from Table 11, Appendix E, tells us the number of positions to be dropped from each end of the ordered data. The remaining extreme values become the bounds of the $1 - \alpha$ confidence interval. That is, the lower boundary for the confidence interval is x_{k+1}, the $(k + 1)$th piece of data; the upper boundary is x_{n-k}, the $(n - k)$th piece of data. The following illustration will clarify this procedure.

Illustration 14-4   Suppose that we have 12 pieces of data in ascending order $(x_1, x_2, x_3, \ldots, x_{12})$ and we wish to form a 0.95 confidence interval for the population median. Table 11 shows a critical value of 2 ($k = 2$) for $n = 12$ and $\alpha = 0.05$ for a hypothesis test. This means that the critical region for a two-tailed hypothesis test would contain the last two values on each end (x_1 and x_2 on the left; x_{11} and x_{12} on the right). The noncritical region is then from x_3 to x_{10}, inclusive. The 0.95 confidence interval is then x_3 to x_{10}, expressed as

$$x_3 \quad \text{to} \quad x_{10}, \qquad 0.95 \text{ confidence interval for } M \qquad ∎$$

In general, the two pieces of data that bound the confidence interval will occupy positions $k + 1$ and $n - k$, where k is the table value read from Table 11. Thus

$$x_{k+1} \quad \text{to} \quad x_{n-k}, \qquad 1 - \alpha \text{ confidence interval for } M$$

If the normal approximation is to be used (including the continuity correction), the position numbers become

$$(\tfrac{1}{2})n \pm (\tfrac{1}{2} + \tfrac{1}{2} \cdot z(\alpha/2) \cdot \sqrt{n})$$ (14-2)

The interval is

$$x_L \quad \text{to} \quad x_U, \qquad 1 - \alpha \text{ confidence interval for } M$$

where

$$L = \frac{n}{2} - \frac{1}{2} - \frac{z(\alpha/2)}{2} \cdot \sqrt{n}$$

$$U = \frac{n}{2} + \frac{1}{2} + \frac{z(\alpha/2)}{2} \cdot \sqrt{n}$$

(L should be rounded down and U should be rounded up to be sure that the level of confidence is at least $1 - \alpha$.)

Illustration 14-5   Estimate the population median with a 0.95 confidence interval for a given set of 60 pieces of data: $x_1, x_2, x_3, \ldots, x_{59}, x_{60}$.

Solution   When we use formula (14-2), the position numbers L and U are

$$(\tfrac{1}{2})(60) \pm [\tfrac{1}{2} + \tfrac{1}{2}(1.96)\sqrt{60}]$$
$$30 \pm 0.50 + 7.59$$
$$30 \pm 8.09$$

Thus

$$L = 30 - 8.09 = 21.91 \quad \text{(21st piece of data)}$$
$$U = 30 + 8.09 = 38.09 \quad \text{(39th piece of data)}$$

Therefore,

$$x_{21} \quad \text{to} \quad x_{39}, \qquad 0.95 \text{ confidence interval for } M \qquad \blacksquare$$

EXERCISES

14.1 Use the sign test to test the hypothesis that the median daily high temperature in the city of Rochester, New York, during the month of December is 48 degrees. The following daily highs were recorded on 20 randomly selected days during December.

$$47 \quad 46 \quad 40 \quad 40 \quad 46 \quad 35 \quad 34 \quad 59 \quad 54 \quad 33$$
$$65 \quad 39 \quad 48 \quad 47 \quad 46 \quad 46 \quad 42 \quad 36 \quad 45 \quad 38$$

a. State the null hypothesis for the test.

b. State specifically what it is that you are actually testing when using the sign test.

c. Complete the test for $\alpha = 0.05$, and carefully state your findings.

14.2 In order to test the null hypothesis that there is no difference in the ages of husbands and wives, the following data were collected.

Husbands	28	45	40	37	25	42	21	22	54	47	35	62	29	44	45	38	59
Wives	26	36	40	42	28	40	20	24	50	54	37	60	25	40	34	42	49

Does the sign test show that there is a significant difference in the ages of husbands and wives at $\alpha = 0.05$?

14.3 A taste test was conducted with a regular beef pizza. Each of 133 individuals was given two pieces of pizza, one with a whole wheat crust and the other with a white crust. Each person was then asked whether he or she preferred whole wheat or white crust. The results were:

65 preferred whole wheat to white crust

53 preferred white to whole wheat crust

15 had no preference

Is there sufficient evidence to verify the hypothesis that whole wheat crust is preferred to white crust at the $\alpha = 0.05$ level of significance?

14.4 Sixteen students were given elementary statistics test on a hot day. Eight randomly selected students took the test in a room that was not air-conditioned. Then, after a short break, they completed a similar test in an air-conditioned room. The procedure was reversed for the other eight students.

Student	1	2	3	4	5	6	7	8	9	10	11	12	13	14	15	16
Not Air-Conditioned	52	90	63	74	87	77	92	72	77	94	67	86	78	84	57	55
Air-Conditioned	49	94	60	78	93	77	93	74	78	93	78	89	92	83	49	68

Does the sample provide sufficient reason to conclude that the use of air conditioning on a hot day has an effect on test grades? Use $\alpha = 0.05$.

 14.5 Determine the 0.95 confidence interval for the median daily high temperature in Rochester during December based on the sample given in Exercise 14.1.

14.6 Find the 0.75 confidence interval estimate for the median swim time for a swimmer whose recorded times are:

24.7 24.7 24.6 25.5 25.7 25.8 26.5 24.5 25.3
26.2 25.5 26.3 24.2 25.3 24.3 24.2 24.2

14.3 THE MANN-WHITNEY *U* TEST

Mann-Whitney *U* test   The **Mann-Whitney *U* test** is a nonparametric alternative for the *t* test for the **difference between two independent means**. It can be applied when we have two independent random samples (independent within each sample as well as between samples) in which the random variable is continuous. This test is often applied in situations in which the two samples are drawn from the same population but different "treatments" are used on each set. We will demonstrate the procedure in the following illustration.

Illustration 14-6   In a large lecture class, when the instructor gives a one-hour exam, she gives two "equivalent" examinations. Students in even-numbered seats take exam A and those in the odd-numbered seats take exam B. It is reasonable to ask "Are these two different exam forms equivalent?" Assuming that the odd- or even-numbered seats had no effect, we would want to test the hypothesis "the test forms yielded scores that had identical distributions."

To test this hypothesis the following two random samples were taken.

A	52	78	56	90	65	86	64	90	49	78
B	72	62	91	88	90	74	98	80	81	71

The size of the individual samples will be called n_a and n_b; actually it makes no difference which way these are assigned. In our illustration they both have the value 10.

The first thing that must be done with the entire sample (all $n_a + n_b$ pieces of data) is to order it into one sample, smallest to largest:

$$49 \quad 52 \quad 56 \quad 62 \quad 64 \quad 65 \quad 71 \quad 72 \quad 74 \quad 78$$
$$78 \quad 80 \quad 81 \quad 86 \quad 88 \quad 90 \quad 90 \quad 90 \quad 91 \quad 98$$

rank   Each piece of data is then assigned a **rank number**. The smallest (49) is assigned rank 1, the next smallest (52) is assigned rank 2, and so on, up to the largest, which is assigned rank $n_a + n_b$ (20). Ties are handled by assigning to each of the tied observations the mean rank of those rank positions that they occupy. For example, in our illustration there are two 78s; they are the 10th and 11th pieces of data. The mean rank for each is then $(10 + 11)/2 = 10.5$. In the case of the three 90s, the 16th, 17th, and 18th pieces of data, each is assigned 17, since $(16 + 17 + 18)/3 = 17$. The rankings are shown in Table 14-2.

Table 14-2
Ranked data for
Illustration 14-6

Ranked Data	Rank	Source	Ranked Data	Rank	Source
49	1	A	78	10.5	A
52	2	A	80	12	B
56	3	A	81	13	B
62	4	B	86	14	A
64	5	A	88	15	B
65	6	A	90	17	A
71	7	B	90	17	A
72	8	B	90	17	B
74	9	B	91	19	B
78	10.5	A	98	20	B

The calculation of the test statistic U is a two-step procedure. We first determine the **sum of the ranks** for each of the two samples. Then, using the two sums of ranks, we calculate two U **scores** for each sample. The **smaller U score is the test statistic**.

The sum of ranks R_a for sample A is computed as

$$R_a = 1 + 2 + 3 + 5 + 6 + 10.5 + 10.5 + 14 + 17 + 17 = 86$$

The sum of ranks R_b for sample B is

$$R_b = 4 + 7 + 8 + 9 + 12 + 13 + 15 + 17 + 19 + 20 = 124$$

The U score for each sample is obtained by using the following pair of formulas:

$$U_a = n_a \cdot n_b + \frac{(n_b)(n_b + 1)}{2} - R_b \qquad (14\text{-}3)$$

$$U_b = n_a \cdot n_b + \frac{(n_a)(n_a + 1)}{2} - R_a \qquad (14\text{-}4)$$

U, the test statistic, will be the smaller of U_a and U_b.

For our illustration, we obtain

$$U_a = (10)(10) + \frac{(10)(10 + 1)}{2} - 124 = 31$$

$$U_b = (10)(10) + \frac{(10)(10 + 1)}{2} - 86 = 69$$

Therefore,

$$U = 31$$

Before we carry out the test for this illustration, let's try to understand some of the underlying possibilities. Recall that the null hypothesis is that the distributions are the same and that we will most likely want to conclude from this that the means are approximately equal. Suppose for a moment that they are indeed quite different—say all of one sample comes before the smallest piece of data in the second sample when they are ranked together. This would certainly mean that we want to reject the null hypothesis. What kind of a value can we expect for U in this case? Suppose, in Illustration 14-6, that the ten A values had ranks 1 through 10 and the ten B values had ranks 11 through 20. Then we would obtain

$$R_a = 55 \qquad R_b = 155$$

$$U_a = (10)(10) + \frac{(10)(10 + 1)}{2} - 155 = 0$$

$$U_b = (10)(10) + \frac{(10)(10 + 1)}{2} - 55 = 100$$

Therefore,

$$U = 0$$

Suppose, on the other hand, that both samples were perfectly matched, that is, a score in each set identical to one in the other. Now what would happen?

$$
\begin{array}{cccccccc}
54 & 54 & 62 & 62 & 71 & 71 & 72 & 72 \quad \cdots \\
A & B & A & B & A & B & A & B \quad \cdots \\
1.5 & 1.5 & 3.5 & 3.5 & 5.5 & 5.5 & 7.5 & 7.5 \quad \cdots
\end{array}
$$

$$R_a = R_b = 105$$

$$U_a = U_b = (10)(10) + \frac{(10)(10 + 1)}{2} - 105 = 50$$

Therefore,

$$U = 50$$

If this was the case, we certainly would want to reach the decision "fail to reject the null hypothesis."

Note that the sum of the two U's $(U_a + U_b)$ will always be equal to the product of the two sample sizes $(n_a \cdot n_b)$. For this reason we need to concern ourselves with only one of them, the smaller one.

Now let's return to the solution of Illustration 14-6.

In order to complete our hypothesis test, we need to be able to determine a critical value for U. Table 12 in Appendix E gives us the critical value for some of the more common testing situations, as long as both sample sizes are reasonably small. Table 12 shows only the critical region in the left-hand tail, and the null hypothesis will be rejected if the observed value for U is less than or equal to the value read from the table. For our example, $n_a = n_b = 10$; at $\alpha = 0.05$ in a two-tailed test, the critical value is 23. We observed a value of 31 and therefore we make the decision "fail to reject the null hypothesis." This means that we do not have sufficient evidence to reject the "equivalent" hypothesis. ∎

If the **samples are larger than size 20**, we may make the test decision with the aid of the standard normal variable z. This is possible since the distribution of U is approximately normal with a mean

$$\mu_U = \frac{n_a \cdot n_b}{2} \tag{14-5}$$

and a standard deviation

$$\sigma_U = \sqrt{\frac{n_a n_b (n_a + b_b + 1)}{12}} \tag{14-6}$$

The null hypothesis is then tested by using

$$z = \frac{U - \mu_U}{\sigma_U} \tag{14-7}$$

in the typical fashion. The test statistic z may be used whenever n_a and n_b are both greater than 10.

Illustration 14-7 A dog obedience trainer is training 27 dogs to obey a certain command. The trainer is using two different training techniques, the reward-and-encouragement method (I) and the no-reward method (II). The following table shows the number of obedience sessions that were necessary before the dogs would obey the command. Does the trainer have sufficient evidence to claim that the reward method will, on the average, require less training time? ($\alpha = 0.05$.)

I	29	27	32	25	27	28	23	31	37	28	22	24	28	31	34
II	40	44	33	26	31	29	34	31	38	33	42	35			

Solution

Step 1 H_0: The average amount of training time required is the same for both methods.

H_a: The reward method requires less time on the average.

Step 2 $\alpha = 0.05$; the critical value is shown in the accompanying figure.

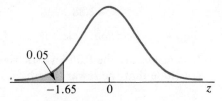

0.05

−1.65   0   *z*

Step 3 The two sets of data are ranked jointly and ranks are assigned, as shown in Table 14-3. Then the sums are

$$R_\mathrm{I} = 1 + 2 + 3 + 4 + 6.5 + \cdots + 20.5 + 23 = 151.0$$

$$R_\mathrm{II} = 5 + 11.5 + \cdots + 26 + 27 = 227.0$$

Table 14-3
Rankings for training methods

Number of Sessions	Group	Rank		Number of Sessions	Group	Rank	
22	I	1		31	II	15 ⌉	14.5
23	I	2		31	II	16 ⌋	14.5
24	I	3		32	I	17	
25	I	4		33	II	18 ⌉	18.5
26	II	5		33	II	19 ⌋	18.5
27	I	6 ⌉	6.5	34	I	20 ⌉	20.5
27	I	7 ⌋	6.5	34	II	21 ⌋	20.5
28	I	8 ⌉	9	35	II	22	
28	I	9	9	37	I	23	
28	I	10 ⌋	9	38	II	24	
29	I	11 ⌉	11.5	40	II	25	
29	II	12 ⌋	11.5	42	II	26	
31	I	13 ⌉	14.5	44	II	27	
31	I	14	14.5				

Now by using formulas (14-3) and (14-4), we can obtain the value of the U scores:

$$U_I = (15)(12) + \frac{(12)(12 + 1)}{2} - 227 = 180 + 78 - 227 = 31$$

$$U_{II} = (15)(12) + \frac{(15)(15 + 1)}{2} - 151 = 180 + 120 - 151 = 149$$

Therefore,

$$U = 31$$

Now we use formulas (14-5), (14-6), and (14-7) to determine the z statistic.

$$\mu_U = \frac{12 \cdot 15}{2} = 90$$

$$\sigma_U = \sqrt{\frac{(12)(15)(12 + 15 + 1)}{12}} = \sqrt{\frac{(180)(28)}{12}} = \sqrt{420} = 20.49$$

$$z = \frac{31 - 90}{20.49} = \frac{-59}{20.49} = -2.879$$

$$z^* = -2.88$$

Step 4   *Decision*: Reject H_0.

Conclusion: At the 0.05 level of significance, the data show sufficient evidence to conclude that the reward method does, on the average, require less training time.

■

Case Study 14-1

HEALTH BELIEFS AND PRACTICES OF RUNNERS VERSUS NONRUNNERS

Valerie Walsh used the Mann-Whitney U statistic to conclude that runners place a greater value on personal health than do nonrunners.

Returns of mailed questionnaires from 77 runner and 63 nonrunner respondents showed that runners placed a statistically higher value on health and performed greater numbers of health-related behaviors. Major differences were found in nutrition, exercise, and medical awareness and self-care. No major differences were found in addictive substance use, stress management, or safety practices. A number of concerns regarding runners' health practices were identified, including running while ill or in pain, incidence of injuries, negative feelings when unable to run, neglect of a conscious cool-down period, low weight levels, and a tendency to increase workouts following perceived dietary indiscretions....

The purpose of this investigation was to explore the differences between runners and nonrunners in terms of specific health beliefs and behaviors. It was hypothesized that there is a difference between runners and nonrunners in the relative value placed on personal health....

RESULTS

The first hypothesis, that there is a difference between runners and nonrunners in the relative value placed on personal health, was tested using the Mann-Whitney U, with alpha set at .05, and was accepted, $p < .019$. The value of U was found to be 1876.5; of *U*, 4990.5. Greater value was placed on personal health by the runners than by the nonrunners....

DISCUSSION

Runners placed a higher value on health and performed more health-related behaviors than did nonrunners. Although these results might have been anticipated in light of findings from other studies (Dawber, 1980; Paffenbarger et al., 1977), the information derived from them regarding health behaviors of runners was limited and inferential at best. The findings from this investigation, however, are congruent with those of Blair et al. (1981), in that runners exert tighter control over the types of nutrients they consume than do nonrunners.

Source: Valerie R. Walsh, *Nursing Research*, November/December 1985, Vol. 34, No. 6.

EXERCISES

14.7 Pulse rates were recorded for 16 men and 13 women. The results are shown in the following table.

Males	61	73	58	64	70	64	72	60
	65	80	55	72	56	56	74	65

Females	83	58	70	56	76	64	80
	68	78	108	76	70	97	

These data were used to test the hypothesis that the distribution of pulse rates differs for men and women. The following MINITAB output printed out the sum of ranks for males ($W = 192.0$) and the *p*-value for z^* of 0.0373. Verify these two values.

```
MTB > SET MALE PULSE RATES IN C1
DATA> 61 73 58 64 70 64 72 60 74 65 65 80 55 72 56 56
DATA> END DATA
MTB > SET FEMALE PULSE RATES IN C2
DATA> 83 58 70 56 76 64 80 76 70 97 68 78 108
DATA> END DATA
MTB > MANN-WHITNEY ON DATA IN C1,C2

Mann-Whitney Confidence Interval and Test

C1            N = 16    MEDIAN =       64.500
C2            N = 13    MEDIAN =       76.000
POINT ESTIMATE FOR ETA1-ETA2 IS       -8.9984
95.4  PCT C.I. FOR ETA1-ETA2 IS  (   -18.0,       0.1)
W =     192.0
TEST OF ETA1 = ETA2 VS. ETA1 N.E. ETA2 IS SIGNIFICANT AT 0.0373
```

14.8 The reaction times for an emergency situation were recorded for individuals from two different age groups. The data are shown in the following table.

| Age Group I (20–30) | 5.1 | 6.2 | 6.0 | 5.5 | 4.9 | 5.0 | 5.5 | 6.0 | 5.7 |
| Age Group II (50–65) | 7.0 | 6.5 | 6.0 | 7.2 | 7.2 | 5.9 | | | |

Do these samples show a significant difference in the reaction times for the two groups at the $\alpha = 0.05$ level of significance?

14.9 To determine whether there is a difference in the breaking strength of lightweight monofilament fishing line produced by two companies, the following test data (pounds of force required to break the line) were obtained. Use the Mann-Whitney U test and $\alpha = 0.05$ to determine whether there is a difference.

| A | 12.4 | 11.9 | 11.8 | 13.5 | 11.6 | 12.0 | 12.9 | 11.3 | 13.8 | 12.3 | |
| B | 12.7 | 10.7 | 13.2 | 11.8 | 11.8 | 12.1 | 12.5 | 11.9 | 12.8 | 11.2 | 11.5 | 11.3 |

14.10 The following set of data represents the ages of drivers involved in automobile accidents. Do these data present sufficient evidence to conclude that there is a difference in the average of men and women drivers involved in accidents? Use a two-tailed test at $\alpha = 0.05$.

Men	70	60	77	39	36	28	19	40
	23	23	63	31	36	55	24	76
Women	62	46	43	28	21	22	27	42
	21	46	33	29	44	29	56	70

   a. State the null hypothesis that is being tested.

   b. Complete the test.

14.11 In a heart study the systolic blood pressure was measured for 24 men of age 25 and for 30 men of age 40. (For your convenience, the data are ranked.) Do these data show sufficient evidence to conclude that the older men have a higher systolic blood pressure, at the 0.02 level of significance?

25-year-olds	95	100	100	105	106	108	110	110		
	115	118	120	122	124	125	130	130		
	130	132	136	138	140	148	150	156		
40-year-olds	108	110	110	114	114	116	118	120	122	124
	126	126	128	130	130	132	136	136	136	140
	142	142	146	148	150	152	154	160	164	176

14.4 THE RUNS TEST

runs test
run

The **runs test** is most frequently used to test the **randomness of data** (or lack of randomness). A **run** is a sequence of data that possesses a common property. One run ends and another starts when a piece of data does not display the property in question. The random variable that will be used in this test is V, the number of runs observed.

Illustration 14-8

To illustrate the idea of runs, let's draw a sample of ten single-digit numbers from the telephone book, using the next-to-last digit from each of the selected telephone numbers.

Sample: 2, 3, 1, 1, 4, 2, 6, 6, 6, 7

Let's consider the property of "odd" (o) or "even" (e). The sample, as it was drawn, becomes $e, o, o, o, e, e, e, e, e, o$, which displays four runs.

$$e \quad o \quad o \quad o \quad e \quad e \quad e \quad e \quad e \quad o$$

Thus $V = 4$. ∎

In Illustration 14-8, if the sample contained no randomness, there would be only two runs—all the evens, then all the odds, or the other way around. We would also not expect to see them alternate—odd, even, odd, even. The maximum number of possible runs would be $n_1 + n_2$, or less (provided n_1 and n_2 are not equal), where n_1 and n_2 are the number of data that have each of the two properties being identified.

We will often want to interpret the maximum number of runs as a rejection of a null hypothesis of randomness, since we often want to test randomness of the data in reference to how they were obtained. For example, if the data alternated all the way down the line, we might suspect that the data had been tampered with. There are many aspects to the concept of randomness. The occurrence of odd and even as discussed in Illustration 14-8 is one aspect. Another aspect of randomness that we might wish to check is the ordering of fluctuations of the data above (a) or below (b) the mean or median of the sample.

Illustration 14-9

Consider the sequence that results from determining whether each of the data points in the sample of Illustration 14-8 is above or below the median value. Test the null hypothesis that this sequence is random. Use $\alpha = 0.05$.

Solution

Step 1 H_0: The numbers in the sample form a random sequence with respect to the two properties "above" and "below" the median value.

H_a: The sequence is not random.

Sample: 2, 3, 1, 1, 4, 2, 6, 6, 6, 7

First we must rank the data and find the median. The ranked data are 1, 1, 2, 2, 3, 4, 6, 6, 6, 7. Since there are ten pieces of data, the median is at the $i = 5.5$ position. Thus $\tilde{x} = (3 + 4)/2 = 3.5$. By comparing each number in the original sample to the value of the median, we obtain the following sequence of a's and b's:

$$b \quad b \quad b \quad b \quad a \quad b \quad a \quad a \quad a \quad a$$

We observe $n_a = 5$, $n_b = 5$, and 4 runs. So $V = 4$.

If n_1 and n_2 are both less than or equal to 20 and a two-tailed test at $\alpha = 0.05$ is desired, then Table 13 of Appendix E will give us the two critical values for the test. For our illustration with $n_a = 5$ and $n_b = 5$, Table 13 shows critical values of 2 and 10. This means that if 2 or fewer, or 10 or more, runs are observed, the null hypothesis will be rejected. If between 3 and 9 runs are observed, we will fail to reject the null hypothesis.

Step 2 $\alpha = 0.05$ and a two-tailed test is used. The critical values for V are found in Table 13 (*see* the accompanying figure). The critical values are 2 and 10.

Reject H_0	Fail to reject H_0	Reject H_0
0	2 3 9	10

V, number of runs

Step 3   Four runs were observed; $V^* = 4$.

Step 4   *Decision*: Fail to reject H_0.

Conclusion: We are unable to reject the hypothesis of randomness at the 0.05 level of significance.   ■

To complete the hypothesis test about randomness when n_1 or n_2 is larger than 20 or when α is other than 0.05, we will use z, the standard normal random variable. V is approximately normally distributed with a mean of μ_V and a standard deviation of σ_V. The formulas are as follows:

$$\mu_V = \frac{2n_1 \cdot n_2}{n_1 + n_2} + 1 \tag{14-8}$$

$$\sigma_V = \sqrt{\frac{(2n_1 \cdot n_2)(2n_1 \cdot n_2 - n_1 - n_2)}{(n_1 + n_2)^2(n_1 + n_2 - 1)}} \tag{14-9}$$

$$z = \frac{V - \mu_V}{\sigma_V} \tag{14-10}$$

Illustration 14-10   Test the null hypothesis that the sequence that results from classifying the sample data in Illustration 14-8 as "odd" or "even" is a random sequence. Use $\alpha = 0.10$.

Solution

Step 1 H_0: The sequence of odd and even occurrences is a random sequence.

H_a: The sequence is not random.

Step 2 $\alpha = 0.10$ and a two-tailed test is to be used. The test criteria are shown in the following figure.

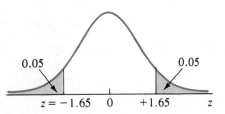

0.05   0.05

$z = -1.65$ 0 $+1.65$ z

Step 3   The sample and the sequence of odd and even properties are shown in Illustration 14-8. $n_1 = n(\text{even}) = 6$ and $n_2 = n(\text{odd}) = 4$. There were 4 runs, so $V = 4$.

$$\mu_V = \frac{2n_1 n_2}{n_1 + n_2} + 1 = \frac{2 \cdot 6 \cdot 4}{6 + 4} + 1 = \frac{48}{10} + 1 = 5.8$$

$$\sigma_V = \sqrt{\frac{(2n_1 n_2)(2n_1 n_2 - n_1 - n_2)}{(n_1 + n_2)^2(n_1 + n_2 - 1)}}$$

$$= \sqrt{\frac{(2 \cdot 6 \cdot 4)(2 \cdot 6 \cdot 4 - 6 - 4)}{(6 + 4)^2(6 + 4 - 1)}}$$

$$= \sqrt{\frac{(48)(38)}{(10)^2(9)}} = \sqrt{\frac{1824}{900}} = \sqrt{2.027} = 1.42$$

$$z = \frac{V - \mu_V}{\sigma_V} = \frac{4.0 - 5.8}{1.42} = \frac{-1.8}{1.42} = -1.268$$

$$z^* = -1.27$$

Step 4   *Decision*: Fail to reject H_0.

Conclusion: We are unable to reject the hypothesis of randomness at the 0.10 level of significance.   ■

EXERCISES

14.12 The following are 24 consecutive downtimes (in minutes) of a particular machine.

20   33   33   35   36   36   22   22   25   27   30   30
30   31   31   32   32   36   40   40   50   45   45   40

The null hypothesis of randomness is to be tested against the alternative that there is a trend. A MINITAB analysis of the number of runs above and below the median follows. Confirm that the number of runs is 4 and compute the value of z^*. Would you reject the hypothesis of randomness?

```
MTB > SET THE FOLLOWING TIMES IN C1
DATA> 20 33 33 35 36 36 22 22 25 27 30 30
DATA> 30 31 31 32 32 36 40 40 50 45 45 40
DATA> END DATA
MTB > RUNS ABOVE OR BELOW 32.5 FOR DATA IN C1

     C1

     K = 32.5000

     THE OBSERVED NO. OF RUNS =   4
     THE EXPECTED NO. OF RUNS = 13.0000
     12 OBSERVATIONS ABOVE K      12 BELOW
                  THE TEST IS SIGNIFICANT AT 0.0002
```

14.13 A manufacturing firm hires both men and women. The following shows the sex of the last 20 individuals hired (M = male, F = female).

M M F M F   F M M M M
M M F M M F M M M M

At the $\alpha = 0.05$ level of significance, are we correct in concluding that this sequence is not random?

14.14 A student was asked to perform an experiment that involved tossing a coin 25 times. After each toss the student recorded the results. The following data were reported (H = heads, T = tails):

H, T, H, T, H, T, H, T, H, H, T, T, H,
H, T, T, H, T, H, T, H, T, H, T, H

Use the runs test at a 5 percent level of significance to test the student's claim that the results reported are random.

14.15 In an attempt to answer the question "Does the husband (h) or wife (w) do the family banking?" the results of a sample of 28 married customers doing the family banking show the following sequence of arrivals at the bank.

w, w, w, w, h, w, h, h, h, h, w, w, w, w,
w, h, h, w, w, w, h, h, h, h, w, h, h, w

Do these data show lack of randomness with regard to whether the husband or wife does the family banking? Use $\alpha = 0.05$.

14.16 The following data were collected in an attempt to show that the number of minutes the city bus is late is steadily growing larger. The data are in order of occurrence.

6,   1,   3,   9,   10,   10,   2,   5,   5,   6,
12,   3,   7,   8,   9,   4,   5,   8,   11,   14

At $\alpha = 0.05$, do these data show sufficient lack of randomness to support the claim?

14.17 The number of absences recorded at a lecture that met at 8 A.M. on Mondays and Thursdays last semester were (in order of occurrence):

5, 16, 6, 9, 18, 11, 16, 21, 14, 17, 12, 14, 10,
6, 8, 12, 13, 4, 5, 5, 6, 1, 7, 18, 26, 6

Do these data show a randomness about the median value, at $\alpha = 0.05$? Complete this test by using (1) critical values from Table 13, and (2) the standard normal distribution.

14.5   RANK CORRELATION

The rank correlation coefficient was developed by C. Spearman in the early 1900s. It is a nonparametric alternative to the linear correlation coefficient that was discussed in Chapters 3 and 13. Only rankings are used in the calculation of this coefficient. If the data are quantitative, each of the two variables must be ranked separately.

Spearman rank correlation coefficient

The **Spearman rank correlation coefficient, r_s,** is found by using the formula

$$r_s = 1 - \frac{6 \sum (d_i)^2}{n(n^2 - 1)} \tag{14-11}$$

where d_i is the difference in the rankings and n is the number of pairs of data. The value of r_s will range from -1 to $+1$ and will be used in much the same manner as the linear correlation coefficient was used previously.

The null hypothesis that we will be testing is "there is no correlation between the two rankings." The alternative hypothesis may be either two-tailed, "there is correlation," or one-tailed, if we anticipate the existence of either positive or negative correlation. The critical region will be on the side(s) corresponding to the specific alternative that is expected. For example, if we suspect negative correlation, the critical region will be in the left-hand tail.

Illustration 14-11   Let's consider a hypothetical situation in which four judges rank five contestants in a contest. Let's identify the judges as A, B, C, and D and the contestants as a, b, c, d, and e. Table 14-4 lists the awarded rankings.

Table 14-4
Rankings for five contestants

Contestant	Judge			
	A	B	C	D
a	1	5	1	5
b	2	4	2	2
c	3	3	3	1
d	4	2	4	4
e	5	1	5	3

When we compare judges A and B, we see that they ranked the contestants in exactly the opposite order—perfect disagreement (see Table 14-5). From our previous work with correlation, we expect the calculated value for r_s to be exactly -1 for these data.

$$r_s = 1 - \frac{6[\sum (d_i)^2]}{n(n^2 - 1)} = 1 - \frac{(6)(40)}{(5)(5^2 - 1)} = 1 - \frac{240}{120} = -1$$

Table 14-5

Rankings of A and B

Contestant	A	B	$d_i = A - B$	$(d_i)^2$
a	1	5	−4	16
b	2	4	−2	4
c	3	3	0	0
d	4	2	2	4
e	5	1	4	16
				40

When judges A and C are compared, we see that their rankings of the contestants are identical (*see* Table 14-6). We would expect to find a calculated correlation coefficient of +1 for these data.

$$r_s = 1 - \frac{(6)(0)}{(5)(5^2 - 1)} = 1 - 0 = 1$$

Table 14-6

Rankings of A and C

Contestant	A	C	$d_i = A - C$	$(d_i)^2$
a	1	1	0	0
b	2	2	0	0
c	3	3	0	0
d	4	4	0	0
e	5	5	0	0
				0

By comparing the rankings of judge A with those of judge B and then with those of judge C, we have seen the extremes: total agreement and total disagreement. Now let's compare the rankings of judge A with those of judge D (*see* Table 14-7). There seems to be no real agreement or disagreement here. Let's compute r_s:

$$r_s = 1 - \frac{(6)(24)}{(5)(5^2 - 1)} = 1 - \frac{144}{120} = 1 - 1.2 = -0.2$$

Table 14-7

Rankings of A and D

Contestant	A	D	$d_i = A - D$	$(d_i)^2$
a	1	5	−4	16
b	2	2	0	0
c	3	1	2	4
d	4	4	0	0
e	5	3	2	4
				24

This is fairly close to zero, which is what we should have suspected since there was no real agreement or disagreement.

The test of significance will result in a failure to reject the null hypothesis when r_s is close to zero, and will result in a rejection of the null hypothesis in cases where r_s is found to be close to $+1$ or -1. The critical values found in Table 14 of Appendix E are positive critical values only. Since the null hypothesis is "the population correlation coefficient is zero" (that is, $\rho = 0$), we have a symmetric test statistic. Hence we need only add a plus or minus sign to the value found in the table, as appropriate. This will be determined by the specific alternative that we have in mind.

In our illustration, the critical values for a two-tailed test at $\alpha = 0.10$ are ± 0.900. (Remember that n represents the number of pairs.) If the calculated value for r_s is between 0.9 and 1.0 or between -0.9 and -1.0, we will reject the null hypothesis in favor of the alternative "there is a correlation." ∎

The Spearman rank coefficient is defined by formula (3-1) [Pearson's product moment r], with rankings used in place of the quantitative x and y values. If there are no ties in the rankings, formula (14-11) is equivalent to formula (3-1). Formula (14-11) provides us with a much easier procedure to use for calculating the r_s statistic. When there are only a few ties, it is common practice to use formula (14-11). The resulting value of r_s is not exactly the same; however, it is generally considered to be an acceptable estimate. Illustration 14-12 shows the procedure for handling ties and uses formula (14-11) for the calculation of r_s. For comparison, Exercise 14.20 asks you to calculate r_s using formula (3-1).

If ties occur in either set of the ordered pairs of rankings, assign each tied observation the mean of the ranks that would have been assigned had there been no ties, as was done in the Mann-Whitney U test.

Illustration 14-12   Students who finish exams more quickly than the rest of the class are often thought to be smarter. The following set of data shows the score and order of finish for 12 students on a recent one-hour exam. At the 0.05 level, do these data support the alternative hypothesis that the first students to complete an exam have higher grades?

Order of Finish	1	2	3	4	5	6	7	8	9	10	11	12
Exam Score	90	74	76	60	68	86	92	60	78	70	78	64

Solution

Step 1 H_0: Order of finish has no relationship to exam score.

H_a: First to finish tend to have higher grades.

Step 2 $\alpha = 0.05$ and $n = 12$; the critical region is shown in the following diagram.

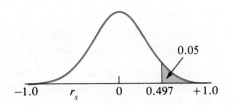

Step 3   Rank the scores from highest to lowest, assigning the highest score the rank number 1, as shown.

92	90	86	78	78	76	74	70	68	64	60	60
1	2	3	4	5	6	7	8	9	10	11	12
			4.5	4.5						11.5	11.5

The rankings and the preliminary calculations are shown in Table 14-8.
    Using formula (14-11), we obtain

$$r_s = 1 - \frac{(6)(235.0)}{(12)(143)} = 1 - \frac{1410}{1716} = 1 - 0.822$$

$$= \mathbf{0.178}$$

Table 14-8
Rankings for test scores
and differences

Test Score Rank	Order of Finish	Difference (d_i)	$(d_i)^2$
1	7	−6	36.00
2	1	1	1.00
3	6	−3	9.00
4.5	9	−4.5	20.25
4.5	11	−6.5	42.25
6	3	3	9.00
7	2	5	25.00
8	10	−2	4.00
9	5	4	16.00
10	12	−2	4.00
11.5	4	7.5	56.25
11.5	8	3.5	12.25
			235.00

Step 4   *Decision*: Fail to reject H_0.

Conclusion: There is not sufficient evidence presented by these sample data to enable us to conclude that the first students to finish have the higher grades, at the 0.05 level of significance. ■

Case Study 14-2

EFFECTS OF THERAPEUTIC TOUCH ON TENSION HEADACHE PAIN

Elizabeth Keller and Virginia Bzdek made use of the Spearman rank correlation coefficient to determine the influence of the subjects' education, age, and several other attributes on the results of therapeutic touch. They reported several observed values and the associated prob-values.

Therapeutic touch (TT) is a modern derivative of the laying on of hands that involves touching with the intent to help or heal. This study investigated the effects of TT on tension headache pain in comparison with a placebo simulation of TT. Sixty volunteer subjects with tension headaches were randomly divided into treatment and placebo groups. The McGill-Melzack Pain Questionnaire was used to measure headache pain levels before each intervention, immediately afterward, and 4 hours later....

HYPOTHESES

I. Tension headache pain will be reduced following therapeutic touch, and the initial reduction will be maintained for a 4-hour period.

II. Subjects who receive therapeutic touch will experience greater tension headache pain reduction than subjects receiving a placebo simulation of therapeutic touch.

III. Subjects who receive therapeutic touch will maintain greater tension headache pain reduction than subjects receiving a placebo simulation of therapeutic touch 4 hours following the intervention....

Findings: Spearman correlation coefficients were calculated to determine the influence of the subject's education, age, sex, practice of meditation, religion, and level of initial skepticism toward TT. The differences in the MMPQ pain scores were not significantly correlated with any belief system or demographic variable in either group. The one exception was the posttest difference on the PPI scale in the placebo group, which was inversely correlated with years of education, $r = -.53$, $p = .002$.

The MMPQ was internally consistent in this study to a statistically significant degree between all three subscales on the pretest, posttest, and delayed posttest. The highest correlation was between the PRI and the NWC, $r = .79$, $p < .0001$. The lowest correlation was between NWC and the PPI, $r = .30$, $p < .02$. When one subscale reported a significant difference between the groups, the other two scales also reported significant differences in the same direction. Based on these observations of internal consistency, the MMPQ was considered to be a reliable instrument in this study.

Source: Elizabeth Keller and Virginia M. Bzdek, *Nursing Research*, March/April 1986, Vol. 35, No. 2.

EXERCISES

 14.18 The following data represent the ages of 12 subjects and the mineral concentration (in parts per million) in their tissue samples.

Age (x)	82	83	64	53	47	50	70	62	34	27	75	28
Mineral Concentration (y)	170	40	64	5	15	5	48	34	3	7	50	10

Refer to the following MINITAB output and verify that the Spearman rank correlation coefficient equals 0.753.

```
MTB > READ AGE IN C1, PPM IN C2
DATA> 82 170
DATA> 83 40
DATA> 64 64
DATA> 53 5
DATA> 47 15
DATA> 50 5
DATA> 70 48
DATA> 62 34
DATA> 34 3
DATA> 27 7
DATA> 75 50
DATA> 28 10
DATA> END DATA
      12 ROWS READ
MTB > RANK C1,PUT RANKS IN C3
MTB > RANK C2,PUT RANKS IN C4
MTB > CORRELATION C3,C4

Correlation of C3 and C4 = 0.753
```

14.19 An advertising company wants to determine whether the number of television commercials (x) are related to the number of sales (y) of a certain product. The following data were obtained.

x	10	12	15	5	7	5	5	15	8	1	13	15
y	30	60	50	12	10	25	10	60	25	10	75	95

a. Compute the Spearman rank correlation coefficient, r_s.

b. At the 0.05 level of significance, is there sufficient reason to conclude that a positive correlation exists?

14.20 Using formula (3-2), calculate the Spearman rank correlation coefficient for the data in Illustration 14-12. Recall that formula (3-2) is equivalent to the definition formula (3-1) and that rank numbers must be used with this formula in order for the resulting statistic to be Spearman's r_s.

14.21 The following sets of air pollution readings were taken from a local newspaper on ten separate days.

Pollution	Day									
	1	2	3	4	5	6	7	8	9	10
Sulfur Dioxide	0.07	0.01	0.01	0.02	0.07	0.02	0.00	0.01	0.05	0.04
Carbon Monoxide	0.9	2.0	0.8	0.8	0.9	1.3	3.8	2.0	1.3	2.5
Soiling or Dust	0.3	0.5	0.3	0.3	0.3	0.5	0.3	0.5	0.3	0.5

a. Draw a scatter diagram showing the sulfur dioxide and carbon monoxide readings as paired data. Use Spearman's rank correlation coefficient to test the correlation between the sulfur dioxide and the carbon monoxide readings. Complete a two-tailed test at $\alpha = 0.10$.

(1) State the null hypothesis being tested.

(2) Complete the test and carefully state the conclusion reached.

b. Repeat (a) using the sulfur dioxide and the soiling readings ($\alpha = 0.05$).

c. Repeat (a) using the carbon monoxide and the soiling readings ($\alpha = 0.01$).

14.22 Refer to the bivariate data shown in the following table.

x	-2	-1	1	2
y	4	1	1	4

a. Construct a scatter diagram.

b. Calculate Spearman's correlation coefficient r_s [formula (14-11)].

c. Calculate Pearson's correlation coefficient r [formula (3-2)].

d. Compare the two results from (b) and (c). Do the two measures of correlation measure the same thing?

14.6 COMPARING STATISTICAL TESTS

Only a few nonparametric tests have been presented in this chapter. Many more nonparametric tests can be found in other books. In addition, most of these other nonparametric tests can be used in place of certain parametric tests. The question is, then, which test do we use, the parametric or the nonparametric? Furthermore, when there is more than one nonparametric test, which one do we use?

The decision about which test to use must be based on the answer to the question "Which test will do the job best?" First, let's agree that we are dealing with two or more tests that are equally qualified to be used. That is, each test has a set of assumptions that must be satisfied before they can be applied. From this starting point we will attempt to define "best" to mean the test that is best able to control the risks of error and at the same time keep the size of the sample to a number that is reasonable to work with. (Sample size means cost, cost to you or your employer.)

Let's look first at the ability to **control the risk of error**. The risk associated with a type I error is controlled directly by the level of significance α. Recall that P (type I error) $= \alpha$ and P (type II error) $= \beta$. Therefore, it is β that we must control. Statisticians like to talk about power (as do others) and the **power of a statistical test** is defined to be $1 - \beta$. Thus the power of a test, $1 - \beta$, is the probability that we reject the null hypothesis when we should have rejected it. If two tests with the same α are equal candidates for use, the one with the greater power is the one you would want to choose.

power of a statistical test

The other factor is the **sample size required** to do a job. Suppose that you set the levels of risk that you can tolerate, α and β, and then are able to determine the sample efficiency

size that it would take to meet your specified challenge. The test that required the smaller sample size would then seem to have the edge. Statisticians usually use the term "efficiency" to talk about this concept. **Efficiency** is defined to be the ratio of the sample size of the best parametric test to the sample size of the best nonparametric test, when compared under a fixed set of risk values. For example, the efficiency rating for the sign test is approximately 0.63. This means that a sample of size 63 with a parametric test will do the same job as a sample of size 100 will do with the sign test.

The power and the efficiency of a test cannot be used alone to determine the choice of test. Sometimes you will be forced to use a certain test because of the data you are given. When there is a decision to be made, the final decision rests in a trade-off of three factors: (1) the power of the test, (2) the efficiency of the test, and (3) the data (and the number of data) available. Table 14-9 shows how the nonparametric tests discussed in this chapter compare with the parametric tests covered in previous chapters.

Table 14-9
Comparison of parametric and nonparametric tests

Test Situation	Parametric Test	Nonparametric Test	Efficiency of Nonparametric Test
One mean	t test (page 328)	Sign test (page 517)	0.63
Two independent means	t test (page 382)	U test (page 523)	0.95
Two dependent means	t test (page 391)	Sign test (page 518)	0.63
Correlation	Pearson's (page 481)	Spearman test (page 535)	0.91
Randomness		Runs test (page 530)	Not meaningful; there is no parametric test for comparison

In Retrospect

In this chapter you have become acquainted with some of the basic concepts of nonparametric statistics. While learning about the use of nonparametric methods and specific nonparametric tests of significance, you should have also come to realize and understand some of the basic assumptions that are needed when the parametric techniques of the earlier chapters are encountered. You now have seen a variety of tests, many of which somewhat duplicate the job done by others. What you must keep in mind is that you should use the best test for your particular needs. The power of the test and the cost of sampling, as related to size and availability of the desired response variable, will play important roles in determining the specific test to be used.

The news article at the beginning of this chapter illustrates only one of many situations in which a nonparametric test can be used. Spearman's rank correlation test may be used to statistically compare the two rankings. They appear to be quite different, but are they significantly different? The answer to this is left for you to determine in Exercise 14.36.

Chapter
Exercises

14.23 Nonparametric tests are also called "distribution-free" tests. However, the normal and the chi-square distributions are used in the inference-making procedures.

a. To what does the "distribution-free" term apply? (The population? The sample? The sampling distribution?) Explain.

b. What is it that has the normal or the chi-square distribution? Explain.

14.24 Track coaches, runners, and fans talk a lot about the "speed of the track." The surface of the track is believed to have a direct effect on the amount of time that it takes a runner to cover the required distance. To test this effect, ten runners were asked to run a 220-yard sprint on each of two tracks. Track A is a cinder track and track B is made of a new synthetic material. The running times are given in the following table. Test the claim that the surface on track B is conducive to faster running times.

Track	Runner									
	1	2	3	4	5	6	7	8	9	10
A	27.7	26.8	27.0	25.5	26.6	27.4	27.2	27.4	25.8	25.1
B	27.0	26.7	25.3	26.0	26.1	25.3	26.7	27.1	24.8	27.1

a. State the null and alternative hypotheses being tested. Complete the test by using the sign test with $\alpha = 0.05$.

b. State your conclusions.

14.25 A test that measures computer anxiety was administered to students in a statistics course that uses statistical packages on the computer. The test was given at the beginning and end of the course. The hypothesis being studied was that computer anxiety would be reduced as the students used computers in the course. (A high score on the test indicates high computer anxiety.) Can we conclude that computer anxiety was reduced during this course? Use the sign test and $\alpha = 0.05$.

Beginning	68	79	49	34	60	87	68	70	60	67	39	59	42	40	62	80	62	63	102	42	68	70	44	63	55
End	77	78	32	40	60	68	87	88	62	68	40	60	31	39	62	63	58	59	110	49	54	67	43	54	41

14.26 A candy company has developed two new chocolate-covered candy bars. Six randomly selected people all preferred candy bar I. Is this statistical evidence, at $\alpha = 0.05$, that the general public will prefer candy bar I?

14.27 While trying to decide on the best time to harvest his crop, a commercial apple farmer recorded the day on which the first apple on the top half and the first apple on the bottom half of 20 randomly selected trees were ripe. The variable x was assigned a value of 1 on the first day that the first ripe apple appeared on 1 of the 20 trees. The days were then numbered sequentially. The observed data are shown in the following table. Do these data provide convincing evidence that the apples on the top of the trees start to ripen before the apples on the bottom half? Use $\alpha = 0.05$.

Position	1	2	3	4	5	6	7	8	9	10	11	12	13	14	15	16	17	18	19	20
Top	5	6	1	4	5	3	6	7	8	5	8	6	4	7	8	10	3	2	9	7
Bottom	6	5	5	7	3	6	6	8	9	4	10	7	5	11	6	11	5	6	9	8

Table header: Tree

14.28 A sample of 32 students received the following grades on an exam:

$$
\begin{array}{cccccccc}
41 & 42 & 48 & 46 & 50 & 54 & 51 & 42 \\
51 & 50 & 45 & 42 & 32 & 45 & 43 & 56 \\
55 & 47 & 45 & 51 & 60 & 44 & 57 & 57 \\
47 & 28 & 41 & 42 & 54 & 48 & 47 & 32
\end{array}
$$

a. Does this sample show that the median score for the exam differs from 50? Use $\alpha = 0.05$.

b. Does this sample show that the median score for the exam is less than 50? Use $\alpha = 0.05$.

14.29 Twenty students were randomly divided into two equal groups. Group 1 was taught an anatomy course using a standard lecture approach. Group 2 was taught using a computer-assisted approach. The test scores on a comprehensive final exam were as follows:

Group 1	75	83	60	89	77	92	88	90	55	70
Group 2	77	92	90	85	72	59	65	92	90	79

Use the Mann-Whitney U test to test the claim that a computer-assisted approach produces higher achievement (as measured by final exam scores) in anatomy courses than does a lecture approach. Use $\alpha = 0.05$.

14.30 A firm is currently testing two different procedures for adjusting the cutting machines used in the production of greeting cards. The results of two samples show the following recorded adjustment times.

Method 1	17	15	14	18	16	15	17	18	15	14	14	16	15		
Method 2	14	14	13	13	15	12	16	14	16	13	14	13	12	15	17	13

Is there sufficient reason to conclude that method 2 requires less time (on the average) than method 1 at the 0.05 level of significance?

14.31 Consider the following sequence of defective parts (d) and nondefective parts (n) produced by a machine.

$$
\begin{array}{ccccccccccccc}
n & n & n & d & n & n & n & n & n & d & n & n \\
n & n & n & n & n & d & n & d & n & n & n & n
\end{array}
$$

Can we reject the hypothesis of randomness at $\alpha = 0.05$?

14.32 A patient was given two different types of vitamin pills, one containing iron and one iron-free. The patient was instructed to take the pills on alternate days. To free himself from remembering which pill he needed to take, he mixed all of the pills together in a large bottle. Each morning he took the first pill that came out of the bottle. To see whether this was a random process, for 25 days he recorded an "I" each morning that he took a vitamin with iron and an "N" for no iron.

Day	1	2	3	4	5	6	7	8	9	10	11	12	13
Type	I	I	N	I	I	N	N	I	N	N	N	N	N

Day	14	15	16	17	18	19	20	21	22	23	24	25
Type	I	I	I	N	I	I	I	I	N	I	I	N

Is there sufficient reason to reject the null hypothesis that the vitamins were taken in a random order, at the 0.05 level of significance?

14.33 Mrs. Brown attended a special reading course that claimed to improve reading speed and comprehension. She took a pretest (on which she scored 106) and a test at the end of each of 15 sessions. Her test scores are shown in the following table. Do these data support the claim of improvement, at the $\alpha = 0.025$ level?

Session	1	2	3	4	5	6	7	8	9	10	11	12	13	14	15
Score	109	108	110	112	108	111	112	113	110	112	115	114	116	116	118

14.34 A computer-science aptitude test score, x, and a mathematical competency score, y, were determined for 15 students. The results are shown in the following table.

x	4	16	20	13	22	21	15	20	19	16	18	17	8	6	5
y	28	35	42	41	44	42	36	44	39	36	40	40	33	27	44

a. Find r_s.

b. Is there sufficient evidence to conclude that the two test scores are correlated? Use $\alpha = 0.05$.

14.35 Can today's high temperature be effectively predicted using yesterday's high? Pairs of yesterday's and today's high temperatures were randomly selected. The results are shown in the following table. Do the data present sufficient evidence to justify the statement "Today's high temperature tends to correlate with yesterday's high temperature"? Use $\alpha = 0.05$.

									Pairs of Days									
Reading	1	2	3	4	5	6	7	8	9	10	11	12	13	14	15	16	17	18
Yesterday's	40	58	46	33	40	51	55	81	85	83	89	64	73	63	46	58	28	69
Today's	40	56	34	59	46	51	74	77	83	84	85	68	65	60	54	62	34	66

14.36 The news article at the beginning of this chapter shows the rankings assigned to ten components of job satisfaction. Do the rankings assigned by the workers and the boss show a significant difference in what each thinks is important? Test by using $\alpha = 0.05$.

14.37 In a study to see whether spouses are consistent in their preferences for television programs, a market research firm asked several married couples to rank a list of 12 programs (1 represents the highest score; 12 represents the lowest). The average ranks for the programs, rounded to the nearest integer, were as follows.

| Rank | \multicolumn{12}{c}{Program} | | | | | | | | | | | |
	1	2	3	4	5	6	7	8	9	10	11	12
Husbands	12	2	6	10	3	11	7	1	9	5	8	4
Wives	5	4	1	9	3	12	2	8	6	10	7	11

Is there significant evidence of negative correlation at the 0.01 level of significance?

14.38 Two table tennis ball manufacturers have agreed that the quality of their products can be measured by the height to which the balls rebound. A test is arranged, the balls are dropped from a constant height, and the rebound heights are measured. The results (in inches) are shown in the following table. Manufacturer A claims, "The results show my product to be superior." Manufacturer B replies, "I know of no statistical test that supports this claim." Can you find a test that supports A's claim?

A	14.0	12.5	11.5	12.2	12.4	12.3	11.8	11.9	13.7	13.2
B	12.0	12.5	11.6	13.3	13.0	13.0	12.1	12.8	12.2	12.6

a. Does the appropriate parametric test show that A's product is superior? (What parametric test (or tests) is appropriate and what exactly does it show?)

b. Does the appropriate nonparametric test show that A's product is superior?

Vocabulary List

Be able to define each term. In addition, describe in your own words and give an example of each term. Your examples should not be ones given in class or in the textbook.

The bracketed numbers indicate the chapters in which the term previously appeared, but you should define the terms again to show increased understanding of their meaning.

binomial random variable [5, 9]
continuity correction [6]
correlation [3, 13]
distribution [2, 5, 6]
distribution-free test

efficiency
independent sample [10]
Mann-Whitney U test
median [2]
nonparametric test

normal approximation [6]   run
paired data [3, 10, 13]   runs test
parametric test   sign test
power   Spearman's rank
randomness [2]   correlation coefficient
rank

Quiz   *Answer "True" if the statement is always true. If the statement is not always true,*
        replace the boldface words with words that make the statement always true.

14.1 One of the advantages that the nonparametric tests have is the necessity for **less restrictive** assumptions.

14.2 The sign test is a possible replacement for the *F* **test**.

14.3 The **sign test** can be used to test the randomness of a set of data.

14.4 If a tie occurs in a set of ranked data, the data that form the tie are **removed from the set**.

14.5 Two dependent **means** can be compared nonparametrically by using the sign test.

14.6 The sign test is a possible alternative to the Student's *t* test for **one mean value**.

14.7 The **runs test** is a nonparametric alternative to the difference between two independent means.

14.8 The **confidence level** of a statistical hypothesis test is measured by $1 - \beta$.

14.9 Spearman's rank correlation coefficient is an alternative to using the **linear correlation coefficient**.

14.10 The **efficiency** of a nonparametric test is the probability that a false null hypothesis is rejected.

Working with Your Own Data

Many variables in everyday life can be treated as bivariate. Often two such variables have a mathematical relationship that can be approximated by means of a straight line. The following illustration demonstrates such a situation.

A | THE AGE AND VALUE OF PEGGY'S CAR

Peggy would like to sell her 1983 Camaro, and she needs to know what price to ask for it in order to write a newspaper advertisement. The car is in average condition and Peggy expects to get an average price for it. She must answer the question "What is an average selling price for a 1983 Camaro?"

Inspection of many classified sections of newspapers turned up only four advertisements for 1983 Camaro. The prices listed varied a great deal. Peggy finally decided that, in order to determine an accurate selling price she would define two variables and collect several pairs of values based on the following definitions.

Population   Used Chevrolet Camaros advertised for sale by individual owners, dealers not included.

Independent variable, x   The age of the car as measured in years and defined by

$$x = \text{(present calendar year)} - \text{(year of manufacture)} + 1.$$

Example: During 1987, Peggy's 1983 Camaro is considered to be 3 years old.

$$x = (1987 - 1983) + 1 = 4 + 1 = 5.$$

Dependent variable, y   The advertised asking price.

The following table lists the data collected in September 1987.

Year of Manufacture:			1981	1981	1981	1981	1987	1982
Asking Price:			5500	4500	3600	4800	9000	6200
Year:	1985	1986	1982	1984	1985	1984	1986	1983
Price:	6300	7000	4800	8000	8700	7500	8000	5200
Year:	1984	1987	1983	1985	1983	1986	1982	1987
Price:	5500	8200	5800	8000	7700	7000	5400	8000
Year:	1986	1984	1983	1986	1982	1984	1986	1982
Price:	9000	6500	4700	8700	7200	7800	6500	5000
Year:	1985	1983						
Price:	7200	6300						

1. Construct and label a scatter diagram of Peggy's data.
2. Determine the equation for the line of best fit.
3. Draw the line of best fit on the scatter diagram.
4. Test the equation of the line of best fit to see whether the linear model is appropriate for the data. Use $\alpha = 0.05$.
5. Construct a 0.95 confidence interval estimate for the mean advertised price for 1983 Camaros.
6. Draw a line segment on the scatter diagram that represents the interval estimate found for question 5.
7. What does the value of the slope, b_1, represent? Explain.
8. What does the value of the y-intercept, b_0, represent? Explain.

B | YOUR OWN INVESTIGATION

Identify a situation of interest to you that can be investigated statistically using bivariate data. (Consult your instructor for specific guidance.)

1. Define the population, the independent variable, the dependent variable, and the purpose for studying these two variables as a regression analysis.
2. Collect 15 to 20 ordered pairs of data.
3. Construct and label a scatter diagram of your data.
4. Determine the equation for the line of best fit.
5. Draw the line of best fit on the scatter diagram.
6. Test the equation of the line of best fit to see whether the linear model is appropriate for the data. Use $\alpha = 0.05$.
7. Construct a 0.95 confidence interval estimate for the mean value of the dependent variable at the following value of x:

Let x be equal to one-third the sum of the lowest value of x in your sample and twice the largest value. That is:

$$x = \frac{L + 2H}{3}$$

8. Draw a line segment on the scatter diagram that represents the interval estimate found for question 7.
9. What does the value of the slope, b_1, represent? Explain.
10. What does the value of the y-intercept, b_0, represent? Explain.

Appendixes

Appendix A

Summation Notation

The Greek capital letter sigma (Σ) is used in mathematics to indicate the summation of a set of addends. Each of these addends must be of the form of the variable following Σ. For example:

1. Σx means sum the variable x.

2. $\Sigma(x - 5)$ means sum the set of addends that are each 5 less than the values of each x.

When large quantities of data are collected, it is usually convenient to index the response variable so that at a future time its source will be known. This indexing is shown on the notation by using i (or j or k) and affixing the index of the first and last addend at the bottom and top of the Σ. For example,

$$\sum_{i=1}^{3} x_i$$

means to add all the consecutive values of x's starting with source number 1 and proceeding to source number 3.

Illustration A-1   Consider the inventory in the following table concerning the number of defective stereo tapes per lot of 100.

Lot Number (i)	1	2	3	4	5	6	7	8	9	10
Number of Defective Tapes per Lot (x)	2	3	2	4	5	6	4	3	3	2

a. Find $\sum_{i=1}^{10} x_i$.   **b.** Find $\sum_{i=4}^{8} x_i$.

552

Solution

a. $\displaystyle\sum_{i=1}^{10} x_i = x_1 + x_2 + x_3 + x_4 + \cdots + x_{10}$

$$= 2 + 3 + 2 + 4 + 5 + 6 + 4 + 3 + 3 + 2 = \mathbf{34}$$

b. $\displaystyle\sum_{i=4}^{8} x_i = x_4 + x_5 + x_6 + x_7 + x_8 = 4 + 5 + 6 + 4 + 3 = \mathbf{22}$ ∎

This index system must be used whenever only part of the available information is to be used. In statistics, however, we will usually use all the available information, and to simplify the formulas we will make an adjustment. This adjustment is actually an agreement that allows us to do away with the index system in situations where all values are used. Thus in our previous illustration, $\displaystyle\sum_{i=1}^{10} x_i$ could have been written simply as $\sum x$.

Note   The lack of the index indicates that all data are being used.

Illustration A-2   Given the following six values for x, 1, 3, 7, 2, 4, 5, find $\sum x$.

Solution

$$\sum x = 1 + 3 + 7 + 2 + 4 + 5 = \mathbf{22}$$ ∎

Throughout the study and use of statistics you will find many formulas that use the $\sum$ symbol. Care must be taken so that the formulas are not misread. Symbols like $\sum x^2$ and $\left(\sum x\right)^2$ are quite different. $\sum x^2$ means "square each x value and then add up the squares," while $\left(\sum x\right)^2$ means "sum the x values and then square the sum."

Illustration A-3   Find (a) $\sum x^2$ and (b) $\left(\sum x\right)^2$ for the sample in Illustration A-2.

Solution

a.

x	1	3	7	2	4	5
x^2	1	9	49	4	16	25

$$\sum x^2 = 1 + 9 + 49 + 4 + 16 + 25 = \mathbf{104}$$

b. $\sum x = 22$, as found in Illustration A-2. Thus,

$$\left(\sum x\right)^2 = (22)^2 = \mathbf{484}$$

As you can see, there is quite a difference between $\sum x^2$ and $\left(\sum x\right)^2$.   ∎

Likewise, $\sum xy$ and $\sum x \sum y$ are different. These forms will appear only when there are paired data, as shown in the following illustration.

Illustration A-4   Given the five pairs of data shown in the following table, find (a) $\sum xy$ and (b) $\sum x \sum y$.

x	1	6	9	3	4
y	7	8	2	5	10

Solution

a. $\sum xy$ means to sum the products of the corresponding x and y values. Therefore, we have

x	1	6	9	3	4
y	7	8	2	5	10
xy	7	48	18	15	40

$$\sum xy = 7 + 48 + 18 + 15 + 40 = \mathbf{128}$$

b. $\sum x \sum y$ means the product of the two summations, $\sum x$ and $\sum y$. Therefore, we have

$$\sum x = 1 + 6 + 9 + 3 + 4 = 23$$
$$\sum y = 7 + 8 + 2 + 5 + 10 = 32$$
$$\sum x \sum y = (23)(32) = \mathbf{736}$$

     There are three basic rules for algebraic manipulation of the $\sum$ notation.

Note c represents any constant value.

Rule 1

$$\sum_{i=1}^{n} c = nc$$

To prove this rule, we need only write down the meaning of $\sum_{i=1}^{n} c$:

$$\sum_{i=1}^{n} c = \underbrace{c + c + c + \cdots + c}_{n \text{ addends}}$$

Therefore,

$$\sum_{i=1}^{n} c = n \cdot c$$

Illustration A-5   Show that $\sum\limits_{i=1}^{5} 4 = (5)(4) = 20$.

Solution

$$\sum_{i=1}^{5} 4 = \underbrace{4_{(\text{when } i=1)} + 4_{(\text{when } i=2)} + 4_{(i=3)} + 4_{(i=4)} + 4_{(i=5)}}$$

five 4s added together

$$= (5)(4) = \mathbf{20}$$

∎

Rule 2

$$\sum_{i=1}^{n} cx_i = c \cdot \sum_{i=1}^{n} x_i$$

To demonstrate the truth of Rule 2, we will need to expand the term $\sum\limits_{i=1}^{n} cx_i$ and then factor out the common term c.

$$\sum_{i=1}^{n} cx_i = cx_1 + cx_2 + cx_3 + \cdots + cx_n$$

$$= c(x_1 + x_2 + x_3 + \cdots + x_n)$$

Therefore,

$$\sum_{i=1}^{n} cx_i = c \cdot \sum_{i=1}^{n} x_i$$

Rule 3

$$\sum_{i=1}^{n} (x_i + y_i) = \sum_{i=1}^{n} x_i + \sum_{i=1}^{n} y_i$$

The expansion and regrouping of $\sum\limits_{i=1}^{n} (x_i + y_i)$ is all that is needed to show this rule.

$$\sum_{i=1}^{n} (x_i + y_i) = (x_1 + y_1) + (x_2 + y_2) + \cdots + (x_n + y_n)$$

$$= (x_1 + x_2 + \cdots + x_n) + (y_1 + y_2 + \cdots + y_n)$$

Therefore,

$$\sum_{i=1}^{n} (x_i + y_i) = \sum_{i=1}^{n} x_i + \sum_{i=1}^{n} y_i$$

Illustration A-6   Show that $\sum\limits_{i=1}^{3} (2x_i + 6) = 2 \cdot \sum\limits_{i=1}^{3} x_i + 18$.

Solution

$$\sum_{i=1}^{3} (2x_i + 6) = (2x_1 + 6) + (2x_2 + 6) + (2x_3 + 6)$$

$$= (2x_1 + 2x_2 + 2x_3) + (6 + 6 + 6)$$
$$= (2)(x_1 + x_2 + x_3) + (3)(6)$$
$$= 2\sum_{i=1}^{3} x_i + 18 \quad\blacksquare$$

Illustration A-7   Let $x_1 = 2$, $x_2 = 4$, $x_3 = 6$, $f_1 = 3$, $f_2 = 4$, and $f_3 = 2$. Find $\sum_{i=1}^{3} x_i \cdot \sum_{i=1}^{3} f_i$.

Solution

$$\sum_{i=1}^{3} x_i \cdot \sum_{i=1}^{3} f_i = (x_1 + x_2 + x_3) \cdot (f_1 + f_2 + f_3)$$

$$= (2 + 4 + 6) \cdot (3 + 4 + 2)$$
$$= (12)(9) = \mathbf{108} \quad\blacksquare$$

Illustration A-8   Using the same values for the x's and f's as in Illustration A-7, find $\sum (xf)$.

Solution   Recall that the use of no index numbers means "use all data."

$$\sum (xf) = \sum_{i=1}^{3} (x_i f_i) = (x_1 f_1) + (x_2 f_2) + (x_3 f_3)$$

$$= (2 \cdot 3) + (4 \cdot 4) + (6 \cdot 2) = 6 + 16 + 12 = \mathbf{34} \quad\blacksquare$$

EXERCISES

A.1 Write each of the following in expanded form (without the summation sign):

a. $\displaystyle\sum_{i=1}^{4} x_i$ **b.** $\displaystyle\sum_{i=1}^{3} (x_i)^2$ **c.** $\displaystyle\sum_{i=1}^{5} (x_i + y_i)$

d. $\displaystyle\sum_{i=1}^{5} (x_i + 4)$ **e.** $\displaystyle\sum_{i=1}^{8} x_i y_i$ **f.** $\displaystyle\sum_{i=1}^{4} x_i^2 f_i$

A.2 Write each of the following expressions as summations, showing the subscripts and the limits of summation:

a. $x_1 + x_2 + x_3 + x_4 + x_5 + x_6$

b. $x_1 y_1 + x_2 y_2 + x_3 y_3 + \cdots + x_7 y_7$

c. $x_1^2 + x_2^2 + \cdots + x_9^2$

d. $(x_1 - 3) + (x_2 - 3) + \cdots + (x_n - 3)$

A.3 Show each of the following to be true:

a. $\sum_{i=1}^{4} (5x_i + 6) = 5 \cdot \sum_{i=1}^{4} x_i + 24$

b. $\sum_{i=1}^{n} (x_i - y_i) = \sum_{i=1}^{n} x_i - \sum_{i=1}^{n} y_i$

A.4 Given $x_1 = 2$, $x_2 = 7$, $x_3 = -3$, $x_4 = 2$, $x_5 = -1$, and $x_6 = 1$, find each of the following:

a. $\sum_{i=1}^{6} x_i$ b. $\sum_{i=1}^{6} x_i^2$ c. $\left(\sum_{i=1}^{6} x_i \right)^2$

A.5 Given $x_1 = 4$, $x_2 = -1$, $x_3 = 5$, $f_1 = 4$, $f_2 = 6$, $f_3 = 2$, $y_1 = -3$, $y_2 = 5$, and $y_3 = 2$, find each of the following:

a. $\sum x$ b. $\sum y$ c. $\sum f$ d. $\sum (x - y)$

e. $\sum x^2$ f. $(\sum x)^2$ g. $\sum xy$ h. $\sum x \cdot \sum y$

i. $\sum xf$ j. $\sum x^2 f$ k. $(\sum xf)^2$

A.6 Suppose that you take out a \$12,000 small-business loan. The terms of the loan are that each month for 10 years (120 months) you will pay back \$100 plus accrued interest. The accrued interest is calculated by multiplying 0.005 (6 percent/12) times the amount of the loan still outstanding. That is, the first month you pay \$12,000 × 0.005 in accrued interest, the second month (\$12,000 − 100) × 0.005 in interest, the third month [\$12,000 − (2)(100)] × 0.005, and so forth. Express the total amount of interest paid over the life of the loan by using summation notation.

The answers to these exercises can be found in the answer section.

Appendix B

Using the Random Number Table

The random number table is a collection of "random" digits. The term *random* means that each of the ten digits $(0, 1, 2, 3, \ldots, 9)$ has an equal chance of occurrence. The digits in Table 1 of Appendix E can be thought of as single-digit numbers $(0-9)$, as two-digit numbers $(00-99)$, as three-digit numbers $(000-999)$, or as numbers of any desired size. The digits presented in Table 1 are arranged in pairs and grouped into blocks of five rows and five columns. This format is used for convenience. Tables in other books may be arranged differently.

Random numbers are used primarily for one of two reasons: (1) to identify the source element of a population (the source of data) or (2) to simulate an experiment.

Illustration B-1   A simple random sample of 10 people is to be drawn from a population of 7564 people. Each person will be assigned a number, using the numbers from 0001 to 7564. We will view Table 1 as a collection of four-digit numbers (two columns used together), where the numbers $0001, 0002, 0003, \ldots, 7564$ identify the 7564 people. The numbers $0000, 7565, 7566, \ldots, 9999$ represent no one in our population; that is, they will be discarded if selected.

Now we are ready to select our 10 people. Turn to Table 1 (page 567). We need to select a starting point and a "path" to be followed. Perhaps the most common way to locate a starting point is to look away and arbitrarily point to a starting point. The number we located this way was 3909. (It is located in the upper left corner of the block that is in the fourth large block from the left and the second large block down.) From here we will proceed down the column, then go to the top of the next set of columns, if necessary. The person identified by number 3909 is the first source of data selected. Proceeding down the column, we find 8869 next. This number is discarded. The number 2501 is next. Therefore, the person identified by 2501 is the second source of data to be selected. Continuing down this column, our sample will be obtained from those people identified by the numbers 3909, 2501, 7485, 0545, 5252, 5612, 0997, 3230, 1051, 2712. (The numbers 8869, 8338, and 9187 were discarded.) ◼

Illustration B-2   Let's use the random number table and simulate 100 tosses of a coin. The simulation is accomplished by assigning numbers to each of the possible outcomes of a particular experiment. The assignment must be done in such a way as to preserve the probabilities. Perhaps the simplest way to make the assignment for the coin toss is to let the even digits (0, 2, 4, 6, 8) represent heads and the odd digits (1, 3, 5, 7, 9) represent tails. The correct probabilities are maintained: $P(H) = P(0, 2, 4, 6, 8) = \frac{5}{10} = 0.5$ and $P(T) = P(1, 3, 5, 7, 9) = \frac{5}{10} = 0.5$. Once this assignment is complete, we are ready to obtain our sample.

Since the question asked for 100 tosses and there are 50 digits to a "block" in Table 1, let's select two blocks as our sample of random one-digit numbers (instead of a column 100 lines long). Let's look away and point to one block on page 566 and then do the same to select one block from page 567. We picked the sixth block down in the first column of blocks on page 566 (24 even and 26 odd numbers) and the sixth block down in the third column of blocks on page 567 (23 even and 27 odd numbers). Thus we obtain a sample of 47 heads and 53 tails for our 100 simulated tosses.

◼

There are, of course, many ways to use the random number table. You must use your good sense in assigning the numbers to be used and in choosing the "path" to be followed through the table. One bit of advice is to make the assignments in as simple and easy a method as possible to avoid errors.

EXERCISES

B.1 A random sample of size 8 is to be selected from a population that contains 75 elements. Describe how the random sample of the 8 objects could be made with the aid of the random number table.

B.2 A coin-tossing experiment is to be simulated. Two coins are to be tossed simultaneously and the number of heads appearing is to be recorded for each toss. Ten such tosses are to be observed. Describe two ways to use the random number table to simulate this experiment.

B.3 Simulate five rolls of three dice by using the random number table.

The answers to these exercises can be found in the answer section.

Appendix C

Round-off Procedure

When rounding off a number, we use the following procedure.

Step 1   Identify the position where the round-off is to occur. This is shown by using a vertical line that separates the part of the number to be kept from the part to be discarded. For example,

125.267	to the nearest tenth is written as	125.2|67
7.8890	to the nearest hundredth is written as	7.88|90

Step 2   Step 1 has separated all numbers into one of four cases. (*X*'s will be used as placeholders for number values in front of the vertical line. These *X*'s can represent any number value.)

Case I: $X\ X\ X\ X|000\ldots$

Case II: $X\ X\ X\ X|----$ (any value from $000\ldots1$ to $499\ldots9$)

Case III: $X\ X\ X\ X|5000\ldots0$

Case IV: $X\ X\ X\ X|----$ (any value from $5000\ldots1$ to $999\ldots9$)

Step 3   Perform the rounding off.

Case I requires no round-off. It is exactly $X\ X\ X\ X$.

Illustration C-1   Round 3.5000 to the nearest tenth.

$$3.5|000 \quad \text{becomes} \quad \textbf{3.5} \quad \blacksquare$$

Case II requires rounding. We will round down for this case. That is, just drop the part of the number that is behind the vertical line.

Illustration C-2   Round 37.6124 to the nearest hundredth.

37.61|24   becomes   **37.61**   ■

Case III requires rounding. This is the case that requires special attention. **When a 5 (exactly a 5) is to be rounded off, round to the even digit.** In the long run, half of the time the 5 will be preceded by an even digit (0, 2, 4, 6, 8) and you will round down, while the other half of the time the 5 will be preceded by an odd digit (1, 3, 5, 7, 9) and you will round up.

Illustration C-3   Round 87.35 to the nearest tenth.

87.3|5   becomes   **87.4**

Round 93.445 to the nearest hundredth.

93.44|5   becomes   **93.44**

(Note: 87.35 is 87.35000... and 93.445 is 93.445000....)   ■

Case IV requires rounding. We will round up for this case. That is, we will drop the part of the number that is behind the vertical line and we will increase the last digit in front of the vertical line by one.

Illustration C-4   Round 7.889 to the nearest tenth.

7.8|89   becomes   **7.9**   ■

Note   **Case I, II, and IV describe what is commonly done. Our guidelines for Case III are the only ones that are different typical procedure.**

If the typical round-off rule (0, 1, 2, 3, 4 are dropped; 5, 6, 7, 8, 9 are rounded up) is followed, then $(n + 1)/(2n + 1)$ of the situations are rounded up. (n is the number of different sequences of digits that fall into each of Case II and Case IV.) That is more than half. You (as many others have) may say, "So what?" In today's world that tiny, seemingly insignificant amount becomes very significant when applied repeatedly to large numbers.

EXERCISES

C.1 Round each of the following to the nearest integer:
   a. 12.94   **b.** 8.762   **c.** 9.05   **d.** 156.49
   e. 45.5   **f.** 42.5   **g.** 102.51   **h.** 16.5001

C.2 Round each of the following to the nearest tenth:
   a. 8.67   **b.** 42.333   **c.** 49.666   **d.** 10.25   **e.** 10.35
   f. 8.4501   **g.** 27.35001   **h.** 5.65   **i.** 3.05   **j.** $\frac{1}{4}$

C.3 Round each of the following to the nearest hundredth:

    a. 17.6666   **b.** 4.444   **c.** 54.5454   **d.** 102.055   **e.** 93.225

    f. 18.005   **g.** 18.015   **h.** 5.555   **i.** 44.7450   **j.** $\frac{2}{3}$

The answers to these exercises can be found in the answer section.

Appendix D

Interpolation Procedure for F Distribution

ONE df VALUE IN TABLE

Illustration D-1   Find the critical value $F(22, 11, 0.05)$.

Solution   Critical values of F are found in Table 8a of Appendix E. Turn to Table 8a and you will see that there is no column for $df_n = 22$. There is a row for $df_d = 11$. Therefore the values closest to $F(22, 11, 0.05)$ are $F(20, 11, 0.05) = 2.65$ and $F(24, 11, 0.05) = 2.61$. Since $df = 22$ is halfway between $df = 20$ and $df = 24$, it seems logical to assume that a good estimate for $F(22, 11, 0.05)$ is 2.63, the value halfway between 2.65 and 2.61.   ■

Illustration D-1 is a simple illustration of linear interpolation. The method of **linear interpolation** assumes that if a number of degrees of freedom lies between two of those listed in the table of critical values, then the F-value for that number of degrees of freedom is a value with the same proportional difference between the F-values listed, and vice versa. In Illustration D-1 the proportion was one-half. The following illustration shows the general procedure used to interpolate when one of the two degrees of freedom is in the table.

Illustration D-2   Find the value of $F(75, 15, 0.05)$.

Solution   In Table 8a there is no column for $df_n = 75$, but there is a row for $df_d = 15$. Therefore we have two nearby values: $F(60, 15, 0.05) = 2.16$ and

563

$F(120, 15, 0.05) = 2.11$. Since degrees of freedom for the denominator is 15 for both, we need to interpolate between $df_n = 60$ and $df_n = 120$ in order to find $df_n = 75$. The following schematic arrangement shows the information we have and the proportional differences that form the proportion that we must solve to find d, the proportional difference.

$$
\begin{array}{c}
\boxed{\;60\;\boxed{15}\;
\begin{array}{l}
F(60, 15, 0.05) = 2.16 \\
F(75, 15, 0.05) = \;? \qquad \boxed{d}\,0.05 \\
F(120, 15, 0.05) = 2.11
\end{array}\;}
\end{array}
$$

The corresponding differences are in proportion, therefore:

$$\frac{15}{60} = \frac{d}{0.05}$$

$$d = \frac{(15)(0.05)}{60}$$

$$d = 0.0125$$

Notice that the value of F decreased as the degrees of freedom increased, therefore the value of d is subtracted from 2.16:

$$F(75, 15, 0.05) = 2.16 - 0.0125 = 2.1475 \text{ or } \mathbf{2.15}$$

to the nearest hundredth. ■

NEITHER df VALUE IN TABLE

When neither of the two degrees of freedom are in the table we will need to interpolate **three** times in order to find the desired value. This interpolation process will follow an "H" pattern as demonstrated by the following illustration.

Illustration D-3   Find the value of $F(28, 72, 0.05)$.

Solution   Table 8a has no column for $df_n = 28$ and no row for $df_d = 72$. Therefore we will use the two columns and the two rows that have df values immediately smaller and immediately larger than 28 and 72, respectively. We find the following four F-values:

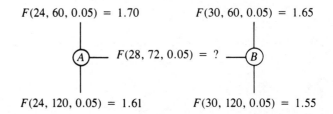

The first two interpolations will be to find F-values in positions A $[F(24, 72, 0.05)]$ and B $[F(30, 72, 0.05)]$.

For A: For B:

$$\frac{12}{60} = \frac{d}{0.09}$$ $$\frac{12}{60} = \frac{d}{0.10}$$

$$d = \frac{(12)(0.09)}{60}$$ $$d = \frac{(12)(0.10)}{60}$$

$$d = 0.018$$ $$d = 0.020$$

Therefore,

$$A = F(24, 72, 0.05)$$ $$B = F(30, 72, 0.05)$$

$$= 1.70 - 0.018$$ $$= 1.65 - 0.020$$

$$= \mathbf{1.682}$$ $$= \mathbf{1.630}$$

The final step is to interpolate between A and B to find our answer.

$$
6 \left\lfloor \boxed{4} \begin{array}{l} F(24, 72, 0.05) = 1.682 \\ F(28, 72, 0.05) = \ ? \\ F(30, 72, 0.05) = 1.630 \end{array} \ \right\rceil \overset{d}{\underset{0.052}{}}
$$

The corresponding differences are in proportion, therefore,

$$\frac{4}{6} = \frac{d}{0.052}$$

$$d = \frac{(4)(0.052)}{6}$$

$$d = 0.03467 = \mathbf{0.035}$$

$F(28, 72, 0.05) = 1.682 - 0.035 = 1.647$ or **1.65** to the nearest hundredth.

Appendix E

Tables

10	09	73	25	33	76	52	01	35	86	34	67	35	48	76	80	95	90	91	17	39	29	27	49	45
37	54	20	48	05	64	89	47	42	96	24	80	52	40	37	20	63	61	04	02	00	82	29	16	65
08	42	26	89	53	19	64	50	93	03	23	20	90	25	60	15	95	33	47	64	35	08	03	36	06
99	01	90	25	29	09	37	67	07	15	38	31	13	11	65	88	67	67	43	97	04	43	62	76	59
12	80	79	99	70	80	15	73	61	47	64	03	23	66	53	98	95	11	68	77	12	17	17	68	33
66	06	57	47	17	34	07	27	68	50	36	69	73	61	70	65	81	33	98	85	11	19	92	91	70
31	06	01	08	05	45	57	18	24	06	35	30	34	26	14	86	79	90	74	39	23	40	30	97	32
85	26	97	76	02	02	05	16	56	92	68	66	57	48	18	73	05	38	52	47	18	62	38	85	79
63	57	33	21	35	05	32	54	70	48	90	55	35	75	48	28	46	82	87	09	83	49	12	56	24
73	79	64	57	53	03	52	96	47	78	35	80	83	42	82	60	93	52	03	44	35	27	38	84	35
98	52	01	77	67	14	90	56	86	07	22	10	94	05	58	60	97	09	34	33	50	50	07	39	98
11	80	50	54	31	39	80	82	77	32	50	72	56	82	48	29	40	52	42	01	52	77	56	78	51
83	45	29	96	34	06	28	89	80	83	13	74	67	00	78	18	47	54	06	10	68	71	17	78	17
88	68	54	02	00	86	50	75	84	01	36	76	66	79	51	90	36	47	64	93	29	60	91	10	62
99	59	46	73	48	87	51	76	49	69	91	82	60	89	28	93	78	56	13	68	23	47	83	41	13
65	48	11	76	74	17	46	85	09	50	58	04	77	69	74	73	03	95	71	86	40	21	81	65	44
80	12	43	56	35	17	72	70	80	15	45	31	82	23	74	21	11	57	82	53	14	38	55	37	63
74	35	09	98	17	77	40	27	72	14	43	23	60	02	10	45	52	16	42	37	96	28	60	26	55
69	91	62	68	03	66	25	22	91	48	36	93	68	72	03	76	62	11	39	90	94	40	05	64	18
09	89	32	05	05	14	22	56	85	14	46	42	75	67	88	96	29	77	88	22	54	38	21	45	98
91	49	91	45	23	68	47	92	76	86	46	16	28	35	54	94	75	08	99	23	37	08	92	00	48
80	33	69	45	98	26	94	03	68	58	70	29	73	41	35	54	14	03	33	40	42	05	08	23	41
44	10	48	19	49	85	15	74	79	54	32	97	92	65	75	57	60	04	08	81	22	22	20	64	13
12	55	07	37	42	11	10	00	20	40	12	86	07	46	97	96	64	48	94	39	28	70	72	58	15
63	60	64	93	29	16	50	53	44	84	40	21	95	25	63	43	65	17	70	82	07	20	73	17	90
61	19	69	04	46	26	45	74	77	74	51	92	43	37	29	65	39	45	95	93	42	58	26	05	27
15	47	44	52	66	95	27	07	99	53	59	36	78	38	48	82	39	61	01	18	33	21	15	94	66
94	55	72	85	73	67	89	75	43	87	54	62	24	44	31	91	19	04	25	92	92	92	74	59	73
42	48	11	62	13	97	34	40	87	21	16	86	84	87	67	03	07	11	20	59	25	70	14	66	70
23	52	37	83	17	73	20	88	98	37	68	93	59	14	16	26	25	22	96	63	05	52	28	25	62
04	49	35	24	94	75	24	63	38	24	45	86	25	10	25	61	96	27	93	35	65	33	71	24	72
00	54	99	76	54	64	05	18	81	59	96	11	96	38	96	54	69	28	23	91	23	28	72	95	29
35	96	31	53	07	26	89	80	93	54	33	35	13	54	62	77	97	45	00	24	90	10	33	93	33
59	80	80	83	91	45	42	72	68	42	83	60	94	97	00	13	02	12	48	92	78	56	52	01	06
46	05	88	52	36	01	39	09	22	86	77	28	14	40	77	93	91	08	36	47	70	61	74	29	41

* For specific details on the use of this table, see page 558.

TABLE 1   Random Numbers   567

Table 1
(Continued)

32 17 90 05 97	87 37 92 52 41	05 56 70 70 07	86 74 31 71 57	85 39 41 18 38	
69 23 46 14 06	20 11 74 52 04	15 95 66 00 00	18 74 39 24 23	97 11 89 63 38	
19 56 54 14 30	01 75 87 53 79	40 41 92 15 85	66 67 43 68 06	84 96 28 52 07	
45 15 51 49 38	19 47 60 72 46	43 66 79 45 43	59 04 79 00 33	20 82 66 95 41	
94 86 43 19 94	36 16 81 08 51	34 88 88 15 53	01 54 03 54 56	05 01 45 11 76	
98 08 62 48 26	45 24 02 84 04	44 99 90 88 96	39 09 47 34 07	35 44 13 18 80	
33 18 51 62 32	41 94 15 09 49	89 43 54 85 81	88 69 54 19 94	37 54 87 30 43	
80 95 10 04 06	96 38 27 07 74	20 15 12 33 87	25 01 62 52 98	94 62 46 11 71	
79 75 24 91 40	71 96 12 82 96	69 86 10 25 91	74 85 22 05 39	00 38 75 95 79	
18 63 33 25 37	98 14 50 65 71	31 01 02 46 74	05 45 56 14 27	77 93 89 19 36	
74 02 94 39 02	77 55 73 22 70	97 79 01 71 19	52 52 75 80 21	80 81 45 17 48	
54 17 84 56 11	80 99 33 71 43	05 33 51 29 69	56 12 71 92 55	36 04 09 03 24	
11 66 44 98 83	52 07 98 48 27	59 38 17 15 39	09 97 33 34 40	88 46 12 33 56	
48 32 47 79 28	31 24 96 47 10	02 29 53 68 70	32 30 75 75 46	15 02 00 99 94	
69 07 49 41 38	87 63 79 19 76	35 58 40 44 01	10 51 82 16 15	01 84 87 69 38	
09 18 82 00 97	32 82 53 95 27	04 22 08 63 04	83 38 98 73 74	64 27 85 80 44	
90 04 58 54 97	51 98 15 06 54	94 93 88 19 97	91 87 07 61 50	68 47 66 46 59	
73 18 95 02 07	47 67 72 62 69	62 29 06 44 64	27 12 46 70 18	41 36 18 27 60	
75 76 87 64 90	20 97 18 17 49	90 42 91 22 72	95 37 50 58 71	93 82 34 31 78	
54 01 64 40 56	66 28 13 10 03	00 68 22 73 98	20 71 45 32 95	07 70 61 78 13	
08 35 86 99 10	78 54 24 27 85	13 66 15 88 73	04 61 89 75 53	31 22 30 84 20	
28 30 60 32 64	81 33 31 05 91	40 51 00 78 93	32 60 46 04 75	94 11 90 18 40	
53 84 08 62 33	81 59 41 36 28	51 21 59 02 90	28 46 66 87 95	77 76 22 07 91	
91 75 75 37 41	61 61 36 22 69	50 26 39 02 12	55 78 17 65 14	83 48 34 70 55	
89 41 59 26 94	00 39 75 83 91	12 60 71 76 46	48 94 97 23 06	94 54 13 74 08	
77 51 30 38 20	86 83 42 99 01	68 41 48 27 74	51 90 81 39 80	72 89 35 55 07	
19 50 23 71 74	69 97 92 02 88	55 21 02 97 73	74 28 77 52 51	65 34 46 74 15	
21 81 85 93 13	93 27 88 17 57	05 68 67 31 56	07 08 28 50 46	31 85 33 84 52	
51 47 46 64 99	68 10 72 36 21	94 04 99 13 45	42 83 60 91 91	08 00 74 54 49	
99 55 96 83 31	62 53 52 41 70	69 77 71 28 30	74 81 97 81 42	43 86 07 28 34	
33 71 34 80 07	93 58 47 28 69	51 92 66 47 21	58 30 32 98 22	93 17 49 39 72	
85 27 48 68 93	11 30 32 92 70	28 83 43 41 37	73 51 59 04 00	71 14 84 36 43	
84 13 38 96 40	44 03 55 21 66	73 85 27 00 91	61 22 26 05 61	62 32 71 84 23	
56 73 21 62 34	17 39 59 61 31	10 12 39 16 22	85 49 65 75 60	81 60 41 88 80	
65 13 85 68 06	87 60 88 52 61	34 31 36 58 61	45 87 52 10 69	85 64 44 72 77	
38 00 10 21 76	81 71 91 17 11	71 60 29 29 37	74 21 96 40 49	65 58 44 96 98	
37 40 29 63 97	01 30 47 75 86	56 27 11 00 86	47 32 46 26 05	40 03 03 74 38	
97 12 54 03 48	87 08 33 14 17	21 81 53 92 50	75 23 76 20 47	15 50 12 95 78	
21 82 64 11 34	47 14 33 40 72	64 63 88 59 02	49 13 90 64 41	03 85 65 45 52	
73 13 54 27 42	95 71 90 90 35	85 79 47 42 96	08 78 98 81 56	64 69 11 92 02	
07 63 87 79 29	03 06 11 80 72	96 20 74 41 56	23 82 19 95 38	04 71 36 69 94	
60 52 88 34 41	07 95 41 98 14	59 17 52 06 95	05 53 35 21 39	61 21 20 64 55	
83 59 63 56 55	06 95 89 29 83	05 12 80 97 19	77 43 35 37 83	92 30 15 04 98	
10 85 06 27 46	99 59 91 05 07	13 49 90 63 19	53 07 57 18 39	06 41 01 93 62	
39 82 09 89 52	43 62 26 31 47	64 42 18 08 14	43 80 00 93 51	31 02 47 31 67	
59 58 00 64 78	75 56 97 88 00	88 83 55 44 86	23 76 80 61 56	04 11 10 84 08	
38 50 80 73 41	23 79 34 87 63	90 82 29 70 22	17 71 90 42 07	95 95 44 99 53	
30 69 27 06 68	94 68 81 61 27	56 19 68 00 91	82 06 76 34 00	05 46 26 92 00	
65 44 39 56 59	18 28 82 74 37	49 63 22 40 41	08 33 76 56 76	96 29 99 08 36	
27 26 75 02 64	13 19 27 22 94	07 47 74 46 06	17 98 54 89 11	97 34 13 03 58	
91 30 70 69 91	19 07 22 42 10	36 69 95 37 28	28 82 53 57 93	28 97 66 62 52	
68 43 49 46 88	84 47 31 36 22	62 12 69 84 08	12 84 38 25 90	09 81 59 31 46	
48 90 81 58 77	54 74 52 45 91	35 70 00 47 54	83 82 45 26 92	54 13 05 51 60	
06 91 34 51 97	42 67 27 86 01	11 88 30 95 28	63 01 19 89 01	14 97 44 03 44	
10 45 51 60 19	14 21 03 37 12	91 34 23 78 21	88 32 58 08 51	43 66 77 08 83	
12 88 39 73 43	65 02 76 11 84	04 28 50 13 92	17 97 41 50 77	90 71 22 67 69	
21 77 83 09 76	38 80 73 69 61	31 64 94 20 96	63 28 10 20 23	08 81 64 74 49	
19 52 35 95 15	65 12 25 96 59	86 28 36 82 58	69 57 21 37 98	16 43 59 15 29	
67 24 55 26 70	35 58 31 65 63	79 24 68 66 86	76 46 33 42 22	26 65 59 08 02	
60 58 44 73 77	07 50 03 79 92	45 13 42 65 29	26 76 08 36 37	41 32 64 43 44	
53 85 34 13 77	36 06 69 48 50	58 83 87 38 59	49 36 47 33 31	96 24 04 36 42	
24 63 73 97 36	74 38 48 93 42	52 62 30 79 92	12 36 91 86 01	03 74 28 38 73	
83 08 01 24 51	38 99 22 28 15	07 75 95 17 77	97 37 72 75 85	51 97 23 78 67	
16 44 42 43 34	36 15 19 90 73	27 49 37 09 39	85 13 03 25 52	54 84 65 47 59	
60 79 01 81 57	57 17 86 57 62	11 16 17 85 76	45 81 95 29 79	65 13 00 48 60	

From tables of the RAND Corporation. Reprinted from Wilfred J. Dixon and Frank J. Massey, Jr., *Introduction to Statistical Analysis*. 3rd ed. (New York: McGraw-Hill, 1969), pp. 446–447. Reprinted by permission of the RAND Corporation.

Table 2

Factorials

n	$n!$
0	1
1	1
2	2
3	6
4	24
5	120
6	720
7	5,040
8	40,320
9	362,880
10	3,628,800
11	39,916,800
12	479,001,600
13	6,227,020,800
14	87,178,291,200
15	1,307,674,368,000
16	20,922,789,888,000
17	355,687,428,096,000
18	6,402,373,705,728,000
19	121,645,100,408,832,000
20	2,432,902,008,176,640,000

* For specific details on the use of this table, see page 205.

TABLE 3   Binomial Coefficients   569

Table 3
Binomial coefficients

n	$\binom{n}{0}$	$\binom{n}{1}$	$\binom{n}{2}$	$\binom{n}{3}$	$\binom{n}{4}$	$\binom{n}{5}$	$\binom{n}{6}$	$\binom{n}{7}$	$\binom{n}{8}$	$\binom{n}{9}$	$\binom{n}{10}$
0	1										
1	1	1									
2	1	2	1								
3	1	3	3	1							
4	1	4	6	4	1						
5	1	5	10	10	5	1					
6	1	6	15	20	15	6	1				
7	1	7	21	35	35	21	7	1			
8	1	8	28	56	70	56	28	8	1		
9	1	9	36	84	126	126	84	36	9	1	
10	1	10	45	120	210	252	210	120	45	10	1
11	1	11	55	165	330	462	462	330	165	55	11
12	1	12	66	220	495	792	924	792	495	220	66
13	1	13	78	286	715	1287	1716	1716	1287	715	286
14	1	14	91	364	1001	2002	3003	3432	3003	2002	1001
15	1	15	105	455	1365	3003	5005	6435	6435	5005	3003
16	1	16	120	560	1820	4368	8008	11440	12870	11440	8008
17	1	17	136	680	2380	6188	12376	19448	24310	24310	19448
18	1	18	153	816	3060	8568	18564	31824	43758	48620	43758
19	1	19	171	969	3876	11628	27132	50388	75582	92378	92378
20	1	20	190	1140	4845	15504	38760	77520	125970	167960	184756

If necessary, use the identity $\binom{n}{k} = \binom{n}{n-k}$

From John E. Freund, *Statistics, A First Course*, Prentice-Hall, Inc., Englewood Cliffs, N. J., 1970, p. 313. Reprinted by permission.

* For specific details on the use of this table, see page 205.

Table 4

Binomial probabilities

$$\left[\binom{n}{x} \cdot p^x q^{n-x} \right]$$

n	x	0.01	0.05	0.10	0.20	0.30	0.40	p 0.50	0.60	0.70	0.80	0.90	0.95	0.99	x
2	0	980	902	810	640	490	360	250	160	090	040	010	002	0+	0
	1	020	095	180	320	420	480	500	480	420	320	180	095	020	1
	2	0+	002	010	040	090	160	250	360	490	640	810	902	980	2
3	0	970	857	729	512	343	216	125	064	027	008	001	0+	0+	0
	1	029	135	243	384	441	432	375	288	189	096	027	007	0+	1
	2	0+	007	027	096	189	288	375	432	441	384	243	135	029	2
	3	0+	0+	001	008	027	064	125	216	343	512	729	857	970	3
4	0	961	815	656	410	240	130	062	026	008	002	0+	0+	0+	0
	1	039	171	292	410	412	346	250	154	076	026	004	0+	0+	1
	2	001	014	049	154	265	346	375	346	265	154	049	014	001	2
	3	0+	0+	004	026	076	154	250	346	412	410	292	171	039	3
	4	0+	0+	0+	002	008	026	062	130	240	410	656	815	961	4
5	0	951	774	590	328	168	078	031	010	002	0+	0+	0+	0+	0
	1	048	204	328	410	360	259	156	077	028	006	0+	0+	0+	1
	2	001	021	073	205	309	346	312	230	132	051	008	001	0+	2
	3	0+	001	008	051	132	230	312	346	309	205	073	021	001	3
	4	0+	0+	0+	006	028	077	156	259	360	410	328	204	048	4
	5	0+	0+	0+	0+	002	010	031	078	168	328	590	774	951	5
6	0	941	735	531	262	118	047	016	004	001	0+	0+	0+	0+	0
	1	057	232	354	393	303	187	094	037	010	002	0+	0+	0+	1
	2	001	031	098	246	324	311	234	138	060	015	001	0+	0+	2
	3	0+	002	015	082	185	276	312	276	185	082	015	002	0+	3
	4	0+	0+	001	015	060	138	234	311	324	246	098	031	001	4
	5	0+	0+	0+	002	010	037	094	187	303	393	354	232	057	5
	6	0+	0+	0+	0+	001	004	016	047	118	262	531	735	941	6
7	0	932	698	478	210	082	028	008	002	0+	0+	0+	0+	0+	0
	1	066	257	372	367	247	131	055	017	004	0+	0+	0+	0+	1
	2	002	041	124	275	318	261	164	077	025	004	0+	0+	0+	2
	3	0+	004	023	115	227	290	273	194	097	029	003	0+	0+	3
	4	0+	0+	003	029	097	194	273	290	227	115	023	004	0+	4
	5	0+	0+	0+	004	025	077	164	261	318	275	124	041	002	5
	6	0+	0+	0+	0+	004	017	055	131	247	367	372	257	066	6
	7	0+	0+	0+	0+	0+	002	008	028	082	210	478	698	932	7
8	0	923	663	430	168	058	017	004	001	0+	0+	0+	0+	0+	0
	1	075	279	383	336	198	090	031	008	001	0+	0+	0+	0+	1
	2	003	051	149	294	296	209	109	041	010	001	0+	0+	0+	2
	3	0+	005	033	147	254	279	219	124	047	009	0+	0+	0+	3
	4	0+	0+	005	046	136	232	273	232	136	046	005	0+	0+	4
	5	0+	0+	0+	009	047	124	219	279	254	147	033	005	0+	5
	6	0+	0+	0+	001	010	041	109	209	296	294	149	051	003	6
	7	0+	0+	0+	0+	001	008	031	090	198	336	383	279	075	7
	8	0+	0+	0+	0+	0+	001	004	017	058	168	430	663	923	8

* For specific details on the use of this table, see page 208.

TABLE 4   Binomial Probabilities   571

Table 4
(Continued)

n	x	0.01	0.05	0.10	0.20	0.30	0.40	0.50	0.60	0.70	0.80	0.90	0.95	0.99	x
9	0	914	630	387	134	040	010	002	0+	0+	0+	0+	0+	0+	0
	1	083	299	387	302	156	060	018	004	0+	0+	0+	0+	0+	1
	2	003	063	172	302	267	161	070	021	004	0+	0+	0+	0+	2
	3	0+	008	045	176	267	251	164	074	021	003	0+	0+	0+	3
	4	0+	001	007	066	172	251	246	167	074	017	001	0+	0+	4
9	5	0+	0+	001	017	074	167	246	251	172	066	007	001	0+	5
	6	0+	0+	0+	003	021	074	164	251	267	176	045	008	0+	6
	7	0+	0+	0+	0+	004	021	070	161	267	302	172	063	003	7
	8	0+	0+	0+	0+	0+	004	018	060	156	302	387	299	083	8
	9	0+	0+	0+	0+	0+	0+	002	010	040	134	387	630	914	9
10	0	904	599	349	107	028	006	001	0+	0+	0+	0+	0+	0+	0
	1	091	315	387	268	121	040	010	002	0+	0+	0+	0+	0+	1
	2	004	075	194	302	233	121	044	011	001	0+	0+	0+	0+	2
	3	0+	010	057	201	267	215	117	042	009	001	0+	0+	0+	3
	4	0+	001	011	088	200	251	205	111	037	006	0+	0+	0+	4
	5	0+	0+	001	026	103	201	246	201	103	026	001	0+	0+	5
	6	0+	0+	0+	006	037	111	205	251	200	088	011	001	0+	6
	7	0+	0+	0+	001	009	042	117	215	267	201	057	010	0+	7
	8	0+	0+	0+	0+	001	011	044	121	233	302	194	075	004	8
	9	0+	0+	0+	0+	0+	002	010	040	121	268	387	315	091	9
	10	0+	0+	0+	0+	0+	0+	001	006	028	107	349	599	904	10
11	0	895	569	314	086	020	004	0+	0+	0+	0+	0+	0+	0+	0
	1	099	329	384	236	093	027	005	001	0+	0+	0+	0+	0+	1
	2	005	087	213	295	200	089	027	005	001	0+	0+	0+	0+	2
	3	0+	014	071	221	257	177	081	023	004	0+	0+	0+	0+	3
	4	0+	001	016	111	220	236	161	070	017	002	0+	0+	0+	4
	5	0+	0+	002	039	132	221	226	147	057	010	0+	0+	0+	5
	6	0+	0+	0+	010	057	147	226	221	132	039	002	0+	0+	6
	7	0+	0+	0+	002	017	070	161	236	220	111	016	001	0+	7
	8	0+	0+	0+	0+	004	023	081	177	257	221	071	014	0+	8
	9	0+	0+	0+	0+	001	005	027	089	200	295	213	087	005	9
	10	0+	0+	0+	0+	0+	001	005	027	093	236	384	329	099	10
	11	0+	0+	0+	0+	0+	0+	0+	004	020	086	314	569	895	11
12	0	886	540	282	069	014	002	0+	0+	0+	0+	0+	0+	0+	0
	1	107	341	377	206	071	017	003	0+	0+	0+	0+	0+	0+	1
	2	006	099	230	283	168	064	016	002	0+	0+	0+	0+	0+	2
	3	0+	017	085	236	240	142	054	012	001	0+	0+	0+	0+	3
	4	0+	002	021	133	231	213	121	042	008	001	0+	0+	0+	4
	5	0+	0+	004	053	158	227	193	101	029	003	0+	0+	0+	5
	6	0+	0+	0+	016	079	177	226	177	079	016	0+	0+	0+	6
	7	0+	0+	0+	003	029	101	193	227	158	053	004	0+	0+	7
	8	0+	0+	0+	001	008	042	121	213	231	133	021	002	0+	8
	9	0+	0+	0+	0+	001	012	054	142	240	236	085	017	0+	9
	10	0+	0+	0+	0+	0+	002	016	064	168	283	230	099	006	10
	11	0+	0+	0+	0+	0+	0+	003	017	071	206	377	341	107	11
	12	0+	0+	0+	0+	0+	0+	0+	002	014	069	282	540	886	12

Table 4

(Continued)

n	x	0.01	0.05	0.10	0.20	0.30	0.40	p 0.50	0.60	0.70	0.80	0.90	0.95	0.99	x
13	0	878	513	254	055	010	001	0+	0+	0+	0+	0+	0+	0+	0
	1	115	351	367	179	054	011	002	0+	0+	0+	0+	0+	0+	1
	2	007	111	245	268	139	045	010	001	0+	0+	0+	0+	0+	2
	3	0+	021	100	246	218	111	035	006	001	0+	0+	0+	0+	3
	4	0+	003	028	154	234	184	087	024	003	0+	0+	0+	0+	4
	5	0+	0+	006	069	180	221	157	066	014	001	0+	0+	0+	5
	6	0+	0+	001	023	103	197	209	131	044	006	0+	0+	0+	6
	7	0+	0+	0+	006	044	131	209	197	103	023	001	0+	0+	7
	8	0+	0+	0+	001	014	066	157	221	180	069	006	0+	0+	8
	9	0+	0+	0+	0+	003	024	087	184	234	154	028	003	0+	9
	10	0+	0+	0+	0+	001	006	035	111	218	246	100	021	0+	10
	11	0+	0+	0+	0+	0+	001	010	045	139	268	245	111	007	11
	12	0+	0+	0+	0+	0+	0+	002	011	054	179	367	351	115	12
	13	0+	0+	0+	0+	0+	0+	0+	001	010	055	254	513	878	13
14	0	869	488	229	044	007	001	0+	0+	0+	0+	0+	0+	0+	0
	1	123	359	356	154	041	007	001	0+	0+	0+	0+	0+	0+	1
	2	008	123	257	250	113	032	006	001	0+	0+	0+	0+	0+	2
	3	0+	026	114	250	194	085	022	003	0+	0+	0+	0+	0+	3
	4	0+	004	035	172	229	155	061	014	001	0+	0+	0+	0+	4
	5	0+	0+	008	086	196	207	122	041	007	0+	0+	0+	0+	5
	6	0+	0+	001	032	126	207	183	092	023	002	0+	0+	0+	6
	7	0+	0+	0+	009	062	157	209	157	062	009	0+	0+	0+	7
	8	0+	0+	0+	002	023	092	183	207	126	032	001	0+	0+	8
	9	0+	0+	0+	0+	007	041	122	207	196	086	008	0+	0+	9
	10	0+	0+	0+	0+	001	014	061	155	229	172	035	004	0+	10
	11	0+	0+	0+	0+	0+	003	022	085	194	250	114	026	0+	11
	12	0+	0+	0+	0+	0+	001	006	032	113	250	257	123	008	12
	13	0+	0+	0+	0+	0+	0+	001	007	041	154	356	359	123	13
	14	0+	0+	0+	0+	0+	0+	0+	001	007	044	229	488	869	14
15	0	860	463	206	035	005	0+	0+	0+	0+	0+	0+	0+	0+	0
	1	130	366	343	132	031	005	0+	0+	0+	0+	0+	0+	0+	1
	2	009	135	267	231	092	022	003	0+	0+	0+	0+	0+	0+	2
	3	0+	031	129	250	170	063	014	002	0+	0+	0+	0+	0+	3
	4	0+	005	043	188	219	127	042	007	001	0+	0+	0+	0+	4
	5	0+	001	010	103	206	186	092	024	003	0+	0+	0+	0+	5
	6	0+	0+	002	043	147	207	153	061	012	001	0+	0+	0+	6
	7	0+	0+	0+	014	081	177	196	118	035	003	0+	0+	0+	7
	8	0+	0+	0+	003	035	118	196	177	081	014	0+	0+	0+	8
	9	0+	0+	0+	001	012	061	153	207	147	043	002	0+	0+	9
	10	0+	0+	0+	0+	003	024	092	186	206	103	010	001	0+	10
	11	0+	0+	0+	0+	001	007	042	127	219	188	043	005	0+	11
	12	0+	0+	0+	0+	0+	002	014	063	170	250	129	031	0+	12
	13	0+	0+	0+	0+	0+	0+	003	022	092	231	267	135	009	13
	14	0+	0+	0+	0+	0+	0+	0+	005	031	132	343	366	130	14
	15	0+	0+	0+	0+	0+	0+	0+	0+	005	035	206	463	860	15

From Frederick Mosteller, Robert E. K. Rourke, and George B. Thomas, Jr., *Probability with Statistical Applications*, 2nd ed., ©1970, Addison-Wesley Publishing Company, Reading, Mass., pp. 475–477. Reprinted with permission.

TABLE 5   Areas of the Standard Normal Distribution   573

Table 5
Areas of the standard normal
distribution

The entries in this table are the probabilities that a random variable having the standard normal distribution assumes a value between 0 and z; the probability is represented by the area under the curve shaded in the accompanying figure. Areas for negative values of z are obtained by symmetry.

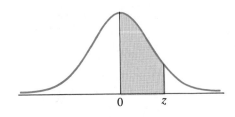

z	0.00	0.01	0.02	0.03	0.04	0.05	0.06	0.07	0.08	0.09
					Second Decimal Place in z					
0.0	0.0000	0.0040	0.0080	0.0120	0.0160	0.0199	0.0239	0.0279	0.0319	0.0359
0.1	0.0398	0.0438	0.0478	0.0517	0.0557	0.0596	0.0636	0.0675	0.0714	0.0753
0.2	0.0793	0.0832	0.0871	0.0910	0.0948	0.0987	0.1026	0.1064	0.1103	0.1141
0.3	0.1179	0.1217	0.1255	0.1293	0.1331	0.1368	0.1406	0.1443	0.1480	0.1517
0.4	0.1554	0.1591	0.1628	0.1664	0.1700	0.1736	0.1772	0.1808	0.1844	0.1879
0.5	0.1915	0.1950	0.1985	0.2019	0.2054	0.2088	0.2123	0.2157	0.2190	0.2224
0.6	0.2257	0.2291	0.2324	0.2357	0.2389	0.2422	0.2454	0.2486	0.2517	0.2549
0.7	0.2580	0.2611	0.2642	0.2673	0.2704	0.2734	0.2764	0.2794	0.2823	0.2852
0.8	0.2881	0.2910	0.2939	0.2967	0.2995	0.3023	0.3051	0.3078	0.3106	0.3133
0.9	0.3159	0.3186	0.3212	0.3238	0.3264	0.3289	0.3315	0.3340	0.3365	0.3389
1.0	0.3413	0.3438	0.3461	0.3485	0.3508	0.3531	0.3554	0.3577	0.3599	0.3621
1.1	0.3643	0.3665	0.3686	0.3708	0.3729	0.3749	0.3770	0.3790	0.3810	0.3830
1.2	0.3849	0.3869	0.3888	0.3907	0.3925	0.3944	0.3962	0.3980	0.3997	0.4015
1.3	0.4032	0.4049	0.4066	0.4082	0.4099	0.4115	0.4131	0.4147	0.4162	0.4177
1.4	0.4192	0.4207	0.4222	0.4236	0.4251	0.4265	0.4279	0.4292	0.4306	0.4319
1.5	0.4332	0.4345	0.4357	0.4370	0.4382	0.4394	0.4406	0.4418	0.4429	0.4441
1.6	0.4452	0.4463	0.4474	0.4484	0.4495	0.4505	0.4515	0.4525	0.4535	0.4545
1.7	0.4554	0.4564	0.4573	0.4582	0.4591	0.4599	0.4608	0.4616	0.4625	0.4633
1.8	0.4641	0.4649	0.4656	0.4664	0.4671	0.4678	0.4686	0.4693	0.4699	0.4706
1.9	0.4713	0.4719	0.4726	0.4732	0.4738	0.4744	0.4750	0.4756	0.4761	0.4767
2.0	0.4772	0.4778	0.4783	0.4788	0.4793	0.4798	0.4803	0.4808	0.4812	0.4817
2.1	0.4821	0.4826	0.4830	0.4834	0.4838	0.4842	0.4846	0.4850	0.4854	0.4857
2.2	0.4861	0.4864	0.4864	0.4871	0.4875	0.4878	0.4881	0.4884	0.4887	0.4890
2.3	0.4893	0.4896	0.4898	0.4901	0.4904	0.4906	0.4909	0.4911	0.4913	0.4916
2.4	0.4918	0.4920	0.4922	0.4925	0.4927	0.4929	0.4931	0.4932	0.4934	0.4936
2.5	0.4938	0.4940	0.4941	0.4943	0.4945	0.4946	0.4948	0.4949	0.4951	0.4952
2.6	0.4953	0.4955	0.4956	0.4957	0.4959	0.4960	0.4961	0.4962	0.4963	0.4964
2.7	0.4965	0.4966	0.4967	0.4968	0.4969	0.4970	0.4971	0.4972	0.4973	0.4974
2.8	0.4974	0.4975	0.4976	0.4977	0.4977	0.4978	0.4979	0.4979	0.4980	0.4981
2.9	0.4981	0.4982	0.4982	0.4983	0.4984	0.4984	0.4985	0.4985	0.4986	0.4986
3.0	0.4987	0.4987	0.4987	0.4988	0.4988	0.4989	0.4989	0.4989	0.4990	0.4990
3.1	0.4990	0.4991	0.4991	0.4991	0.4992	0.4992	0.4992	0.4992	0.4993	0.4993
3.2	0.4993	0.4993	0.4994	0.4994	0.4994	0.4994	0.4994	0.4995	0.4995	0.4995
3.3	0.4995	0.4995	0.4995	0.4996	0.4996	0.4996	0.4996	0.4996	0.4996	0.4997
3.4	0.4997	0.4997	0.4997	0.4997	0.4997	0.4997	0.4997	0.4997	0.4997	0.4998
3.5	0.4998									
4.0	0.49997									
4.5	0.499997									
5.0	0.4999997									

* For specific details on the use of this table, see page 224.

Table 6

Critical values of Student's t distribution

The entries in this table are the critical values for Student's t for an area of α in the right-hand tail. Critical values for the left-hand tail are found by symmetry

df	Amount of α in One-tail					
	0.25	0.10	0.05	0.025	0.01	0.005
1	1.000	3.08	6.31	12.7	31.8	63.7
2	0.816	1.89	2.92	4.30	6.97	9.92
3	0.765	1.64	2.35	3.18	4.54	5.84
4	0.741	1.53	2.13	2.78	3.75	4.60
5	0.727	1.48	2.02	2.57	3.37	4.03
6	0.718	1.44	1.94	2.45	3.14	3.71
7	0.711	1.42	1.89	2.36	3.00	3.50
8	0.706	1.40	1.86	2.31	2.90	3.36
9	0.703	1.38	1.83	2.26	2.82	3.25
10	0.700	1.37	1.81	2.23	2.76	3.17
11	0.697	1.36	1.80	2.20	2.72	3.11
12	0.695	1.36	1.78	2.18	2.68	3.05
13	0.694	1.35	1.77	2.16	2.65	3.01
14	0.692	1.35	1.76	2.14	2.62	2.98
15	0.691	1.34	1.75	2.13	2.69	2.95
16	0.690	1.34	1.75	2.12	2.58	2.92
17	0.689	1.33	1.74	2.11	2.57	2.90
18	0.688	1.33	1.73	2.10	2.55	2.88
19	0.688	1.33	1.73	2.09	2.54	2.86
20	0.687	1.33	1.72	2.09	2.53	2.85
21	0.686	1.32	1.72	2.08	2.52	2.83
22	0.686	1.32	1.72	2.07	2.51	2.82
23	0.685	1.32	1.71	2.07	2.50	2.81
24	0.685	1.32	1.71	2.06	2.49	2.80
25	0.684	1.32	1.71	2.06	2.49	2.79
26	0.684	1.32	1.71	2.06	2.48	2.78
27	0.684	1.31	1.70	2.05	2.47	2.77
28	0.683	1.31	1.70	2.05	2.47	2.76
29	0.683	1.31	1.70	2.05	2.46	2.76
z	0.674	1.28	1.65	1.96	2.33	2.58

NOTE: For df $\geq$ 30, the critical value $t(\text{df}, \alpha)$ is approximated by $z(\alpha)$, given in the bottom row of table.

Adapted from E. S. Pearson and H. O. Hartley, *Biometrika Tables for Statisticians*, vol. I (1966), p. 146. Reprinted by permission of the Biometrika Trustees. The two columns headed "0.10" and "0.01" are taken from Table III (adapted) on p. 46 of Fisher and Yates, *Statistical Tables for Biological, Agricultural and Medical Research*, 6th ed., published by Longman Group Ltd., London, 1974 (previously published by Oliver and Boyd, Edinburgh), and by permission of the authors and publishers.

* For specific details on the use of this table, see page 329.

TABLE 7   Critical Values of the χ^2 Distribution   575

Table 7
Critical values of the χ^2
distribution

The entries in this table are the critical values for chi square for which the area to the right under the curve is equal to α.

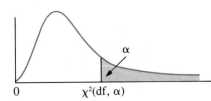

$$\chi^2(df, \alpha)$$

	Amount of α in Right-hand Tail									
df	0.995	0.990	0.975	0.950	0.900	0.100	0.050	0.025	0.010	0.005
1	0.0000393	0.000157	0.000982	0.00393	0.0158	2.71	3.84	5.02	6.65	7.88
2	0.0100	0.0201	0.0506	0.103	0.211	4.61	6.00	7.38	9.21	10.6
3	0.0717	0.115	0.216	0.352	0.584	6.25	7.82	9.35	11.4	12.9
4	0.207	0.297	0.484	0.711	1.0636	7.78	9.50	11.1	13.3	14.9
5	0.412	0.554	0.831	1.15	1.61	9.24	11.1	12.8	15.1	16.8
6	0.676	0.872	1.24	1.64	2.20	10.6	12.6	14.5	16.8	18.6
7	0.990	1.24	1.69	2.17	2.83	12.0	14.1	16.0	18.5	20.3
8	1.34	1.65	2.18	2.73	3.49	13.4	15.5	17.5	20.1	22.0
9	1.73	2.09	2.70	3.33	4.17	14.7	17.0	19.0	21.7	23.6
10	2.16	2.56	3.25	3.94	4.87	16.0	18.3	20.5	23.2	25.2
11	2.60	3.05	3.82	4.58	5.58	17.2	19.7	21.9	24.7	26.8
12	3.07	3.57	4.40	5.23	6.30	18.6	21.0	23.3	26.2	28.3
13	3.57	4.11	5.01	5.90	7.04	19.8	22.4	24.7	27.7	29.8
14	4.07	4.66	5.63	6.57	7.79	21.1	23.7	26.1	29.1	31.3
15	4.60	5.23	6.26	7.26	8.55	22.3	25.0	27.5	30.6	32.8
16	5.14	5.81	6.91	7.96	9.31	23.5	26.3	28.9	32.0	34.3
17	5.70	6.41	7.56	8.67	10.1	24.8	27.6	30.2	33.4	35.7
18	6.26	7.01	8.23	9.39	10.9	26.0	28.9	31.5	34.8	37.2
19	6.84	7.63	8.91	10.1	11.7	27.2	30.1	32.9	36.2	38.6
20	7.43	8.26	9.59	10.9	12.4	28.4	31.4	34.2	37.6	40.0
21	8.03	8.90	10.3	11.6	13.2	29.6	32.7	35.5	39.0	41.4
22	8.64	9.54	11.0	12.3	14.0	30.8	33.9	36.8	40.3	42.8
23	9.26	10.2	11.0	13.1	14.9	32.0	35.2	38.1	41.6	44.2
24	9.89	10.9	12.4	13.9	15.7	33.2	36.4	39.4	43.0	45.6
25	10.5	11.5	13.1	14.6	16.5	34.4	37.7	40.7	44.3	46.9
26	11.2	12.2	13.8	15.4	17.3	35.6	38.9	41.9	45.6	48.3
27	11.8	12.9	14.6	16.2	18.1	36.7	40.1	43.2	47.0	49
28	12.5	13.6	15.3	16.9	18.9	37.9	41.3	44.5	48.3	51.0
29	13.1	14.3	16.1	17.7	19.8	39.1	42.6	45.7	49.6	52.3
30	13.8	15.0	16.8	18.5	20.6	40.3	43.8	47.0	50.9	53.7
40	20.7	22.2	24.4	26.5	29.1	51.8	55.8	59.3	63.7	66.8
50	28.0	29.7	32.4	34.8	37.7	63.2	67.5	71.4	76.2	79.5
60	35.5	37.5	40.5	43.2	46.5	74.4	79.1	83.3	88.4	92.0
70	43.3	45.4	48.8	51.8	55.3	85.5	90.5	95.0	100.0	104.0
80	51.2	53.5	57.2	60.4	64.3	96.6	102.0	107.0	112.0	116.0
90	59.2	61.8	65.7	69.1	73.3	108.0	113.0	118.0	124.0	128.0
100	67.3	70.1	74.2	77.9	82.4	114.0	124.0	130.0	136.0	140.0

Adapted from E. S. Pearson and H. O. Hartley, *Biometrika Tables for Statisticians*, vol. I (1962), pp. 130–131. Reprinted by permission of the Biometrika Trustees.

* For specific details on the use of this table, see page 346.

Table 8a
Critical values of the *F*
distribution ($\alpha = 0.05$)

The entries in this table are critical values of *F* for which the area under the curve to the right is equal to 0.05.

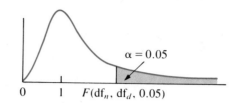

	Degrees of Freedom for Numerator									
	1	2	3	4	5	6	7	8	9	10
1	161	200	216	225	230	234	237	239	241	242
2	18.5	19.0	19.2	19.2	19.3	19.3	19.4	19.4	19.4	19.4
3	10.1	9.55	9.28	9.12	9.01	8.94	8.89	8.85	8.81	8.79
4	7.71	6.94	6.59	6.39	6.26	6.16	6.09	6.04	6.00	5.96
5	6.61	5.79	5.41	5.19	5.05	4.95	4.88	4.82	4.77	4.74
6	5.99	5.14	4.76	4.53	4.39	4.28	4.21	4.15	4.10	4.06
7	5.59	4.74	4.35	4.12	3.97	3.87	3.79	3.73	3.68	3.64
8	5.32	4.46	4.07	3.84	3.69	3.58	3.50	3.44	3.39	3.35
9	5.12	4.26	3.86	3.63	3.48	3.37	3.29	3.23	3.18	3.14
10	4.96	4.10	3.71	3.48	3.33	3.22	3.14	3.07	3.02	2.98
11	4.84	3.98	3.59	3.36	3.20	3.09	3.01	2.95	2.90	2.85
12	4.75	3.89	3.49	3.26	3.11	3.00	2.91	2.85	2.80	2.75
13	4.67	3.81	3.41	3.18	3.03	2.92	2.83	2.77	2.71	2.67
14	4.60	3.74	3.34	3.11	2.96	2.85	2.76	2.70	2.65	2.60
15	4.54	3.68	3.29	3.06	2.90	2.79	2.71	2.64	2.59	2.54
16	4.49	3.63	3.24	3.01	2.85	2.74	2.66	2.59	2.54	2.49
17	4.45	3.59	3.20	2.96	2.81	2.70	2.61	2.55	2.49	2.45
18	4.41	3.55	3.16	2.93	2.77	2.66	2.58	2.51	2.46	2.41
19	4.38	3.52	3.13	2.90	2.74	2.63	2.54	2.48	2.42	2.38
20	4.35	3.49	3.10	2.87	2.71	2.60	2.51	2.45	2.39	2.35
21	4.32	3.47	3.07	2.84	2.68	2.57	2.49	2.42	2.37	2.32
22	4.30	3.44	3.05	2.82	2.66	2.55	2.46	2.40	2.34	2.30
23	4.28	3.42	3.03	2.80	2.64	2.53	2.44	2.37	2.32	2.27
24	4.26	3.40	3.01	2.78	2.62	2.51	2.42	2.36	2.30	2.25
25	4.24	3.39	2.99	2.76	2.60	2.49	2.40	2.34	2.28	2.24
30	4.17	3.32	2.92	2.69	2.53	2.42	2.33	2.27	2.21	2.16
40	4.08	3.23	2.84	2.61	2.45	2.34	2.25	2.18	2.12	2.08
60	4.00	3.15	2.76	2.53	2.37	2.25	2.17	2.10	2.04	1.99
120	3.92	3.07	2.68	2.45	2.29	2.18	2.09	2.02	1.96	1.91
∞	3.84	3.00	2.60	2.37	2.21	2.10	2.01	1.94	1.88	1.83

Degrees of Freedom for Denominator (vertical axis label)

* For specific details on the use of this table, see page 375.

Table 8a (Continued)		Degrees of Freedom for Numerator								
		12	15	20	24	30	40	60	120	∞
	1	244	246	248	249	250	251	252	253	254
	2	19.4	19.4	19.4	19.5	19.5	19.5	19.5	19.5	19.5
	3	8.74	8.70	8.66	8.64	8.62	8.59	8.57	8.55	8.53
	4	5.91	5.86	5.80	5.77	5.75	5.72	5.69	5.66	5.63
	5	4.68	4.62	4.56	4.53	4.50	4.46	4.43	4.40	4.37
	6	4.00	3.94	3.87	3.84	3.81	3.77	3.74	3.70	3.67
	7	3.57	3.51	3.44	3.41	3.38	3.34	3.30	3.27	3.23
	8	3.28	3.22	3.15	3.12	3.08	3.04	3.01	2.97	2.93
	9	3.07	3.01	2.94	2.90	2.86	2.83	2.79	2.75	2.71
Degrees of Freedom for Denominator	10	2.91	2.85	2.77	2.74	2.70	2.66	2.62	2.58	2.54
	11	2.79	2.72	2.65	2.61	2.57	2.53	2.49	2.45	2.40
	12	2.69	2.62	2.54	2.51	2.47	2.43	2.38	2.34	2.30
	13	2.60	2.53	2.46	2.42	2.38	2.34	2.30	2.25	2.21
	14	2.53	2.46	2.39	2.35	2.31	2.27	2.22	2.18	2.13
	15	2.48	2.40	2.33	2.29	2.25	2.20	2.16	2.11	2.07
	16	2.42	2.35	2.28	2.24	2.19	2.15	2.11	2.06	2.01
	17	2.38	2.31	2.23	2.19	2.15	2.10	2.06	2.01	1.96
	18	2.34	2.27	2.19	2.15	2.11	2.06	2.02	1.97	1.92
	19	2.31	2.23	2.16	2.11	2.07	2.03	1.98	1.93	1.88
	20	2.28	2.20	2.12	2.08	2.04	1.99	1.95	1.90	1.84
	21	2.25	2.18	2.10	2.05	2.01	1.96	1.92	1.87	1.81
	22	2.23	2.15	2.07	2.03	1.98	1.94	1.89	1.84	1.78
	23	2.20	2.13	2.05	2.01	1.96	1.91	1.86	1.81	1.76
	24	2.18	2.11	2.03	1.98	1.94	1.89	1.84	1.79	1.73
	25	2.16	2.09	2.01	1.96	1.92	1.87	1.82	1.77	1.71
	30	2.09	2.01	1.93	1.89	1.84	1.79	1.74	1.68	1.62
	40	2.00	1.92	1.84	1.79	1.74	1.69	1.64	1.58	1.51
	60	1.92	1.84	1.75	1.70	1.65	1.59	1.53	1.47	1.39
	120	1.83	1.75	1.66	1.61	1.55	1.50	1.43	1.35	1.25
	∞	1.75	1.67	1.57	1.52	1.46	1.39	1.32	1.22	1.00

From E. S. Pearson and H. O. Hartley, *Biometrika Tables for Statisticians*, vol. I (1958), pp. 159–163. Reprinted by permission of the Biometrika Trustees.

Table 8b

Critical values of the *F*
distribution ($\alpha = 0.025$)

The entries in this table are critical values of F for which the area under the curve to
the right is equal to 0.025.

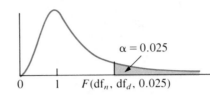

$\alpha = 0.025$

$0 \qquad 1 \qquad F(\mathrm{df}_n, \mathrm{df}_d, 0.025)$

		Degrees of Freedom for Numerator								
	1	2	3	4	5	6	7	8	9	10
1	648	800	864	900	922	937	948	957	963	969
2	38.5	39.0	39.2	39.2	39.3	39.3	39.4	39.4	39.4	39.4
3	17.4	16.0	15.4	15.1	14.9	14.7	14.6	14.5	14.5	14.4
4	12.2	10.6	9.98	9.60	9.36	9.20	9.07	8.98	8.90	8.84
5	10.0	8.43	7.76	7.39	7.15	6.98	6.85	6.76	6.68	6.62
6	8.81	7.26	6.60	6.23	5.99	5.82	5.70	5.60	5.52	5.46
7	8.07	6.54	5.89	5.52	5.29	5.12	4.99	4.90	4.82	4.76
8	7.57	6.06	5.42	5.05	4.82	4.65	4.53	4.43	4.36	4.30
9	7.21	5.71	5.08	4.72	4.48	4.32	4.20	4.10	4.03	3.96
10	6.94	5.46	4.83	4.47	4.24	4.07	3.95	3.85	3.78	3.72
11	6.72	5.26	4.63	4.28	4.04	3.88	3.76	3.66	3.59	3.53
12	6.55	5.10	4.47	4.12	3.89	3.73	3.61	3.51	3.44	3.37
13	6.41	4.97	4.35	4.00	3.77	3.60	3.48	3.39	3.31	3.25
14	6.30	4.86	4.24	3.89	3.66	3.50	3.38	3.28	3.21	3.15
15	6.20	4.77	4.15	3.80	3.58	3.41	3.29	3.20	3.12	3.06
16	6.12	4.69	4.08	3.73	3.50	3.34	3.22	3.12	3.05	2.99
17	6.04	4.62	4.01	3.66	3.44	3.28	3.16	3.06	2.98	2.92
18	5.98	4.56	3.95	3.61	3.38	3.22	3.10	3.01	2.93	2.87
19	5.92	4.51	3.90	3.56	3.33	3.17	3.05	2.96	2.88	2.82
20	5.87	4.46	3.86	3.51	3.29	3.13	3.01	2.91	2.84	2.77
21	5.83	4.42	3.82	3.48	3.25	3.09	2.97	2.87	2.80	2.73
22	5.79	4.38	3.78	3.44	3.22	3.05	2.93	2.84	2.76	2.70
23	5.75	4.35	3.75	3.41	3.18	3.02	2.90	2.81	2.73	2.67
24	5.72	4.32	3.72	3.38	3.15	2.99	2.87	2.78	2.70	2.64
25	5.69	4.29	3.69	3.35	3.13	2.97	2.85	2.75	2.68	2.61
30	5.57	4.18	3.59	3.25	3.03	2.87	2.75	2.65	2.57	2.51
40	5.42	4.05	3.46	3.13	2.90	2.74	2.62	2.53	2.45	2.39
60	5.29	3.93	3.34	3.01	2.79	2.63	2.51	2.41	2.33	2.27
120	5.15	3.80	3.23	2.89	2.67	2.52	2.39	2.30	2.22	2.16
∞	5.02	3.69	3.12	2.79	2.57	2.41	2.29	2.19	2.11	2.05

Degrees of Freedom for Denominator

* For specific details on the use of this table, see page 375.

Table 8b
(Continued)

		Degrees of Freedom for Numerator								
		12	15	20	24	30	40	60	120	∞
	1	977	985	993	997	1,001	1,006	1,010	1,014	1,018
	2	39.4	39.4	39.4	39.5	39.5	39.5	39.5	39.5	39.5
	3	14.3	14.3	14.2	14.1	14.1	14.0	14.0	13.9	13.9
	4	8.75	8.66	8.56	8.51	8.46	8.41	8.36	8.31	8.26
	5	6.52	6.43	6.33	6.28	6.23	6.18	6.12	6.07	6.02
	6	5.37	5.27	5.17	5.12	5.07	5.01	4.96	4.90	4.85
	7	4.67	4.57	4.47	4.42	4.36	4.31	4.25	4.20	4.14
	8	4.20	4.10	4.00	3.95	3.89	3.84	3.78	3.73	3.67
	9	3.87	3.77	3.67	3.61	3.56	3.51	3.45	3.39	3.33
	10	3.62	3.52	3.42	3.37	3.31	3.26	3.20	3.14	3.08
Degrees of Freedom for Denominator	11	3.43	3.33	3.23	3.17	3.12	3.06	3.00	2.94	2.88
	12	3.28	3.18	3.07	3.02	2.96	2.91	2.85	2.79	2.72
	13	3.15	3.05	2.95	2.89	2.84	2.78	2.72	2.66	2.60
	14	3.05	2.95	2.84	2.79	2.73	2.67	2.61	2.55	2.49
	15	2.96	2.86	2.76	2.70	2.64	2.59	2.52	2.46	2.40
	16	2.89	2.79	2.68	2.63	2.57	2.51	2.45	2.38	2.32
	17	2.82	2.72	2.62	2.56	2.50	2.44	2.38	2.32	2.25
	18	2.77	2.67	2.56	2.50	2.44	2.38	2.32	2.26	2.19
	19	2.72	2.62	2.51	2.45	2.39	2.33	2.27	2.20	2.13
	20	2.68	2.57	2.46	2.41	2.35	2.29	2.22	2.16	2.09
	21	2.64	2.53	2.42	2.37	2.31	2.25	2.18	2.11	2.04
	22	2.60	2.50	2.39	2.33	2.27	2.21	2.14	2.08	2.00
	23	2.57	2.47	2.36	2.30	2.24	2.18	2.11	2.04	1.97
	24	2.54	2.44	2.33	2.27	2.21	2.15	2.08	2.01	1.94
	25	2.51	2.41	2.30	2.24	2.18	2.12	2.05	1.98	1.91
	30	2.41	2.31	2.20	2.14	2.07	2.01	1.94	1.87	1.79
	40	2.29	2.18	2.07	2.01	1.94	1.88	1.80	1.72	1.64
	60	2.17	2.06	1.94	1.88	1.82	1.74	1.67	1.58	1.48
	120	2.05	1.95	1.82	1.76	1.69	1.61	1.53	1.43	1.31
	∞	1.94	1.83	1.71	1.64	1.57	1.48	1.39	1.27	1.00

From E. S. Pearson and H. O. Hartley, *Biometrika Tables for Statisticians*, vol. I (1958), pp. 159–163. Reprinted by permission of the Biometrika Trustees.

Table 8c
Critical values of the *F*
distribution ($\alpha = 0.01$)

The entries in the table are critical values of F for which the area under the curve to the right is equal to 0.01.

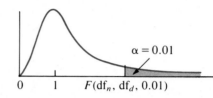

$\alpha = 0.01$

$0 \quad 1 \quad F(df_n, df_d, 0.01)$

		Degrees of Freedom for Numerator								
	1	**2**	**3**	**4**	**5**	**6**	**7**	**8**	**9**	**10**
1	4,052	5,000	5,403	5,625	5,764	5,859	5,928	5,982	6,023	6,056
2	98.5	99.0	99.2	99.2	99.3	99.3	99.4	99.4	99.4	99.4
3	34.1	30.8	29.5	28.7	28.2	27.9	27.7	27.5	27.3	27.2
4	21.2	18.0	16.7	16.0	15.5	15.2	15.0	14.8	14.7	14.5
5	16.3	13.3	12.1	11.4	11.0	10.7	10.5	10.3	10.2	10.1
6	13.7	10.9	9.78	9.15	8.75	8.47	8.26	8.10	7.98	7.87
7	12.2	9.55	8.45	7.85	7.46	7.19	6.99	6.84	6.72	6.62
8	11.3	8.65	7.59	7.01	6.63	6.37	6.18	6.03	5.91	5.81
9	10.6	8.02	6.99	6.42	6.06	5.80	5.61	5.47	5.35	5.26
10	10.0	7.56	6.55	5.99	5.64	5.39	5.20	5.06	4.94	4.85
11	9.65	7.21	6.22	5.67	5.32	5.07	4.89	4.74	4.63	4.54
12	9.33	6.93	5.95	5.41	5.06	4.82	4.64	4.50	4.39	4.30
13	9.07	6.70	5.74	5.21	4.86	4.62	4.44	4.30	4.19	4.10
14	8.86	6.51	5.56	5.04	4.70	4.46	4.28	4.14	4.03	3.94
15	8.68	6.36	5.42	4.89	4.56	4.32	4.14	4.00	3.89	3.80
16	8.53	6.23	5.29	4.77	4.44	4.20	4.03	3.89	3.78	3.69
17	8.40	6.11	5.19	4.67	4.34	4.10	3.93	3.79	3.68	3.59
18	8.29	6.01	5.09	4.58	4.25	4.01	3.84	3.71	3.60	3.51
19	8.19	5.93	5.01	4.50	4.17	3.94	3.77	3.63	3.52	3.43
20	8.10	5.85	4.94	4.43	4.10	3.87	3.70	3.56	3.46	3.37
21	8.02	5.78	4.87	4.37	4.04	3.81	3.64	3.51	3.40	3.31
22	7.95	5.72	4.82	4.31	3.99	3.76	3.59	3.45	3.35	3.26
23	7.88	5.66	4.76	4.26	3.94	3.71	3.54	3.41	3.30	3.21
24	7.82	5.61	4.72	4.22	3.90	3.67	3.50	3.36	3.26	3.17
25	7.77	5.57	4.68	4.18	3.86	3.63	3.46	3.32	3.22	3.13
30	7.56	5.39	4.51	4.02	3.70	3.47	3.30	3.17	3.07	2.98
40	7.31	5.18	4.31	3.83	3.51	3.29	3.12	2.99	2.89	2.80
60	7.08	4.98	4.13	3.65	3.34	3.12	2.95	2.82	2.72	2.63
120	6.85	4.79	3.95	3.48	3.17	2.96	2.79	2.66	2.56	2.47
∞	6.63	4.61	3.78	3.32	3.02	2.80	2.64	2.51	2.41	2.32

Degrees of Freedom for Denominator

Table 8c (Continued)		Degrees of Freedom for Numerator								
		12	15	20	24	30	40	60	120	∞
	1	6,106	6,157	6,209	6,235	6,261	6,287	6,313	6,339	6,366
	2	99.4	99.4	99.4	99.5	99.5	99.5	99.5	99.5	99.5
	3	27.1	26.9	26.7	26.6	26.5	26.4	26.3	26.2	26.1
	4	14.4	14.2	14.0	13.9	13.8	13.7	13.7	13.6	13.5
	5	9.89	9.72	9.55	9.47	9.38	9.29	9.20	9.11	9.02
	6	7.72	7.56	7.40	7.31	7.23	7.14	7.06	6.97	6.88
	7	6.47	6.31	6.16	6.07	5.99	5.91	5.82	5.74	5.65
	8	5.67	5.52	5.36	5.28	5.20	5.12	5.03	4.95	4.86
	9	5.11	4.96	4.81	4.73	4.65	4.57	4.48	4.40	4.31
Degrees of Freedom for Denominator	10	4.71	4.56	4.41	4.33	4.25	4.17	4.08	4.00	3.91
	11	4.40	4.25	4.10	4.02	3.94	3.86	3.78	3.69	3.60
	12	4.16	4.01	3.86	3.78	3.70	3.62	3.54	3.45	3.36
	13	3.96	3.82	3.66	3.59	3.51	3.43	3.34	3.25	3.17
	14	3.80	3.66	3.51	3.43	3.35	3.27	3.18	3.09	3.00
	15	3.67	3.52	3.37	3.29	3.21	3.13	3.05	2.96	2.87
	16	3.55	3.41	3.26	3.18	3.10	3.02	2.93	2.84	2.75
	17	3.46	3.31	3.16	3.08	3.00	2.92	2.83	2.75	2.65
	18	3.37	3.23	3.08	3.00	2.92	2.84	2.75	2.66	2.57
	19	3.30	3.15	3.00	2.92	2.84	2.76	2.67	2.58	2.49
	20	3.23	3.09	2.94	2.86	2.78	2.69	2.61	2.52	2.42
	21	3.17	3.03	2.88	2.80	2.72	2.64	2.55	2.46	2.36
	22	3.12	2.98	2.83	2.75	2.67	2.58	2.50	2.40	2.31
	23	3.07	2.93	2.78	2.70	2.62	2.54	2.45	2.35	2.26
	24	3.03	2.89	2.74	2.66	2.58	2.49	2.40	2.31	2.21
	25	2.99	2.85	2.70	2.62	2.53	2.45	2.36	2.27	2.17
	30	2.84	2.70	2.55	2.47	2.39	2.30	2.21	2.11	2.01
	40	2.66	2.52	2.37	2.29	2.20	2.11	2.02	1.92	1.80
	60	2.50	2.35	2.20	2.12	2.03	1.94	1.84	1.73	1.60
	120	2.34	2.19	2.03	1.95	1.86	1.76	1.66	1.53	1.38
	∞	2.18	2.04	1.88	1.79	1.70	1.59	1.47	1.32	1.00

Table 9
Critical values of *r* when
p = 0

The entries in this table are the critical values of *r* for a two-tailed test at α.
For simple correlation, df $= n - 2$, where *n* is the number of pairs of data in the sample. For a one-tailed test, the value of α shown at the top of the table is double the value of α being used in the hypothesis test.

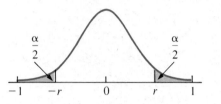

df	α 0.10	0.05	0.02	0.01
1	0.988	0.997	1.000	1.000
2	0.900	0.950	0.980	0.990
3	0.805	0.878	0.934	0.959
4	0.729	0.811	0.882	0.917
5	0.669	0.754	0.833	0.874
6	0.662	0.707	0.789	0.834
7	0.582	0.666	0.750	0.798
8	0.549	0.632	0.716	0.765
9	0.521	0.602	0.685	0.735
10	0.497	0.576	0.658	0.708
11	0.476	0.553	0.634	0.684
12	0.458	0.532	0.612	0.661
13	0.441	0.514	0.592	0.641
14	0.426	0.497	0.574	0.623
15	0.412	0.482	0.558	0.606
16	0.400	0.468	0.542	0.590
17	0.389	0.456	0.528	0.575
18	0.378	0.444	0.516	0.561
19	0.369	0.433	0.503	0.549
20	0.360	0.423	0.492	0.537
25	0.323	0.381	0.445	0.487
30	0.296	0.349	0.409	0.449
35	0.275	0.325	0.381	0.418
40	0.257	0.304	0.358	0.393
45	0.243	0.288	0.338	0.372
50	0.231	0.273	0.322	0.354
60	0.211	0.250	0.295	0.325
70	0.195	0.232	0.274	0.302
80	0.183	0.217	0.256	0.283
90	0.173	0.205	0.242	0.267
100	0.164	0.195	0.230	0.254

* For specific details on the use of this table, see page 481.

Table 10

Confidence belts for the
correlation coefficient
$(1 - \alpha) = 0.95$

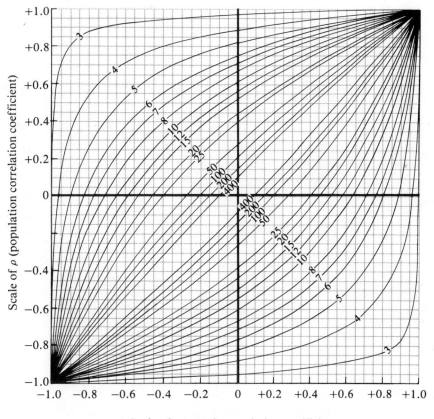

Scale of r (sample correlation coefficient)

The numbers on the curves are sample sizes.

* For specific details on the use of this table, see page 482.

From E. S. Pearson and H. O. Hartley, *Biometrika Tables for Statisticians*, vol. I (1962), p. 140. Reprinted by permission of the Biometrika Trustees.

Table 11

Critical values of the sign test

The entries in this table are the critical values for the number of the least frequent sign for a two-tailed test at α for the binomial $p = 0.5$. For a one-tailed test, the value of α shown at the top of the table is double the value of α being used in the hypothesis test.

	α					α			
n	0.01	0.05	0.10	0.25	n	0.01	0.05	0.10	0.25
1					51	15	18	19	20
2					52	16	18	19	21
3				0	53	16	18	20	21
4				0	54	17	19	20	22
5			0	0	55	17	19	20	22
6		0	0	1	56	17	20	21	23
7		0	0	1	57	18	20	21	23
8	0	0	1	1	58	18	21	22	24
9	0	1	1	2	59	19	21	22	24
10	0	1	1	2	60	19	21	23	25
11	0	1	2	3	61	20	22	23	25
12	1	2	2	3	62	20	22	24	25
13	1	2	3	3	63	20	23	24	26
14	1	2	3	4	64	21	23	24	26
15	2	3	3	4	65	21	24	25	27
16	2	3	4	5	66	22	24	25	27
17	2	4	4	5	67	22	25	26	28
18	3	4	5	6	68	22	25	26	28
19	3	4	5	6	69	23	25	27	29
20	3	5	5	6	70	23	26	27	29
21	4	5	6	7	71	24	26	28	30
22	4	5	6	7	72	24	27	28	30
23	4	6	7	8	73	25	27	28	31
24	5	6	7	8	74	25	28	29	31
25	5	7	7	9	75	25	28	29	32
26	6	7	8	9	76	26	28	30	32
27	6	7	8	10	77	26	29	30	32
28	6	8	9	10	78	27	29	31	33
29	7	8	9	10	79	27	30	31	33
30	7	9	10	11	80	28	30	32	34
31	7	9	10	11	81	28	31	32	34
32	8	9	10	12	82	28	31	33	35
33	8	10	11	12	83	29	32	33	35
34	9	10	11	13	84	29	32	33	36
35	9	11	12	13	85	30	32	34	36
36	9	11	12	14	86	30	33	34	37
37	10	12	13	14	87	31	33	35	37
38	10	12	13	14	88	31	34	35	38
39	11	12	13	15	89	31	34	36	38
40	11	13	14	15	90	32	35	36	39
41	11	13	14	16	91	32	35	37	39
42	12	14	15	16	92	33	36	37	39
43	12	14	15	17	93	33	36	38	40
44	13	15	16	17	94	34	37	38	40
45	13	15	16	18	95	34	37	38	41
46	13	15	16	18	96	34	37	39	41
47	14	16	17	19	97	35	38	39	42
48	14	16	17	19	98	35	38	40	42
49	15	17	18	19	99	36	39	40	43
50	15	17	18	20	100	36	39	41	43

* For specific details on the use of this table, see pages 517–522.

From Wilfred J. Dixon and Frank J. Massey, Jr., *Introduction to Statistical Analysis*, 3d ed., (New York: McGraw-Hill, 1969), p. 509. Reprinted by permission.

TABLE 12   Critical Values of U in the Mann-Whitney Test   585

Table 12
Critical values of U in the
Mann-Whitney test

A. The entries are the critical values of U for a one-tailed test at 0.025 or for a two-tailed test at 0.05.

n_2 \ n_1	1	2	3	4	5	6	7	8	9	10	11	12	13	14	15	16	17	18	19	20
1																				
2								0	0	0	0	1	1	1	1	1	2	2	2	2
3					0	1	1	2	2	3	3	4	4	5	5	6	6	7	7	8
4				0	1	2	3	4	4	5	6	7	8	9	10	11	11	12	13	13
5			0	1	2	3	5	6	7	8	9	11	12	13	14	15	17	18	19	20
6			1	2	3	5	6	8	10	11	13	14	16	17	19	21	22	24	25	27
7			1	3	5	6	8	10	12	14	16	18	20	22	24	26	28	30	32	34
8		0	2	4	6	8	10	13	15	17	19	22	24	26	29	21	34	36	38	41
9		0	2	4	7	10	12	15	17	20	23	26	28	31	34	37	39	42	45	48
10		0	3	5	8	11	14	17	20	23	26	29	33	36	39	42	45	48	52	55
11		0	3	6	9	13	16	19	23	26	30	33	37	40	44	47	51	55	58	62
12		1	4	7	11	14	18	22	26	29	33	37	41	45	49	53	57	61	65	69
13		1	4	8	12	16	20	24	28	33	37	41	45	50	54	59	63	67	72	76
14		1	5	9	13	17	22	26	31	36	40	45	50	55	59	64	67	74	78	83
15		1	5	10	14	19	24	29	34	39	44	49	54	59	64	70	75	80	85	90
16		1	6	11	15	21	26	31	37	42	47	53	59	64	70	75	81	86	92	98
17		2	6	11	17	22	28	34	39	45	51	57	63	67	75	81	87	93	99	105
18		2	7	12	18	24	30	36	42	48	55	61	67	74	80	86	93	99	106	112
19		2	7	13	19	25	32	38	45	52	58	65	72	78	85	92	99	106	113	119
20		2	8	13	20	27	34	41	48	55	62	69	76	83	90	98	105	112	119	127

B. The entries are the critical values of U for a one-tailed test at 0.05 or for a two-tailed test at 0.10.

n_2 \ n_1	1	2	3	4	5	6	7	8	9	10	11	12	13	14	15	16	17	18	19	20
1																			0	0
2					0	0	0	1	1	1	1	2	2	2	3	3	3	4	4	4
3			0	0	1	2	2	3	3	4	5	5	6	7	7	8	9	9	10	11
4			0	1	2	3	4	5	6	7	8	9	10	11	12	14	15	16	17	18
5		0	1	2	4	5	6	8	9	11	12	13	15	16	18	19	20	22	23	25
6		0	2	3	5	7	8	10	12	14	16	17	19	21	23	25	26	28	30	32
7		0	2	4	6	8	11	13	15	17	19	21	24	26	28	30	33	35	37	39
8		1	3	5	8	10	13	15	18	20	23	26	28	31	33	36	39	41	44	47
9		1	3	6	9	12	15	18	21	24	27	30	33	36	39	42	45	48	51	54
10		1	4	7	11	14	17	20	24	27	31	34	37	41	44	48	51	55	58	62
11		1	5	8	12	16	19	23	27	31	34	38	42	46	50	54	57	61	65	69
12		2	5	9	13	17	21	26	30	34	38	42	47	51	55	60	64	68	72	77
13		2	6	10	15	19	24	28	33	37	42	47	51	56	61	65	70	75	80	84
14		2	7	11	16	21	26	31	36	41	46	51	56	61	66	71	77	82	87	92
15		3	7	12	18	23	28	33	39	44	50	55	61	66	72	77	83	88	94	100
16		3	8	14	19	25	30	36	42	48	54	60	65	71	77	83	89	95	101	107
17		3	9	15	20	26	33	39	45	51	57	64	70	77	83	89	96	102	109	115
18		4	9	16	22	28	35	41	48	55	61	68	75	82	88	95	102	109	116	123
19	0	4	10	17	23	30	37	44	51	58	65	72	80	87	94	101	109	116	123	130
20	0	4	11	18	25	32	39	47	54	62	69	77	84	92	100	107	115	123	130	138

* For specific details on the use of this table, see page 526.

Table 13
Critical values for total number of runs (V)

The entries in this table are the critical values* for a two-tailed test using $\alpha = 0.05$. For a one-tailed test, $\alpha = 0.025$ and use only one of the critical values: the smaller critical value for a left-hand critical region, the larger for a right-hand critical region.

The smaller of n_1 and n_2 (rows) vs. *The larger of n_1 and n_2* (columns). Each cell gives the smaller critical value (upper) and the larger critical value (lower).

smaller \ larger	5	6	7	8	9	10	11	12	13	14	15	16	17	18	19	20
2								2/6	2/6	2/6	2/6	2/6	2/6	2/6	2/6	2/6
3		2/8	2/8	2/8	2/8	2/8	2/8	2/8	2/8	3/8	3/8	3/8	3/8	3/8	3/8	3/8
4	2/9	2/9	2/10	3/10	3/10	3/10	3/10	3/10	3/10	3/10	3/10	4/10	4/10	4/10	4/10	4/10
5	2/10	3/10	3/11	3/11	3/12	3/12	4/12	4/12	4/12	4/12	4/12	4/12	4/12	5/12	5/12	5/12
6		3/11	3/12	3/12	4/13	4/13	4/13	4/13	5/14	5/14	5/14	5/14	5/14	5/14	6/14	6/14
7			3/13	4/13	4/14	5/14	5/14	5/14	5/15	5/15	5/15	6/16	6/16	6/16	6/16	6/16
8				4/14	5/14	5/15	5/15	6/16	6/16	6/16	6/16	6/17	7/17	7/17	7/17	7/17
9					5/15	5/16	6/16	6/16	6/17	7/17	7/18	7/18	7/18	8/18	8/18	8/18
10						6/16	6/17	7/17	7/18	7/18	7/18	8/19	8/19	8/19	8/20	9/20
11							7/17	7/18	7/19	8/19	8/19	8/20	9/20	9/20	9/21	9/21
12								7/19	8/19	8/20	8/20	9/21	9/21	9/21	10/22	10/22
13									8/20	9/20	9/21	9/21	10/22	10/22	10/23	10/23
14										9/21	9/22	10/22	10/23	10/23	11/23	11/24
15											10/22	10/23	11/23	11/24	11/24	12/25
16												11/23	11/24	11/25	12/25	12/25
17													11/25	12/25	12/26	13/26
18														12/26	13/26	13/27
19															13/27	13/27
20																14/28

* See page 532 in regard to critical values.

From C. Eisenhart and F. Swed, "Tables for testing randomness of grouping in a sequence of alternatives," *The Annals of Statistics*, vol. 14 (1943): 66–87. Reprinted by permission.

For $n_1 > 20$ or $n_2 > 20$, treat V as a normal variable with a mean and a standard deviation of

$$\mu_v = \frac{2n_1 n_2}{n_1 + n_2} + 1$$

$$\sigma_v = \sqrt{\frac{2n_1 n_2 (2n_1 n_2 - n_1 - n_2)}{(n_1 + n_2)^2 (n_1 + n_2 - 1)}}$$

Table 14
Critical values of Spearman's
rank correlation coefficient The entries in this table are the critical values of r_s for a two-tailed test at α. For a one-tailed test, the value of α shown at the top of the table is double the value of α being used in the hypothesis test.

n	$\alpha = 0.10$	$\alpha = 0.05$	$\alpha = 0.02$	$\alpha = 0.01$
5	0.900	—	—	—
6	0.829	0.886	0.943	—
7	0.714	0.786	0.893	—
8	0.643	0.738	0.833	0.881
9	0.600	0.683	0.783	0.833
10	0.564	0.648	0.745	0.794
11	0.523	0.623	0.736	0.818
12	0.497	0.591	0.703	0.780
13	0.475	0.566	0.673	0.745
14	0.457	0.545	0.646	0.716
15	0.441	0.525	0.623	0.689
16	0.425	0.507	0.601	0.666
17	0.412	0.490	0.582	0.645
18	0.399	0.476	0.564	0.625
19	0.388	0.462	0.549	0.608
20	0.377	0.450	0.534	0.591
21	0.368	0.438	0.521	0.576
22	0.359	0.428	0.508	0.562
23	0.351	0.418	0.496	0.549
24	0.343	0.409	0.485	0.537
25	0.336	0.400	0.475	0.526
26	0.329	0.392	0.465	0.515
27	0.323	0.385	0.456	0.505
28	0.317	0.377	0.448	0.496
29	0.311	0.370	0.440	0.487
30	0.305	0.364	0.432	0.478

From E. G. Olds, "Distribution of sums of squares of rank differences for small numbers of individuals," *Annals of Statistics*, vol. 9 (1938), pp. 138–148, and amended, vol. 20 (1949), pp. 117–118. Reprinted by permission.

* For specific details on the use of this table, see page 537.

Answers to Selected Exercises

CHAPTER 1

Section 1.2, page 8

1.1 a. The amount of each type of throwaway discarded by each person in the neighborhood
b. Weight of the throwaways, in pounds
c. It appears that they collected the amount of each type of throwaway for each person for a given day.
d. Weight measured in pounds
e. Percentage of total. Since pounds are mentioned, we assume that the percentages are based on pounds.

Section 1.3, page 12

1.4 a. Those individuals who have the condition called hypertension (a very large but unknown number)
b. The sample is the 5000 people in the study.
c. The parameter is the proportion in the population for which the drug is effective.
d. The statistic is the proportion in the sample for which the drug is effective. The 80 percent is the value of the statistic.
e. The value of the parameter is unknown, but the 80 percent could be used to estimate its value.

1.6 (1) attribute   (2) continuous   (3) discrete

1.8 a. Population—all students enrolled at a certain college this semester
b. Variable—the amount of money each student spent to purchase textbooks for this semester

1.9 a. Population parameter—average cost of textbooks per student for all students
b. Sample statistic—average cost of textbooks per student in the sample of size 50
c. The average cost could be found by adding the cost of textbooks for all 50 students and then dividing the sum by 50.

1.11 a. continuous   **b.** attribute   **c.** discrete
d. attribute   **e.** discrete   **f.** continuous

Section 1.4, page 15

1.14 Group 2

Section 1.5, page 20

1.16 a. The sampling frame is that set of elements from which the sample is actually drawn.
b. A computer list of this semester's full-time enrollment
1.19 Stratified sampling—each supermarket chain is a strata and each strata is sampled

Section 1.6, page 21

1.21 a. statistics   **b.** probability   **c.** statistics
d. probability

Section 1.7, page 23

1.23 Several large computer programs (called *statistical packages*) have been developed that perform many of the statistical computations and tests you will study in this text. In order to have the statistical package perform the computation or run the statistical test, simply enter your data into the computer and it does the rest. This saves time and, therefore, money.

Chapter Exercises, page 23

1.24 Each student's answers will differ. Some possibilities are:
a. color of hair, major, sex, marital status, hometown
b. number of courses currently enrolled in, number of semesters at college, number of roommates
c. height, weight, distance from hometown, cost of textbooks

1.27 Each of the numbers reported in *A Liberty Blast* is a value that represents the amount of tape, wire, etc., used to

589

put on the fireworks display at the Statue of Liberty. The values reported do not result from repeated observations of variables.

1.30 a. $T = 3$ is a piece of data—the result obtained from one person
b. What is the average number of times per week that the people in the sample go grocery shopping?
c. What is the average number of times per week that the people in the sampled population go grocery shopping?

1.32 Cluster sampling. The blocks are clusters, or strata. Some, but not all, of the clusters are selected to be part of the sample. In this example, cluster sampling, as compared with simple random sampling, would save on expense (time and travel).

CHAPTER 2

Section 2.1, page 35

2.1 "I enjoy using computers"

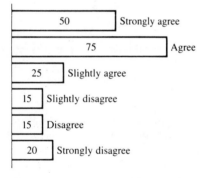

50	Strongly agree
75	Agree
25	Slightly agree
15	Slightly disagree
15	Disagree
20	Strongly disagree

2.4 a. 15 patients **b.** 11 days **c.** 50 days
d. 20 days

2.6   Number of Detectable Emissions
                          in 10 Seconds

0	8
1	8 5 8 9 2 6
2	3 2 2 1 3 5 4 2 1 2 2 7 6
3	7 2

Section 2.2, page 46

2.10 a. 1, 3, 16, 10, 12, 5, 1, 2
b. 0.02, 0.06, 0.32, 0.20, 0.24, 0.10, 0.02, 0.04
c. 0.02, 0.08, 0.40, 0.60, 0.84, 0.94, 0.96, 1.00

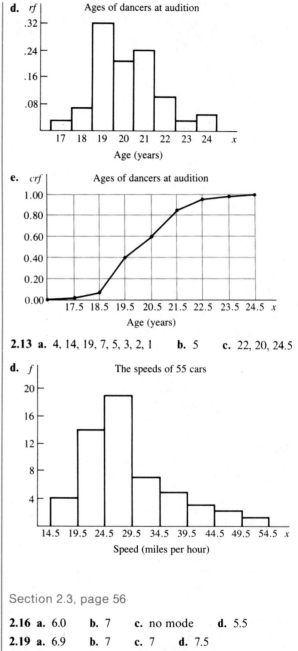

d. Ages of dancers at audition

e. Ages of dancers at audition

2.13 a. 4, 14, 19, 7, 5, 3, 2, 1   **b.** 5   **c.** 22, 20, 24.5

d. The speeds of 55 cars

Section 2.3, page 56

2.16 a. 6.0   **b.** 7   **c.** no mode   **d.** 5.5
2.19 a. 6.9   **b.** 7   **c.** 7   **d.** 7.5
2.21 35,400; 33,375; 31,500; 39,750
[*Note*: There is a typographical error in the data as listed in the article. See the paragraph, the midrange, and specific reference made to the largest data.]
2.23 a. 13.1   **b.** 13.1   **c.** 13.1   **d.** 13.15
2.26 28.5

Section 2.4, page 65

2.29 a. 7   **b.** 8.5   **c.** 2.9

2.32 a. 3.1   **b.** 3.1   **c.** 1.8

2.33 $\bar{x}_1 = 2.000$ and $s_1 = 0.0027$
$\bar{x}_2 = 2.000$ and $s_2 = 0.0074$
Even though both machines produced shafts of the same mean diameter, machine 2 had a standard deviation that was over 2.5 times the standard deviation of machine 1. Machine 2 will produce more shafts that do not meet specifications.

2.35 1.2, 1.1

2.39 7.9

2.41 a. 22,153.6   **b.** 22,153.6

2.43 The statement must be incorrect. The standard deviation can never be negative. The computations need to be checked; they have to be in error.

Section 2.5, page 75

2.45 a. 3.8, 5.6   **b.** 4.7   **c.** 3.5, 4.0, 6.9

2.46 a. 8.3   **b.** 9.25   **c.** 10.85   **d.** 14.8

e.

7.1	8.3	9.25	10.85	15.5
L	Q_1	$\tilde{x}$	Q_3	H

f.

```
     ┌──────┬────────┐
─────┤      │        ├────────────────
     └──────┴────────┘
  7.1 8.3 9.25  10.85            15.5
```

2.49 a. 1.0   **b.** 0.0   **c.** 2.25   **d.** −1.25

2.50 680

2.53 A has the higher relative position.

Section 2.6, page 81

2.55 a. Since almost all the data fall between $\bar{x} - 3s$ and $\bar{x} + 3s$, then $(\bar{x} + 3s) - (\bar{x} - 3s) = 6s$ should be approximately equal to the range.
b. Since $6s$ and the range are approximately equal, s could be approximated by range divided by 6.

2.56 a. at most 11%   **b.** at most 6.25%

2.57 a. 7, 10, 22, 8, 7, 2, 3, 0, 1
b. 3.4, 1.7   **c.** 1.7, 5.1   **d.** 47, 78%
e. 0.0, 6.8   **f.** 56, 93%   **g.** −1.7, 8.5
h. 98.3%
i. 93% is at least 75% and 98.3% is at least 89%, which agrees with the claims of Chebyshev's theorem.

j. 78%, 93%, and 98.3% are not approximately equal to the 68%, 95%, and 99.7% claimed by the empirical rule.

2.60 a. 6, 9, 8, 10, 6, 4, 4, 2, 1

b.

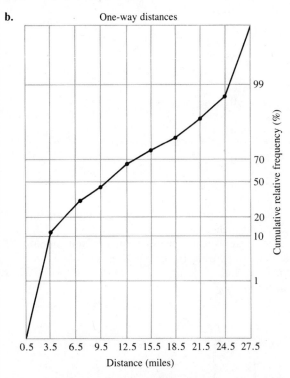

One-way distances

Cumulative relative frequency (%)

Distance (miles)

c. Yes. The line segments (disregarding the end segments) form a path that is nearly straight. This straight-line appearance is what indicates the near normality of the distribution.

d. $P_1 = 3.41$, $Q_1 = 5.7$, $\tilde{x} = 10.1$, $Q_3 = 14.7$, $P_9 = 19.7$

Chapter Exercises, page 87

2.61 a. The mean increases; the sum increases when one piece of data increases.
b. The median is unchanged; the value of the median is affected only by the middle value(s).
c. The mode is unchanged.
d. The midrange increases; an increase in the value of one of the extremes increases the sum, $H + L$.
e. The range increases; the difference between the lowest- and highest-valued data has increased.
f, g. The variance and the standard deviation have both increased, since the data are now more spread out.

2.63 a. 243.5   **b.** 256   **c.** 238.25

2.65 a. 11.8%   **b.** 11.9%   **c.** 1.9

2.68   Hours of TV Watched

0.	00000 00000 0
0.	55
1.	0000
1.	55
2.	0000
2.	55555
3.	0
3.	
4.	0
4.	
5.	0
5.	
6.	0

b. 1.45   **c.** 1   **d.** 0
e. 3
f. The mode represents the most common amount of television watched.
g. The mean includes the concept of total amount of television watched by the people in the sample.
h. 6   **i.** 2.41   **j.** 1.55

2.70   79.88, 12.4

2.72

x	f
3	1
4	2
5	3
6	4
7	5
8	5
9	6
10	7
11	8
12	9
13	6
14	2
sum	58

b. 58 litters
c. $\sum f$ is the number of litters, 58

2.74 a.

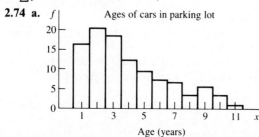

Ages of cars in parking lot

b. 4.0, 3, 2, 6, 3.75   **c.** 2, 5.5
d. 1, 1
e. 10, 6.7, 2.6

2.77   3000, 127,400

2.79 a.

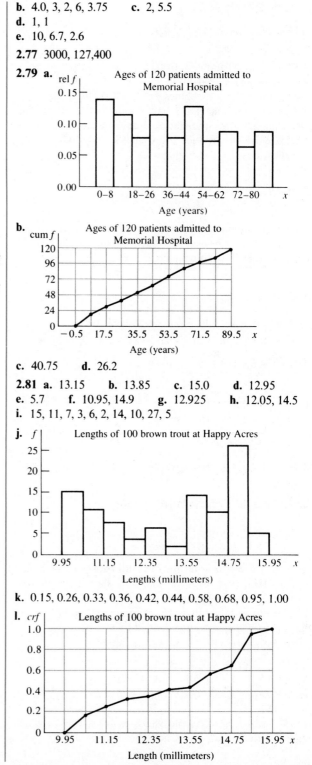

Ages of 120 patients admitted to Memorial Hospital

b.

Ages of 120 patients admitted to Memorial Hospital

c. 40.75   **d.** 26.2

2.81 a. 13.15   **b.** 13.85   **c.** 15.0   **d.** 12.95
e. 5.7   **f.** 10.95, 14.9   **g.** 12.925   **h.** 12.05, 14.5
i. 15, 11, 7, 3, 6, 2, 14, 10, 27, 5
j.

Lengths of 100 brown trout at Happy Acres

k. 0.15, 0.26, 0.33, 0.36, 0.42, 0.44, 0.58, 0.68, 0.95, 1.00
l.

Lengths of 100 brown trout at Happy Acres

m. 13.15   **n.** 1.91

2.85 98th, since there is approximately 97.5% of the distribution to the left of the z-score $+2$

2.91 The empirical rule states that 99.7% of a normal distribution lies within three standard deviations of the mean, that is, between $z = -3$ and $z = 3$.

2.92 The interval 22,500 to 37,500 represents the mean plus or minus three standard deviations.

a. If the distribution is normal, then approximately 99.7% of the distribution is contained within the interval.

b. If nothing is known about the shape of the distribution, we can be sure that at least 89% of the distribution is contained within the interval.

CHAPTER 3

Section 3.1, page 103

3.2

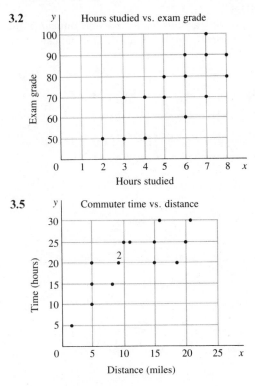

3.5

Section 3.2 page 111

3.7 Impossible. The correlation coefficient must be a value between -1.0 and $+1.0$.

3.9 b. 0.741

3.12 0.453

3.14 a. A moderate correlation might be interpreted to mean that there appears to be a slight linear pattern showing that communities with higher viewership rates tend to also have higher rates of reported rape.

b. No cause and effect is shown by this information. It says only that when one rate is higher, the other tends to be higher.

3.16 Numerator of formula (3-1):

$$\text{numerator} = \sum(x - \bar{x})(y - \bar{y})$$
$$= \sum(xy - \bar{x}y - x\bar{y} + \bar{x}\bar{y})$$
$$= \sum xy - \bar{x}\sum y - \bar{y}\sum x + n\bar{x}\bar{y}$$
$$= \sum xy - \left(\frac{\sum x}{n}\right)\sum y - \left(\frac{\sum y}{n}\right)\sum x$$
$$+ n\left(\frac{\sum x}{n}\right)\left(\frac{\sum y}{n}\right)$$
$$= \sum xy - \frac{\sum x \sum y}{n} = SS(xy)$$

Denominator of formula (3-1):

$$\text{denominator} = (n-1)s_x s_y$$
$$= (n-1)\sqrt{\frac{SS(x)}{n-1}}\sqrt{\frac{SS(y)}{n-1}} = \sqrt{SS(x)SS(y)}$$

Therefore, formula (3-1) is equivalent to formula (3-2).

Section 3.3, page 122

3.17

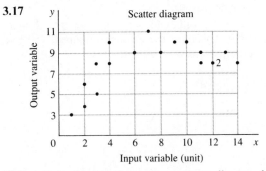

This scatter diagram does not suggest a linear relationship between the two variables. Therefore, we would not be justified in using techniques of linear regression. If we were to calculate the line of best fit, we would get an equation that would be worthless.

3.18 The vertical scale shown on Figure 3-21 is located at $x = 50$ and is therefore not the y-axis. The y-axis is the vertical line located at $x = 0$.

3.19 68,100

3.21

a.

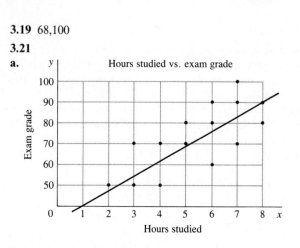

Estimation: $\hat{y} = 33 + 7x$

b. $\hat{y} = 3.96 + 0.625x$, which translates to $\hat{y} = 39.6 + 6.25x$

d. 77

e. $\hat{y} = 77$ is the average score expected for all those who studied for 6 hours.

3.23 a. 0.937   **b.** $\hat{y} = -0.21 + 0.043x$

Chapter Exercises, page 124

3.26 a, c, e. correlation   **b, d.** regression

3.27 a. The purpose of a correlation analysis is to determine whether or not two variables are linearly related. The product of correlation is the numerical value of r.

b. The purpose of regression analysis is to determine the equation of the line of best fit. The product of regression is the equation.

3.31 a. 2150.00, 3950.00

b. These extreme values of x are outside the domain of the study. Therefore the answers obtained, 350 and 30,500, respectively, are unrealistic.

3.34 a.

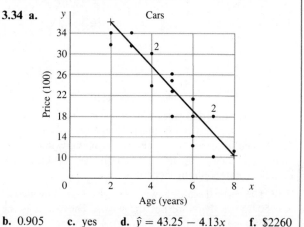

b. 0.905   **c.** yes   **d.** $\hat{y} = 43.25 - 4.13x$ **f.** \$2260

CHAPTER 4

Section 4.1, page 137

4.4 a. Relative frequency for $1 = 0.14$, $2 = 0.22$, $3 = 0.12$, $4 = 0.18$, $5 = 0.22$, $6 = 0.12$

b. Relative frequency for $H - 0.50$ and $T - 0.50$

Section 4.2, page 144

4.10 The 31.4 is the percentage of households tuned to "Family Ties" the week of the poll. This is an observed relative frequency, the proportion of the sampled households.

Section 4.3, page 148

4.11 Let J = jack, Q = queen, K = king, S = spade, D = diamond, H = heart, C = club

S = {JS, JD, JH, JC, QS, QD, QH, QC, KS, KD, KH, KC}

4.14 a. S = {HH, HT, TH, TT}

b. S = {HH, HT, TH, TT}

Note: The sample spaces listing the possible outcomes are identical. The listed sample spaces do not indicate relative probabilities.

4.16

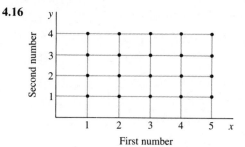

4.18 D = defective, N = nondefective

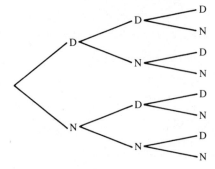

Section 4.4, page 151

4.19 $P(R) = x$, $P(Y) = x$, and $P(G) = 2x$

$x + x + 2x = 4x = 1$ and therefore $x = \frac{1}{4}$

$P(R) = 0.25$, $P(Y) = 0.25$, $P(G) = 0.50$

4.23 The three success ratings (highly successful, successful, and not successful) appear to be nonintersecting and their union appears to be the sample space. If this is true, none of the three sets of probabilities is appropriate.
a. A has a total probability of 1.2. The total must be exactly 1.0.
b. B has a negative probability of -0.1. All probability numbers are values between 0 and 1.
c. C has a total probability of 0.9. The total must be exactly 1.0.

4.25 a. 0.55   **b.** 0.40

Section 4.5, page 159

4.27 a, b, c. not mutually exclusive
d. mutually exclusive

4.29 a. 0.7   **b.** 0.6   **c.** 0.7   **d.** 0.0

4.31 No. Female students can be working students. Also, if the given probabilities are correct, there must be an intersection; otherwise the total probability is more than 1.0.

4.33 a. 0.375   **b.** 0.85   **c.** 0.35

Section 4.6, page 167

4.36 a. independent   **b.** not independent
c. independent   **d.** independent
e. not independent   **f.** not independent

4.39 a. 0.5   **b.** 0.666...   **c.** no

4.41 a, b. independent   **c.** dependent

4.43 a. $\frac{64}{343}$ **b.** $\frac{4}{35}$

4.46 a. 0.01   **b.** 0.22   **c.** 0.77

Section 4.7, page 175

4.49 a. 0.25   **b.** 0.2   **c.** 0.6   **d.** 0.8
e. 0.7   **f.** no   **g.** no

4.51 a. 0.0   **b.** 0.7   **c.** 0.6   **d.** 0.0   **e.** 0.5
f. no

4.53 a. $\frac{1}{2}$ **b.** $\frac{1}{4}$ **c.** $\frac{1}{8}$
b. $\frac{9}{16}, \frac{9}{32}, \frac{9}{64}$

4.56 a. 0.2   **b.** 0.7   **c.** 0.5

4.59 0.0085

Chapter Exercises, page 177

4.61 a, b. true

4.63 a. $\frac{89}{216}$ **b.** $\frac{111}{216}$ **c.** $\frac{64}{113}$

4.65 a, c, d. false,   **b.** true

4.68 0.080

4.70 a. 0.3168   **b.** 0.4659   **c.** no   **d.** no

4.74 a. $\frac{6}{12} = 0.5$ **b.** $\frac{1}{6} = 0.167$ **c.** $\frac{5}{12} = 0.417$

4.75 $P((\text{med or sh}) \text{ and } (\text{mod or sev})) =$
$(90 + 121 + 35 + 54)/1000 = 0.300$

4.77 a. 0.30   **b.** 0.60   **c.** 0.10   **d.** 0.60
e. 0.333   **f.** 0.25

4.80 0.56

4.84 p^6 **a.** 0.531   **b.** 0.262   **c.** 0.047

4.87 0.592

4.88 a. 0.80   **b.** 0.96   **c.** 0.64

4.90 0.061

CHAPTER 5

Section 5.1, page 189

5.1 number of children per family
$x = 0, 1, 2, 3, \ldots$

5.3 $x = n(\text{bull's-eye shots})$
$x = 0, 1, 2, 3, \ldots, 7, 8$

Section 5.2, page 195

5.5

x	0	1	2	3
$P(x)$	0.20	0.30	0.40	0.10

5.8

x	$P(x)$
1	$\frac{4}{10}$
2	$\frac{3}{10}$
3	$\frac{2}{10}$
4	$\frac{1}{10}$
sum	$\frac{10}{10} = 1.0$

It is a probability function: (a) Each $P(x)$ is a value between zero and one and (b) the sum of the $P(x)$s is one.

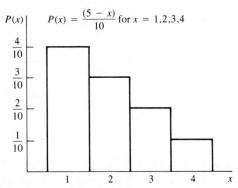

$P(x) = \dfrac{(5 - x)}{10}$ for $x = 1, 2, 3, 4$

5.10

x	$R(x)$
0	0.2
1	0.2
2	0.2
3	0.2
4	0.2
sum	1.0

It is a probability function: (a) Each $R(x)$ is a value between zero and one and (b) the sum of the $R(x)$s is one.

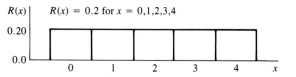

$R(x) = 0.2$ for $x = 0,1,2,3,4$

5.12 a. The percentages are based on observed data. The accidents are reported for 7,458 athletes, not all athletes. **b.** Using age as the variable and skiing as the identifying characteristic for the population of athletes, the percentage of injured for each age becomes the probability. Each injured person can belong to only one age class. **c.** One injured person could have multiple responses to the anatomic-site-of-injury variable. For example, one person could injure a knee, an ankle, and break a leg, all in one accident.

Section 5.3, page 199

5.13 2.0, 1.0

5.15 2.0, 1.1

5.18

a.

$P(x) = 0.1$ for $x = 0,1,2,3, \ldots ,9$

Random digits

b. 4.5, 2.87

5.19
$$\sigma^2 = \sum (x - \mu)^2 P(x)$$
$$\sum (x^2 - 2x + \mu^2) P(x)$$
$$\sum x^2 P(x) - 2\mu \sum x P(x) + \mu^2 \sum P(x)$$
$$\sum x^2 P(x) - 2\mu^2 + \mu^2$$
$$\sigma^2 = \sum x^2 P(x) - \mu^2 \quad \text{or} \quad \sum x^2 P(x) - \left\{ \sum x P(x) \right\}^2$$

Section 5.4, page 209

5.20 a. 24   **b.** 5,040   **c.** 1   **d.** 360

e. 10   **f.** 15   **g.** 0.0081   **h.** 35   **i.** 10
j. 1   **k.** 0.4096   **l.** 0.16807

5.21 a. x is not a binomial random variable because the trials are not independent. The probability of success (get an ace) changes from trial to trial. On the first trial it is $\frac{4}{52}$. The probability of an ace on the second trial can be either $\frac{4}{51}$ or $\frac{3}{51}$, depending on whether or not the first trial was a success. The probability of success on the third and fourth trials depends on what has occurred on the previous trials. The probability of success does not remain constant throughout the experiment because we do not have independence of trials.
b. x is a binomial random variable because the trials are independent. The probability of success on any one trial will be $\frac{1}{13}$, since each card that is drawn is replaced before the next drawing. This way the probabilities remain constant trial after trial.

5.23 a.

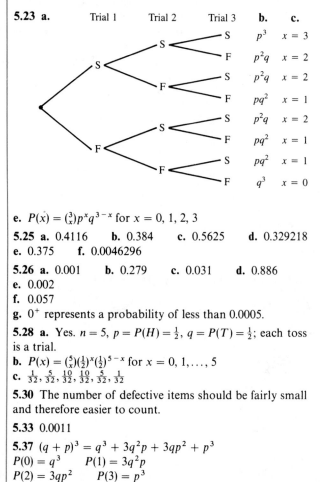

e. $P(x) = \binom{3}{x} p^x q^{3-x}$ for $x = 0, 1, 2, 3$

5.25 a. 0.4116   **b.** 0.384   **c.** 0.5625   **d.** 0.329218
e. 0.375   **f.** 0.0046296

5.26 a. 0.001   **b.** 0.279   **c.** 0.031   **d.** 0.886
e. 0.002
f. 0.057
g. 0^+ represents a probability of less than 0.0005.

5.28 a. Yes. $n = 5$, $p = P(H) = \frac{1}{2}$, $q = P(T) = \frac{1}{2}$; each toss is a trial.
b. $P(x) = \binom{5}{x}(\frac{1}{2})^x(\frac{1}{2})^{5-x}$ for $x = 0, 1, \ldots, 5$
c. $\frac{1}{32}, \frac{5}{32}, \frac{10}{32}, \frac{10}{32}, \frac{5}{32}, \frac{1}{32}$

5.30 The number of defective items should be fairly small and therefore easier to count.

5.33 0.0011

5.37 $(q + p)^3 = q^3 + 3q^2 p + 3qp^2 + p^3$
$P(0) = q^3$ $P(1) = 3q^2 p$
$P(2) = 3qp^2$ $P(3) = p^3$

Section 5.5, page 213

5.38 a. and b. 2.5, 1.2

5.39 a. 7.7, 2.7   **b.** 24.0, 4.7   **c.** 44.0, 2.3
d. 2.5, 1.6

5.40 400, 0.5

Chapter Exercises, page 214

5.43 i. Each probability, $P(x)$, has a value between zero and one.
ii. The sum of the individual probabilities is exactly one.

5.45 a. yes   **b.** no   **c.** no

5.47 0.95   **b.** 0.6

5.51 2600

5.53 1 branch has 0 successes, 4 branches have 1 success, 6 branches have 2 successes, 4 branches have 3 successes, 1 branch has 4 successes.

5.54 $n = 4$, $x = n$(males), $p = \frac{2}{3}$, $q = \frac{1}{3}$
$P(x) = \binom{4}{x}(\frac{2}{3})^x(\frac{1}{3})^{4-x}$ for $x = 0, 1, 2, 3, 4$

x	0	1	2	3	4
$P(x)$	0.012	0.099	0.296	0.395	0.198

5.57 0.1035

5.60 0.930

5.62 0.668

5.64 a. 0.930   **b.** 0.264

5.68 two: 0.997; four: 1.000

5.70 Let success be the event that an item is defective on a given draw. The probability of success on the first draw is $\frac{3}{10}$, or 0.3. The probability of success on the second draw is dependent on the results of the first draw. If the first was defective, then the probability is $\frac{2}{9}$. If the first is not defective, then the probability is $\frac{3}{9}$. Since the trials are not independent, x is not a binomial random variable.

5.71 a. 0.914   **b.** 0.625

5.74 a. 28
b. 12 23 34 45 56 67 78
    13 24 35 46 57 68
    14 25 36 47 58
    15 26 37 48
    16 27 38
    17 28
    18
c. $\frac{1}{28}$

d.

x	$P(x)$
3	$\frac{1}{28}$
4	$\frac{1}{28}$
5	$\frac{2}{28}$
6	$\frac{2}{28}$
7	$\frac{3}{28}$
8	$\frac{3}{28}$
9	$\frac{4}{28}$
10	$\frac{3}{28}$
11	$\frac{3}{28}$
12	$\frac{2}{28}$
13	$\frac{2}{28}$
14	$\frac{1}{28}$
15	$\frac{1}{28}$

9.0, 3.0

5.76
$$\mu = \sum xP(x)$$
$$= (1)\left(\frac{1}{n}\right) + (2)\left(\frac{1}{n}\right) + \cdots + (n)\left(\frac{1}{n}\right)$$
$$= \left(\frac{1}{n}\right)(1 + 2 + \cdots + n)$$
$$= \left(\frac{1}{n}\right)\left(\frac{(n)(n+1)}{2}\right) = \frac{n+1}{2}$$

CHAPTER 6

Section 6.2, page 228

6.1 A bell-shaped distribution with a mean of zero and a standard deviation of one

6.2 a. 0.4032   **b.** 0.3997   **c.** 0.4993   **d.** 0.4761

6.3 a. 0.4821   **b.** 0.4949   **c.** 0.3849   **d.** 0.4418

6.4 a. 0.4394   **b.** 0.0606   **c.** 0.9394   **d.** 0.8788

6.7 a. 0.5000   **b.** 0.1469   **c.** 0.9893   **d.** 0.9452
e. 0.0548

6.10 a. 1.14   **b.** 0.47   **c.** 1.66   **d.** 0.86
e. 1.74   **f.** 2.23

6.12 a. 1.65   **b.** 1.96   **c.** 2.33

6.15 a. +0.84   **b.** +1.04

6.18 a. 1.28   **b.** 1.65   **c.** 2.33

Section 6.3, page 234

6.19 a. 0.5000   **b.** 0.3849   **c.** 0.6072   **d.** 0.2946
e. 0.9502   **f.** 0.0139

6.23 a. 89.6   **b.** 79.2   **c.** 57.3

6.26 a. 0.1131   **b.** 0.0505   **c.** 4.64

6.29 0.12

Section 6.4, page 239

6.32 a. $z(0.03)$ **b.** $z(0.14)$ **c.** $z(0.75)$
d. $z(0.13)$ **e.** $z(0.91)$ **f.** $z(0.82)$
6.35 a. 1.28, 1.65, 1.96, 2.05, 2.33, 2.58
b. $-2.58, -2.33, -2.05, -1.96, -1.65, -1.28$
6.36 a. 0.925   **b.** 3.61

Section 6.5, page 245

6.37 a, b. not appropriate   **c, d.** appropriate
6.39 0.1822, 0.177
6.41 0.9429, 0.943
6.43 0.8078
6.46 0.0009

Chapter Exercises, page 246

6.50 a.

x	f	x	f	x	f
-7	4	3	48	13	9
-6	2	4	52	14	1
-5	2	5	62	15	2
-4	4	6	46	16	6
-3	2	7	40	17	2
-2	9	8	34	18	2
-1	18	9	22	19	0
0	24	10	22	20	. 2
1	39	11	9	21	1
2	55	12	13		

b. 74.8%, 90.6%, 94.9%
c. Empirical rule: 68%, 95%, 99.7%
These data have: 74.8%, 90.6%, 94.9% (not very close)
d. 44.2% compared to 43.32%
e. The histogram looks normal, but not all of the percentages compare very closely.
6.52 a. 1.26   **b.** 2.16   **c.** 1.13
6.55 a. 0.0930   **b.** 0.9684
6.57 a. 1.175 or 1.18   **b.** 0.58   **c.** -1.04
d. -2.33
6.59 18.8
6.61 10.9
6.64 10.033
6.67 a. The normal approximation is reasonable since both $np = 7.5$ and $nq = 17.5$ are greater than 5.

b. 7.5, 2.29
6.69 a. 0.001   **b.** 1.000   **c.** 0.995   **d.** 0.0668
6.71 a. 0.1170   **b.** 0.7824   **c.** 0.0008
6.73

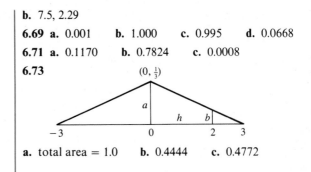

a. total area $= 1.0$ **b.** 0.4444   **c.** 0.4772

CHAPTER 7

Section 7.1, page 259

7.1 a. A sampling distribution of sample means is the distribution formed by the means from all possible samples of a fixed size that can be taken from a population.
b. It is one element in the distribution of the means of all samples of size 3.
c. There were 25 equally likely samples. Therefore, each has a probability of $\frac{1}{25}$, or 0.04.
7.2 a. 11   31   51   71   91
      13   33   53   73   93
      15   35   55   75   95
      17   37   57   77   97
      19   39   59   79   99

b.

$\bar{x}$	$P(\bar{x})$
1	0.04
2	0.08
3	0.12
4	0.16
5	0.20
6	0.16
7	0.12
8	0.08
9	0.04

c.

R	0	2	4	6	8
$P(R)$	0.20	0.32	0.24	0.16	0.08

7.5 $P(x)$ Binomial probability distribution
$n = 4$ and $p = 0.1$

7.7 a. Each boss was asked to report the average number of hours per day he spends on the phone.

b.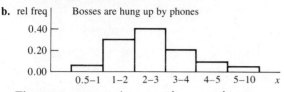

c. The average reported may not be a sample mean.

Section 7.2, page 267

7.8 a. one   **b.** $\sigma_{\bar{x}} = \sigma/\sqrt{n}$

7.10 6.25, 5.00, 3.54, 2.50

Section 7.3, page 272

7.14 a. 0.9974   **b.** 0.8849   **c.** 0.9282

7.15 0.7888

7.17 a. 0.3830   **b.** 0.9938   **c.** 0.3085   **d.** 0.0031

7.20 0.0009

Chapter Exercises, page 275

7.23 a. 0.9544   **b.** 0.9974

7.24 a. 0.6826   **b.** 0.9544   **c.** 0.9974

7.25 a. 0.6826   **b.** 0.9544   **c.** 0.3830

7.28 a. Individual score x: the distribution of xs is normal, with a mean of 720 and a standard deviation of 60.
b. Mean scores $\bar{x}$: The distribution of $\bar{x}$s is normal, with a mean of 720 and a standard deviation of $6(60/\sqrt{100})$.
c. 0.5359   **d.** 0.8238

7.30 0.49

7.33 0.0004

7.35 a. 0.9759   **b.** 69 to 75

7.38 0.0000+. The chances are virtually zero; the manufacturer's figures of 35,000 and 5,000 should be in doubt.

CHAPTER 8

Section 8.1, page 291

8.1 a. Type A correct decision: The accused is indeed innocent and is acquitted. Type I error: The accused is in fact innocent but is convicted. Type II error: The accused is actually guilty but is acquitted. Type B correct decision: The accused is guilty and is convicted.
b. no   **c.** no

8.4 A type I error occurs if the company concludes that the additive increases average coverage when in fact it does not. A type II error occurs if the company concludes that the additive does not increase the coverage when in fact it does.

8.6 We are willing to allow the type I error to occur with a probability of (a) 0.001, (b) 0.05, (c) 0.10.

8.9 The null hypothesis states that the teaching techniques have no effect (change is equal to zero), whereas the alternative (what they want to show) is the statement that the teaching techniques do have an effect on student's exam scores. They are looking for improvement, therefore using a one-tailed test.

8.13 a. The critical region is the set of all values of the test statistic that will cause us to reject H_0.
b. The critical value(s) is the value that forms the boundary between the critical region and the noncritical region. The critical value is in the critical region.

8.14 Because α and β are interconnected. If we reduce α, then β will become larger.

Section 8.2, page 302

8.16 a. $H_0: \mu = 26$ yrs $(<, =)$; $H_a: \mu > 26$
c. $H_0: \mu = 1600$ hrs $(>, =)$; $H_a: \mu < 1600$
e. $H_0: \mu = 4.7$ mi $(>, =)$; $H_a: \mu < 4.7$
g. $H_0: \mu = 20$ deg $(>, =)$; $H_a: \mu < 20$

8.18

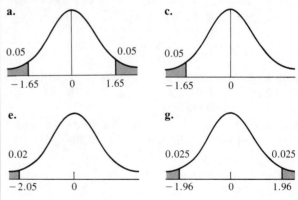

8.19 a. 3   **b.** yes

8.21 20.825

8.22 $H_a: \mu > 4.9$; $z^* = 1.33$; fail to reject H_0

8.25 $H_a: \mu < 1600$; $z^* = -2.51$; reject H_0

8.28 a. $H_a: r > a$. Failure to reject H_0 will result in the drug being marketed. Because of the high current mortality rate, burden of proof is on the old, ineffective drug.

b. $H_a: r < a$. Failure to reject H_0 will result in the new drug not being marketed. Because of the low current mortality rate, burden of proof should be on the new drug.

Section 8.3, page 309

8.29 a. 0.0694   **c.** 0.2420   **e.** 0.3524

8.30 a. 1.57   **b.** -2.13 **c.** $+2.87$ or -2.87

8.31 a. fail to reject H_0 **b.** reject H_0

8.33 $H_a: \mu > 21.5; P = P(z > 1.60) = 0.0548$

Section 8.4, page 318

8.35 width $= (\bar{x} - z(\alpha/2)(\sigma/\sqrt{n})) - (\bar{x} - z(\alpha/2)(\sigma/\sqrt{n}))$
$= 2z(\alpha/2)\sigma/\sqrt{n}$

a. The higher the level of confidence, the larger the value of $z(\alpha/2)$, and the greater the width. Therefore, as level of confidence increases, the width of the confidence interval increases.
b. The larger the sample size n is, the smaller the standard error gets, and the smaller the width. Therefore, as the sample size n increases, the width of the confidence interval decreases.
c. The more variable the characteristic measured is, the larger the standard deviation will be, the larger the standard error will be, and the greater the width. Therefore, as the standard deviation increases, the width of the confidence interval increases.

8.37 a. 7280.00   **b.** 6888 to 7672   **c.** 6764 to 7796

8.39 a. 25.3   **b.** 24.29 to 26.31   **c.** 23.97 to 26.63

8.41 959

Chapter Exercises, page 320

8.45 a. $H_0: \mu = 100$ **b.** $H_a: \mu \neq 100$ **c.** 0.01
d. ± 2.58 **e.** 100   **f.** 96   **g.** 12   **h.** 1.7
i. -2.35 **j.** fail to reject H_0

8.46 a. 32.0   **b.** 2.4   **c.** 64   **d.** 0.90   **e.** 1.65
f. 0.3   **g.** 0.495   **h.** 32.495   **i.** 31.505

8.48 a. 22.61   **b.** 22.45, no

8.51 a. $H_a: \mu > 9; z^* = 3.14$; reject H_0
b. $H_a: \mu > 9; P = P(z > 3.14) = 0.0008$

8.55 The level of confidence determines the $z(\alpha/2)$ to be used. z is the number of multiples of the standard error used to determine the maximum error.

8.58 a. 0.86 to 2.18   **b.** 1.00 to 2.04
c. larger n causes narrower interval

8.60 a. 92.44 to 97.56   **b.** 93.72 to 96.28
c. 93.72 to 96.28

8.62 a. 124.85 to 159.15   **b.** 2,497,000 to 3,183,000

8.64 60

CHAPTER 9

Section 9.1, page 334

9.1 a. 1.71   **b.** 1.37   **c.** 2.60   **d.** 2.08
e. -1.72 **f.** -2.06 **g.** -2.47 **h.** 1.96

9.2 a. 1.73   **b.** $-3.18, 3.18$ **c.** -2.55
d. 1.33   **e.** $-1.89, 1.89$

9.3 7

9.5 a. 0.89   **b.** 0.945

9.7 a. symmetric about mean; mean equals zero
b. Standard deviation of t distribution is larger than one; t distribution is different for each different sample size while there is only one z distribution.

9.9 a. $H_a: \mu < 25$ (less than); $t^* = -3.50$; reject H_0
b. $H_a: \mu < 25$;
$P = P(t < -3.50, \text{with } df = 35) = P(z < -3.50)$
$= 0.0002$
reject H_0

9.12 11.57 to 13.63

9.14 79.15 to 92.85

Section 9.2, page 342

9.16 a. 0.090   **b.** 0.029   **c.** 0.007

9.18 a. correctly fail to reject H_0 **b.** 0.0143
c. commit a type II error   **d.** 0.0384

9.19 a. $H_a: p < 0.90$ (less than); $z^* = -4.77$; reject H_0
b. $H_a: p < 0.90; P = P(z < -4.77) < 0.000003$

9.23 a. 0.162, 0.438   **c.** 0.24, 0.76   **e.** 0.474, 0.526
g. as n increases, interval becomes narrower

9.25 a. 0.019   **b.** 0.078   **c.** 0.139

9.27 a. 0.21   **b.** 0.043

9.30 522

Section 9.3, page 351

9.33 a. 34.8   **b.** 28.9   **c.** 13.4   **d.** 48.3
e. 12.3   **f.** 3.25   **g.** 37.7   **h.** 10.9

9.34 a. 30.1   **c.** 7.56   **e.** 11.6, 32.7   **g.** 1.24

9.37 0.94

9.38 a. $H_a: \sigma > 0.25$ (increase); $\chi^{2*} = 37.24$; reject H_0
b. $H_a: \sigma > 0.25$ (increase);
$P = P(\chi^2$ with $df = 19 > 37.24)$; $0.005 < P < 0.010$

9.41 30.2 to 80.8

Chapter Exercises, page 354

9.43 a. $H_a: \mu \neq 80$ (different); $t^* = -4.47$; reject H_0
b. $H_a: \mu \neq 80$; $P = 2P(t > 4.47) < 0.001$; reject H_0

9.46 a. 11.49, 0.47   **b.** 11.23 to 11.75

9.48 a. $0.025 < P < 0.05$ **b.** $0.025 < P < 0.05$
c. $0.05 < P < 0.10$ **d.** $0.05 < P < 0.10$

9.51 a. 8.782, 0.71   **b.** 8.782   **c.** 8.643 to 8.921
d. 0.71   **e.** 0.622 to 0.825

9.54 0.0261

9.57 0.533 to 0.579

9.60 a. 0.126 to 0.340
b. The dealer has overestimated the percentage of satisfied customers; it appears to be less than 90%.

9.63 9604

9.67 $s^2 > 25$

9.69 a. 3.44, 0.653
b. $H_a: \mu < 4.9$; $t^* = -7.75$; reject H_0
c. $H_a: \sigma^2 > 0.25$ (more than); $\chi^{2*} = 18.8$; fail to reject H_0

9.72 Formula (9-5): $E = z(\alpha/2)\sqrt{\dfrac{pq}{n}}$

$$E = \frac{z(\alpha/2)\sqrt{pq}}{\sqrt{n}}$$

$$E\sqrt{n} = z(\alpha/2)\sqrt{pq}$$

$$\sqrt{n} = \frac{z(\alpha/2)\sqrt{pq}}{E}$$

$$n = \left\{ \frac{z(\alpha/2)\sqrt{pq}}{E} \right\}^2$$

Formula (9-6): $n = \dfrac{\{z(\alpha/2)\}^2 pq}{E^2}$

CHAPTER 10

Section 10.1, page 366

10.1 dependent samples

10.4 independent samples

Section 10.2, page 372

10.7 a. 0.5028   **b.** 0.3830

10.8 a. 0.0322   **b.** 0.0198   **c.** 0.0029

10.10 a. $H_a: \mu_I - \mu_{II} \neq 0$; $z^* = -1.31$; fail to reject H_0
b. $H_a: \mu_A - \mu_B \neq 0$; $P = P(|z| > 1.31) = 0.1902$

10.13 a. $H_a: \mu_c - \mu_e = <0$; $z^* = -10.0$; reject H_0
b. $H_a: \mu_c - \mu_e = <0$;
$P = P(z < -10.00) < 0.000001$; reject H_0

10.15 0.6 to 1.8

Section 10.3, page 380

10.17 a. $F(15, 17, 0.99)$ and $F(15, 17, 0.01)$
c. $F(7, 19, 0.05)$ **e.** $F(24, 34, 0.95)$

10.18 a. 2.51   **b.** 2.20   **c.** 2.91   **d.** 4.10
e. 2.67   **f.** 3.77   **g.** 1.79   **h.** 2.99

10.22 a. $H_a: \sigma_k^2 \neq \sigma_m^2$; $F^* = \dfrac{3.2}{2.4} = 1.33$;

fail to reject H_0
b. $P(F > 1.33) > 0.10$

10.24 0.47 to 2.41

Section 10.4, page 388

10.27 a. Formula (10-11) is used when it is assumed that $\sigma_1 = \sigma_2$, and formula (10-12) is used when it is assumed that the standard deviations are not equal. Notice that a different formula is used to estimate the standard error.
b. The critical region becomes larger in the case where it is assumed that the standard deviations are not equal.
c. The number of degrees of freedom is smaller in the case where it is assumed that the standard deviations are not equal.

10.28 1st: $H_a: \sigma_R^2 \neq \sigma_S^2$; $F^* = 1.54$;
fail to reject H_0, therefore assume $\sigma_R^2 = \sigma_S^2$ and use Case I methods.
2nd: $H_a: \mu_R - \mu_S > 0$; $t^* = 3.348$; reject H_0

10.31 $P = 0.394$ means that if the null hypothesis is rejected, there is a high probability of committing a type I error.

10.32 1st: $F^* = 2.0$; therefore assume $\sigma_A = \sigma_B$ and use Case 1 methods
2nd: $H_a: \mu_s - \mu_n \neq 0$; $t^* = -50.00$; reject H_0
b. $P = P(|t| > 50.00) = 0.0+$

10.34 a. $H_a: \sigma_1^2 \neq \sigma_2^2$; $F^* = 12.8$; reject H_0
b. $H_a: \mu_2 - \mu_1 > 0$; $t^* = 2.84$; reject H_0

10.37 1.90 to 8.35

10.40 $s_p\sqrt{\dfrac{1}{n_1}+\dfrac{1}{n_2}} = \sqrt{\dfrac{(n-1)s_1^2 + (n-1)s_2^2}{n+n-2}}\sqrt{\dfrac{1}{n}+\dfrac{1}{n}}$

$\qquad = \sqrt{\dfrac{(n-1)(s_1^2 + s_2^2)}{2(n-1)}}\sqrt{\dfrac{2}{n}}$

$\qquad = \sqrt{\dfrac{2(n-1)(s_1^2 + s_2^2)}{2(n-1)(n)}}$

$\qquad = \sqrt{\dfrac{s_1^2 + s_2^2}{n}}$

Section 10.5, page 395

10.42 a. H_a: $\mu_d > 0$ (beneficial); $t^* = 3.06$; reject H_0
b. H_a: $\mu_d > 0$ (beneficial); $P = P(t > 3.06) = 0.0011$
10.45 -1.35 to 8.85

Section 10.6, page 405

10.47 a. H_a: $p_A - p_B \neq 0$; $z^* = -1.09$; fail to reject H_0
b. $P = P(|z| > 1.09) = 0.2758$
10.49 a. H_a: $p_c - p_m \neq 0$; $z^* = -1.42$; fail to reject H_0
b. $P = P(|z| > 1.42) = 0.1556$; fail to reject H_0
10.51 a. 0.0405
b. 0.0506; Table C shows 6, the calculated value rounded up.
c. 14%—anything 6% or over is significant
d. 2%—anything less than 6% is not significant
10.52 -0.091 to 0.291

Chapter Exercises, page 407

10.55 a. H_a: $\mu_t - \mu_1 \neq 0$; $z^* = -1.80$; fail to reject H_0
10.58 0.0524
10.59 1.02 to 4.38
10.60 H_a: $\sigma_2^2 \neq \sigma_1^2$; $F^* = 1.65$; fail to reject H_0
10.63 a. 0.6944   **b.** 0.2846 to 1.8749
c. 0.533 to 1.369
10.65 a. 1st: H_a: $\sigma_R^2 \neq \sigma_P^2$; $F^* = 1.57$; fail to reject H_0
2nd: H_a: $\mu_R - \mu_P \neq 0$; $t^* = 4.41$; reject H_0
b. $P = P(|t| > 4.68$, with $df = 13)$; $P < 0.01$
10.66 a. H_a: $\sigma_A^2 \neq \sigma_B^2$; $F^* = 2.15$; reject H_0; therefore assume $\sigma_L^2 \neq \sigma_H^2$ and use Case 2 methods
b. H_a: $\mu_A - \mu_B > 0$; $t^* = 6.02$; reject H_0
c. 2.18 to 3.82

10.70 1st: $F^* = \dfrac{6.38}{4.02} = 1.59$; therefore assume $\sigma_1 = \sigma_2$ and use Case 1 methods

a. H_a: $\mu_m - \mu_f > 0$ (males larger); $t^* = 8.4$; reject H_0
10.74 a. H_a: $\mu_d \neq 0$ (is diff.); $t^* = 2.26$; reject H_0
b. $P = P(|t| > 2.26$, with $df = 11)$; $0.02 < P < 0.05$
10.76 a. H_a: $\mu_d > 0$ (improvement); $t^* = 3.82$; reject H_0
b. $P = P(t > 3.82$, with $df = 9) < 0.005$
10.78 -8.85 to 16.02
10.80 a. H_a: $p_A - p_B \neq 0$; $z^* = -1.71$; reject H_0
b. $P = P(|z| > 1.71) = 0.0872$
10.83 -0.116 to 0.216
b. No; the interval estimate contains the value zero.

CHAPTER 11

Section 11.2, page 429

11.1 a. $P(A) = P(B) = P(C) = P(D) = P(E) = \frac{1}{5}$
b. χ^2
c. (i) H_0: equal preference; $\chi^{2*} = 4.40$; fail to reject H_0
(ii) $P = P(\chi^2 > 4.40$, with $df = 4)$; $0.10 < P < 0.50$
11.4 a. H_0: this year's distribution is the same as the previous year's; $\chi^{2*} = 7.66$; fail to reject H_0
b. $P = P(\chi^2 > 7.66$, with $df = 4)$; $0.10 < P < 0.50$
11.6 The student leaders were able to give multiple answers and therefore the percentages reported total more than 100%. The multinomial experiment requires exactly one answer from each student.

Section 11.3, page 439

11.8 H_0: the number of defective items is independent of the day of the week; $\chi^{2*} = 8.548$; fail to reject H_0
11.11 a. The information compares several distributions, a distribution for each region of New York State.
b. A test of homogeneity compares several distributions.
11.12 a. H_0: the distribution of reactions is the same for both groups. $\chi^{2*} = 22.56$; reject H_0
b. $P = P(\chi^2 > 22.56$, with $df = 2)$; $P < 0.005$
11.15 a. H_0: fear darkness and do not fear darkness are proportioned the same for each age group; $\chi^{2*} = 80.957$; reject H_0
b. $P = P(\chi^2 > 80.96$, with $df = 4)$; $P < 0.005$

Chapter Exercises, page 442

11.16 a. H_0: distribution is 10, 20, 40, 20, 10 percent; $\chi^{2*} = 8.637$; fail to reject H_0
b. $P = P(\chi^2 > 8.64$, with $df = 4)$; $0.05 < P < 0.10$

11.19 H_0: the weights are normally distributed with a mean of 160 pounds and a standard deviation of 15 pounds; $P = P(\chi^{2*} > 5.812$, with $df = 5) > 0.10$; fail to reject H_0

11.22 H_0: independence; $P = P(\chi^2 > 23.339) < 0.005$; reject H_0

11.24 a. H_0: proportion of popcorn that popped is the same for all brands; $\chi^{2*} = 2.839$; fail to reject H_0
b. $P = P(\chi^2 > 2.839$, with $df = 3)$; $0.10 < P < 0.50$

11.27 a. H_0: rate of absenteeism is the same for all groups; $\chi^{2*} = 27.2369$; reject H_0

11.29 a. $z^* = 0.7918$ **b.** $\chi^{2*} = 0.6268$

CHAPTER 12

Section 12.3, page 464

12.1 a. The mean levels of the test factor are not all the same.
b. The mean levels of the test factor are all the same.
c. The mean levels of the test factor are all the same.

12.2 df(factor) appears first in the critical-number notation since MS(factor) is the numerator for the calculated value of the test statistic F.

12.3 a. 0   **b.** 3   **c.** 16   **d.** 60   **e.** 1232

12.5 a. H_0: The mean values for each of the levels of the tested factor are all equal.
H_a: The mean values for each of the levels of the tested factor are not all equal. That is, at least one is different in value from the rest.
b. We would conclude that the alternative hypothesis is correct.
c. We would conclude that the evidence found was not sufficient to contradict the null hypothesis.
d. The decision is made using an F test by comparing the calculated F^* with the critical value of F obtained from Table 8.

12.9 H_0: The mean values for workers are all equal.

Source	SS	df	MS
Workers	17.73	2	8.87
Error	25.20	12	2.10
Total	42.93	14	

$F^* = 4.22$

Reject H_0

Chapter Exercises, page 467

12.11 H_0: The mean typing speeds on both types of typewriters are equal.

Source	SS	df	MS
Typewriter	280.333	1	280.333
Error	806.333	10	80.633
Total	1086.666	11	

$F^* = 3.477$

Fail to reject H_0

12.14 H_0: There is no difference in the yield using the three concentrations.

Source	SS	df	MS
Concent.	691.60	2	345.80
Error	114.40	12	9.53
Total	806.00	14	

$F^* = 36.27$

Reject H_0

12.17 H_0: The mean stopping distance is not affected by the brand of tire.

Source	SS	df	MS
Brand	95.359	3	31.7865
Error	126.467	19	6.6562
Total	221.826	22	

$F^* = 4.78$

Reject H_0

12.20 H_0: The mean amounts dispensed by the machines are all equal.

Source	SS	df	MS
Machine	20.998	4	5.2495
Error	2.158	13	0.166
Total	23.156	17	

$F^* = 31.6$

Reject H_0

CHAPTER 13

Section 13.1, page 479

13.1 Refer to the definition and development of the measures of central tendency, Section 2.3, and the measures of dispersion, Section 2.4.

13.3 a. y

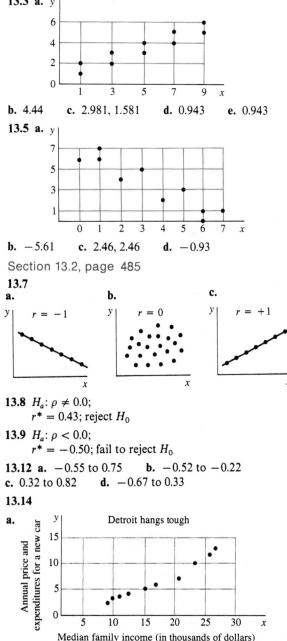

b. 4.44   **c.** 2.981, 1.581   **d.** 0.943   **e.** 0.943

13.5 a. y

b. −5.61   **c.** 2.46, 2.46   **d.** −0.93

Section 13.2, page 485

13.7
a.   **b.**   **c.**

13.8 $H_a: \rho \neq 0.0$;
$r^* = 0.43$; reject H_0

13.9 $H_a: \rho < 0.0$;
$r^* = -0.50$; fail to reject H_0

13.12 a. −0.55 to 0.75   **b.** −0.52 to −0.22
c. 0.32 to 0.82   **d.** −0.67 to 0.33

13.14

a.

Detroit hangs tough

b. 0.9   Yes, these variables seem to be linearly correlated. The two graphs shown in Case Study 13-2 have almost identical patterns, and the scatter diagram drawn in **a.** shows the ordered pairs following a strong upward pattern.

Section 13.3, page 491

13.18 $\hat{y} = 16.34 + 0.4245x$, 21.088

13.21

a. y

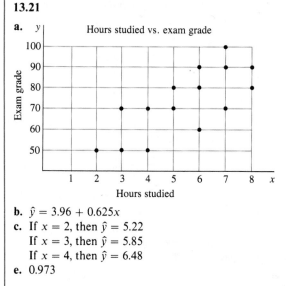

Hours studied vs. exam grade

b. $\hat{y} = 3.96 + 0.625x$
c. If $x = 2$, then $\hat{y} = 5.22$
If $x = 3$, then $\hat{y} = 5.85$
If $x = 4$, then $\hat{y} = 6.48$
e. 0.973

Section 13.4, page 495

13.23

a. y

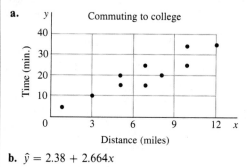

Commuting to college

b. $\hat{y} = 2.38 + 2.664x$
c. $H_a: \beta_1 > 0.0$; $t^* = 6.54$; reject H_0
d. 1.48 to 3.84

Section 13.5, page 504

13.25 a. 15.4 to 19.1   **b.** 11.6 to 22.8

13.27 The standard deviation for $\bar{x}$s is much smaller than for individual xs (central limit theorem). Thus, in accordance with this, the confidence interval will be narrower.

13.28 Yes. $s\sqrt{1/n} = s/\sqrt{n}$, and that is the estimate for the standard error of the mean.

Chapter Exercises, page 506

13.29 a. always   **b.** never   **c.** sometimes
d. sometimes   **e.** always

13.32

a.

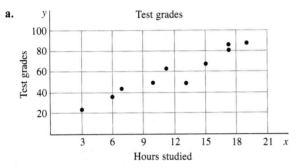

b. 0.961
c. $H_a: \rho > 0.0$; $r^* = 0.961$; reject H_0 **d.** 0.85 to 0.98
13.35 a. -0.912 **b.** 0.912
c. They are numerically equal but of opposite sign.

13.38 a.

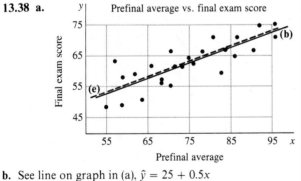

b. See line on graph in (a), $\hat{y} = 25 + 0.5x$
d. $\hat{y} = 25.82 + 0.492x$ **f.** 0.82
g. $H_a: \rho \neq 0.0$; $r^* = 0.824$; reject H_0
h. 0.65 to 0.92   **i.** 4.26   **j.** 0.346 to 0.638
k. $H_a: \beta_1 > 0.0$; $t^* = 6.97$; reject H_0
l. 65.30 to 69.96   **m.** 55.18 to 73.20

13.41

x	y	$\hat{y}$	$y - \bar{y}$	$(y - \bar{y})^2$	$y - \hat{y}$	$(y - \hat{y})^2$	$\hat{y} - \bar{y}$	$(\hat{y} - \bar{y})^2$
0	1	1.5	1.0	1.00	-0.5	0.25	-0.5	0.25
1	3	2	1.0	1.00	1.0	1.00	0.0	0.00
2	2	2.5	0.0	0.00	-0.5	0.25	0.5	0.25
Sum				2.00		1.50		0.50

Therefore, $\sum(y - \bar{y})^2 = \sum(y - \hat{y})^2 + \sum(\hat{y} - \bar{y})^2$

CHAPTER 14

Section 14.2, page 522

14.1 a. $H_0: M = 48$
b. $H_0: P(+) = 0.5$ ($+$ stands for above 48)
c. $H_a: M \neq 48$; $x = 3$; reject H_0
14.3 H_0: Equal preference for whole wheat and white crust
H_a: whole wheat crust is preferred
$(+)$ = whole wheat preferred; $x' = 64.5$
$z^* = 1.01$; fail to reject H_0
14.5 39 to 47

Section 14.3, page 529

14.8 H_0: no difference in the reaction times; $U = 5$; reject H_0
14.11 H_0: no difference between the average systolic blood pressure of men of age 40 and men of age 25; $U = 480$; $z^* = 2.09$; reject H_0

Section 14.4, page 533

14.13 H_0: the hiring sequence is random; $V = 9$; fail to reject H_0
14.17 H_0: randomness in number of absences (about median); $V = 9$; fail to reject H_0

Section 14.5, page 539

14.19 a. $r_s = 0.87$ **b.** $H_a: \rho_s > 0$; reject H_0
14.22 a.

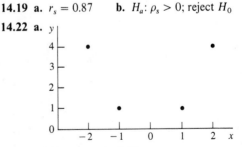

b. $r_s = 0.10$ **c.** $r = 0.00$
d. The two results are not identical, but both coefficients are near or at zero. The rank correlation measures the correlation between the rank numbers, whereas the Pearson coefficient uses the numerical values.

Chapter Exercises, page 543

14.24 a. H_0: no difference between average times (no faster)

H_a: average time on B is less than on A (B is faster)
$x = 2$
b. Fail to reject H_0; the evidence is not sufficient to justify the claim that track B is faster.

14.27 H_0: no difference
H_a: top ripens first; $(+) =$ if top ripens first $x = 4$; reject H_0

14.29 H_0: no effect
H_a: computer-assisted instruction produced higher achievement $U = 47.5$; fail to reject H_0

14.31 H_0: random order
H_a: lack of randomness $V = 9$; fail to reject H_0

14.34 H_a: $\rho_s \neq 0$ (correlated); $r_s = 0.39$; fail to reject H_0

Answers to Chapter Quizzes

Only the replacement for the word(s) in boldface type is given. (If the statement is true, no answer is shown. If the statement is false, a replacement is given.)

CHAPTER 1 QUIZ, PAGE 25

1.1 Descriptive
1.2 Inferential
1.4 sample
1.5 population
1.6 attribute data
1.7 discrete
1.8 continuous
1.9 random

CHAPTER 2 QUIZ, PAGE 96

2.1 median
2.2 dispersion
2.3 never
2.4 zero
2.5 higher than

CHAPTER 3 QUIZ, PAGE 128

3.1 Regression
3.2 strength of the
3.3 $+1$ or -1
3.5 positive
3.7 positive
3.8 -1 and $+1$
3.9 output or predicted value

CHAPTER 4 QUIZ, PAGE 184

4.1 any number value between 0 and 1, inclusive

4.4 simple
4.5 seldom
4.6 sum to 1.0
4.7 dependent
4.8 complementary
4.9 mutually exclusive or dependent
4.10 multiplication rule

CHAPTER 5 QUIZ, PAGE 219

5.1 continuous
5.3 one
5.5 exactly two
5.6 binomial
5.7 one success occurring on 1 trial
5.8 population
5.9 population parameters

CHAPTER 6 QUIZ, PAGE 250

6.1 its mean
6.4 one standard deviation
6.6 right
6.7 zero, 1
6.8 some (many)
6.9 mutually exclusive
6.10 normal

CHAPTER 7 QUIZ, PAGE 278

7.1 is not
7.2 some (many)
7.3 population
7.4 divided by $\sqrt{n}$
7.5 decreases
7.6 approximately normal

7.7 sampling
7.8 means
7.9 random

CHAPTER 8 QUIZ, PAGE 325

8.1 Alpha
8.2 Alpha
8.3 sample distribution of the mean
8.7 type II error
8.8 beta or $1 - 2$
8.9 correct decision
8.10 critical (rejection) region

CHAPTER 9 QUIZ, PAGE 359

9.2 Student's t
9.3 chi-square
9.4 to be rejected
9.6 t score
9.7 $n - 1$
9.9 $\sqrt{pq/n}$
9.10 z(normal)

CHAPTER 10 QUIZ, PAGE 415

10.1 two independent means
10.3 F distribution
10.4 Student's t distribution
10.7 is not

CHAPTER 11 QUIZ, PAGE 447

11.1 one less than
11.3 expected

11.4 contingency table
11.6 test of homogeneity
11.8 approximated by chi-square

CHAPTER 12 QUIZ, PAGE 471

12.2 mean square
12.3 SS(factor) or MS(factor)
12.5 Reject H_0
12.7 the number of factors less one
12.8 mean
12.9 need to
12.10 does not indicate

CHAPTER 13 QUIZ, PAGE 511

13.2 need not be
13.3 does not prove
13.4 need not be
13.6 the linear correlation coefficient
13.8 Regression
13.9 $n - 2$

CHAPTER 14 QUIZ, PAGE 547

14.2 t-test
14.3 runs test
14.4 assigned equal ranks
14.7 Mann-Whitney U test
14.8 power
14.10 power

Index